Springer Texts in Education

Springer Texts in Education delivers high-quality instructional content for graduates and advanced graduates in all areas of Education and Educational Research. The textbook series is comprised of self-contained books with a broad and comprehensive coverage that are suitable for class as well as for individual self-study. All texts are authored by established experts in their fields and offer a solid methodological background, accompanied by pedagogical materials to serve students such as practical examples, exercises, case studies etc. Textbooks published in the Springer Texts in Education series are addressed to graduate and advanced graduate students, but also to researchers as important resources for their education, knowledge and teaching. Please contact Natalie Rieborn at textbooks.education@springer.com for queries or to submit your book proposal.

More information about this series at http://www.springer.com/series/13812

Ben Akpan · Teresa J. Kennedy
Editors

Science Education in Theory and Practice

An Introductory Guide to Learning Theory

 Springer

Editors
Ben Akpan
Science Teachers Association of Nigeria
Abuja, Nigeria

Teresa J. Kennedy
University of Texas at Tyler
Tyler, TX, USA

ISSN 2366-7672   ISSN 2366-7680   (electronic)
Springer Texts in Education
ISBN 978-3-030-43619-3   ISBN 978-3-030-43620-9   (eBook)
https://doi.org/10.1007/978-3-030-43620-9

Preface

The field of science education is a relatively broad and dynamic area. Theories and applications of science teaching strategies are at the very core of the success of any science education program. While various books abound in support of international science education programs, there is a dearth of books that provide a collection of applicable learning theories and their applications to science teaching in a single source. *Science Education in Theory and Practice* fills this gap.

Science education specialists from 14 countries (Canada, China, Estonia, Germany, Ireland, Mauritius, New Zealand, Nigeria, Norway, South Africa, Sweden, Turkey, UK, and USA) anchored the 32 chapters of the book. This broad geographical coverage consisting of 44 authors provides applicable generational and regional perspectives from around the world.

This book provides a synthesis of historical theories while also providing practical implications for the improvement of pedagogical practices aimed at advancing the field into the future. The 32 chapters are divided among five significant areas:

- humanistic theories;
- behaviourist theories;
- cognitivist theories;
- constructivist theories; and the
- intellectually oriented and skill-based theories.

The theoretical viewpoints included span cognitive and social human development, address theories of learning, as well as describe approaches to teaching and curriculum development. In addition, wider issues are also addressed related to philosophical positions supporting science education. With a global readership in mind, each chapter follows a reader-motivated approach beginning with an introduction/background, followed by the primary or related issues through historical and/or theoretical background and reference to current debate and practice. Each chapter also provides recommended resources for extended reading and ends with a summary or list of main ideas. The genre of the writing is the narrative form.

Science Education in Theory and Practice, is intended for use by undergraduate and post-graduate students and their teachers, as well as researchers in the field of science education. It also serves as a guide for those aiming at creating optional learning experiences to prepare the next generation STEM (science, technology, engineering, and mathematics) workforce. The theories and practical applications highlighted serve as an active framework driving teaching and learning; and supporting pedagogical practices that facilitate student learning in the classroom.

We are immensely grateful to all the contributing authors to this book and to Springer International Publishing AG for facilitating its publication.

Abuja, Nigeria   Ben Akpan
Tyler, USA   Teresa J. Kennedy

Contents

Contents   ix

Chapter 1
Introduction—Theory into Practice

Ben Akpan and Teresa J. Kennedy

What Role Do Theories Play in Teaching and Learning?

In everyday parlance, the term theory has several meanings: 1. A coherent body of knowledge that is widely accepted as an explanation for some phenomena; 2. An insight into the natural world which is tentative but which is capable of providing explanations for natural phenomena if true; 3. Principles on which the practice of an activity is based; and 4. An idea that can guide behaviour. This last definition helps in understanding everyday behaviour of individuals. The first two definitions fit well within the fields of natural and applied sciences. In education, especially in teaching and learning, it is the third definition that prevails. In this sense, a theory becomes a versatile tool for understanding certain processes especially how teaching might result in effective learning. According to Woolfolk (2014):

> Given a number of established principles, educational psychologists have developed explanations for the relationships among many variables and even whole systems of relationships. There are theories to explain how … people learn. Theories are based on systematic research and they are the beginning and ending points of the research cycle. In the beginning, theories provide the research *hypotheses* to be tested or the questions examined. A hypothesis is a prediction of what will happen in a research study based on theory and previous research. For example, two different theories might suggest two competing predictions that could be tested. Piaget's theory might suggest that instruction cannot teach young children to think more abstractly, whereas Vygotsky's theory might suggest that this is possible. Of course, at times, psychologists don't know enough to make predictions, so they just ask *research questions*. (p. 30)

B. Akpan (✉)
Science Teachers Association of Nigeria, Abuja, Nigeria
e-mail: ben.b.akpan@gmail.com

T. J. Kennedy
College of Education and Psychology, College of Engineering, University of Texas, Tyler, USA
e-mail: tkennedy@uttyler.edu

© Springer Nature Switzerland AG 2020   1
B. Akpan and T. Kennedy (eds.), *Science Education in Theory and Practice*,
Springer Texts in Education, https://doi.org/10.1007/978-3-030-43620-9_1

Science Education in Theory and Practice presents 31 theories and describes how they may be used in science teaching and learning. These theories are categorised into five groups: humanistic, behaviourist, cognitivist, constructivist, and intellectually oriented and skill-based theories. In what follows we examine each group in turn.

Humanistic Approach

There are various, but related, definitions of humanism. We adopt, here, the definition by *The Humanist Magazine* (American Humanist Association, 2018):

> Humanism is a rational philosophy informed by science, inspired by art, and motivated by compassion. Affirming the dignity of each human being, it supports the maximization of individual liberty and opportunity consonant with social and planetary responsibility. It advocates the extension to participatory democracy and the expansion of the open society, standing for human rights and social justice. Free of supernaturalism, it recognizes human beings as part of nature and holds that values - be they religious, ethical, social or political – have their source in human experience and culture. Humanism thus derives the goals of life from human need and interest rather than from theological or ideological abstractions, and asserts that humanity must take responsibility for its own destiny. (p. 1)

Prior to the advent of the humanistic approach, psychodynamic theories such as psychoanalysis prevailed. Central to psychodynamics is the interrelation between conscious and unconscious processes as well as emotions that determine personality and motivation. Unfortunately, psychodynamics could not address issues such as the meaning of behaviour, as well as the nature of healthy growth. These concerns gave rise to the humanistic approach which places emphasis on subjective meaning, rejects determinism (a philosophical theory which posits that all events are inevitable consequences of antecedent sufficient causes; often denying the possibility of free will), and shows concern for positive growth rather than pathology. Pathology here refers to deviations from healthy or normal conditions. To humanists, individuals are capable of understanding their own behaviour as the meaning of behaviour is essentially personal and subjective. Ultimately, humanists posit, all individuals are subjective. Also, individuals are good, active, and creative persons who live in the present (Carducci, 2009).

In the humanistic theory of learning, learners observe what others do and the result of their actions. The teachers are role models in the class, and for that reason, they have to ensure inappropriate behaviours are avoided. When presenting learning tasks, teachers are expected to provide reasons and motivations as well. It is in the nature of humanism that learners have to take responsibility for their own learning including setting goals that are realistic for themselves. In science learning and teaching, the approach lays a high premium on collaborative and discussion groups which create the desired environment for learners to observe their peers. In this book, we have devoted four chapters to humanistic theories encompassing the ideas, among others, of Abraham Maslow, William Glasser, Thomas Malone, and von Humbolt. Some of the highpoints of the humanistic approaches to learning include placing emphasis on free will of learners, the placement of learners' subjective experiences as well as meanings at the centre, the focus on whole individual learners, and treatment of learners with high level of respect, ultimately viewing learners as very active agents in the teaching and learning process. On the flip side, some challenges exist. These include laying too much emphasis on the free will of learners, and the fact that the tenets of the humanistic approach are not subject to rigorous and objective methods of research thus making it difficult to make predictions that can either be ascertained or falsified.

Behaviourist Approach

Some psychologists' views do differ from the humanistic approach to learning. One such group are the proponents of behaviourism—a theory that assumes that the behaviour of humans and animals should be explained in terms of conditioning, without recourse to thoughts and feelings. The group also maintains that psychological disorders can be treated by altering behaviour patterns. At its core, behaviourists see behaviours as responses to stimuli. It is assumed that human behaviour is determined by the environment where the person resides. These environments provide stimuli to which the person responds. In contrast to other psychologists, behaviourists say that it is not necessary to consider internal mental processes in explaining behaviour. What is important to them is finding out which stimuli bring about particular responses. To them, therefore, complex behaviours of humans are a consequence of learning as a result of interaction with the environment. In this book, we dedicate three chapters to the behaviourist theories, encompassing the views of Ivan Pavlov, Albert Bandura, and Edward Thorndike.

The behavioural approach has some advantages. Unlike the humanistic approach, it uses rigorous experimental methodologies, makes arguments in support of nurture in learning and has demonstrated instances where the use of the approach is recommended. Shuell (2013) maintains that operant conditioning is better than any other theory in explaining how information is acquired, and how physical and mental skills are learned. According to Shuell, classical conditioning provides the:

> best explanation of how and why people including students, respond emotionally to a wide variety of stimuli and situations. The many types of emotional reactions acquired through classical conditioning include: anger toward or hatred for a particular person or group, phobias to a particular subject area or to school itself, and infatuation with another person. (p. 2)

These advantages have positive implications in science teaching and learning. For example, students will not like to repeat behaviours that have been disapproved; instead students tend to repeat actions that lead to positive consequences.

However, there are some disadvantages as well. The approach is overly deterministic, placing control of behaviour on nurture, thus ignoring the role of nature. Humans are therefore regarded as passive learners.

Cognitivist Approach

Cognitivism is the psychology of learning that is concerned with several mental processes such as thinking, perception, recall of information, learning, and problem-solving. This approach is as old as the field of psychology itself, but was at some stage, dwarfed by the rise of behaviourism. In recent decades, psychologists have redirected their interest to the area, especially as cognitive science (the study of memory and cognition) is gaining prominence. There are marked differences between the cognitivist approach and the behaviourist approach to learning as shown in Table 1.1.

Following are major points to note about cognitive science in relation to learning (Woolfolk, 2014):

- Whenever learning takes place the brain is involved. The brain shapes and is itself being shaped by activities involving cognition.
- Both knowing and knowledge result from learning. Previous knowledge is of overarching importance for future learning. Some knowledge may be general while some are specific, or to use cognitive terminology, *domain-specific*.
- What people already know, or what they need to know, determines how they pay attention. People with attention-deficit disorder experience difficulty in paying attention or ignoring competing information. Learners cannot process information which they don't recognise; so, attention is crucial for learning.
- People's working memory serves as the workbench of conscious thought. It holds information that is current. In contrast, long-term memory holds information that is properly learned. Information that is lost from the working memory has definitely disappeared but given the right conditions information in long-term memory can be retrieved. However, when neural connections become weak due to non-use (*time decay*), information might be lost from long-term memory.
- Learners differ in how well and how fast they learn some tasks due to differences in their metacognitive knowledge and skills. Metacognition means *thinking about*

Table 1.1 Differences between behavioural and cognitive approaches

	Behavioural approach	Cognitive approach
1.	New behaviours are learned	Knowledge and strategies are learned; changes in behaviour are due to changes in knowledge and strategies
2.	Reinforcement strengthens responses	Reinforcement provides information about what could occur when behaviours change or are repeated
3.	Learning is influenced by events in the environment	Learners actively choose and take decisions in pursuance of their learning goals
4.	Identification of general laws of learning which apply to all humans is of great interest	No concern with general laws. Consequently, no one theory of learning is representative of the cognitive view

thinking. It regulates thinking and learning through planning, monitoring, and evaluation.

- Learning strategies facilitate the cognitive engagement of learners.

Seven chapters anchor the cognitivist approach in this book. Expectedly, the various positions canvassed by the theorists are very disparate. Jean Piaget in his stage theory of cognitive development presents a four-stage model showing how new information is processed in the mind. Children, he claims, progress through sensorimotor, preoperational, concrete operational, and formal operational stages and they all do so in that same order. Benjamin Bloom's mastery learning is based on the need to give each student enough time, help, and support through individualised and differentiated instruction in order to attain mastery of a learning task. For David Ausubel, meaningful learning takes place when new information is related to previous knowledge, an idea that gave rise to concept maps. Jerome Bruner and Robert Gagné both subscribe to discovery learning but while Bruner goes for pure discovery, Gagné supports guided discovery. For Bruner, learning is a process. For Gagné, learning is both a process and a product. According to Bruner:

> To instruct someone… is not a matter of getting him to commit results to mind. Rather, it is to teach him to participate in the process that makes possible the establishment of knowledge. We teach a subject not to produce little living libraries on that subject, but rather to get a student to think mathematically for himself, consider matters as an historian does, to take part in the process of knowledge-getting. Knowing is a process not a product (Smith, 2002 p. 4).

Elsewhere, Bruner also reiterated:

> It is my hunch that it is only through the exercise of problem-solving and the effort of discovery that one learns the working heuristics of discovery, and the more one has practice, the more likely is one to generalise what one has learned into a style of problem-solving or inquiry …I think the matter is self-evident, but what is unclear is what kinds of training and teaching produce the best effects. How do we teach a child to, say, cut his losses but at the same time be persistent in trying out an idea; to risk forming an early hunch without at the same time formulating the one so early and with so little evidence as to be stuck with it waiting for the appropriate evidence to materialize…to pose good testable guesses that are neither too brittle nor too sinuously incorrigible. Practice in inquiry, in trying to figure out things for oneself, is indeed what is needed, but in what form? Of only one thing I am convinced. I have never seen anybody improve in the art and technique of inquiry by any other means other than engaging in inquiry. (Bruner, 1975, p. 87)

However, Gagné (1975) retorts:

> Broad, generalizable knowledge is a prerequisite for the successful practice of enquiry, whether as part of the total instructional process or as a terminal capability. How does the student acquire this broad knowledge? … such knowledge cannot be attained by the student by the use of the method of enquiry itself. Were we to follow this suggestion, we should have to put the student back in the original situation that Newton found himself in, and ask the student to invent a solution, as Newton did. It would be difficult to achieve this situation, in the first place, and presumably not all students would achieve what Newton did, even then. But the major difficulty with this suggestion is that it would be a most terrible waste of time. Are we going to have students rediscover the laws of motion, the periodic table, the structure of the atom, the circulation of blood, and all the other achievements of science simply in order

to ensure that instructional conditions are "pure", in the sense that they demand enquiry? (pp. 95-96).

Taken together, cognitive approaches are very popular and are applicable in many areas of science teaching and learning. Overall, they demonstrate the function of the human brain in the learning and thought processes. The approaches can be combined with other theoretical models based on the human and material resources available. However, reliance on cognitive science has precluded several aspects of humanity such as genetic, biological, and chemical features and imbalances. Also, where people are observed in controlled environments, the findings may differ from real-world settings where several stimuli compete for attention at the same time.

Constructivist Approaches

The constructivist approaches are closely related to the cognitivist approaches. Constructivism is a philosophy of teaching which maintains that students perform mental construction in the process of learning. By using personal experiences and relating these to new knowledge, students are able to construct meanings for themselves. Thus, students create their mental models (also called schemas) in a bid to understand new subject matter. New knowledge is accommodated through the adjustment of the schemas. All students thus actively search for meanings in constructivist learning approaches. In general, therefore, constructivists are of the view that learners construct their own knowledge and that knowledge construction processes are greatly enhanced by social interactions (Woolfolk, 2014). Constructivism has provided a solid foundation for the learning sciences—an interdisciplinary research area that focuses on fields of learning such as neuroscience, computer studies, psychology, sociology, philosophy, and anthropology.

There are three major ways knowledge is constructed: (1) Knowledge acquisition is externally directed through reconstruction of external reality as in the case of information processing models of learning; (2) Knowledge acquisition occurs through internal direction by transformation or reorganisation of past knowledge as in Jean Piaget's theory; and (3) Knowledge acquisition is attained through both knowledge of the outside world and previous knowledge as in the case of Lev Vygotsky's theory. Additionally, knowledge may be situated, a form of enculturation or adoption of norms.

Although there are differences in the various cognitivist positions as exemplified in eleven chapters in this book, some learning activities typify the approaches. These, according to Windschitl (2002) include complex, meaningful, problem-based activities; obtaining students' ideas on specific topics and organising suitable learning experiences to help them improve on the current knowledge; task-oriented collaborative activities involving many students; asking students to apply knowledge and experiences in explaining concepts, interpreting phenomena, and constructing coherent arguments based on evidence; and using diverse assessment methods to find out how students are progressing and to give feedback to the students. In addition, Woolfolk (2014) maintains that the approaches often involve *scaffolding*—situations where 'teachers and students make meaningful connections between what the teacher knows and what the students know and need in order to help the students learn more'. (p. 393)

As a group, the constructivist approaches are very effective in hands-on environments and in helping learners to relate subject matter to lived experiences. Science classes implementing these approaches enable teachers to identify and place emphasis on topics that learners tend to like. In addition, by working in groups, learners acquire the much-needed social skills, are able to assist one another, and indirectly learn to respect the point of views of other persons. However, implementing the approaches may be expensive in terms of materials and professional development of science teaching personnel. With variations in previous knowledge of students, difficulties may arise in agreeing on the operational curriculum for a class; and as standardised testing and grading is downplayed, comparisons of achievements across states and regions, for example, become problematic.

Intellectually Oriented and Skill-Based Theories

There are six theories in this book which are collectively grouped under intellectually oriented and skill-based theories. These are multiple intelligences, systems thinking, gender/sexuality, indigenous knowledge systems, STEAM education, and twenty-first-century skills. It is important to note that where there is an overarching need to take care of diversity among students by designing appropriate and suitable teaching and learning strategies, these theories will be very useful.

Multiple Intelligences Theory

The multiple intelligences theory proposed by Howard Gardner in 1983 (Edutopia, 2016) takes the view that the time-honoured view of intelligence based on IQ tests is highly limited. The theory thus proposes many different intelligences in humans: verbal-linguistic intelligence (words), logical-mathematical intelligence (logic, numbers), visual-spatial intelligence (pictures), bodily-kinesthetic intelligence (physical experience), musical-rhythmic intelligence (music), interpersonal intelligence (social experience), intrapersonal intelligence (self-reflection), naturalist intelligence (experience in the natural world), and existential intelligence (using values and intuition to generate understanding). Thus, if a science teacher is having difficulty in reaching a learner in the more traditional linguistic or logical ways of instruction, the theory provides many other approaches through which the learning task could be presented to facilitate effective science teaching and learning.

Systems Thinking

Systems thinking is a theory that explores an understanding of a system by examining the linkages and interactions between the different parts that make up the entire system that is being considered. It looks at things as a whole rather than various components. Basically, a system, whether human-made or natural, exists and functions as a whole through the dynamic interaction of its component parts. A change in one part of the system affects other parts of the system, and subsequently affects the stability and sustainability of the system. In applying systems thinking in science teaching and learning, learners: try to understand the big picture, figure out patterns and trends in scientific systems, become aware of how a scientific system's structure causes its behaviour, identify and explore cause and effect relationships, test scientific assumptions, and search for clues as to where unintended consequences may arise.

Gender/Sexuality Theory

Gender/Sexuality is a dedicated area of interdisciplinary study on gender representation and identity with respect to women and men. Such studies are conducted in several fields of human endeavour such as science, technology, engineering, mathematics, education, medicine, politics, languages, and anthropology. Across these fields, variations as to how and why these studies are conducted are discernible. In science teaching and learning, part of the focus is on gender neutrality and equity in instructional materials development and implementation as well as in planning, streaming, resource allocation, assessment procedures, guidance and counselling, and careers in related fields. A challenge facing science educators is to vigilantly monitor their classroom activities and adjust pedagogical practices to promote gender parity (Kennedy & Sundberg, 2017).

Indigenous Knowledge Systems

Blades and Mcivor (2017) are of the view that indigenous people are those who first settle in an area in any part of the world. The knowledge developed by such people is termed indigenous knowledge and is used as a basis for making decisions and in other personal and communal pursuits. According to McCallum (2012), indigenous knowledge should be integrated into the science curriculum for the following reasons: (1) Since our planet is encountering ecological problems and indigenous knowledge has very strong links to global sustainability, such knowledge should be helpful in ensuring efficient use of land; (2) Ways of conservation of resources as well as energy can be imparted to students by using indigenous knowledge; (3) Indigenous knowledge is useful in imparting knowledge about intellectual property rights of indigenous peoples; and (4) Both indigenous knowledge and western knowledge can be taught together thereby fostering an awareness of the culture of indigenous peoples and promotion of peace and tolerance across the world.

The STEAM Framework

STEAM is a teaching strategy created by the Rhode Island School of Design in the USA. The STEAM framework adds the *Arts* to the original STEM framework. The approach demonstrates how interdisciplinarity can contribute to the understanding and knowledge of scientific principles to solve societal challenges (Akpan, 2016). According to the European Union (2015), the approach involves:

- learning science through other disciplines and learning about other disciplines through science;
- strengthening connections and synergies between science, creativity, entrepreneurship, and innovation; and
- placing more emphasis on ensuring all citizens are equipped with the skills and competences needed in the digitalised world starting from preschool.

As noted by The Vision Board (2017), the STEAM approach uses science, technology, engineering, and mathematics as access points for guiding student inquiry, dialogue, and critical thinking. Indeed, STEAM enables teachers to use project-based learning that crosses all 5 disciplines (University of San Diego, 2017). It thus provides an inclusive learning environment such that all learners are able to engage and contribute. The STEAM framework is obviously not an easy one but as The Vision Board (2017) stated:

> the benefits to students and entire school community are tremendous. Students and teachers engaged in STEAM make more real-life connections so that school is not a place where you go to learn but instead becomes the entire experience of learning itself. We are always learning, always growing, always experimenting. School doesn't have to be a place, but rather a frame of mind that uses the Arts as a lever to explosive growth, social-emotional connections, and the foundation for the innovators of tomorrow… today! (p. 1)

Twenty-First-Century Skills

Twenty-first-century skills, are a group of skills and capabilities which are considered necessary for a successful participation in learning, business, work, and other societal responsibilities in this modern age. These skills include innovation, creativity, curiosity, health and awareness literacy, critical thinking, environmental literacy, scientific literacy, problem-solving, perseverance, analysis, imagination, listening, collaboration, media literacy, ethics, entrepreneurialism, humanitarianism, scientific method, communication, interpretation, planning, synthesising information, teamwork, and information and communications technology. There are numerous classifications of these skills but for this discourse we align with the classification provided by Crockett (2016) because it tends to cover all the areas of concern with respect to science teaching and learning in the emerging world: problem-solving, creativity, analytic thinking, collaboration, communication; and ethics, action, and accountability. The task for teachers of science is how to use suitable approaches earlier described in ensuring students acquire these skills.

How to Use This Book

In the foregoing discourse, we have provided an overview of the various groups of theories presented in this book. It should be noted that some chapters deal with theories of human development (either cognitive or social), some address theories of learning, others deal with approaches to teaching and/or curriculum construction, and some address wider issues relating to philosophical positions. As a result, it is helpful for readers to appreciate that chapter authors are likely to take significantly disparate approaches. Similarly, although there are measures that have standardised the various chapters—for example, each author provides an abstract, an introduction, and a closing summary and spends some time outlining the biographical details of the major theoreticians being discussed—there are, nonetheless, striking differences. These differences actually cater to the special and specific needs of the various chapters and so are a source of strength for the book rather than an indication of a lack of coherence and organisation. Thus, each of the 31 chapters anchoring these theories has highlighted how the theory may be applied in teaching and learning science. We also think it is appropriate here to state that the grouping of the theories in the book is not cast in stone. The theories could be grouped in various other ways. For example, in terms of cognitive and social development of humans, etc. Also, even with the present groupings, some theories may fit into groups other than where they are currently placed. Having said that, we want to take a close look at how to use this book, and in particular, how to relate the various theories to classroom practice. Shuell (2013) has given some viewpoints regarding the relationship between theory and practice: (1) Ordinarily, instructional practices should be based on the best theories that one can find. However, in practice this is not necessarily the case as schools are more or less based on philosophical beliefs rather than on learning theories. Indeed, establishment of schools globally is based on people's cultural beliefs. In every educational system, there is a theory of learning but these theories are usually highly nuanced and so go unrecognised; (2) Contrasting philosophical views result in very disparate classrooms. What is consistent is that theoretical beliefs have a direct impact on practices in classrooms. There is, however, the chicken and egg dilemma here; namely, whether or not it is the theory that comes before the practice or whether the theory may be generated from classroom practice. In a way, there is the view that although theories of learning are expected to provide direction for educational practice (theories of teaching), it is not always the case. There is, indeed, a two-way relationship between theory (learning theories) and practice (theories of teaching); and (3) In the teaching and learning process, the teacher and the students are important but ultimately it is whether the students have learnt effectively that determines the success or otherwise of the teacher. To that extent, the student plays an overarching role, way ahead of the teacher, in determining what is actually learnt.

In the end, which theory do we choose for instruction? Here, we align our thoughts with those of Morrison (2014):

> Is any one set of instructional method better than the other? No … there is a variety of methods that serve different needs. It's the skilled and intuitive educator that analyses a learning situation, leverages the resources at his or her disposal … and is able to analyse the situation and design the very best learning experience for his or her student. (p. 5)

Summary

In this chapter, we have discussed the following:

- The theories in this book are categorised into five groups: humanistic theories, behaviourist theories, cognitivist theories, constructivist theories, and intellectually oriented and skill-based theories.
- In science learning and teaching, the humanistic approach lays a high premium on collaboration and discussion groups that create the desired environment for learners to observe their peers.
- At its core, behaviourists see behaviours as responses to stimuli. It is assumed that human behaviour is determined by the environment where the person resides. These environments provide stimuli to which the person responds.
- Cognitivism is the psychology of learning that is concerned with several mental processes such as those of thinking, perception, recall of information, learning, and problem-solving.
- Constructivist approaches are very effective in hands-on environments and in helping learners to relate subject matter to lived experiences. Science classes implementing these approaches enable teachers to identify and place emphasis on topics that learners tend to like.
- The multiple intelligences theory has proposed many different intelligences in humans aimed at promoting the implementation of diverse teaching methods.
- Systems thinking looks at things as a whole rather than various components.
- In science education, gender/sexuality theory focuses on gender neutrality and equity in instructional materials development and implementation as well as in planning, streaming, resource allocation, assessment procedures, guidance and counselling, and careers in related fields.
- The knowledge developed by indigenous people is termed indigenous knowledge and is used as a basis for making decisions and in other personal and communal pursuits.

- The STEAM framework adds the *Arts* to the original STEM framework.
- Twenty-first-century skills, are a group of skills and capabilities which are considered necessary for successful participation in learning, business, work, and other societal responsibilities at this modern age.

Further Readings

Cruciun, B., & Dumitru, S. B. (2011). Knowledge management—The importance of learning theory. Journal of Knowledge Management, Economics, and Information Technology. Retrieved June 6, 2016, from http://www.scientificpapers.org/wp-content/files/1209_Craciun_Bucur_Matei_Knowledge_Management_the_importance_of_Learning_Theory.pdf.

Learning theory (education). 2016, April, 30. In Wikipedia. The Free Encyclopedia. Retrieved June 6, 2016, from https://en.wikipedia.org/w/index.php?title=Learning_theory_(education)&oldid=717947183.

Greeno, J. G. (2006). Learning in activity. In R. K. Sawyer (Ed.), The Cambridge handbook of the learning sciences (pp. 79–96). New York: Cambridge University Press.

Post, T. (1988). Some notes on the nature of mathematics learning. In T. Post (Ed.), Teaching Mathematics in Grades K-8: Research Based Methods (pp. 1–19). Boston: Allyn & Bacon. Retrieved 19 March, 2018, from http://www.cehd.umn.edu/ci/rationalnumberproject/88_9.html.

Salomon, G. (Ed.). (1993). Distributed cognitions: psychological and educational considerations. New York: Cambridge University Press.

References

Akpan, B. B. (2016). Science education research and national development. *Journal of the Science Teachers Association of Nigeria, 51*(1), 105–116.

American Humanist Association. (2018). Definition of humanism. Retrieved from https://americanhumanist.org/what-is-humanism/definition-of-humanism/.

Blades, D., & Mcivor, O. (2017). Science education and indigenous learners. In K. S. Taber & B. Akpan (Eds.), *Science education: An international course companion* (pp. 465–478). Rotterdam: Sense Publishers.

Bruner, J. S. (1975). The act of discovery. In E. Victor & M. S. Lerner (Eds.), *Readings in science education for the elementary school* (3rd ed., pp. 77–89). New York: Macmillan Publishing Company Inc.

Carducci, B. J. (2009). *The psychology of personality* (2nd ed.). Chichester: Wiley-Blackwell.

Crockett, L. W. (2016). The critical 21st century skills every student needs and why. Global Digital Citizen Foundation. Retrieved from https://globaldigitalcitizen.org/21st-century-skills-every-student-needs.

Edutopia. (2016). Multiple intelligences: What does the research say? Retrieved from https://www.edutopia.org/multiple-intelligences-research.

European Union. (2015). *Science education for responsible citizenship.* Luxemburg: Publications Office of the European Union

Gagné, R. M. (1975). The learning requirements of enquiry. In E. Victor & M. S. Lerner (Eds.), *Readings in science education for the elementary school* (3rd ed., pp. 89–103). New York: Macmillan Publishing Company Inc.

Kennedy, T. J., & Sundberg, C. (2017). International perspectives and recommendations on equity and gender: Development studies in science education. In B. Akpan (Ed.), *Science education: A global perspective: international addition* (Chapter 15, pp. 295–311). Switzerland: Springer International Publishing. ISBN 978-3-319-32350-3; ISBN 978-3-319-32351-0 (eBook).

McCallum, D. (2012). Seven Reasons to integrate indigenous knowledge into science curriculum. Working effectively with indigenous peoples [Blog]. Indigenous Corporate Training Inc. Retrieved from https://www.ictinc.ca/blog/7-reasons-to-integrate-indigenous-knowledge-into-science-curriculum.

Morrison, D. (2014, January 31). Why educators need to know learning theory [Blog]. Retrieved fromhttps://onlinelearninginsights.wordpress.com/2014/01/31/why-educators-need-to-know-learning-theory/.

Shuell, T. (2013). Theories of learning. Retrieved from http://www.education.com/reference/article/theories-of-learning/.

Smith, M. K. (2002). Jerome S. Bruner and the process of education. *The encyclopedia of informal education*. Retrieved from http://infed.org/mobi/jerome-bruner-and-the-process-of-education/.

The Vision Board. (2017). STEAM education. Retrieved from https://educationcloset.com/steam-education/.

University of San Diego. (2017). STEAM education: A 21st century approach to learning. Retrieved from https://onlinedegrees.sandiego.edu/steam-education-in-schools/.

Windschitl, M. (2002). Framing constructivism in practice as the negotiation of dilemmas: An analysis of the conceptual, pedagogical, cultural, and political challenges facing teachers. *Review of Educational Research, 72,* 131–175.

Woolfolk, A. (2014). *Educational psychology*. Noida, India: Dorling Kindersley India Pvt. Ltd.

Ben Akpan Ph.D., a professor of science education, is the Executive Director of the Science Teachers Association of Nigeria (STAN). He served as President of the International Council of Associations for Science Education (ICASE) from 2010 to 2013 and currently serves on the Executive Committee of ICASE as the Chair of World Conferences Standing Committee. Ben's areas of interest include chemistry, science education, environmental education, and support for science teacher associations. He is the editor of *Science Education: A Global Perspective* published by Springer and co-editor (with Keith S. Taber) of *Science Education: An International Course Companion* published by Sense Publishers. Ben is a member of the Editorial Boards of the Australian Journal of Science and Technology (AJST), Journal of Contemporary Educational Research (JCER), and Action Research and Innovation in Science Education (ARISE) Journal.

Teresa J. Kennedy Ph.D., holds a joint appointment as Professor of International STEM and Bilingual/ELL Education in the College of Education and Psychology and in the College of Engineering at the University of Texas at Tyler, United States of America. She served as President of the International Council of Associations for Science Education (ICASE) from 2014 to 2017 and currently serves on the Executive Committee of ICASE as the Representative to UNESCO. Teresa is a two-time Fulbright Scholar, first in 1993 in Ecuador, and again during the 2014–15 academic year focusing on engineering education in Argentina. Her research interests include STEM Education, international comparative studies, gender equity, and brain research in relation to second language acquisition and bilingualism. She is a member of the Editorial Boards of the *Journal of Educational Research and Review (JERR)*, the open-access journal *Education Sciences*, and ISCI Publishing *Arts, Humanities and Social Sciences*.

Part I
Humanistic Theories

Chapter 2
Theory of Human Motivation—Abraham Maslow

Shannon L. Navy

Introduction and Chapter Map

When I was a high school science teacher in the United States (US), I often taught students who were in remedial classes. The students were in these classes because of previous grades or test scores. Some of the students, ages 15–19, were staying in school until they were able to drop out or attend an alternative education program. Some of them would come to school for the food at breakfast and lunch (80% of the students were on free and reduced lunch, an indicator of poverty in the US), and/or to socialize with friends. At this stage in their schooling, many of them were seemingly no longer motivated to learn.

The more I worked with my students, the more I witnessed their capabilities of achieving success when they were motivated to learn. A question I often asked myself was, "How can I motivate my students to learn science?" To help answer this question, I found myself reflecting to Abraham Maslow's hierarchy of needs and theory of human motivation. Once I began to apply some of this theory to my science teaching, I saw the benefit and value of its application to classroom practice and student learning.

This chapter is intended to highlight the main components of Maslow's theory of human motivation and how it applies to science teaching. It begins with a brief biography of Maslow, including the influences on his ideas and theory. Next, the theory is explained through the hierarchy of needs beginning with the basic physiological needs and advancing to self-actualization. Beyond the hierarchy of needs, additional components of the theory are described including: hierarchy reversal of needs, degrees of relative satisfaction, and multiple motivations of behavior. The controversies associated with the theory are presented based on findings in the literature. The second part of the chapter discusses the applications of Maslow's theory

S. L. Navy (✉)
Kent State University, Kent, OH, USA
e-mail: snavy@kent.edu

© Springer Nature Switzerland AG 2020
B. Akpan and T. Kennedy (eds.), *Science Education in Theory and Practice*,
Springer Texts in Education, https://doi.org/10.1007/978-3-030-43620-9_2

to science instruction. Practical considerations for science teaching are suggested for each of the levels in the hierarchy of needs.

Biography

Abraham Maslow (1908–1970) is considered the father of humanistic psychology. He was born in Brooklyn, New York as the first of seven children to parents who were Russian-Jewish immigrants. His father wanted him to be a lawyer, so he tried law school for two weeks before deciding that he was not interested in becoming a lawyer (Hall, 1968). Although studying law did not interest him, he had a quest for learning and continued pursuing areas of interest.

In 1928, Maslow decided to study psychology at the University of Wisconsin where he earned a B.A., M.A., and Ph.D. by 1934. In 1935, Maslow transitioned to a postdoctoral position at Colombia University where he conducted research with Edward Thorndike, a distinguished behaviorist. Although Maslow was initially impressed with behaviorism, he soon realized the limitations to the strict behaviorist approach. He believed that not everything about human nature could be studied by behaviorist approaches, which try to reduce elements of being human to variables and numbers.

Maslow moved on from Columbia University to Brooklyn College where he taught for 14 years. He encountered Ruth Benedict and Max Wertheimer, two friends he became fascinated with because of their brilliance, creativity, and caring natures (Frager, 1987). He wondered what made them admirable human beings and brilliant scholars. Slowly, he started to realize that Benedict and Wertheimer were a type of human being with comparable characteristics. This began to guide his thinking about human nature and motivation.

In the 1940s, Maslow encountered Kurt Goldstein, a neurologist who created a holistic theory of organisms. Goldstein introduced Maslow to the concept of self-actualization. Goldstein considered self-actualization to be an individual's "tendency to actualize, as much as possible, its individual capacities" (Goldstein, 1939, p. 196). Goldstein considered this tendency to be the basic drive of human life. As Maslow continued studying individuals he believed to be healthy, highly fulfilled people, he made many connections to Goldstein's concept of self-actualization. In his 1943 paper, *A Theory of Human Motivation*, Maslow described how each person has a set of basic needs which, once satisfied, will no longer motivate behavior. Motivation is then driven by human fulfillment needs which he described as Goldstein's concept of self-actualization.

From 1951 to 1969, Maslow worked as the chairman of the psychology department at Brandeis University in Boston. He continuously refined his ideas toward a more comprehensive theory of human nature. Up until his death of a heart attack in 1970, he was studying and observing human nature to more fully understand the complicated connections of motivation, learning, and being.

A Theory of Human Motivation

The basis of this chapter is Maslow's (1943) paper, *A Theory of Human Motivation*. According to Maslow, humans are motivated by needs and these needs are hierarchically organized by priority. Unsatisfied needs are what motivate human behavior. The hierarchy of needs in Maslow's theory is most often represented as a pyramid (see Fig. 2.1).

The Basic Needs

The needs are categorized into five levels, from highest priority to lowest priority: physiological, safety, love and belonging, esteem, and self-actualization. Once one level of needs is satisfied or gratified in the hierarchy, the next level of needs becomes the focal center of motivation for an individual. For example, individuals with inadequate food (a physiological need) must meet that need before seeking to establish stability (a need at the safety level). If and when individuals have met their physiological needs, safety needs will then become a priority for motivation. This principle continues through the hierarchy of needs (Maslow, 1943, 1970).

The physiological needs. The starting point for Maslow's motivation theory is the physiological needs of hunger, thirst, health, and sleep. These are the greatest priority of all the needs. This means that if an individual is lacking anything in an extreme fashion, then the physiological needs are the major motivation (Maslow, 1943). As Maslow (1943) wrote, "For the man who is extremely and dangerously hungry, no other interests exist but food. He dreams food, he thinks about food, he emotes only about food, he perceives only food and he wants only food" (p. 374). For

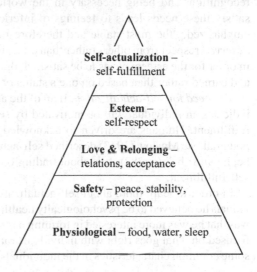

Fig. 2.1 The pyramid representation of Maslow's theory of human motivation

Self-actualization – self-fulfillment

Esteem – self-respect

Love & Belonging – relations, acceptance

Safety – peace, stability, protection

Physiological – food, water, sleep

the extremely hungry person, aspects of life or learning not related to food remain in the background until the food need is satisfied.

The safety needs. If the physiological needs are satisfied, then an individual's new center of focus becomes safety. Safety needs include feelings of peace, security, stability, and protection (Maslow, 1970). They ensure an individual does not feel threatened or endangered (Maslow, 1943). For children, characteristics of their upbringing and their parents/guardians are important components to satisfying this need. Children who are raised in loving homes without quarreling, assault, abuse, and separation often feel safe and secure in the world. Maslow (1943) indicated, "we may generalize and say that the average child in our society generally prefers a safe, orderly, predictable, organized world, which he can count on" (p. 378). Children without a home or family security often lack this sense of organization, structure, and safety. Adults do not often experience the same safety concerns of children. However, some adults may experience feelings of danger from wild animals, extreme weather, criminals, and/or abuse.

The love and belonging needs. Once the physiological and safety needs are fulfilled, an individual becomes motivated by love and belonging needs. These include loving and being loved, belonging in a community, and having friends and family. In this level of the hierarchy, an individual will strive for relationships with people. It is also important to note that this level involves both giving and receiving love. Maslow (1970) connected this level in the hierarchy to basic animal tendencies when he remarked that humans have a "deeply animal tendency to herd, to flock, to join, to belong" (p. 44). Maslow believed that the increase in frequency and popularity of many training, personal growth, and community groups were related to humans' motivations to belong and connect with people.

The esteem needs. If the physiological, safety, and love and belonging needs are met, esteem needs become the focal center of motivation. The esteem needs include feelings of self-respect, confidence, achievement, success, self-worth, reputation, recognition, and being necessary in the world (Maslow, 1943, 1970). Failing to satisfy these needs leads to feelings of inferiority or uselessness. Maslow (1970) emphasized, "the most stable and therefore most healthy self-esteem is based on *deserved* respect from others rather than on … unwarranted adulation" (p. 46). Thus, in order for the esteem needs to be satisfied, the respect from others must be genuine and earned rather than based on one's status or fame in society.

The need for self-actualization. If all of the above needs are met, Maslow's theory indicates an individual will be motivated by self-actualization, or a desire for self-fulfillment. Humans are driven to acknowledge, become, and fulfill their human potential. As Maslow (1943) described self-actualization he wrote, "what a man *can* be, he *must* be" (p. 382). It is about finding one's calling in life in order to achieve self-fulfillment.

In order to learn more about self-actualization in humans, Maslow studied individuals he believed to be psychologically healthy adults. His justification for doing so was that he was more interested in forming a positive account of human behavior that focused on what goes right with individuals rather than what goes wrong. He found some common characteristics in the individuals he studied who he believed achieved self-actualization. These characteristics included: accurate perception, spontaneity, detachment, autonomy, interpersonal relations, and creativeness (Maslow, 1970). Although the individuals Maslow studied were psychologically healthy adults, they were far from perfect human beings. They discussed their failings, which Maslow believed helped them remain psychologically healthy and achieve self-actualization. Maslow discovered that self-actualizing people also had qualities that helped them resolve conflicting dichotomies, such as: heart versus head, duty versus pleasure, and selfishness versus unselfishness (Maslow, 1970). The values which motivated them were *being values* (B-values) rather than *deficiency values* (D-values) (Maslow, 1968).

The five-level model of Maslow's theory of human motivation moves from physiological needs to self-actualization. Lower level needs must be satisfied before progressing to the higher levels and achieving self-actualization. However, Maslow believed that not all individuals achieved self-actualization because society often rewards motivation based on esteem and belonging. Yet, he remained hopeful that humans could achieve their potential and indicated that growth toward self-actualization was a dynamic process.

Additional Characteristics

In addition to the hierarchical model of human needs that drive motivation, Maslow also explained some important components of the theory. These included: hierarchy reversal of needs, degrees of relative satisfaction, and multiple motivations of behavior. Understanding these aspects of the theory helps to understand Maslow's intentions with the extent of the theory.

Hierarchy reversal. Maslow created the hierarchy of needs based on his examinations of the humans he studied. Although the order of needs seems to be accurate for most people he studied, there are some exceptions. Therefore, the hierarchy is not intended to be a fixed entity, but rather a less rigid structure and order.

There are five hierarchy reversal exceptions to consider in Maslow's theory. One reversal is that, for some individuals, self-esteem is more important than love. Another is that creativity is the most important motivating factor for some individuals. As such, some revised models of Maslow's hierarchy include creativity as a separate component. A third reversal is that for some individuals who are in a chronically devastating condition, such as extreme hunger or chronic abuse, aspirations and motivations may become permanently lowered. For such instances, having food and water and safety might be sufficient in determining one's satisfaction or gratification. A fourth reversal is with individuals who have lacked love from a very early age. In such circumstances, the desire and ability to love and be loved is nonexistent. A final hierarchy reversal is the underestimated value of needs that have been sufficiently satisfied for a long time. Maslow explained this latter reversal using the example of hunger. People who have never experienced chronic hunger will likely deem food as the most important need if they ever do experience chronic hunger, even if other needs in their lives are currently satisfied or dominating (Maslow, 1943). These reversal exceptions illustrate some of the fluidity in the hierarchy of needs proposed by Maslow.

Relative satisfaction. In further explaining the concepts in his theory, Maslow indicated that needs in the hierarchy do not require 100% satisfaction for an individual to move on to the next level. In fact, Maslow explained that most normal humans are partially satisfied and partially unsatisfied in all their basic needs at any one time (Maslow, 1943). He believed it would be more accurate to portray the hierarchy of needs in relative decreasing percentages going up the hierarchy. For instance, an individual would have to be at least 85% satisfied in the physiological needs, 70% satisfied in the safety needs, and so on as one moves up the hierarchy. Therefore, complete 100% gratification at any one level may be unnecessary for determining a person's motivation.

Multiple motivations of behavior. Although the needs are described and classified in different levels of a hierarchy, Maslow cautioned that most behavior is multi-motivated and cannot be isolated to a single factor. Therefore, human behavior is simultaneously motivated by many levels of the hierarchy. In the multi-motivated view of behavior, some levels may highly motivate an individual while others weakly motivate an individual at any given time or in any given situation.

Controversies

Maslow's (1943) theory of human motivation has generated sustained interest and support since its inception. It is still one of the most referenced and remembered theories of human motivation. However, as with most grand theories, there exist some criticisms and controversies. It has been criticized for focusing on individualism and elitism, and relating to primarily Western cultures.

Individualism. One of the main criticisms of Maslow's theory is its excessive individualism, which is also a critique of the larger branch of humanistic psychology (Pearson, 1999). Critics argue that a tension exists between the individual and society, between the self and others. Buss (1979) indicated that Maslow primarily focused on individual efforts, freedom, and development rather than on society's development needs. Although Maslow did give some recognition to societal and cultural forces, at the center of his work were assumptions of individual capacity, human-centeredness, autonomy, and responsibility. Pearson (1999) synthesized the individualism in Maslow's theory into the concepts of self, growth, responsibility, and capability to influence social progress. Indeed, Maslow believed individuals have a responsibility to grow and fulfill their potentials, which he considered to be self-actualization.

The individualistic nature of Maslow's work was criticized by Marxists and Post-modernists. Marxists emphasized society's influence in shaping individuals, thereby rejecting the notion of autonomy. In order to reshape human nature, they argued, society must be reshaped first (Pearson, 1999). Postmodernists critiqued the notion of human agency and the concept of normal in Maslow's theory. In this view, humans are completely constructed by practices of power so the idea of the human self in Maslow's theory is rejected (Pearson, 1999).

Maslow constantly thought about his work and considered the critiques of his theory. His journal writings indicated he was focusing more and more on social and political factors (Lowry, 1979). In his later writings, his view of the world was one where the individual and society developed in synergy (Pearson, 1999). These adjustments were likely in response to the criticisms of individualism in the theory.

Elitism. Maslow's theory of human motivation has also been critiqued for being elitist. The premise of this critique is that not everyone in society can be self-actualized given various societal circumstances. This puts those individuals who are self-actualized as elite members of society (Cooke, Mills, & Kelley, 2005; Pearson, 1999). In this way, the hierarchy of needs is essentially a social hierarchy (Buss, 1979; Cooke et al., 2005; Cullen, 1997).

Cooke et al. (2005) explained that a tension is created between democracy and elitism since not everyone can be self-actualized even though it is described as a basic human condition in Maslow's theory. For individuals who do not reach a level of self-actualization, they may blame themselves, or others may blame them for their hardships, rather than recognizing the social injustices that created the hardships (Shaw & Colimore, 1988). This social hierarchy perspective of the theory illuminates the criticism that larger questions of societal structure can remain hidden with an elitist stance.

Culture. Maslow's theory of human motivation has also been criticized for being primarily applicable to Western cultures. The theory itself was developed based on Maslow's research on US subjects. Gambrel and Cianci (2003) indicated that the hierarchy represented Maslow's values and those of the US middle class. In a critique of Maslow's work, Bouzenita and Boulanouar (2016) indicated that any hierarchy of needs created will be dependent upon the degree of individualism and/or collectivism in the society. Since Maslow's hierarchy was developed from an individualistic perspective, which is pervasive in US culture, Nevis (1983) developed a hierarchy of needs based on Chinese culture, which is known as being collectivist. In this hierarchy, the basic need is belonging rather than physiological, and there is no self-esteem need. Additionally, self-actualization is achieved by meeting the developmental needs of society.

In his original explanation of the theory, Maslow (1943) claimed that it was not intended to be universal for all cultures. Yet, he believed the types of needs would cross cultural boundaries as he felt there are certain characteristics of human nature that are similar from culture to culture. In a large number of cultural contexts, Maslow's theory has received empirical support (Davis-Sharts, 1986; Taormina & Gao, 2013). However, the cultural criticism based on collectivist views and approaches to human nature remains a present part of the controversy today.

Applications to Science Teaching

As mentioned in the introduction to this chapter, I routinely found myself referencing Maslow's theory of human motivation when I was trying to determine how to motivate my students to learn science. Given empirical support for the theory and its relevance to my students and teaching context, I applied a number of teaching strategies which I believe captured the essence of the levels in Maslow's hierarchy of needs. I believe these approaches, combined with an unwavering belief in my students, helped increase my students' motivation to learn science and scientific practices and, ultimately, their achievement in the high school science class. In the sections below, I describe applications to science teaching suitable for learning science and scientific practices for each of the five levels in Maslow's theory.

Physiological. Ensuring students' physiological needs, such as food, water, and sleep are met requires institutional support and teacher awareness. Many schools in the US have a free and reduced lunch program for students to receive food for breakfast and/or lunch at free or reduced prices. At the school level, this is a supportive program to ensure children are nourished for breakfast and lunch, which likely provides them some nutrients and energy to learn throughout the day.

Teachers also have to be aware of the physiological needs of the students in their classes. They should reach out to the appropriate support resources (e.g., school counselors) if there are any concerns with a student's nourishment, sleep, and/or health. For example, if a student is regularly coming to class and trying to sleep, a teacher should talk one-on-one with the student to decipher a cause for the behavior. However, if it is a repeated behavior, additional support resources may be needed to find out if the student is obtaining a healthy amount of sleep and nourishment at home. Many times when I was teaching, my students would come to class tired from working late shifts or taking care of family members at home. I would always check in with the students. If a student is showing symptoms of fatigue, allowing the student to stand up and get a drink of water might help awaken the mind and body.

Safety. Maslow's theory of human motivation indicates that humans prefer familiarity, consistency, and comfort to feel safe and reduce anxiety. In the science classroom, it is important to establish a welcoming, comfortable, and respectful environment where students can contribute to the construction of scientific knowledge. This begins on the first day of school with clear and consistent rules and expectations for behavior and assignments. The consistency helps students understand what is expected of them as they progress through their learning. Especially with laboratories in science, safety rules need to be strictly followed to ensure students do not find themselves in any type of danger.

Setting up a welcoming classroom environment also involves explicitly teaching and modeling respect. One of the bulletin boards in the front of my classroom read "Give Respect to Get Respect." On the first day of class, I had students share their ideas about what a respectful classroom and a disrespectful classroom looks and sounds like. These ideas were discussed and written on post-it notes which were then displayed on the bulletin board for the entire year. A culture of respect is necessary in order for students to feel safe and learn science concepts and practices. I also made sure that my students' work was displayed on the walls throughout the year. This gives the students a sense of ownership in the classroom. Students always enjoyed seeing their work on the classroom walls.

Belonging. Building a sense of belonging and acceptance in the science classroom begins with a respectful and welcoming environment discussed in the safety level section above. Once this is established, peer interaction and teacher rapport with students foster the fulfillment of the belonging needs. Collaborative learning activities and peer dialogue help to build a sense of community in the classroom, especially when groups are working toward a common goal. Many of the scientific practices in the *Next Generation Science Standards* (NGSS Lead States, 2013) incorporate collaborative learning and dialogue. For example, when students are working on engineering design solutions or asking questions to solve problems, they are often discussing and collaborating with peers. When they are engaging in argumentation, they are communicating their ideas which can involve small or whole group discussions. In a respectful environment, these collaborations and conversations help ensure a sense of belonging and acceptance of one's ideas and views. It also creates a sense of community and interdependence, important features of a supportive learning environment.

Likewise, important in meeting students' belonging needs is a teacher's rapport with students. This begins by getting to know each student as an individual in the class. There are many ways to do this at the beginning of the year, including student interest surveys or questionnaires. Reading through student responses on these helps a teacher get to know his or her students, which helps not only with rapport but also making the science content relevant to students' lives. I would have my students make a "My Biology" poster at the beginning of the year, which included any information they were willing to share about their cultures, interests, family, hobbies, etc. They could also include a photograph, which I would take and get developed if a student did not have one available. I would laminate these posters and hang them on the front sidewall of my classroom. It helped me get to know my students, and it created a sense of belonging in our classroom.

Esteem. The esteem needs involve feelings of self-respect, confidence, achievement, success, and recognition. Meeting these needs in the science classroom across the year involves student-centered instructional approaches, productive questioning strategies, and recognition of effort and success. To build esteem in science, instructional approaches should build on students' prior knowledge and guide them to accurate understandings. Instructional approaches or assessments where students are seeking to find a correct answer may reduce self-esteem in science if the student's answer is incorrect. Rather, teachers should guide students to deeper understandings in science by building on their prior knowledge. Questioning strategies can also be open-ended so students can explain their understandings rather than recalling facts from memorization. Open-ended responses provide a teacher with more information to guide student thinking and increase student confidence in science rather than close-ended responses which can often be marked as right or wrong.

Recognition of effort and success also can help build students' self-esteem in science. Praising student effort or progress rather than product can help create a growth-oriented nature toward learning (Dweck, 2006). Recognizing success in the classroom and awarding student achievement can also build self-esteem in science. There are many ways this can be done. Some teachers use achievement stars for every test score above 80%. Other teachers recognize positive group work and collaboration. Yet, others use raffle tickets for prizes which students can earn in various ways. Positive comments and feedback on work also helps encourage students and increase their self-esteem in science. However a teacher decides to recognize success and effort will depend on the approach toward recognition and/or rewards in the classroom.

Self-Actualization. Helping students toward self-actualization in the science classroom builds on meeting the needs from the previous levels in the hierarchy. Self-actualization involves the urge to grow and fulfill one's calling in the world. It certainly builds on having a strong sense of self and self-esteem. Teachers want to help students achieve their dreams and career goals. Through using a facilitative orientation toward teaching science and encouraging self-directed learning, teachers can help students progress toward attaining self-actualization. Inquiry or practice-based science instruction are more student-centered approaches to learning. Teachers can also have students set short- and long-term goals to have them work toward self-actualization. For example, when I was teaching, my students collectively wrote a whole class goal, an individual career goal, and a personal goal for their time in the class. This helped them consider their contribution to the overall collective class goal, monitor their progress in the class, and remain focused on their career goal.

In addition to the above strategies and suggestions for the five levels of the hierarchy, a teacher's overall awareness of Maslow's theory and student needs can help interpret student behavior. For instance, if students are misbehaving in class (e.g., causing disruptions, or not turning in assignments), they can be sent to the office, or the teacher may call home to parents/guardians as punishment for the misbehavior. However, these actions may be useless if the cause of the behavior is unknown. If the student lacks food to eat or a place to sleep, a visit to the main office or a phone call to a parent/guardian is not going to help remedy the situation. If students are misbehaving because of a safety need, then the teachers need to do what they can to help the students feel safe. If the student is not turning in assignments, it might be because of low esteem. Using the levels of needs in Maslow's theory can help teachers understand the cause of student behavior and, therefore, help determine appropriate actions to correct misbehavior.

The ideas and teaching applications in the above sections represent a sampling of possible ways to incorporate Maslow's theory of human motivation into science teaching. They are based on my experiences as a high school biology teacher in the US. Although the ideas are not a comprehensive list of every possible strategy, my intention is to spark ideas to implement in your own teaching context.

Summary

- Maslow believed not everything in human nature can be studied by behaviorist approaches. He studied individuals he considered to be caring and brilliant in order to learn more about the connections of motivation, learning, and being.
- The theory of human motivation explains that humans are motivated by needs which are organized into a hierarchy. Once one level of needs in the hierarchy is met, a new level becomes the focal center of motivation for an individual.
- The basic needs in Maslow's theory and hierarchy are: physiological, safety, love and belonging, esteem, and self-actualization. These are arranged in order from greatest to least priority, according to Maslow. For instance, for a person who faces extreme hunger (a physiological need), food is the primary motivating factor.
- Self-actualization, the highest tier in Maslow's theory, is a desire or motivation toward self-fulfillment. To understand more about self-actualization, Maslow studied individuals he considered to be physiologically healthy adults.
- In addition to the hierarchy of needs, Maslow's theory also contains important characteristics, such as: hierarchy reversal of needs, degrees of relative satisfaction, and multiple motivations of behavior. For instance, at times, there are exceptions to the order of the tiers in the hierarchy. Additionally, needs do not have to be 100% satisfied before progressing toward higher levels of motivation.
- The main criticisms of Maslow's theory are that it focuses on individualism and elitism and relates primarily to Western cultures.
- Teaching strategies can be applied for each of the levels in the hierarchy. The main strategies presented in this paper are:
  - Physiological—free and reduced lunch programs, teacher awareness, support resources
  - Safety—respectful and welcoming classroom, clear rules and expectations, laboratory safety
  - Belonging—collaborative learning groups/activities, peer dialogue, teacher rapport with students
  - Esteem—inquiry and student-centered instructional approaches, productive questioning strategies, recognition of effort and success
  - Self-actualization—inquiry and student-centered instructional approaches, goal setting strategies
- Maslow's theory of human motivation remains one of the most referenced and remembered theories to this day. Maslow continued to ponder questions about human motivation and nature until his death in 1970.

Acknowledgements The author would like to thank Melissa Jurkiewicz and Ryan Nixon for providing thoughtful feedback on drafts of this chapter.

References

Bouzenita, A. I., & Boulanouar, A. W. (2016). Maslow's hierarchy of needs: An Islamic critique. *Intellectual Discourse, 24*(1), 59–81.

Buss, A. R. (1979). Humanistic psychology as liberal ideology: The socio-historical roots of Maslow's theory of self-actualization. *Journal of Humanistic Psychology, 19*(3), 43–55.

Cooke, B., Mills, A. J., & Kelley, E. S. (2005). Situating Maslow in cold war America: A recontextualization of management theory. *Group and Organization Management, 32*(2), 129–152.

Cullen, D. (1997). Maslow, monkeys, and motivation theory. *Organization, 4*(3), 355–373.

Davis-Sharts, J. (1986). An empirical test of Maslow's theory of need hierarchy using hologeistic comparison by statistical sampling. *Advances in Nursing Science, 9*(1), 58–72.

Dweck, C. (2006). *Mindset: The new psychology of success*. New York: Random House.

Frager, R. (1987). Foreward: The influence of Abraham Maslow. In A. H. Maslow, R. Frager, J. Fadiman, C. Reynolds, & R. Cox (Eds.), *Motivation and personality* (3rd ed., Rev., pp. 33–41). Noida, India: Pearson.

Gambrel, P. A., & Cianci, R. (2003). Maslow's hierarchy of needs: Does it apply in a collectivist culture. *Journal of Applied Management and Entrepreneurship, 8*(2), 143–161.

Goldstein, K. (1939). *The organism: A holistic approach to biology derived from pathological data in man*. New York: American Book Company.

Hall, M. H. (1968). A conversation with Abraham H. Maslow. *Psychology Today, 35–37,* 54–57.

Lowry, R. (Ed.). (1979). *The journals of A.H. Maslow*. Monterey, CA: Brooks/Cole.

Maslow, A. H. (1943). A theory of human motivation. *Psychological Review, 50*(4), 370–396.

Maslow, A. H. (1968). *Toward a psychology of being* (2nd ed.). New York: Van Nostrand.

Maslow, A. H. (1970). *Motivation and personality* (2nd ed.). New York: Harper and Row.

Nevis, E. C. (1983). Using an American perspective in understanding another culture: Toward a hierarchy of needs for the People's Republic of China. *The Journal of Applied Behavioral Science, 19*(3), 249–264.

NGSS Lead States. (2013). *Next Generation Science Standards: For states, by states*. Washington, DC: The National Academies Press.

Pearson, E. M. (1999). Humanism and individualism: Maslow and his critics. *Adult Education Quarterly, 50*(1), 41–55.

Shaw, R., & Colimore, K. (1988). Humanistic psychology as ideology. An analysis of Maslow's contradictions. *Journal of Humanistic Psychology, 28*(3), 51–74.

Taormina, R. J., & Gao, J. H. (2013). Maslow and the motivation hierarchy: Measuring satisfaction of the needs. *The American Journal of Psychology, 126*(2), 155–177.

Shannon L. Navy is an Assistant Professor of Science Education at Kent State University in the United States. She taught high school biology prior to working in teacher education. Her research focuses on science teacher learning and development, particularly during the induction years. Related to human motivation, she is interested in teachers' motivations to engage in life-long learning opportunities. Her current work investigates teachers' mindsets and trajectories of learning.

Chapter 3
Glasser's Choice Theory and Science Education in British Columbia

Todd M. Milford and Robert B. Kiddell

Introduction

William Glasser developed *choice theory* (CT), which provides the foundations of *reality therapy* (RT), in the 1960s in an effort to explain both human behavior and motivation. In teacher education programs, and particularly in the areas of classroom management and special education, CT is relatively common and is often addressed along with other theories such as Skinner's *behaviorism*, Bandura's *self-efficacy,* and Adler's *individual psychology*. The first author was introduced to Glasser and CT within his own course work as an early career teacher while doing a Special Education Diploma at the University of British Columbia in the mid-1990s. In these classes, Glasser's ideas around motivation and student choice were appealing as they could be directly applied to the elementary classroom, the classroom in which the author planned to focus their teaching career. Glasser's writings offer a clear and straightforward explanation describing how a science classroom might be set up to best function; that is, how one can live their life in a way that works for them while getting along well with those they need to get along with (Glasser, 1998). Integrated with classroom functioning, building and maintaining positive relationships are a key area of emphasis within CT, much as relationships between students and teacher and student are emphasized in the British Columbia (BC) Science Curriculum. Glasser recognized that important human relationships were critical to success in life. This recognition of how students interact with each other, as well as with the teacher in the classroom, was also appealing as it countered the external control approach. The external control approach argued for punishing students who are doing wrong so they might do right and rewarding students for doing right so they will continue

T. M. Milford (✉) · R. B. Kiddell
University of Victoria, Victoria, BC, Canada
e-mail: tmilford@uvic.ca

R. B. Kiddell
e-mail: rkiddell@uvic.ca

© Springer Nature Switzerland AG 2020
B. Akpan and T. Kennedy (eds.), *Science Education in Theory and Practice*,
Springer Texts in Education, https://doi.org/10.1007/978-3-030-43620-9_3

to do so and was common in teacher education at the time. The recent revision of the Kindergarten (K) to Grade 12 curriculum in British Columbia (BC) emphasizes more personalized learning in science education, which better meets the needs of individual students and offers a place for Glasser and his ideas to provide guidance.

This chapter outlines the basic ideas of Glasser's CT, its relationship to additional theories and theorists, and explores how CT can be applied to the science classroom. As both authors are from BC, where the curriculum has recently undergone substantial revision to better meet the needs of twenty-first-century students, the discussion of CT and classroom applications—particularly in science—will explore the new BC curriculum in some detail.

Choice Theory and Reality Therapy

William Glasser

William Glasser was an American psychiatrist and the creator of both RT and CT. He was an anti-*Freudian* and anti-*Behaviorist* who focused on personal responsibility and personal transformation as a way to mental health and success in life. Glasser did not believe in, nor did he promote, the dominant paradigm in traditional psychiatry that the common goal was to diagnose a patient with a mental illness and prescribe medications to treat the particular illness. Instead, he believed the patient was typically acting out of unhappiness, not some kind of mental illness. Glasser notably deviated from conventional psychiatrists by warning the general public about the potential detriments caused by the profession of psychiatry. In fact, he was denied a teaching position early in his career because of his efforts to counter the teachings of Freud (Henderson & Thompson, 2010).

Glasser publicized his approaches to psychiatry and mental health through a number of single and coauthored books across a variety of topics including mental health, counseling, school improvement, and teaching. Several publications advocated a public health approach, which emphasized mental health versus the prevailing "medical" model that focused on illness and medication. He founded the Institute for Reality Therapy in 1968, which offered both introductory and advanced courses for professionals working in the areas of mental health services. Information on William Glasser can be found easily on the internet through the William Glasser Institute (n.d.; http://www.wglasser.com/) which offers research, training, journals, counseling, membership, and conference information.

Reality Therapy. RT, developed by Glasser prior to the further detailing of its theoretical foundation CT, is a person-centered approach to counseling that primarily addresses the present instead of dwelling in the past. Fundamental to RT is the suggestion that psychological problems are not the result of a mental illness, but instead human psychological problems are the result of one's inability to meet basic needs. The modern science curriculum considers how basic needs affect elementary children's learning. Glasser detailed five basic needs (i) *love and belonging* (to a family, other loved one, and/or community); (ii) *power and achievement* (a sense of winning or a sense of self-worth); (iii) *survival* (basic needs of survival, nourishment, and shelter); (iv) *freedom* (to be independent or maintain personal autonomy); and, (v) *fun* (to achieve satisfaction, enjoyment, and pleasure).

Glasser believed that an individual has control over their behavior. When an individual makes choices to change their behavior, rather than attempting to change someone else's behavior, they will more successfully meet their needs. Life becomes problematic for people when they engage in *irresponsible behaviors*; these irresponsible behaviors are defined as any effort to satisfy one's own needs that infringe upon the rights of others to meet their needs (Henderson & Thompson, 2010). RT emphasizes individual efforts to meet basic needs, and at the same time facilitates individuals (clients in this case) to become aware of, and change negative thoughts and actions. Under this approach, when an individual, or elementary science student, is feeling poorly it is because one or more of the five basic needs is not being met. The goal becomes to help the individual recognize that changing their actions may have a positive effect on the way the individual feels as well as on their ability to meet their needs, which has implications for the elementary science classroom.

According to RT, the source of almost all human problems is unsatisfactory or nonexistent connections with people. RT works by helping the person in therapy focus on the present and on what needs can be satisfied (William Glasser Institute, 2010). In this way, a specific issue or concern becomes the focus of what they can actually change. Adapting Glasser's approach helps students and teachers create connections, the teacher who follows Glasser's work likely will (a) focus on the present; (b) avoid discussing complaints; (c) avoid blaming or criticizing; (d) offer a nonjudgmental perspective; (e) avoid excuses; (f) focus on the specifics; and, (g) help students to make a tangible and workable plan to reconnect with the people they need in their lives. RT offers teachers and students a self-help tool, which effectively improves the science classroom and boosts their confidence and self-esteem.

RT has been presented as an effective approach to dealing with challenging individuals who exhibit both resistive and uncooperative behaviors (Wubbolding, 1991). Before his death in 2013, Glasser had a good deal of success applying the ideas of RT and CT (see below) at the Ventura School for Girls in California where he reduced the recidivism rate from over 90% to lower than 20% in a short period of time (Henderson & Thompson, 2010). To achieve this outcome, Glasser assigned each girl personal responsibility for her actions, favored praise over punishment, and demonstrated personal interest in each of the girls' well-being. Essentially, he applied the tenets of both RT and CT with positive results.

Choice Theory. CT primarily focusses on the idea of external versus internal control and indicates how these factors influence behavior. Glasser (1998) suggests that from birth we begin to understand how we externally control others to meet our needs (e.g., crying to gain attention or to be fed). As we age and mature, however, continuing to try and externally control others' behavior to meet our needs actually leads to unhappiness (Glasser, 1998). Glasser saw the efforts to get others to do things as the ultimate cause of relationship break downs. Central to the idea of CT is that we are truly internally motivated and that external influences never force us to do anything. Instead people are responsible for their choices, decisions, goals, and the general degree of happiness in their lives (Henderson & Thompson, 2010). He believed the best way to improve human relationships was for all people to embed the ideas of CT in their own lives. This responsibility is not to say that we have unlimited choice or that the external world is unimportant, just that individuals have a good deal of control over and are responsible for the choices they make in life. Student choice and responsibility is the core element of the BC Science Curriculum.

What motivates us internally is the personal image of a *quality world* we have created for ourselves. Whenever we feel good it is because we are "choosing to behave so that someone, something, or some belief in the real world has come close to matching a picture of that person, thing, or belief in our quality world" (Glasser, 1998, p. 45). Based upon this ideal, individuals evaluate their behavior and determine if it is the best choice to move them toward their *quality world*. CT is about understanding that one can only control their own behavior and that individuals have the ability to make choices to improve their lives (Henderson & Thompson, 2010).

The most germane idea from CT for science teachers is that educators should empower instead of control students. If, as a teacher, you deem external control and punishment as inappropriate and ineffective in helping your students to succeed, then the idea of bringing students to the point of cooperation in the classroom is appealing. Glasser believes the teacher should play the role of a manager by motivating students to make their own choices and by empowering them to take responsibility for their own learning. Thus, it is not difficult to see how Glasser's ideas have the potential to fit nicely into a science curriculum focused on personalized learning.

Relationship to Other Behavior Theories

Realizing that there are chapters in this volume on both Bandura's *social learning theory* (SCT) (see Chap. 7 this volume) as well as Vygotsky's *sociocultural theory* (see Chap. 20 this volume), these will not be dealt with to any great detail here. However, some exploration of how these theories overlap with parts of Glasser's CT provides a clearer understanding of CT as well as transitioning readers to the second part of this chapter which is associated with the application of CT in the classroom and science classroom.

Social Learning Theory. Albert Bandura's SCT suggests learning occurs in a social environment, can be acquired by observing and replicating what others do, and can also occur through the observations of behavioral reward and punishment (Bandura, 1971). Like *behaviorism* (see Chap. 6 this volume), SCT maintains that when a behavior is reinforced it will tend to continue and if not reinforced then the behavior will tend to diminish. From a comparative perspective, SCT and CT share the belief that individuals control their own lives and actions, despite some language differences (Malone, 2002). For example, Bandura speaks of an individual's self-regulatory capabilities while Glasser frames the choice or lack of choice in a behavior by asking what need was potentially being fulfilled. Self-efficacy, one's belief in their ability to succeed in specific situations or accomplish a task (Bandura, 1977), is also relevant as any discussion of CT relates to the "importance of human relationships, and, to have feelings of worth, individuals need to feel a sense of competence (self-efficacy)" (Malone, 2002, p. 11).

Sociocultural Theory. Another useful theory, when trying to understand CT and its applications to the classroom, is Vygotsky's sociocultural theory. Vygotsky emphasized the role of the environment and social interaction on cognitive development (Crain, 2011). He approached learning as a social experience, promoting social interaction as a key theme in an individual's cognitive development. The similarities between CT and sociocultural theory in terms of the importance of social relationships are obvious. Two key contributions to the understanding of cognitive development within Vygotsky's theory were the zone of proximal development (ZPD) and that of internalization. The ZPD describes tasks that a learner is unable to complete on their own but are appropriate when some assistance is provided from a more knowledgeable person (Louis, 2009). Cognitive development occurs when learners are confronted with tasks within this zone. The other contribution, internalization, suggests that social forces are key to learning and that much of what children learn is through interactions they have with the environment (Crain, 2011). Apparent in sociocultural theory is that effective social interactions are a necessary foundation for cognitive development. Glasser's CT contributes to sociocultural theory through the tools required to set this foundation for social interactions in a classroom (Louis, 2009).

Choice Theory and Teaching

The ideas that Glasser forwarded in CT, namely that humans have five basic needs (i.e., survival, freedom, power, love and belonging, and fun) they seek to satisfy, have applicability within the K to Grade 12 science classroom. Glasser felt students did their best learning when they were happy and to realize this, he felt schools needed to be places where "students can attain a sense of belonging, maintain the belief that they have some control over their academic achievement, make developmentally appropriate and meaningful choices, and appreciate school as a joyful place" (Wubbolding, 2007, p. 254). A classroom that reflects CT is one where social interactions are paramount (Irvine, 2015) recognizing children learn best when positive relationships between students, teachers, administration, and parents are actively fostered (Wubbolding, 2007). The practice of teaching in a classroom, with such positive interaction, is more about leading than about demanding. This sort of classroom is also where teachers openly demonstrate they have students' best interests at heart, allowing students to place teachers into their *quality world*. Before discussing how CT translates and informs teaching in a science classroom, a brief discussion of the BC science curriculum and its recent focus on personalized learning provides a context for why this change is warranted.

BC Science Curriculum

The public as well as the private sector provide education in Canada although funding and control are situated primarily at the provincial level. Up until the 1990s, provinces developed their own curricula without considering the rest of the landscape in Canada. In the 1990s, the Council of Ministers of Education with representation from across the country formed the Pan-Canadian Protocol (PCP) for Collaboration on School Curriculum to help each province develop their own curricula within the larger context of Canada. The first development project initiated by the PCP was the *Common Framework of Science Learning Outcomes, K to 12* (Council of Ministers of Education, Canada [CMEC], 1997), a project that focused on science education across the country.

The framework provided a vision and foundation statements for scientific literacy in Canada, outlined general and specific learning outcomes, and included illustrative examples for some of those outcomes. The framework created common ground for the development of curriculum within each participating jurisdiction, with the intent to provide greater consistency in the learning outcomes for K to Grade 12 science across jurisdictions. Other benefits included a greater harmonization of science curriculum for increased student mobility, the development of quality pan-Canadian learning resources, and collaboration for professional development activities by teachers of science. Each jurisdiction determined how the framework was to be used. The *Common Framework* had a large impact on the science curriculum across Canada and despite its relatively advanced age and calls for its revision (Milford, Jagger, Yore, & Anderson, 2010), current science curriculum from across the country continues to reflect the common framework.

The British Columbia Ministry of Education (BCME) drew heavily from the *Common Framework* when it created the integrated resource package (IRPs), the school curriculum, which was implemented in 2005 and is now replaced with a revised curriculum. The goals for science education in BC, as stipulated in the IRPs, were to provide students with scientific literacy through: understanding the nature of science, technology, and the environment; skills for inquiry; knowledge and understanding across the major domains of science; and the development of responsible attitudes toward scientific and technological knowledge (BCME, 2005).

Science curriculum developers were informed that these four goals were critical to students' scientific literacy and that the science curriculum must adhere to three principles of learning: (i) learning requires the active participation of the student; (ii) people learn in a variety of ways and at different rates; and (iii) learning is both an individual and a group process. The IRPs were broken down by subject and grade level; there was one science IRP for all of K (5 years old) to Grade 7 (12-year old), additional science IRPs for Grades 8, 9, and 10 (13–15-year old) and then subject-specific ones for upper level (16–18-year old) science (e.g., Biology 11 or Physics 12). The standards within the IRPs were called *prescribed learning outcomes* (PLOs), which set the required attitudes, skills, and knowledge students were expected to know and be able to do for each subject and grade level.

New K-12 BC Science Curriculum

In 2010, the BCME initiated the Learning Modernization Project (LMP) with the goal of helping to transform education to better meet the needs of all learners (Milford, Hawkey, Glickman, & Anderson, 2017). The LMP was a consultative process involving stakeholders, provincial partners, and school districts that took the form of local sessions, provincial and regional conferences and meetings, conversations with international experts, and online dialogue. In addition, explorations into best practices in education within BC and a review of transformation plans from other parts of Canada and from around the world helped advise this transformation.

From this process, a direction materialized reflecting the conviction that the province needed a flexible curriculum that was less prescriptive than the IRP curriculum. The new curriculum should enable teachers and students by providing choice, encouraging collaboration, and empowering innovation. The LMP offered the vision of a K to Grade 12 school system focused on competencies best suited to prepare students for their futures based upon a new curriculum that was less prescriptive, allowing for greater focus on important outcomes (and individual needs) and providing more flexibility to innovate. This vision of education has similar goals to those proposed by Glasser in CT.

The BC Education Plan (BCME, 2015a), is the most recent revised articulation of this new vision for education in BC. The Plan's vision is further informed by the understanding that capable young people should thrive in this rapidly changing world and the education system, and therefore curriculum, must better engage students in their own learning and allow students to foster the skills and competencies they will need to succeed. Much like CT, the BCME (2015b), proposes

> The best outcomes are achieved through learner-centered approaches that are sensitive to individual and group differences, that promote inclusive and collaborative learning, that harness students' passions and interests, and that deliver tailored feedback and coaching. (p. 3)

The BCME intends to achieve their vision forwarded by the BC Education Plan through a focus on personalized learning and encouraging students to learn by exploring their own interests and passions. This personalized approach supports student-initiated, self-directed, and interdisciplinary learning.

The curriculum (https://curriculum.gov.bc.ca/) that emerged from this cooperative undertaking was implemented for K to Grade 9 (5–14-year old) students in the 2016–2017 school year and was implemented for Grades 10–12 (15–18-year old) students beginning in the 2018–2019 school year. Learners in science, as in every discipline in the new curriculum, are expected to develop three core competencies: *communication, thinking,* and *personal and social* (see Table 3.1). The communication competency encompasses the set of abilities that students use to impart and exchange information. The thinking competency encompasses the knowledge, skills, and processes associated with intellectual development. The personal and social competency is the set of abilities that relate to students' identity in the world, both as individuals and as members of their community and society. The core competencies

Table 3.1 Description of the core competencies in the revised BC curriculum (BCME, 2017)

Competency	Description
Communication	The set of abilities that students use to impart and exchange information, experiences, and ideas, to explore the world around them, and to understand and effectively engage in the use of digital media
Thinking	Encompasses the knowledge, skills, and processes we associate with intellectual development. It is through their competency as thinkers that students take subject-specific concepts and content and transform them into a new understanding. Thinking competence includes specific thinking skills as well as habits of mind and metacognitive awareness
Personal and social	The set of abilities that relate to students' identity in the world, both as individuals and as members of their community and society. Personal and social competency encompasses the abilities students need to thrive as individuals, to understand and care about themselves and others, and to find and achieve their purposes in the world

are intended for use in everyday school, designed to become an integral part of the learning in all curriculum areas, and theorized to generalize to aspects of the student's life outside of school. The core competencies, while developed separately by the BCME, relate very closely to Glasser's CT.

The new BC science curriculum's curricular competencies describe further and delve deeper into the BCEP's competencies (Fig. 3.1).

The new provincial curriculum resulted in a new science curriculum for students in K to Grade 9 in the 2016–2017 school year. The redesigned science curriculum indicates for teachers the content to be addressed and embeds the core competencies of critical thinking, innovation, and collaboration. These core competencies are further informed by science curricular competencies (see Fig. 3.1) which necessitates, for example learners in Grade 4 to assume different perspectives, designing and planning their own investigation and opportunities for expressing and reflecting on personal or shared experiences of place.

The BC science curriculum also provides elaborations for the curricular competencies at each grade level. These elaborations are not science content but suggest entry points through which students may investigate concepts related to the curricular competencies. The content, with many fewer descriptors than the competencies, appears as a medium through which each individual can further their understanding of the competencies. Thus, as Glasser suggests, the students become involved in their learning and are expected to make choices that meet their needs. The curriculum was specifically transformed to encourage this sort of student success. As one tenet of that transformation states:

> Personalized learning focuses on enhancing student engagement in learning and giving students choices—more of a say in what and how they learn—leading to lifelong, self-directed learning. Students and teachers develop learning plans to build on student's interests, goals, and learning needs. Involving students in reflecting on their work and setting new goals based on their reflections allows them to take more control of their learning. Personalized

Big Ideas			
All living things sense and respond to their environment	Matter has Mass, takes up space, and can change phase	Energy can be transformed	The motions of earth and the moon cause observable patterns that affect living and non-living things

Curricular Competencies					
Questioning and Predicting: Demonstrate curiosity about the natural world. Observe objects and events in familiar contexts. Identify questions about familiar objects and events that can be investigated scientifically. Make predictions based on prior knowledge	Planning and Conducting Suggest ways to plan and conduct an inquiry to find answers to their questions. Consider ethical responsibilities when deciding how to conduct an experiment. Safely use appropriate tools to make observations and measurements. Make observations about living and non-living things in the local environment. Collect simple data.	Processing and analyzing data and information Experience and interpret the local environment. Identify First People's perspectives and knowledge as sources of information. Sort and classify data and information using drawings or provided tables. Use tables, simple bar graphs, or other formats to represent data and show simple patterns and trends. Compare results with predictions, suggesting possible reasons for findings.	Communicating Represent and communicate ideas and findings in a variety of ways, such as diagrams and simple reports, using digital technologies as appropriate. Express and reflect on personal or shared experiences of place.	Evaluating Make simple inferences based on their results and prior knowledge. Reflect on whether an investigation was a fair test. Demonstrate an understanding and appreciation of evidence. Identify some simple environmental implications of their and others' actions.	Applying and Innovating Contribute to care for self, others, school, and neighbourhood through individual or collaborative approaches. Cooperatively design projects. Transfer and apply learning to new situations. Generate and introduce new or refined ideas when engaged in problem-solving

Content
Sensing and Responding: humans, other animals, plants Biomes as large regions with similar environmental features Phases of matter The effect of temperature on particle movement Energy: has various forms, is conserved Devices that transform energy Local changes caused by Earth's axis, rotation, and orbit The effects of the relative positions of the sun, moon and Earth including local First People's perspectives

Fig. 3.1 Big ideas, curricular competencies, and content from the Grade 4 science BC revised curriculum (BCME, 2017)

learning also encompasses place-based learning, where learning experiences are adapted to
the local environment or an individual context. (BCME, 2015b, p. 2)

The result is a science curriculum that is much more personalized than the previous
curriculum in the province. The New BC Science Curriculum from its vision, through
to its goals, big ideas and competencies, and content, is designed to engage the
individual learner. The vision and goals provide a clear direction toward individual
learning while the design of the big ideas, curricular competencies, and content lead
to the development of the whole child intellectually, personally, and socially. The
science curriculum prepares students for their lives as individuals and moves teaching
from simply helping students master the knowledge and skills acquired through the
standard subject areas to applying it in their lives. BC's science curriculum design is
intended to enable a personalized, flexible, and innovative approach at all levels of
the education system.

CT and Science Teaching

The transition to a more personalized, innovating, and flexible curriculum in BC as
outlined in the previous section has the potential to address concerns Glasser voiced
with the education system. Glasser had two major difficulties with the institution
that he labeled *schooling*. First, the entire enterprise was designed to make students
acquire facts and information that have limited value in the real world; and second,
that in order to make these acquisitions occur, the system needed to emphasize
external control and punishment. Instead, he argued for *education* that encouraged
students to meet their basic needs of love and belonging, power and achievement,
survival, freedom, and fun. For Glasser, a successful education system could be
designed to meet these different needs in students.

However, the literature associated with CT in the classroom is sparse. One area
where CT has some overlap is with the education transformation currently underway
in BC. The BCME has stated that it is transforming the education system in BC
to one that "better engages students in their own learning and fosters the skills and
competencies students will need to succeed" (BCME, 2019, para. 2). The BCME
clearly emphasized *personalized learning* in its most recent curriculum revisions
and further states that to achieve personalization, "educators on the ground will
need to innovate and identify successful approaches" (BCME, 2015b, p. 8). We
believe one potential approach would be for teachers to employ CT in their efforts
to bring core and curricular competencies to the science classroom. For example, a
Grade 4 classroom that seeks to encourage students in the personal and social core
competency would fit nicely with the basic ideas of CT. The personal and social
core competency relates to how a student as an individual and as a member of a
group is associated with the larger community, as well as how students learn to care
about themselves and others. The basic principle within CT of empowering students
by offering choice within their educational experience and encouraging them to take

personal control over their interactions in the classroom would clearly align to support this competency. In this way, students can bring both science as a subject as well as their teachers into their *quality world*.

Basic needs. As a way to explore what *education* as opposed to *schooling* might look like in a science classroom, each of Glasser's basic needs (*love and belonging*; *power and achievement*; *survival*; *freedom*; and *fun*) is used as a basis for discussion.

Love and belonging. The first of Glasser's stated needs is for relationships, social connections, affection, and group membership as there is a positive association between the relationships in our lives and our welfare. Kohn (2006) suggests that students are unsuccessful, not because of cognitive deficits, but because they feel unwelcome, detached, or alienated from significant others in the educational environment. A classroom environment that counters this alienation and supports positive relationships is one of mutual trust, where students are identified positively for actions and achievements, and learning activities occur along a continuum from individual to whole group instruction. For example, the 5E model of engage, explore, explain, elaborate, and evaluate (Bybee, 1997) easily becomes a place for teachers to set students up in learning groups for explorations, ensuring that each student has a key role to play in the activity and also understands the direction and learning in the group. Additionally, the *Applying and Innovating* Curricular Competency for Grade 4 Science indicates that students are expected to (i) Contribute to care for self, others, school, and neighborhood through individual or collaborative approaches and to (ii) Cooperatively design projects.

Power and achievement. This need is based on the personal recognition and understanding of achievement, competence, and skill. Rich performance assessment tasks built upon formative assessment offer pathways for students to achieve this need. Assessment tasks, tied intimately with instruction, help teachers and students to understand what learning needs to have taken place before assessment occurs. In this way, students better understand where they have strength and weakness, which they are then able to address while teachers offer support and guidance on a more individual and differentiated level. Irvine (2015) also identified opportunities for this need in a science classroom where students became "scientists for a day" with the opportunity to understand a scientific concept and present this to the class. He suggested that this opportunity supported student choice, competence, and self-efficacy. At all levels, students of science are adept at forming rubrics or other tools on which they wish their work to be assessed. This is common practice in many schools with which we have worked.

Survival. Survival is the basic need for food, shelter, and safety. How it relates to or manifests in a science classroom is probably the most general of the needs as it easily generalizes to all classrooms. To help students feel safe and secure in the classroom, teachers can do a number of things from encouraging student self-care to maintaining a safe classroom. Allowing students some autonomy and responsibility to take care of themselves with such things as personal water bottles, nutrition, or bathroom breaks during class can help them meet these needs. Within the classroom environment, developing and maintaining consistent behavioral expectations, as well as procedures and routines can add to a sense of security. These procedures and routines may be the most suitable for the science classroom as not only does it help maintain safety while encouraging learning but also helps extend to students that they are important, valued, and cared for. Safety in a science classroom needs to be paramount but this does not mean that the "rules" must always be predetermined.

Freedom. The fourth need addresses independence, autonomy, choice, and control over one's life. On a simple level, this need can be addressed by allowing some choice on things like classroom seating and how groups are established for learning opportunities. On a larger level, we have worked in schools that offered one hour each day for "choice work." This choice time was for students to focus on one area of interest to explore it in depth. The teacher's role was as facilitator, guiding students through goal-setting, check-ins, and assistance. The science student and class benefit, for example, when a science demonstration conducted by a student is confidently discussed and explored; especially if the students are well prepared in their chosen area.

Fun. The final need deals with pleasure, play, and laughter. One way to meet this need in the science classroom is through games for learning. Herr (2008) suggests that games are alternatives to the passive recitation type classroom activities and allows students to be active and engaged participants. Although Herr looks at these primarily as review sessions, the opportunity to use games of various forms, across many aspects of science instruction, is possible as they promote "teamwork, an attitude and skill that is invaluable in every aspect of life" (2008, p. 244). One game that we have seen and played in our science methods course is Science Taboo. In this game, students are broken into teams and in turn, team members select words to describe to their peers without using the words that are identified as being taboo. For example, the described word could be telescope and the taboo words might be star, observatory, Galileo, and astronomy. Students are motivated to do well for themselves and for their peers and the ability to explain concepts without using taboo terms is an indicator of good understanding; besides, it is also quite fun.

WDEP system. Lastly, the acronym WDEP (i.e., wants and needs, direction and doing, self-evaluation, and planning) has been employed with the practice of CT to describe key procedures (Corey, 2005) and can be extended to an application in the science classroom. As previously stated, RT considers unsatisfactory relationships as the basis of human problems and therapy focuses on how to help patients make more effective choices. One of these choices for an individual is change, and the realization that they can control their own behavior.

The new curriculum in BC encourages more student choice in that it focuses on personalized learning and encouraging students to learn by exploring their own interests and passions. In a science classroom, the WDEP acronym might operationalize into the following:

- *Wants and needs*. Teachers can be encouraged to ask students what they want out of a science class or unit as well as what they are expecting. Questions to guide this process might include ones such as "What kind of science class might your idea one look like?" and "What is it that you want to get out of science class?"

- *Direction and doing.* Direction and doing focuses on current behavior and asks the question "What are you doing now?" The focus here is for teachers to encourage students to take action in changing what they are thinking and doing.
- *Self*-evaluation. The core of this approach is for students to self-evaluate and asks students if their current behavior is taking them in the direction they wish to go. Questions such as "Is what you are doing now what you want to be doing?" can be used to guide this evaluation process. CT is based upon the idea that individuals will not change until they determine a change will be beneficial.
- *Planning and action.* Once students have determined what they want from their science class, what they are currently doing, and if what they are doing is helping them meet these needs, they are then in a position to determine what they want to change (if anything) and to develop an action plan. This involves creating and carrying out a plan as well as working with the teacher to devise a different plan if the initial one does not work (Corey, 2005).

Because the new curriculum in BC is built upon student collaboration, investigation, problem-solving, communication, innovation, and discovery—all in an effort to increase understanding through hands-on science—the steps that make up the WDEP acronym could be used as a template for teachers and students to create a plan for their science learning that is doable, positive, independent, realistic, and ultimately, attainable.

Conclusion

In this chapter, we have offered a brief overview of Glasser's ideas associated with RT and CT, both counseling theories that seek to address individual's unhappiness or dissatisfaction by improving their abilities to establish relationships. Additionally, we argued for the potential application of these theories to the classroom—in particular how they might easily mesh with the new and revised curriculum in BC. The possibility of employing parts of CT and combining it with a personalized and student-centered curriculum has some distinct advantages for science teachers.

Summary

- This chapter explores the work of William Glasser, an American psychiatrist and the creator of both Reality Therapy and Choice Theory.
- Reality Therapy is a person-centered approach to counseling that primarily addresses the present instead of dwelling in the past.
- Choice Theory attempts to explain human behavior and motivation and maintains that humans have five basic needs (i.e., survival, freedom, power, love and belonging, and fun) they seek to satisfy.

- These five needs have general application from the Kindergarten to Grade 12 classroom.
- The British Columbia Ministry of Education in Canada has recently revised its Kindergarten to Grade 12 curriculum to better meet the needs of learners in the twenty-first century.
- One of the major areas of the revised British Columbia curriculum includes personalized learning.
- Teacher efforts to meet Glasser's five needs easily overlap with ideas of personalization in the classroom.

Recommended Resources

http://wglasser.com/our-approach/reality-therapy/.
http://wglasser.com/our-approach/choice-theory/.
http://www.assessmentforlearning.edu.au/default.asp.
https://curriculum.gov.bc.ca/.
http://www2.gov.bc.ca/gov/content/education-training/k-12/support/bcs-education-plan.
http://www2.gov.bc.ca/assets/gov/education/kindergarten-to-grade12/teach/pdfs/curriculum/
    curriculum-transformation-overview.pdf.

References

Bandura, A. (1971). *Social learning theory*. New York: General Learning Corporation.
Bandura, A. (1977). Self-efficacy: Towards a unifying theory of behavioral change. *Psychological Review, 84*(2), 191–215.
British Columbia Ministry of Education. (2005). *Science K to 7: Integrated resource package 2005*. Retrieved from www2.gov.bc.ca/assets/gov/education/kindergarten-to-grade.../sciences/2005scik7.pdf.
British Columbia Ministry of Education. (2015a). *British Columbia education plan*. Retrieved from http://www2.gov.bc.ca/gov/content/education-training/k-12/support/bcs-education-plan.
British Columbia Ministry of Education. (2015b). *Introduction to curriculum redesign*. Retrieved from https://curriculum.gov.bc.ca/curriculum-info.
British Columbia Ministry of Education. (2017). *Core competencies*. Retrieved from https://curriculum.gov.bc.ca/competencies.
British Columbia Ministry of Education. (2019). *Curriculum overview*. Retrieved from https://curriculum.gov.bc.ca/curriculum/overview.
Bybee, R. W. (1997). *Achieving scientific literacy*. Portsmouth, NH: Heinemann.
Corey, G. (2005). *Theory and practice of counseling and psychotherapy*. Belmont, CA: Brooks/Cole.
Council of Ministers of Education, Canada. (1997). *Common framework of science learning outcomes, K to 12: Pan-Canadian protocol for collaboration on school curriculum for use by curriculum developers*. Toronto, ON: Author. Retrieved from https://archive.org/details/commonframework00coun.
Crain, W. (2011). *Theories of development: Concepts and applications*. New York, NY: Routledge.
Glasser, W. (1998). *Choice theory*. New York, NY: Harper Collins.

Henderson, D. A., & Thompson, C. L. (2010). *Counseling Children*. Belmont, CA: Thomson Brooks/Cole.

Herr, N. (2008). *The sourcebook for teaching science: Strategies, activities, and instructional resources*. San Francisco, CA: Jossey-Bass.

Irvine, J. (2015). Encting Glasser's (1998) Choice theory in a grade 3 classroom: A case study. *Journals of Case studies in Education, 7*, 1–14.

Kohn, A. (2006). *Beyond discipline from compliance to community*. Alexandria: Virginia: Association for Supervision and Curriculum Development.

Louis, G. W. (2009). Using Glasser's choice theory to understand Vygotsky. *International Journal of Reality Therapy, 28*(2), 20–23.

Malone, Y. (2002). Social cognitive theory and choice theory: A comparative analysis. *International Journal of Reality Therapy, 22*(1), 10–13.

Milford, T. M., Hawkey, C., Glickman, V., & Anderson, J. O. (2017). Documentation sources for the evaluation of the British Columbia Learning Modernization Project (SRFP No. MED-GAD-003). Research Report. May 2017. Ministry of Education, Victoria, BC, Canada.

Milford, T. M., Jagger, S., Yore, L. D., & Anderson, J. O. (2010). National influences on science education reform in Canada. *Canadian Journal of Science, Mathematics and Technology Education, 10*(4), 370–381. https://doi.org/10.1080/14926156.2010.528827.

William Glasser Institute. (n.d.). *Welcome to the William Glasser institute—US*. Retrieved from http://www.wglasser.com/.

William Glasser Institute (2010). *Realty therapy*. Retrieved from http://www.wglasser.com/the-glasser-approach/reality-therapy.

Wubbolding, R. E. (1991). Using reality therapy with difficult and resistant people. *The Journal of Correctional Training*, 12.

Wubbolding, R. E. (2007). Glasser quality school. *Group Dynamics: Theory, Research and Practice, 11*(4), 253–261.

Todd M. Milford (BSc-1994, BEd-1997, Dip SpecEd-2000, MEd-2004, Ph.D.-2009; University of Victoria) is an associate professor in science education and research methodologies at the University of Victoria. He was a lecturer in the Art, Law, and Education Group at Griffith, University in Brisbane Australia. He has mathematics, science, and special education teaching experience at the elementary and secondary level (Victoria and Vancouver, BC) as well as in the online environment (SIDES, Saanich, BC). He has been teaching at the postsecondary level since 2005 primarily in the areas of science education, mathematics education, and classroom assessment. His research has been and continues to be varied; however, the constant theme is on using data and data analysis to help teachers and students in the classroom.

Robert B. Kiddell (BSc-1978, Cert. Ed.-1979, MA-1987; UBC) is a Ph.D. candidate in science education at the University of Victoria. He was an elementary and middle school principal in Canada and Singapore for 23 years. He has taught science education at the University of British Columbia and been Chief Judge of the Canada Wide Science Fair. Since 2016, he has been teaching science methods courses and supervising science education students at the University of Victoria.

Chapter 4
Intrinsically Motivating Instruction—Thomas Malone

Bodil Svendsen, Tony Burner, and Fredrik Mørk Røkenes

> *No compulsory learning can remain in the soul.... In teaching children, train them by a kind of game, and you will be able to see more clearly the natural bent of each (Plato, The Republic, Book VII).*

Introduction

This chapter is about the *Intrinsically Motivating Instruction Theory* by Thomas Malone. This is a manifestation which covers a wide range of in-service activities, from the acquisition of functional information to deep philosophical reflection about teaching and learning. A number of features of intrinsically motivating environments will be described further in this chapter, such as curiosity, challenge, and fantasy.

Simply put, an activity bringing about intrinsic motivation is satisfying in itself, and it is not influenced by what others will think about performance, or what kind of reward that awaits when the task is completed. Imagine children playing and how unaffected they are of extrinsic motivation, i.e., some kind of external reward in order to be engaged in playing. According to Olson, Malone, and Smith (2001, pp. 334–335) an activity is said to be intrinsically motivating if people engage in it for its own sake and if they do not engage in the activity to receive some external reward such as money, prizes, or status.

B. Svendsen (✉) · F. M. Røkenes
Norwegian University of Science and Technology, Trondheim, Norway
e-mail: bodil.svendsen@ntnu.no

F. M. Røkenes
e-mail: fredrik.rokenes@ntnu.no

T. Burner
University of Southeast, Drammen, Norway
e-mail: tony.burner@usn.no

© Springer Nature Switzerland AG 2020
B. Akpan and T. Kennedy (eds.), *Science Education in Theory and Practice*,
Springer Texts in Education, https://doi.org/10.1007/978-3-030-43620-9_4

Intrinsic Motivation

The concept of intrinsic motivation was introduced in the 1950s in the field of animal psychology and further developed by others within human psychology. People are motivated to bring to their cognitive structures three of the characteristics of well-formed scientific theories: completeness, consistency, and parsimony (Malone, 1981). The way to engage learners' curiosity is to present them with just enough information to make their existing knowledge seem incomplete, inconsistent, or unparsimonious. The learners will then be motivated to learn more, to make their cognitive constructions better. Intrinsic motivation is an impetus for behavior that the individual will perform, although it does not cause any external rewards or any external consequences (Bailey et al., 2012; Skogen, 2014).

Extrinsic reinforcement such as external rewards may destroy the intrinsic motivation if a person should engage in an activity, and degrade the quality of certain kinds of task performances (Condry, 1977; Lepper & Greene, 1979). An example given by Lepper, Greene, and Nisbett (1973) regards nursery school children who liked to play with marking pens, receiving a promised reward for doing so. Consequently, they later played less with the marking pens compared to children in a control group who received no promised reward.

Both Piaget (1951) and Bruner (1962) argued for the significance of intrinsically motivated play-like activities for many kinds of deep learning. If students are intrinsically motivated to learn something, they may spend more time and effort learning, feel better about what they learn, and use it more in the future. Shulman and Keislar (1966) argued that learners may learn "better" in the sense that more fundamental cognitive structures are modified, including the development of such skills as "learning how to learn." Papert (1980) discusses the "power principle," which is the notion that the knowledge being learned should "… empower the learner to perform personally meaningful projects that could not be done without it" (p. 54). Cognitive curiosity can be thought of as a desire to bring better "form" to one's knowledge structures (Olson et al., 2001, p. 363). Finally, praise can play an important role in enhancing intrinsic motivation (Henderlong & Lepper, 2002). This is where the teacher and the peers become important in laying the foundation for intrinsically motivating contexts. According to a review of what type of praise enhances intrinsic motivation, Henderlong and Lepper (2002, p. 787) point out that praise which is perceived as sincere, encourages adaptive performance attributions, promotes perceived autonomy, provides positive information about personal performance without social comparisons, and conveys standards and expectations that are realistic and not disruptive. This is also supported by educational research on student assessment (Hattie & Timperley, 2007).

Designing Instructional Environments

We live in a digitalized world where information and communications technology (ICT) has become deeply embedded into our lives and daily activities. Digital tools such as computers play a significant role at work, in education, and in society in general by modifying and transforming existing practices, and patterns of interaction. The presence of ICT makes it possible to simulate complex systems and processes such as the effects of gravitational forces on astronauts while on board a space shuttle in outer space. As a result, the school curriculum can focus on "authentic" problems parallel to those that students will be facing in real-world settings. For example, modeling and visualization tools can be used to bridge experience and abstraction, and controversial topics may be discussed with experts outside the immediate classroom (Crosier, Cobb, & Wilson, 2002). Previous studies on ICT in schools often focus on issues related to teachers' and students' lack of ICT skills, barriers, and enablers to the uptake of ICT in schools, and the lack of computer technology in schools (Mork, 2011). Today, the focus seems to have shifted somewhat from focusing on technological infrastructure in schools toward educational use of ICT in subject disciplines. Here, an interesting research focus is on studies that can help us understand how game-based learning with the use of ICT or "gamification" may be used to enhance students' learning (Connolly, Boyle, MacArthur, Hainey, & Boyle, 2012).

Studies of the use of ICT in educational settings have focused on issues like design, change of classroom practices and learning outcomes (Clark & Jorde, 2004). Learning about science involves being introduced to natural science concepts, conventions, laws, theories, principles, and how to work in science. This also means recognizing how this knowledge can be used in social, technological, and environmental contexts.

We will further in this chapter provide examples of intrinsically motivating instructions in science, based on different environmental features such as challenge, fantasy, and curiosity as proposed by Malone (1981).

Challenge

Challenge is captivating because it engages a person's self-esteem. Success in an instructional environment, like success in any challenging activity, can make people feel better about themselves. The opposite side of this principle is, however, that failure in a challenging activity can lower a person's self-esteem and, if it is severe enough, decrease the person's interest in the instructional activity.

One simple implication of this relationship is that instructional activities should have a variable difficulty level so learners can work at an appropriate level for their ability. Another implication might be that performance feedback should be presented in a way that minimizes the possibility of self-esteem damage (Henderlong & Lepper, 2002).

Fantasy

Fantasies can make instructional environments more interesting and more educational. Malone (1981, p. 337) defines a fantasy inducing environment as one that evokes "mental images of things not present to the senses or within the actual experience of the person involved." These mental images can be either of physical objects or of social situations, and they may or may not be likely to occur in the learner's environment.

One relatively easy way to try to increase the fun of learning is to take an existing curriculum and overlay it with a game in which the player progresses toward some fantasy goal, or avoids some fantasy catastrophe (like hangman), depending only on whether the player's answers are right or wrong. These are examples of *extrinsic fantasies*, where the fantasy depends on the use of the skill but not vice versa. Other factors such as speed of answering can also affect extrinsic fantasies (like in the game-based digital quiz application Kahoot). In *intrinsic fantasies*, on the other hand, not only does the fantasy depend on the skill, but the skill also depends on the fantasy. This usually means that problems are presented in terms of the elements of the fantasy world, and players receive a natural kind of constructive feedback. In general, intrinsic fantasies are both (a) more interesting and (b) more instructional than extrinsic fantasies (Asgari & Kaufman, 2004). One advantage of intrinsic fantasies is that they often indicate how the skill could be used to accomplish some real-world goal.

Metaphors or analogies of the kind provided by intrinsic fantasies can often help a learner apply old knowledge in understanding new things. Another cognitive advantage of intrinsic fantasies is simply that, by provoking vivid images related to the material being learned, they can improve memory of the material. Fantasies in computer games almost certainly derive some of their appeal from the emotional needs they help to satisfy in the people who play them. It is very difficult to know what emotional needs people have and, for example, how these needs might be partially met by computer games. Computer games that embody emotionally involving fantasies like war, destruction, and competition are likely to be more popular than those with less emotional fantasies (Malone, 1981). One obvious consequence of the importance of emotional aspects of fantasies is that different people will find different fantasies appealing. If instructional designers can create many kinds of fantasies for various kinds of people, their activities are likely to have much broader appeal. For example, one can easily envision a math game where different students see the same problems but can choose which fantasy they want to see. Instructional designers might also create environments into which students can project their own fantasies in a relatively unconstrained way. For instance, one could let students name imaginary participants in a computer game.

It is difficult to predict what kinds of fantasies will be appealing to different people. There are also difficult questions about whether it is sometimes bad to encourage certain fantasies. For example, if a computer game provides an outlet for aggressive fantasies, then that could have detrimental effects.

Curiosity

The educational environment should be neither too complicated nor too simple with respect to the learner's existing knowledge. It should be *novel* and *surprising*, but not completely incomprehensible. In general, an optimally complex environment will be one where the learner knows enough to have expectations about what will happen, but where these expectations are sometimes unmet.

There are several parallels between challenge and curiosity. Both often depend on adjusting the environment to the learner's ability or understanding. Both also depend on feedback to reduce uncertainty. Challenge could be explained as curiosity about one's own ability, or curiosity could be explained as a challenge to one's understanding. While the notion of self-esteem is central to the idea of challenge, self-esteem is not involved in most curiosity.

Sensory curiosity involves the attention-attracting value of changes in the light, sound, or other sensory stimuli of an environment. There is no reason why educational environments have to be impoverished sensory environments. Examples of this are artifacts like colorfully illustrated textbooks and television. Computers provide even more possibilities for graphics, animation, music, and other captivating *audio and visual effects*. These effects can be used: (1) *as decoration* (e.g., music at the beginning of a game), (2) *to enhance fantasy*, (3) *as a reward*, and (4) *as a representation system* that may be more effective than words or numbers.

Cognitive curiosity can be thought of as a desire to bring better "form" to one's knowledge structures. In particular, people are motivated to bring to all their cognitive structures three of the characteristics of well-formed scientific theories: *completeness, consistency*, and *parsimony* (Malone, 1981). Cognitive curiosity is about engaging learners and presenting exactly enough information to make the learner's existing knowledge seem incomplete, inconsistent, or unparsimonious. The learners are then motivated to learn more, to make their cognitive structures better-formed.

Designing Instructional Environment by Using the 5E-Model

An example of a model to fit the features of *challenge, fantasy*, and *curiosity* is found in the 5E-model (Bybee et al., 2006; Svendsen, 2015). The model has its origins in the Biological Sciences Curriculum Study (BSCS), in which American scholars developed educational programs and research on teaching and learning in natural science. The five Es are the initial letters in the words *engage, explore, explain, elaborate*, and *evaluate*.

In the 5E-model, the teachers teach by *engaging* the students with a starter. A startup should be both motivating and related to phenomena that students can relate to (like every day phenomena). The students' prior knowledge is accessed by the teacher or the curriculum, and helps students to become engaged in a concept using short activities, or introduction to phenomena in order to endorse interest and provoke prior knowledge. The activities of this phase make connections to past experiences and expose students' misconceptions; they should serve to ease cognitive imbalance. Activity refers to both mental and physical activity (Bybee et al., 2006).

Once the activities have engaged the students, the students need time to *explore* the ideas. Activities should be designed for the students to have common, concrete experiences upon which they continue formulating concepts, processes, and skills. The students work actively with the material (read, write, investigate, play, observe, etc.) and add knowledge and skills to reach new learning goals. This level is concrete and hands-on, and the use of touchable materials and concrete experiences is essential but not necessary. The aim of creating cognitive curiosity is to establish experiences that teachers and students can use later to introduce and discuss concepts, processes, or skills. Explanation provides openings for teachers to directly introduce a concept, process, or skill. The students *explain* their understanding of the concept. An explanation from the teacher may guide them toward a deeper understanding, which is a critical part of their new understanding. By facilitating activities that build on the knowledge and skills the student already possesses, and allow students to reflect, discuss, read, and write to achieve the learning objectives, the teacher can introduce new concepts that challenge student's conceptual understanding (Bybee et al., 2006).

Teachers have a variety of techniques and strategies at their disposal to stimulate and develop student explanations. Once the students have explanations and terms for their learning tasks, it is important to involve them in further experiences which extend, or *elaborate*, the concepts, processes, or skills. This level facilitates the transfer of concepts to closely related but new situations. Students' theoretical understandings and skills are challenged by their new experiences and by the guidance of their teachers. They develop deeper and extensive understanding, more information, and adequate skills. Students apply their understanding of the concept by conducting supplementary activities. Elaborative activities provide further time and experiences that contribute to learning.

Evaluation should be continuous, varied, and be a part of all levels. Evaluation concerns the activities and has a meta-perspective on them. Assessment is on the individual level and concerns self- and peer assessment, continuous assessment, and final assessment of processes and products. It can be conducted orally, in writing or in a combination. Students consider their own learning and understanding, and the teacher and/or peers will assess student learning in relation to learning objectives in a given subject or in an activity, and in relation to the objectives of the curriculum.

We will now introduce an example of cognitive curiosity, which is stimulated by pointing out inconsistencies or paradoxes in the learner's knowledge. Concept cartoons are drawings that trigger students' fantasy and curiosity about science issues. They often show different characters arguing about an everyday situation. The characters express both scientific viewpoints and common misconceptions. In debating the ideas, students use scaffolding to articulate their thoughts (Wood, Bruner, & Ross, 1976), challenge each other, propose claims and explanations, and justify their reasoning. Concepts can be understood by using ICT and gaming. In lack of ICT opportunities in classrooms, concept cartoons constitute a useful teaching, learning, and assessment tool. For instance, students may be told that plants require healthy soil for the process of growing, but why is that important? It raises curiosity about what makes plants grow, and wherefrom they gain weight. One might also evoke curiosity by giving a number of examples of a general rule before showing how (or letting students discover that) all the examples can be explained more parsimoniously by the new knowledge.

In conclusion, the 5E-model can be supportive in making teaching by inquiry through the use of ICT or gaming explicit and targeted. By shaping clear learning aims for teaching, teachers and students can use the model as a reflection tool for designing, planning, implementing, and evaluating their teaching sequences, and thus expand students' learning processes.

Summary

- Malone's theory of intrinsically motivating instruction is based on three categories:
  - *Challenge* is hypothesized to depend on goals with uncertain outcomes. Several ways of making outcomes uncertain are discussed, including variable difficulty level, multiple level goals, hidden information, and randomness.
  - *Fantasy* has both cognitive and emotional advantages in designing instructional environments. A distinction is made between extrinsic fantasies that depend only weakly on the skill used in a game, and intrinsic fantasies that are intimately related to the use of the skill.
  - *Curiosity* is separated into sensory and cognitive components, and it is suggested that cognitive curiosity can be aroused by making learners believe their knowledge structures are incomplete, inconsistent, or unparsimonious.
- An activity bringing about intrinsic motivation is satisfying in itself, and it is not influenced by what others will think about performance, or what kind of reward that awaits when the task is completed.
- An activity is said to be intrinsically motivating if people engage in it for its own sake.
- A way to engage learners' curiosity is to present them with just enough information to make their existing knowledge seem incomplete, inconsistent, or unparsimonious.

Recommended Resources

Books
Sawyer, R. K. (Ed.). (2014). *The Cambridge Handbook of the Learning Sciences* (2nd ed.).
    Cambridge: Cambridge University Press.
Journals
Instructional Science: https://link.springer.com/journal/11251.
Learning and Instruction: https://www.journals.elsevier.com/learning-and-instruction/.
Internet Source
The MIT Center for Collective Intelligence conduct research on how people and computers can
    work together more intelligently: http://cci.mit.edu/.

References

Asgari, M., & Kaufman, D. (2004). Relationships among computer games, fantasy, and learning.
    Paper presented at the *2nd International Conference on Imagination and Education.*
Bailey, R., Pearce, G., Smith, C., Sutherland, M., Stack, N., Winstanley, C., et al. (2012). Improv-
    ing the educational achievement of gifted and talented students: A systematic review. *Talent
    Development & Excellence, 4*(1), 33–48.
Bruner, J. S. (1962). *On knowing: Essays for the left hand.* Cambridge, MA: Belknap Press of
    Harvard University Press.
Bybee, R., Taylor, J. A., Gardner, A., Van Scotter, P., Carlson, J., Westbrook, A., et al. (2006). *The
    BSCS 5E instructional model: Origins and effectiveness.* Colorado Springs, CO: BSCS.
Clark, D., & Jorde, D. (2004). Helping students revise disruptive experientially supported ideas about
    thermodynamics: Computer visualizations and tactile models. *Journal of Research in Science
    Teaching, 41,* 1–23.
Condry, J. (1977). Enemies of exploration: Self-initiated versus other-initiated learning. *Journal of
    Personality and Social Psychology, 35,* 459–477.
Connolly, T. M., Boyle, E. A., MacArthur, E., Hainey, T., & Boyle, J. M. (2012). A systematic
    literature review of empirical evidence on computer games and serious games. *Computers &
    Education, 59*(2), 661–686.
Crosier, J. K., Cobb, S. V. G., & Wilson, J. R. (2002). Key lessons for the design and integration of
    virtual environments in secondary science. *Computers & Education, 38,* 77–94.
Hattie, J., & Timperley, H. (2007). The power of feedback. *Review of Educational Research, 77*(1),
    81–112.
Henderlong, J., & Lepper, M. R. (2002). The effects of praise on children's intrinsic motivation: A
    review and synthesis. *Psychological Bulletin, 128*(5), 774–795.
Lepper, M. R., & Greene, D. (1979). *The hidden costs of reward.* Morristown, NJ: Lawrence
    Erl-baum Associates.
Lepper, M. R., Greene, D., & Nisbett, R. E. (1973). Undermining children's intrinsic interest with
    extrinsic rewards: A test of the overjustification hypothesis. *Journal of Personality and Social
    Psychology, 28,* 129–137.
Malone, T. (1981). Toward a theory of intrinsically motivating instruction. *Cognitive Science, 4,*
    333–369.
Millgate House of Education: https://www.millgatehouse.co.uk/
Mork, S. M. (2011). An interactive learning environment designed to increase the possibilities
    for learning and communication about radioactivity. *Interactive learning environments, 19*(2),
    163–177.
Olson, G. M., Malone, T. W., & Smith, J. B. (Eds.). (2001). *Coordination theory and collaboration
    technology.* Mahwah, NJ: Erlbaum.

Papert, S. (1980). *Mindstorms: Children, computers, and powerful ideas*. New York, NY: Basic Books.

Piaget, J. (1951). *Play, dreams, and imitation in childhood*. New York, NY: Norton.

Shulman, L. S., & Keislar, E. R. (Eds.). (1966). *Learning by discovery: A critical appraisal*. Chicago, IL: Rand McNally.

Skogen, K. (2014). De evnerikebarnaogspesialpedagogikken. [High ability children and special pedagogy]. I S. Germeten (Red.) (Ed.). *De utenfor: Forskning om spesialpedagogikkogspesialundervisning [Outsiders: Special Needs and Special Education]* (pp. 89–101). Bergen: Fagbokforlaget.

Svendsen, B. (2015). Mediating artifact in teacher professional development. *International Journal of Science Education, 37*(11), 1834–1854.

Wood, D. J., Bruner, J. S., & Ross, G. (1976). The role of tutoring in problem solving. *Journal of Child Psychiatry and Psychology, 17*(2), 89–100.

Bodil Svendsen (Ph.D.) works as an Associate Professor of Natural Science didacticts at the Department of Teacher Education at the Norwegian University of Science and Technology (NTNU), where she has taught in-service and pre-service courses to teachers and students since 2006. She has also been the Head of the Gifted Children Center at Trondheim Science Center in the period form 2016-2019. Bodil has teaching experience from elementary school, middle school, senior high school, and adult education teaching Natural Science, Biology, and Geography since 1997. She has broad international experience with research, among others from Finland, Scotland, Denmark, Sweden, and England. Her main research interests are Natural Science Education, school development, R&D work, teacher mentoring, professional development, and gifted children.

Tony Burner (Ph.D.) works as a Professor of English at the Department of Languages and Literature Studies at the University College of Southeast Norway, where he has taught in-service and pre-service courses to teachers and student teachers the last 11 years. He has broad international experience with research, among others from Finland, Australia, Iraqi Kurdistan, and Vietnam. His main research interests are English education, classroom assessment, research and development work, teacher mentoring, multilingualism, and professional development.

Fredrik Mørk Røkenes (Ph.D.) works as an Associate Professor of English didactics at the Department of Teacher Education at the Norwegian University of Science and Technology, where he has taught in-service and pre-service courses to teachers and student teachers the last 7 years. He has international experience with research, among others from Australia and Belgium, and is currently leading a large-scale national project on the digitalization of teacher education. His main research interests are professional digital competence, English didactics, literature reviews, design-based research, and teacher education.

Chapter 5
The *Bildung* Theory—From von Humboldt to Klafki and Beyond

Jesper Sjöström and Ingo Eilks

Bildung

Bildung is the central theory of education in the German speaking part of Europe and in Scandinavia (Sweden, Denmark, and Norway), and it is also influential on traditions of education in some South American countries, like Brazil (Sjöström, Frerichs, Zuin, & Eilks, 2017). *Bildung* covers a more than 200-year-long central European tradition of education dating back to works of Wilhelm von Humboldt in the late eighteenth century (see a translation of Humboldt's work from 1793 in Westbury, Hopmann, & Riquarts, 2000). Since then it has had an important role in central and northern European educational philosophy and policy.

Bildung is a theory of defining the aims and objectives of any education. It is a complex educational concept that has connections to both the Enlightenment and Romanticism. In the eighteenth century, *Bildung* was mainly connected to humanity and in the end of the nineteenth century it became mainly understood as a value and commodity (Sjöström et al., 2017). There was a decline in the use of the concept during the 1960s and 1970s, due to both the Sputnik shock and the student movement. However, since the 1980s the concept has to some extent reappeared and during the last two decades it has been reconsidered from late/postmodern perspectives (see for example Sjöström, 2018).

J. Sjöström (✉)
Faculty of Education and Society, Department of Science-Mathematics-Society, Malmö University, 205 06 Malmö, Sweden
e-mail: jesper.sjostrom@mau.se

I. Eilks
Department of Biology and Chemistry, Institute for Science Education, University of Bremen, 28359 Bremen, Germany
e-mail: ingo.eilks@uni-bremen.de

© The Author(s) 2020, corrected publication 2021
B. Akpan and T. Kennedy (eds.), *Science Education in Theory and Practice*,
Springer Texts in Education, https://doi.org/10.1007/978-3-030-43620-9_5

Over the past two centuries, various scholars have contributed to clarify the concept of *Bildung*. Some important early *Bildung*-theorists from Germany were Wilhelm von Humboldt (1767–1835) and Johann Gottfried Herder (1744–1803). Examples of *Bildung*-scholars from Scandinavia are Nikolaj Frederik Severin Grundtvig (1783–1872), Carl Adalph Agardh (1785–1859), and Ellen Key (1849–1926). More recent German scholars in the field were Hans-Georg Gadamer (1900–2002), Paul Ricoeur (1913–2005), Erich Weniger (1894–1961), and Wolfgang Klafki (1927–2016).

The concept of *Bildung* is rich and complex. Generally, it consists of two elements: an ideal picture of desirable knowledge and skills, and free learning processes, or in other words both "the process of personal development and the result of this development process" (Fischler, 2011, p. 33). The seminal works leading to our contemporary understanding of *Bildung* stem mainly from the 1950s to the 1970s. Klafki and others defined *Bildung* (or *Allgemeinbildung* meaning *Bildung* for all and in all human capacities; see further below) as the ability to recognize and follow one's own interests in society and to behave within society as a responsible citizen (see the translated and updated contributions of Klafki in Westbury et al., 2000). This was linked to developing the capacity for self-determination, participation, and solidarity within society. Within this debate, *Bildung* was never understood as something one can be taught, but *Bildung*-oriented education is suggested as a way for everyone to support developing *Bildung* on their own. *Bildung* in a theoretical view is more of a concept of achieving capacity and skills than a set of facts and theories to be learned. *Bildung* is viewed more as a process of activating potential than a process of learning (see a translation of Weniger's work from 1952 in Westbury et al., 2000).

Schneider (2012) describes *Bildung* as a reflexive event and its function to design and form the self, a complex meaning-making process that occurs from childhood to advanced age. It is understood as a lifelong challenge and opportunity and is connected to developing critical consciousness, a process of character-formation and self-discovery. It is connected to issues of finding truth, value, and meaning. For Bauer (2003, p. 212), *Bildung* covers "creative, critical and transformative processes which change the relationship of self and world in conjunction with a changing social and material environment." In other words, *Bildung* consists of autonomous self-formation and reflective and responsible action in, and in interaction with, society. As a humanistic theory, *Bildung* theory (or better theories as will be described below) has similarities to some of the theories described in this book in Sect. V, such as systems thinking and transdisciplinary teaching. Contemporary ideas of critical-reflexive *Bildung*, which is in focus in this chapter, adds philosophical as well as political dimensions to the teaching and learning of science. As such, it is a vehicle for promoting socio-political activism, that is, assisting students to become active citizens in addressing science and technology-related issues at both local and global levels.

Because there is no precise English translation, the German term *Bildung* has been used in the international science education literature (see, for example, Elmose & Roth, 2005; Hofstein, Eilks, & Bybee, 2011; Sjöström, 2013). The often-used translation of *Bildung* as only "education" ignores its special roots and the unique philosophical framework behind the concept. See Westbury et al. (2000) for some translated original contributions from the history of *Bildung* and the related concept of *Didaktik*. It is necessary to say that the *Bildung*-connected meaning of the term *Didaktik* in German and Scandinavian languages differs a lot from how the word *didactics* is used in English (Gundem, 2010). *Didaktik* in German and Scandinavian languages means the praxis knowledge about teaching and at the same time it covers the research area about teaching and learning (Kansanen, 2009).

In the German-speaking countries there has for a long time been a debate about what is to be meant by *Bildung* with its both individual and societal implications when it comes to the teaching and learning of science (e.g., Marks, Stuckey, Belova, & Eilks, 2014). Also in Scandinavia, there has been an interest in this debate. For example, in 1998 Svein Sjøberg published the first edition of his teacher education textbook *Science as part of Bildung for all—a critical subject-Didaktik* (our translation). It has become a standard text in science-teacher education in the whole of Scandinavia. In recent years, the concept of *Bildung* has been used to justify new philosophies of science education, like the ideas of critical scientific literacy (Sjöström & Eilks, 2018) or eco-reflexive science education (Sjöström, Eilks, & Zuin, 2016).

Before further applying the concept of *Bildung* on science teaching and learning and connecting it to the concept of "scientific literacy," we will first describe different ideas related to *Bildung* and then also its connection to what is called critical-constructive *Didaktik*.

Different Ideas Related to *Bildung*

With reference to the literature, Sjöström and Eilks (2018) and Sjöström et al. (2017) recently identified five educational traditions directly related to the *Bildung* theory. They can be called: (a) classical *Bildung*, (b) liberal education, (c) Scandinavian *folk-Bildung*, (d) democratic education, and (e) critical-hermeneutic *Bildung*:

(a) *Classical Bildung*: Classical Bildung is based on von Humboldt's way of understanding *Bildung* as a process of individualization, where the human being develops personality in all their human capacity. However, today von Humboldt is often—at least at universities—more associated with free search for knowledge, free from both the state and the market. The works of von Humboldt are also sometimes misused. His idea that *Bildung* manifests itself mainly in languages, led to a long time of devaluing the sciences for developing own worldviews in the individual.

(b) *Liberal education*: The thoughts behind liberal education can also be tracked back to von Humboldt in the means of education as character-formation. The character-formation ideal is emphasized especially in the English version, whereas the canon has been emphasized in the American version. A famous representative for a more critical and cosmopolitical version of liberal education is Martha Nussbaum (1997, 2010). She argues for ethical self-reflection and critical approaches to the own culture and its traditions as essential part of education. This is needed to create enlightened citizens, rather than efficient workers and uncritical consumers. Nussbaum uses typical *Bildung*-type arguments for liberal education, however, without explicitly using the term.

(c) *Scandinavian folk-Bildung*: From the late nineteenth century a unique tradition called *folkbildning* in Swedish (might be translated as "*Bildung* for the whole people") was developed in Scandinavia. *Folk-Bildung* is less academically oriented than the classical *Bildung*. In this tradition, *Bildung* was combined with a pronounced benefit-approach. The political dimension was much more explicit in *folk-Bildung* than in the classical German version, but it was not especially radical.

(d) *Democratic education*: The idea of education for all was also developed in the USA by John Dewey (1859–1952). The connection of democratic education with *Bildung* lies in promoting social-ethical foundations of a society to promote democratic habits. Dewey used the term *Bildung* in his work, although not systematically.

(e) *Critical-hermeneutic Bildung*: This tradition is rooted in the works of Hans-Georg Gadamer and Paul Ricoeur and was developed mainly in the 1950s and 1960s by Erich Weniger and Wolfgang Klafki. They developed a new understanding of *Bildung* connected to educational practices and a democratic and emancipatory view of society. They created the term *Allgemeinbildung*. Within this concept, part of the word, *Allgemein* (which can be translated as "general") has two dimensions. The first dimension means to achieve *Bildung* for all persons (like in the Scandinavian approach of *folkbildning*). The second dimension aims at *Bildung* in all human capacities. Klafki's thinking was based on the thought that responsible life and action of any citizen in a democratic society needs *Bildung* as the capacity to determine one's own life, to be able to participate in society, and to act solidary toward others. This educational philosophy has a clear democratic and critical approach and is the most complex and advanced concept of *Bildung* (Sjöström & Eilks, 2018). It has later been influenced by late/postmodern perspectives in contrast to the other four *Bildung*-traditions, which are mainly based on Western modernism (e.g., Sjöström, 2018).

Klafki's Concepts of Material, Formal, and Categorical *Bildung*

Klafki's *Bildung* theory and its connected ideas of *Didaktik* include both epistemological aspects and practice-oriented concepts for use in lesson planning. Klafki explained his view of *Bildung* with the term categorical *Bildung* (see the contributions of Klafki in Westbury et al., 2000). It was developed based on an analysis of 150 years of views of knowledge and learning in educational theory. Klafki identified two main ideas of thought: (1) material *Bildung* and (2) formal *Bildung*, respectively.

Then he suggested the concept of (3) categorical *Bildung,* which includes elements from both material and formal *Bildung.*

(1) In material *Bildung* theories, content knowledge is prioritized over developing general competences of the learner. In other words, the objective side is prioritized over the subjective. There are two subgroups of material *Bildung* theories: scientific *Bildung* and humanistic *Bildung,* respectively. Scientific *Bildung* is based on a belief in the objectivity of knowledge, that is epistemological positivism. Humanistic *Bildung* focuses on cultural quality. It is about learning about human traditions.

(2) In formal *Bildung* theories, competences of the learner are prioritized over learning of content knowledge. In other words, the subjective side is prioritized over the objective. There are two subgroups of formal *Bildung* theories: functional *Bildung* and method-based *Bildung,* respectively. Functional *Bildung* has its roots in the philosophy of Rousseau and is also the type of *Bildung* emphasized by von Humboldt. Focus is on human powers and potentials. Method-based *Bildung* focuses on the processes of learning methods and ways of thinking to "master life." This line of thinking is connected to ideas of meta-learning and learning strategies. It is connected to the ideas of John Dewey.

Generally, Klafki prioritized formal over material *Bildung.* However, there are several arguments to why formal *Bildung* theories are not enough of their own. The main problem with pure formal *Bildung* theories is that it is hard to develop any competences without having any content to apply them on. How can a teacher motivate students to develop skills without engaging in specific content? Instead of turning back to a content-based curriculum, however, Klafki suggested the concept of categorical *Bildung.*

(3) In categorical *Bildung,* Klafki suggested to connect both views. He suggested that any learning activity should contribute to both material and formal gains in the learner. He suggested selecting content that is elementary and basic for the discipline; that is fundamental for essential experiences of and insights into the world; and that has exemplary significance to offer structure for understanding the field of study.

The relationship between the three different types of *Bildung* is illustrated in Fig. 5.1. Material, formal, and categorical *Bildung* can further be connected to the five *Bildung*-traditions (described above) in the following ways:

		Content orientation	
		Low	High
Skills orientation	Low		Material *Bildung*
	High	Formal *Bildung*	Categorical *Bildung*

Fig. 5.1 Relationships between material, formal, and categorical *Bildung*

- Material *Bildung* emphasizes content knowledge. It is in line with the American version of liberal education, mainly focusing the canon of topics. Important aspects of especially humanistic material *Bildung* can also be found in classical *Bildung*, although von Humboldt's orientation is probably better described as formal *Bildung*.
- Formal *Bildung* emphasizes the development of the person. For example, the character-formation ideal is emphasized in the English tradition of liberal education, which focuses on skills development. Both the Scandinavian "folk-*Bildung*"-tradition and democratic education have many aspects that can be categorized as formal *Bildung*.
- Categorical *Bildung* emphasizes both content and the skills development in the learner. We would claim that critical-hermeneutic *Bildung* is most compatible with categorical *Bildung*.

Except emancipation, Klafki's view of *Bildung* can be described with terms such as autonomy, responsibility, reason, and interdependence, but also humanity, world, and objectivity education. He suggested the following three elements as guiding principles that characterize *Bildung*:

- Self-determination ability (to be able to take up one's own interests as part of society).
- Ability for participation (to be able to actively participate in and contribute to the development of society).
- Solidarity ability (to act responsibly in society with a view on those whose opportunities for self-determination and participation are limited).

In line with the thinking of Klafki, *Bildung* can be suggested as a critical concept in a late/postmodern world (Elmose & Roth, 2005; Sjöström & Eilks, 2018; Sjöström, 2018). It can form the basis for new interpretations of *Bildung* to come up with challenges of our contemporary society as a risk society to make education an eco-reflexive and transformative practice (Sjöström et al., 2016) and to provide relevant education in all its dimensions (Stuckey, Hofstein, Mamlok-Naaman, & Eilks, 2013).

Bildung and Critical-Constructive *Didaktik*

For educational operation Klafki and others developed a tool called *Didaktik* analysis as being part of what has been called critical-constructive *Didaktik* (see the contributions of Klafki in Westbury et al., 2000). According to Duit (2015, p. 325) *Didaktik* "stands for a multifaceted view of planning and performing instruction. It is based on the German concept of *Bildung* [… and it] concerns the analytical process of transposing (and transforming) human knowledge (the cultural heritage) into knowledge for schooling that contributes to Bildung." It is about answering the three fundamental *Didaktik*-questions: *why?* (intentions—aims and objectives), *what*? (topic of instruction—content), and *how*? (methods of instruction and media used) (see also for example Duit, 2012).

Klafki's *Bildung* theory is, as already indicated above, connected to the German tradition of *Didaktik* (also called *Bildung*-centered *Didaktik*). If the German understanding of *Didaktik* is compared to the Anglo-American concepts of curriculum and instruction, *Didaktik* can be understood as teaching based on *Bildung*, focusing on matter and meaning, and autonomy of teaching and learning. Kansanen (2009) compared subject-specific *Didaktik* (in German *Fachdidaktik*) with Lee Shulman's idea of *Pedagogical Content Knowledge* (PCK). According to him, *Didaktik* is much broader and also containing aspects of values and other characteristics related to curriculum theory and pedagogy. *Didaktik* focuses predominantly on the *why*-question (and its implication on practice), while the pragmatic Anglo-American curriculum tradition focuses mainly on the *how*-question (Duit, 2015). *Didaktik* supports the idea that education is not only about teaching methods but also an issue of selecting and justifying content for education (Fischler, 2011).

Didactical analysis, originally suggested by Klafki in 1958 (see the translated and updated contribution of Klafki in Westbury et al., 2000; see also Duit, 2015), offers guidance to reflect whether an issue or topic is relevant enough to be taught. It consists of five questions:

1. What wider or general sense or reality does this content exemplify and open up to the learner? What basic phenomenon or fundamental principal, law, criterion, problem, method, technique, or attitude can be grasped by dealing with this content as an "example"?
2. What significance does the content in question, or the experience, knowledge, ability, or skill to be acquired through this topic, process in the minds of the children in my class? What significance should it have from a pedagogical point of view?
3. What constitutes the topic's significance for the children's future?
4. How is the content structured? How can it be placed in a specifically pedagogical perspective by questions 1, 2, and 3?
5. What are the special cases, phenomena, situations, experiments, persons, elements of aesthetic experience, and so forth, in terms of which the structure of the content in question can become interesting, stimulating, approachable, conceivable, or vivid for children of the stage of development of this class?

These questions try to identify epoch-typical relevant knowledge and key problems to learn about, which are of importance for the individuals and the society the students live in and operate, today and for the future. Except learning the science behind relevant issues such as climate change, students should get "the potential to learn about how such an issue is handled within society and one can learn about the interplay of science with ecology, economics, politics, cultural beliefs and values" (Marks et al. 2014, p. 287).

Connecting *Bildung* to Different Visions of Scientific Literacy

Roberts (2007) has suggested two different visions of scientific literacy to understand science learning. The more traditional Vision I describes science learning as mainly focusing learning science content for later application and further education. This approach is often considered from and driven by the inner structure of the corresponding academic discipline. For a more student-oriented approach to science education, Roberts suggested a Vision II. In Vision II science learning should provide the learner with understanding about the usefulness of scientific knowledge in life and society by starting from meaningful contexts.

Inspired by the ideas of education for sustainability, a Vision III of scientific literacy was recently suggested (Sjöström & Eilks, 2018; Sjöström et al., 2017). Largely inspired by critical versions of *Bildung*, it emphasizes science learning for scientific engagement and "knowing-in-action." This point of view wants to strengthen the learning beyond science content, contexts, and processes. It argues for general skill development via engagement with issues of science that is relevant for a sustainable development of our society and global world. Figure 5.2 provides an organizer to understand the differences between the three visions. Visions I and II focus on individual content knowledge development and how it is applied in everyday-life contexts. In the tradition of critical-hermeneutic *Bildung*, Vision III aims at critical skills development and transformative learning for actively shaping the future society in a sustainable fashion.

Scientific content knowledge and contextual understanding about science might be considered necessary pre-requisites to participate in informed scientific and societal discourses on the technological applications of science and its corresponding effects on the environment and society. However, this is not enough. Contemporary

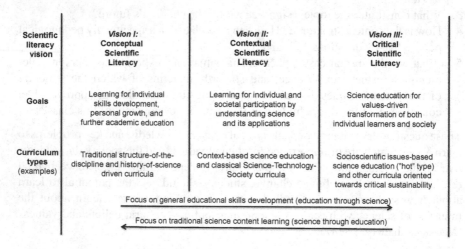

Scientific literacy vision	*Vision I:* Conceptual Scientific Literacy	*Vision II:* Contextual Scientific Literacy	*Vision III:* Critical Scientific Literacy
Goals	Learning for individual skills development, personal growth, and further academic education	Learning for individual and societal participation by understanding science and its applications	Science education for values-driven transformation of both individual learners and society
Curriculum types (examples)	Traditional structure-of-the-discipline and history-of-science driven curricula	Context-based science education and classical Science-Technology-Society curricula	Socioscientific issues-based science education ("hot" type) and other curricula oriented towards critical sustainability

Focus on general educational skills development (education through science) →

← Focus on traditional science content learning (science through education)

Fig. 5.2 Three different visions of scientific literacy (from Sjöström et al., 2017), where vision III can be connected to critical-reflexive *Bildung*

Bildung theory suggests that skills and a critical stance are also needed that promote understanding of the responsibility of any individual and at the same time enables and directs the individual to act accordingly within society.

Applying *Bildung* and *Didaktik* to Science Education

In this section, we will briefly describe selected theoretical and empirical works where the ideas of categorical *Bildung* in a critical and reflexive interpretation according to Sjöström and Eilks (2018) are applied to science education. Critical-hermeneutic *Bildung,* or critical-reflexive *Bildung,* in science education adds philosophical as well as political dimensions to the teaching and learning of science. It focuses on both meta-perspectives and socio-political actions grounded in a problematizing stance toward contemporary society (see for example Elmose & Roth, 2005; Hofstein et al., 2011; Sjöström, 2013).

There is not much written about ideological assumptions that underpin different formulations of science education, but Pedretti and Nazir (2015, p. 934) recently wrote: "a view that science education should be focused on teaching science content (a predominantly transmissive view) rather than focused on social reconstruction and change (a transformative view) can produce radically different experiences and challenges in the science classroom." The latter view includes values, worldviews, politicization, and actions and is connected to critical-reflexive *Bildung,* whereas the first view hardly will be able to open all the learners' corresponding perspectives.

Coming from *Bildung* theory, Stolz, Witteck, Marks, and Eilks (2013) have elaborated a set of five characteristics, including provable criteria, for identifying socio-scientific issues (SSIs) that lead to *Allgemeinbildung*-oriented learning. They suggested SSIs for the promotion of *Bildung* in science education to be: (a) authentic, (b) relevant, (c) undetermined in evaluation in a socio-scientific respect, (d) offering the chance for open debate, and (e) connected to science and technology (see also column two in Fig. 5.3). In their model they suggested clear criteria: (a) concerning authenticity, they ask whether there is an authentic debate in society on any issues, documented in everyday-life media; (b) relevance asks whether there is any decision to be drawn, at the individual or societal levels, that would make a difference to the life of the learner so that any debate is worth pursuing; (c) openness asks whether there are different points of view that are mirrored in positions by different stakeholders in the authentic debate on the individual or societal levels; (d) offering the chance for open debate asks whether debate is possible by exchanging arguments without harming any individual learner; and, finally, (e) connectedness to science and technology asks whether there are arguments from science or technology used in the public debate. Based on this, they suggested implementing understanding of communication and decision-making practices about techno-scientific queries from society into the teaching and learning of science (for example by mimicking corresponding societal practices in role plays and business games).

Concept of the socio-critical and problem-oriented approach to science teaching

Objectives	Criteria for selecting issues and approaches	Methods	Structure of the lesson plans
Allgemeinbildung/ education through science	Authenticity	Authentic media	1. Textual approach and problem analysis
(Multidimensional) Scientific Literacy	Relevance	Student oriented science learning and laboratory work	2. Clarifying the science background, e.g. in a laboratory environment
Promotion of evaluation skills	Evaluation undetermined in a socio-scientific respect	Learner centred instruction and cooperative learning	3. Resuming the socio-scientific dimension
Promotion of communication skills	Allows for open discussion	Methods structuring controversial debating	4. Discussing and evaluating different points of view
Learning science	Deals with questions from science and technology	Methods provoking the explication of individual opinions	5. Meta-reflection

Fig. 5.3 Framework outlining the socio-critical and problem-oriented approach to science teaching (from Marks et al., 2014, based on Marks & Eilks, 2009)

It is obvious that such critical versions of SSI-based teaching are related to *Bildung*, but it is not fully clear how much it is still connected to democratic education rather than with critical-reflexive *Bildung*. The curriculum model by Marks and Eilks (2009), called the socio-critical and problem-oriented approach to science teaching, uses the term *Allgemeinbildung* as the first instance of objectives (column one); it is used in the meaning of Weniger and Klafki.

The curriculum model by Marks and Eilks (2009) is a *Didaktik*-model in the tradition of *Allgemeinbildung* and the German subject-specific *Didaktik*. It suggests science teaching should start with actual and authentic media from everyday life to demonstrate the authenticity and relevance of any SSI for the individual and society. Media is used to provoke questions on a topic and also to demonstrate how any given topic is related to both society and science. Questions in the lesson plans generally cover both issues of science and technology as well as corresponding ecological, economic, and societal impacts. Learning the science behind a technology is justified by allowing students to understand the sources and processes behind any development and the controversy around its scale. It allows an evaluation of the issue from a scientific point of view, but it does not stop there. Science teaching of this type aims at understanding how the individual and the society is communicating and deciding about the issues of science and technology in its multidimensional relations and impacts. Therefore, the model suggests a thorough analysis of which SSI-related questions can be answered by science and which cannot. Science cannot answer any political or ethical questions; it can only contribute to their understanding. In a democratic society, such questions are negotiated and decided in public forums, media, and parliaments. Consequently, the *Didaktik*-model suggests moving over to mimicking authentic societal practices of communication and decision-making as essential parts of SSI-based science lessons with different pedagogies to make the learner skillful for self-determination, participation, and solidarity (Marks et al., 2014).

Typical issues that are authentic, relevant, open-ended, debatable, and science-based often stem from the environmental and sustainability debate. The issues of climate change, renewable energy and materials supply, green engineering, sustainable agriculture, preserving biodiversity, risks of chemicals in the environment, or provision of clean water resources are only a few among many examples. However, health and living issues are also important and highly relevant since questions of the chemicalization of our environment, the provision of organic food, the use of chemicals in consumer products, or questions of genetically manipulated food growth are all authentic, and decisions about these—all on the individual, societal, and/or global levels—are highly relevant for our present and future and the challenge of transforming our contemporary society that is thoroughly impacted by developments in science and technology.

Summary

Bildung as a theory of education is very old. It covers a history of more than 200 years. The meaning and understanding of *Bildung* in theory and practice changed over time. It only slowly found its way into the international literature of science education. With growing ecological and technological challenges in our current societies (and a growing number of fake news about them), however, reflecting the ideas and directions of the concept of *Bildung* for science education might be considered to be more relevant than ever. Categorial *Bildung* in the means of Klafki is needed for the responsible citizen to behave and to react to challenges like climate change, the chemicalization of our world, or the need for more efficient and sustainable use of natural resources. It is also highly relevant in times of needed political decisions on the development and transformation of our today's world for a sustainable future. Both knowledge from science and technology is needed as well as skills to apply this knowledge for a self-determined life, participation in society, and solidarity with others. *Bildung* gives guidance to how to select content and learning objectives for this direction via its tools like *Didaktik* analysis or societal-oriented approaches to science teaching—and in the other way around it also provides criteria to assess teaching practices whether they are of potential to promote *Bildung* to enable the young generation to become responsible citizens.

- *Bildung* is a unique central and northern European tradition of education that has its roots in the late eighteenth century.
- *Bildung* just recently was being recognized in the international science education literature.
- Recognizing contemporary interpretations of *Bildung* involves rethinking science education toward a more critical view to allow transformative learning of science,

which promotes capabilities in the student for self-determined life and responsible citizenry.
- Contemporary interpretations of *Bildung* suggest a more thorough operation of current and controversial socio-scientific issues as drivers for modern science education.

References

Bauer, W. (2003). On the relevance of *Bildung* for democracy. *Educational Philosophy and Theory, 35,* 212–225.

Duit, R. (2012). The model of educational reconstruction—A framework for improving teaching and learning science. In D. Jorde & J. Dillon (Eds.), *Science education research and practice in Europe* (pp. 39–61). Rotterdam: Sense.

Duit, R. (2015). Didaktik. In R. Gunstone (Ed.), *Encyclopedia of science education* (pp. 325–327). Dordrecht: Springer.

Elmose, S., & Roth, W. M. (2005). *Allgemeinbildung*: Readiness for living in risk society. *Journal of Curriculum Studies, 37,* 11–34.

Fischler, H. (2011). Didaktik—An appropriate framework for the professional work of science teachers? In D. Corrigan, J. Dillon, & R. Gunstone (Eds.), *The professional knowledge base of science teaching* (pp. 31–50). Dordrecht: Springer.

Gundem, B. (2010). Didactics–Didaktik–Didactique. In C. Kridel (Ed.), *Encyclopedia of curriculum studies* (pp. 293–294). Thousand Oaks: SAGE.

Hofstein, A., Eilks, I., & Bybee, R. (2011). Societal issues and their importance for contemporary science education: A pedagogical justification and the state of the art in Israel, Germany and the USA. *International Journal of Science and Mathematics Education, 9,* 1459–1483.

Kansanen, P. (2009). Subject-matter didactics as a central knowledge base for teachers, or should it be called pedagogical content knowledge? *Pedagogy, Culture & Society, 17,* 29–39.

Marks, R., & Eilks, I. (2009). Promoting scientific literacy using a socio-critical and problem-oriented approach to chemistry teaching: Concept, examples, experiences. *International Journal of Environmental and Science Education, 4,* 131–145.

Marks, R., Stuckey, M., Belova, N., & Eilks, I. (2014). The societal dimension in German science education—From tradition towards selected cases and recent developments. *Eurasia Journal of Mathematics, Science & Technology Education, 10,* 285–296.

Nussbaum, M. (1997). *Cultivating humanity: A classical defense of reform in liberal education.* Cambridge: Harvard University Press.

Nussbaum, M. (2010). *Not for profit: Why democracy needs the humanities.* Princeton: Princeton University Press.

Pedretti, E., & Nazir, J. (2015). Science, technology and society (STS). In R. Gunstone (Ed.), *Encyclopedia of science education* (pp. 932–935). Dordrecht: Springer.

Roberts, D. A. (2007). Scientific literacy/science literacy. In S. K. Abell & N. G. Lederman (Eds.), *Handbook of research on science education* (pp. 729–780). Mahwah: Lawrence Erlbaum.

Schneider, K. (2012). The subject-object transformations and 'Bildung'. *Educational Philosophy and Theory, 44,* 302–311.

Sjöström, J. (2013). Towards *Bildung*-oriented chemistry education. *Science & Education, 22,* 1873–1890.

Sjöström, J. (2018). Science teacher identity and eco-transformation of science education: Comparing western modernism with confucianism and reflexive *Bildung*. *Cultural Studies of Science Education, 13,* 147–161.

Sjöström, J., & Eilks, I. (2018). Reconsidering different visions of scientific literacy and science education based on the concept of *Bildung*. In Y. Dori, Z. Mevarech, & D. Baker (Eds.), *Cognition, metacognition, and culture in STEM education* (pp. 65–88). Dordrecht: Springer.

Sjöström, J., Eilks, I., & Zuin, V. G. (2016). Towards eco-reflexive science education: A critical reflection about educational implications of green chemistry. *Science & Education, 25,* 321–341.

Sjöström, J., Frerichs, N., Zuin, V. G., & Eilks, I. (2017). Use of the concept of *Bildung* in the international science education literature, its potential, and implications for teaching and learning. *Studies in Science Education, 53,* 165–192.

Stolz, M., Witteck, T., Marks, R., & Eilks, I. (2013). Reflecting socio-scientific issues for science education coming from the case of curriculum development on doping in chemistry education. *Eurasia Journal of Mathematics, Science and Technological Education, 9,* 273–282.

Stuckey, M., Hofstein, A., Mamlok-Naaman, R., & Eilks, I. (2013). The meaning of 'relevance' in science education and its implications for the science curriculum. *Studies in Science Education, 49,* 1–34.

Westbury, I., Hopmann, S., & Riquarts, K. (Eds.). (2000). *Teaching as a reflective practice: The German Didaktik tradition.* Mahwah: Lawrence Erlbaum.

Jesper Sjöström is trained as an upper-secondary school teacher oriented toward chemistry and general science teaching. After a Ph.D. in chemistry in 2002 at Lund University, Sweden, he was a post-doc for three years in the area of Science and Technology Studies. He joined Malmö University in Sweden in 2007 and since 2015 he is an Associate Professor of Science Education at the Department of Science-Mathematics-Society. His current research interests encompass socio-oriented science education, with a particular focus on chemistry and its links to philosophy, media and environmental and health issues. He is also interested in teacher education research and development.

Ingo Eilks studied chemistry, mathematics, philosophy, and education. He is a full-trained grammar school teacher, having a Ph.D. and habilitation in chemistry education. Since 2004, he is a full professor for chemistry education at the Institute for Science Education at the University of Bremen, Germany. His research interests encompass societal-oriented science education, action research in science education, teaching methodology, ICT in education, teacher education, and innovations in higher chemistry teaching. Recently, he led the EU-ERASMUS+ project ARTIST—Action Research to Innovate Science Teaching.

Part II
Behaviourist Theories

Chapter 6
Classical and Operant Conditioning—Ivan Pavlov; Burrhus Skinner

Ben Akpan

Introduction

Learning refers to a process in which a person acquires a relatively permanent change in behaviour as a result of experience (Kuppuswamy, 2013; Driscoll, 2000; Schunk, 2005). This definition excludes changes due to physical growth and development. The focus of attention in any school is on learning. The student, school and the teacher are the most important factors in the learning process. Through all the ages, educationists have been pre-occupied with learning how the learning process itself occurs.

One approach to learning is behaviouristic in outlook. Central to behavioural learning theories is a focus on the ways in which desirable and undesirable consequences of behaviour change people's behaviour over time and ways in which people model their behaviour on that of others (Slavin, 2009). In this chapter, I will be presenting two behavioural theories and their applications in science education. These are classical conditioning and operant conditioning.

Classical Conditioning

Ivan Pavlov (see Box 6.1) is credited with the development of classical conditioning. In his study of the process of digestion in dogs, Pavlov observed that a hungry dog would salivate automatically if powdered meat was placed in its mouth or near to it. The salivation occurred without any previous training. Pavlov and his team of scientists named the powdered meat *unconditioned stimulus*—meaning a stimulus

B. Akpan (✉)
Science Teachers Association of Nigeria, Abuja, Nigeria
e-mail: ben.b.akpan@gmail.com

© Springer Nature Switzerland AG 2020   71
B. Akpan and T. Kennedy (eds.), *Science Education in Theory and Practice*,
Springer Texts in Education, https://doi.org/10.1007/978-3-030-43620-9_6

that naturally brings about a particular response. The salivation was named *uncon-ditioned response*—a behaviour that is an automatic result of a stimulus, in this case salivation occurred automatically in the presence of powdered meat. Pavlov and his team further observed that there were some other stimuli which could not produce salivation automatically.

Box 6.1
Ivan Pavlov Bio

Ivan Pavlov (1849–1936) author of *Lectures on the Work of the Principal Digestive Glands* (1897), *Lectures on Conditioned Reflexes* (1928), and *Conditioned Reflexes and Psychiatry* (1941) hailed from Ryazan, Russia. He was a natural science graduate of the University of St. Petersburg when he got admitted into the Academy of Medical Surgery. After obtaining his doctorate, he joined the Military Medical Academy serving as director in the department of physiology in the Institute of Experimental Medicine. In 1904, his work on the digestive secretions of dogs at the Institute earned him the Nobel Prize. Pavlov passed on in 1936 having suffered from pneumonia.

An example of these is a bell. The bell is said to be a neutral stimulus—a stimulus that does not evoke a particular response. Pavlov demonstrated that by presenting the powdered meat and the bell together salivation was observed. Interestingly, salivation was equally evoked when the bell was later presented alone meaning that the bell (a previously neutral stimulus) when paired with an unconditioned stimulus (powdered meat) became a conditioned stimulus—a previously neutral stimulus that gives rise to a specific response after having been paired with an unconditioned stimulus. This kind of conditioning is commonly referred to as classical (or Pavlovian) conditioning. For a schematic representation, see Fig. 6.1.

Here is an example of classical conditioning:

A particular tone might be paired with a puff of air to the eye. The tone is initially neutral in that it elicits no response from a person. The puff of air to the eye however, does elicit a response—an eye blink. After repeated pairing of the tone with the puff of air, eventually the tone will produce an eye blink even without the puff of air. In this example, the puff of air is referred to as the unconditioned stimulus and the tone as the conditioned stimulus. (Tiwari, 2016, p. 238)

Terminologies Associated with Classical Conditioning

Acquisition

Acquisition refers to the initial stage of learning where a response is elicited and gradually strengthened. According to Cherry (2017), during the acquisition phase a neutral stimulus is repeatedly paired with an unconditioned stimulus following

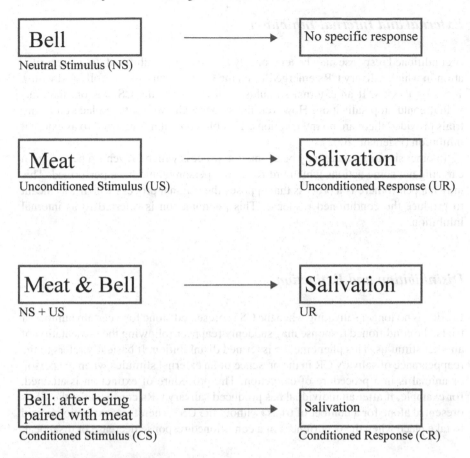

Fig. 6.1 Classical conditioning process

which the subject elicits a behaviour in response to the previously neutral stimulus, which is now known as the conditioned stimulus. Cherry (2017, p. 2) provides the following example:

> …imagine that you are conditioning a dog to salivate in response to the sound of a bell. You repeatedly pair the presentation of food with the sound of the bell. You can say the response has been acquired as soon as the dog begins to salivate in response to the bell tone. Once the response has been established, you can gradually reinforce the salivation response to make sure the behaviour is well learned

External and Internal Inhibition

A conditioned response may be temporarily suppressed or inhibited. Imagine a situation in which salivary CRs emerged when the CS was consistently followed within 1–5 s by the US. If an external stimulus occurred while the CS was on, the individual could stop salivating. However, the salivary CR would take place on future trials provided there are no interruptions. This phenomenon is referred to as external inhibition (Frieman, 2002).

In other situation, there may be an internal process which prevents a person from carrying out some actions which ordinarily the person would have performed. This may happen if there is a process that opposes the original process which was meant to produce the conditioned response. This phenomenon is referred to as internal inhibition.

Disinhibition and Extinction

If a dog is no longer salivating when the CS is presented alone for a certain number of trials, the conditioned response may suddenly reappear following the presentation of another stimulus. This phenomenon is termed disinhibition. It basically refers to the reappearance of salivary CR in the presence of an external stimulus when the person or animal is in a procedure of extinction. The procedure of extinction is attained, for example, if after an individual has produced salivary CRs consistently, the CS is presented alone for a number of trials (without the US). Therefore, extinction is said to take place when the occurrences of a conditioned response decrease or disappear.

Spontaneous Recovery

Pavlov discovered that behaviour that has become extinct may reappear if the person returns to that situation after some time has passed. This is referred to as spontaneous recovery.

It should be noted that Pavlov used the phenomena of disinhibition and spontaneous recovery to demonstrate that extinction could not undo the effects of conditioning.

Figure 6.2 gives a schematic representation of Pavlovian extinction and recovery.

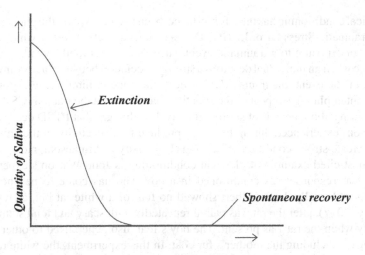

Fig. 6.2 Schematic representation of Pavlovian extinction and recovery

Assessing Pavlov's Contribution

In a critique on Pavlov's work, Frieman (2002) writes:

> His dogs were strapped in harnesses in isolation rooms, and everything the dogs experienced
> was controlled by the experimenter who stayed in an adjoining room. Even the saliva was
> collected in tubes that carried it to the experimenter. What can we possibly learn from such an
> artificial and sterile arrangement that is relevant to us? The answer is that we can learn a great
> deal... Pavlov showed us how we can use systematic observation and experimentation to
> explore psychological processes, and he provided a way to measure behaviour objectively...
> He did not deny that humans have an inner world; his objection was to those who explained
> the behaviour of animals in terms of these mental states. Pavlov believed that we can come
> to understand the behaviour of animals by studying how they react to their environment.
> (pp. 34–35)

Indeed, the principle behind classical conditioning hinged on Pavlov's experimental strategy provided a ground-breaking initiative for the development of psychology as a truly scientific discipline as opposed to its hitherto philosophical orientation. Pavlov was to greatly influence many leading behaviourist psychologists including John Watson and Burrhus Skinner.

Classical conditioning has also led to the understanding of specific areas of the brain that are vital in learning memory. Some sections of the brain such as the cerebellum, prefrontal cortex, hippocampus, and amygdala are said to play vital roles in the conditioning process. Also, certain behavioural therapies make use of classical conditioning. An example is aversion therapy which is programmed to ensure patients stop an unwanted habit by linking such habit with an unconditioned stimulus that is not pleasant. Other uses of classical conditioning are in the areas of drug tolerance, conditioned hunger, and conditioned emotional response. These are, however, outside the scope of this book.

Classical conditioning has thus, for instance, been used to explain the phenomenon of Post-traumatic Stress Disorder (PTSD)—a serious anxiety disorder which develops after an exposure to a traumatic event, such as an automobile accident. PTSD takes place when an individual develops a strong association between factors involved in the traumatic event, the trauma (US). Due to the conditioning, being exposed to, or even contemplating a repeat of the situation that produced the trauma (CS), is sufficient to bring about the CR of serious anxiety. It is thought that PTSD occurs when the emotions experienced during the event produce neural activity in the amygdala and thus create strong conditioned learning (University of Minnesota, n.d.).

The most cited example of classical conditioning is John Watson's experiment where a fear response was conditioned in a boy who has come to be known as Little Albert. Initially, Little Albert showed no fear of a white rat but, as reported by Cherry (2017), after the rat was paired repeatedly with scary and loud sounds, he would cry when the rat was present. The boy's fear also generalised to other fuzzy white objects including the mother's fur coat. In this experiment, the white rat was the neutral stimulus prior to conditioning while the unconditioned stimulus was the scary loud sound. The fear response caused by the noise (unconditioned stimulus) was the conditioned response and resulted from repeated pairing of the white rat (now the conditioned stimulus) with the unconditioned stimulus. The Little Albert experiment demonstrates how phobias can be formed through classical conditioning.

University of Minnesota (n.d.) reports a research by Lewicki (1985) which demonstrated the influence of stimulus generalisation and the speed at which it can occur. In that experiment, students were first made to have a brief interaction with a female experimenter who had short hair and eye glasses. The study was designed in such a way as to ensure students asked the experimenter a question, and (based on random assignment) the experimenter either responded in a neutral or negative manner to the questions. Thereafter, the students were directed to go into another room where two experimenters were present, and to approach any one of them. The researchers ensured that one of the two experimenters looked a lot like the original experimenter, while the other one did not (she was made to appear with longer hair without eye glasses). The results showed that the students were significantly more likely to avoid the experimenter who looked more like the earlier one that responded negatively to them. The students thus showed stimulus generalisation.

In everyday life, it has been shown that forming associations through conditioning can have survival benefits for the individual or organism. If people take a meal that makes them sick, they will learn to avoid such meals in order to prevent the reoccurrence of the sickness. Cherry, (2017) reports one study where researchers injected sheep carcasses with a poison that made coyotes ill but not kill them. This helped sheep ranchers to reduce the number of sheep coyote killing. The experiment worked by lowering the number of sheep that died and also made some of the coyotes to develop such a strong aversion so much that they actually ran away on scenting or sighting a sheep. Even so, it bears to state quite clearly that in many instances people do not respond exactly in the same way as Pavlov's dogs.

By way of implications, Mangal (2015, p. 191) posits that

> In our everyday life, we are usually exposed to simple classical conditioning. Fear, love and hatred towards an object, phenomenon or event are created through conditioning…a teacher with his defective methods of teaching or improper behaviour may condition a child to develop a distaste and hatred toward him, the subject he teaches and even the school environment. On the contrary, affection, a loving attitude and sympathetic treatment given to the child by the parents at home or by the teachers at school may produce a desirable impact on him through the process of conditioning. Most of our learning is associated with the process of conditioning. A child learns to call his father 'daddy', his mother 'mummy'…(and) as a result of stimulus generalization, he may attribute the name of daddy to all adult males, mummy to all adult females…Gradually, he comes to the stage of stimulus discrimination and then learns to discriminate and recognise and attribute different names to different persons…What is termed as abnormality in one's behaviour may…be taken as learned…Thus, much of our behaviour in the shape of interests, attitudes, habits … is fashioned through conditioning.

Systematic desensitisation, a behaviour therapy, draws on principles of classical conditioning. The main principle in this therapy is to teach the patient how to relax as the session progresses. For instance, if students have intense fear for carrying out experiments with toads in the laboratory, they can undergo systemic desensitisation. As a first step, a mild fear situation will be created in relation to toad such as by holding a picture of toad far away from the students. The students are taught to relax in the presence of the distant picture. When it is obvious that they have been able to relax sufficiently without any fear, the picture of the toad is brought nearer so the level of fear is increased. Again, they are made to relax sufficiently to overcome the fear. Next, the therapist can bring in a toad at a distance. This should create more fear and again the students are made to relax sufficiently in order to overcome the fear. This process will go on and at each stage the level of fear is increased followed by relaxation on the part of the students until the students are able to carry out experiments involving toads without any more fear. If there is fear at any stage, the therapist will return to a previous stage before proceeding. The process may be repeated over long intervals of time to assure there is no more fear. Essentially, relaxation has been substituted for phobia for toads as the conditioned response. In practice, there may just be one student as the patient. Where possible also, rather than use physical objects, the patient may be made to imagine various scenarios that will create graduated levels of fear of toad.

Operant Conditioning

As discussed above, classical conditioning is an automatic conditioning of reflex-like responses as in the case of salivation (Woolfolk, 2014). In real-life settings, many human behaviours are not automatic. In several cases, people have to deliberately act on the environment; and such deliberate efforts are termed 'operant', thus giving rise to operant conditioning first proposed by B. F. Skinner (see Box 6.2). This kind of conditioning emphasises the role of external stimuli in determining human behaviour. Essentially, Skinner (1953) contends that humans and animals repeat acts if such produce outcomes that are favourable. Conversely, humans and animals will suppress acts that give rise to outcomes that are not favourable. For an example, if a rat obtains a pellet of food that is delicious by pressing a bar, it may like to repeat the action of pressing the bar. The bar-pressing is referred to as an operant while the pellet of food is known as the reinforcer.

Box 6.2
Burrhus Frederic Skinner Bio

Author of *The Behaviour of Organisms: An Experimental Analysis* (1938), *Walden Two* (1948), *Science and Human Behaviour* (1953), *Verbal Behaviour* (1957), and *Beyond Freedom and Dignity* (1971), Burrhus Frederic Skinner (1904–1920) hailed from Susquehanna, Pennsylvania, USA. An English graduate from Hamilton College, New York, Skinner later enrolled for a psychology programme in Harvard where he earned a doctorate. Skinner later worked in the University of Minnesota (1936–1946), Indiana University (1946–1947), and Harvard (1948–1990). He died of leukaemia in 1990.

Terminologies Associated with Operant Conditioning

Reinforcement

In operant conditioning terms, behaviours are sustained by reinforcers (see Fig. 6.3). Reinforcers are stimuli that strengthen or weaken the behaviours that produced them. This implies that there are two types of reinforcement.

Positive and Negative Reinforcement

In positive reinforcements new stimuli are produced. This is irrespective of whether or not the behaviour in question is appropriate or inappropriate. According to Woolfolk (2014), if the consequence that strengthens a behaviour is the appearance of a new stimulus, positive reinforcement is said to occur. Conversely, in negative reinforcement, the consequence that strengthens a behaviour is the disappearance of a stimulus. In practical terms, positive reinforcement is done by providing desired stimulus after the behaviour is exhibited, whereas removal of irritating or unpleasant stimulus when behaviour occurs will lead to negative reinforcement.

Fig. 6.3 Behaviour
reinforcement is a cyclic
process

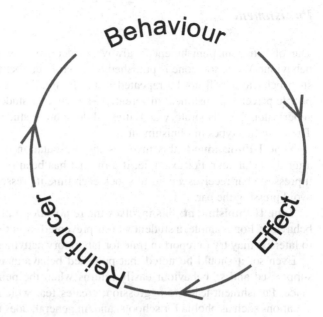

By way of examples, some pupils may develop the habit of being empathetic
towards more disadvantaged peers if their friends consistently reinforce that kind of
behaviour by praising them. Also, pupils may put in more effort in studying particular
science subjects if their efforts result in grades that are high.

Primary and Secondary Reinforcers

There are two types of reinforcers: primary and secondary reinforcers. Primary reinforcers are about basic needs of people such as water, food, warmth, security, and sex.
On the other hand, secondary reinforcers are valued when considered in conjunction
with primary reinforcers or other very important secondary reinforcers. Examples
include money, school grades, smiles, hugs, games, and toys.

Extinction

It is generally observed that the possibility of a specific response occurring increases if
the response is followed by a positive reinforcer. Conversely, if positive reinforcement
does not follow the response, the possibility of it occurring decreases. This means
that responses which are not advantageous tend not to be elicited. This procedure
which results in not reinforcing a specific response is termed extinction.

Punishment

The objective in punishment is always to decrease, suppress, or stop observed behaviour. When someone is punished for a particular behaviour, it is assumed that such behaviour will not be repeated in the future. This is contentious, though, as people perceive punishment differently—a particular student may be unhappy with supervision from laboratory activities while another student may be glad about it. There are two types of punishment.

Type I Punishment: this involves the presentation of an unpleasant stimulus after the behaviour. For example, if a rat that has been reinforced with meat when it presses a bar receives a painful shock each time it presses the bar, it will begin to avoid pressing the bar.

Type II Punishment: this involves the removal of a pleasant stimulus after the behaviour. For example, a student who is prevented from the laboratory activity due to lateness may try to report in time for laboratory activities in future.

Even so, it should be noted that punished behaviour is not forgotten but only suppressed and so behaviour easily returns when the punishment is no longer in place. Punishment leads to aggression; creates fear which can generalise to other situations such as phobia for schools; and, in general, does not provide a good guide toward desired behaviour (McLeod, 2015).

Stimulus Generalisation and Discrimination

It is possible for a response to be obtained in a situation that is a little different from the one in which original reinforcement occurred. In such an instance, the strength of the response will decrease in the new situation. Should the new situation be very different from the original situation, the response may not occur. In everyday life, a situation may be a little different from another and still be possible to produce the same reinforcement for the same response. Thus, it is possible for the original stimulus situation to generalise to new situations. This phenomenon is termed stimulus generalisation. Conversely, it is clearly impossible for an individual to completely generalise from one situation to all other situations. Perfect stimulus generalisation is clearly not possible otherwise inappropriate responses would be the order of the day. As a result, individuals usually demonstrate stimulus discrimination.

Morgan et al. (1993) give the following example for a discrimination experiment:

> The experimenter, on some trials, pairs one stimulus (called the CS^+) with an unconditioned stimulus. On other trials, another stimulus (called CS^-) is presented alone, without the unconditioned stimulus. In other words, while responses to the CS^+ are being conditioned on some trials, extinction of any tendency to respond to the CS^- is occurring on other trials. As a result, the learner forms discrimination; conditioned responses are made to the CS^+ but not to the CS^-. (p. 147)

A Note on Operant Conditioning

Operant conditioning can be used to explain a large variety of human behaviour such as learning, drug addiction and acquisition of new language. According to Mangal (2015, pp. 199–200):

> A response or behaviour is not necessarily dependent...upon a specific known stimulus. It is more correct to think that a behaviour or response is dependent upon its consequences...The individual's behaviour should get the reward and he should in turn, act in such a way that he is rewarded again and so on...The principle of operant conditioning may be successfully applied in behaviour modification...The development of personality can be successfully manipulated through operant conditioning...The theory ... does not attribute motivation to internal processes within the organism. It takes for granted the consequences of a behaviour or response as a source of motivation to further occurrence of that behaviour.

Indeed, its applications in educational settings, prisons and certain hospitals have been established. Nonetheless, operant conditioning is unable to take into account the role played by inherited as well as cognitive factors in the teaching and learning process. Consequently, it offers an incomplete explanation of the process of learning.

The over-justification effect is another issue to be considered in operant conditioning. Studies have reported a drop in performance following a period of reward especially if the tasks were just performed for their own sake and the rewards were given in ways that showed they were purposely for motivating the persons to carry out the tasks. Gray (2007) calls this the *over-justification effect* which arises 'because the reward presumably provides an unneeded justification for engaging in the behaviour. The result...is that the person comes to regard the task as work (the kind of thing one does for external rewards such as money, certificates or improved resume) rather than play'. (p. 117)

Application of Classical and Operant Conditioning in Science Education

The following are guidelines for the implementation of classical and operant conditioning in science teaching and learning:

1. Set goals for student behaviours and ensure those behaviours are reinforced whenever they occur. In science classes, endeavour to reward good work and provide feedback on assignments as well as practical exercises. When new science experiments are conducted, endeavour to provide reinforcement at different stages until completion.
2. Let students know in advance what desirable behaviours you expect from them and how they may be rewarded. In science practical classes, for instance, let them know which aspects will attract more marks and thus where much premium should be placed.

3. Act promptly in reinforcing behaviours. When reinforcements are delayed, they may be counter-productive.

4. If students have submitted their science assignments for review, action should be expedited on the teacher's part and the students are provided with feedback as soon as possible.

5. Teach students to praise themselves and appreciate their strengths especially when they have successfully concluded a science task.

6. Listen carefully to what students tell you in the science class and provide reinforcements where necessary.

7. Lay great emphasis on group work instead of individual competition.

8. Ensure that learning tasks are associated with events that are both positive and pleasant.

9. Explore ways and means of ensuring that students are capable of deciphering similarities and differences in various settings such that they are able to discriminate and generalise from one situation to the other.

10. Publicise good effort in the science class. For example, the best science project can be on display in the classroom. Also, the best written assignment can be displayed on the notice board.

11. Where possible, let parents know about their wards' attainment in science classes. Some occasional notes to parents may be a good way to go.

12. Enlist the help of parents in reinforcing good behaviour at home. This way, the effort of the school is complemented.

13. Create a positive classroom learning environment in order to assist students overcome fear and anxiety. For example, in the science class, experiments that provoke anxiety can be performed in laboratory environments that help the students focus on the subject matter rather than becoming unnecessarily anxious and tensed up. Laboratory demonstrations that are observed by students in relaxed mood will be far more beneficial than those carried out in rowdy situations.

14. Undesirable behaviours such as tardiness and dominating class discussion is easily prevented by being ignored by the teacher instead of being reinforced by having attention drawn to them.

15. It is necessary to vary the type of reinforcement provided in order to ensure desired behaviour is elicited.

16. Use Integrated Learning Systems (ILS): This is a computerised system that combines tutorial programmes with programmes which track the performance of students and give feedback to both the teacher and the student. The system produces some measure of positive results, especially for students achieving below average.

17. Make use of Computer-Based Instruction (CBI) Drill-and-practice programmes, tutorial programmes, and problem-solving programmes such as simulations and games.

18. Shape behaviour by reducing complex behaviours into a sequence of simpler behaviours; reinforcing successive approximations to complex behaviour.

Summary

- In classical conditioning, an individual learns to associate a neutral stimulus (CS) with a stimulus (US) that usually elicits a behaviour (UR). Due to this association, the previously neutral stimulus comes to produce the CR. Extinction occurs if the CS is repeatedly presented in the absence of US such that the CR ultimately disappears, even though it may reappear in future in a process termed spontaneous recovery. Stimulus generalisation takes place if the individual produces the same response as the original stimulus does. Conversely, when the individual learns to differentiate between the CS and another similar stimulus, stimulus discrimination takes place.
- In operant conditioning, it is theorised that if a behaviour is followed by reinforcement, that behaviour is very likely to be elicited in the future but if it is followed by punishment, it is not likely to be elicited. Burrhus Skinner carried out his research on rats and pigeons using positive reinforcement, negative reinforcement, or punishment. He explained operant conditioning on the basis of learned, physical aspects of the world such as the organism's life history and evolution.
- Both classical and operant conditionings have wide applications in psychology, medicine, education, and science education.

Further Reading

McSweeney, F., & Murphy, E. S. (Eds.). (2014). *The Wiley Blackwell handbook of operant and classical conditioning*. New Jersey: Blackwell Publishing. Retrieved September 3, 2017, from http://www.blackwellreference.com/public/book?id=g9781118468180_9781118468180.

Gewirtz, J. L., & Pelaez-Nogueras, M. (1992). Skinner, B. K: Legacy to human infant behavior and development. *American Psychologist, 47,* 1411–1422.

Stricker, J. M., Miltenberger, R. G., Garlinghouse, M. A., Deaver, C. M., & Anderson, C. A. (2001). Evaluation of an awareness enhancement device for the treatment of thump sucking children. *Journal of Applied Behaviour Analysis, 34,* 77–80.

References

Cherry, K. (2017). What is classical conditioning?—A step-by-step guide to how classical conditioning really works. Retrieved September 3, 2017, from https://www.verywell.com/classical-conditioning-2794859.

Driscoll, M. P. (2000). *Psychology of learning for instruction* (2nd ed.). Boston: Allyn & Bacon.

Frieman, J. (2002). *Learning and adaptive behavior*. California: Wadsworth Group.

Gray, P. (2007). *Psychology* (5th ed.). New York: Worth Publishers.

Kuppuswamy, B. (Ed.). (2013). *Advanced educational psychology*. New Delhi: Sterling Publishers Private Limited.

Lewickki, P. (1985). Nonconscious biasing effects of single instances on subsequent judgements. *Journal of Personality and Social Psychology, 48,* 563–574.

Mangal, S. K. (2015). *Advanced educational psychology*. Delhi: PHI Learning Private Limited.

Morgan, C. T., King, R. A., Weisz, J. R., & Schopler, J. (1993). *Introduction to psychology* (7th ed.). New Delhi: McGraw Hill Education (India) Private Limited.

McLeod, S. (2015). Skinner—Operant conditioning. Retrieved September 3, 2017, from https://www.simplypsychology.org/operant-conditioning.html.

Schunk, D. H. (2005). Self-regulated learning: The education legacy of Paul R. Pintrich. *Educational Psychologist, 40*(2), 85–94.

Skinner, B. F. (1953). *Science and human behaviour*. New York: Macmillan.

Slavin, R. E. (2009). *Educational psychology: Theory and practice* (9th ed.). New Jersey: Pearson Education Inc.

Tiwari, N. M. (2016). *Child Psychology*. New Delhi: Saurabh Publishing House.

University of Minnesota. (n.d.). Learning by association: Classical conditioning. Retrieved September 3, 2017, from http://open.lib.umn.edu/intropsyc/chapter/7-1-learning-by-association-classical-conditioning/.

Woolfolk, A. (2014). *Educational psychology*. Noida, India: Dorling Kindersley India Pvt. Ltd.

Ben Akpan a Professor of science education is the Executive Director of the Science Teachers Association of Nigeria (STAN). He served as President of the International Council of Associations for Science Education (ICASE) for 2011–2013 and currently serves on the Executive Committee of ICASE as the Chair of World Conferences Standing Committee. Ben's areas of interest include chemistry, science education, environmental education and support for science teacher associations. He is the editor of *Science Education: A Global Perspective* published by Springer and co-editor (with Keith S. Taber) of *Science Education: An International Course Companion* published by Sense Publishers. Ben is a member of the Editorial Boards of the Australian Journal of Science and Technology (AJST), Journal of Contemporary Educational Research (JCER) and Action Research and Innovation in Science Education (ARISE) Journal.

Chapter 7
Social Learning Theory—Albert Bandura

Anwar Rumjaun and Fawzia Narod

> *We human beings are social beings. We come into the world as the result of others' actions. We survive here in dependence on others. Whether we like it or not, there is hardly a moment of our lives when we do not benefit from others' activities. For this reason, it is hardly surprising that most of our happiness arises in the context of our relationships with others.*
> —Dalai Lama XIV.

Introduction

From the above quote, it is evident that interactions with others play an important role in our lives as social beings. As early as the conception of a being (the formation of zygote) in the mother's body, the zygote is dependent on the mother for growth and development to become a full-fledged baby. Even the initial informal learning of toddlers and pre-school children start through their interactions with others like identifying their body parts, their parents, and siblings. It is thus not surprising that researchers trying to understand about learning have put forward theories which are based upon learners' interactions with other people—teachers, peers, parents, and siblings among others.

Such theories include the Vygotsky's Social Development theory, also called Vygotsky's Sociocultural theory, (Chen, 2015; John-Steiner & Mahn, 1996; McDevitt & Ormrod, 2002; Ormrod, 2008), and the Bandura's Social Learning Theory (Jarvis, Holford, & Griffin, 2003), among others. According to Vygotsky's theory, cognitive development is dependent on the child's social and cultural environments and as such interactions with others impact learning and cognition as would be elaborated in Chap. 19.

A. Rumjaun (✉) · F. Narod
Department of Science Education, Mauritius Institute of Education, Moka, Mauritius
e-mail: a.rumjaun@mie.ac.mu

F. Narod
e-mail: f.narod@mie.ac.mu

© Springer Nature Switzerland AG 2020   85
B. Akpan and T. Kennedy (eds.), *Science Education in Theory and Practice*,
Springer Texts in Education, https://doi.org/10.1007/978-3-030-43620-9_7

On the other hand, Bandura's Social Learning Theory postulates that people learn from each other through observation and modeling. His theory is often referred to as a junction or bridge between cognitive and behaviorist theories (McLeod, 2016). According to his theory, learning is based on a social behavioral approach—people learn from others (social element) by observing and modeling their behavior (behaviorist approach), but Bandura also brings into picture cognitive processes to explain learning. He proposes observational learning as opposed to direct imitation: people learn by observing others' behavior, but their cognitive processes or internal mental states will determine whether they will "imitate" the behavior or not (Boundless Psychology, 2016).

This chapter seeks to document SLT in its historical and educational perspectives. It also discusses the importance of the theory and its relevance in relation to current educational debates and reforms occurring worldwide. Drawing from current practices, the chapter furthermore emphasizes the relevance of the theory in supporting the teaching and learning of science and analyses to what extent the twenty-first-century science curriculum reconciles itself with SLT (Bandura, 1977). Some ideas and examples of science teaching and learning using SLT will also be provided. Finally, the chapter seeks to provide a critical lens of embedding SLT in science classes including the issues and challenges thereof.

Historical Perspective of the Social Learning Theory

The origin of the Social Learning Theory can be traced back to the work of Miller and Dollard (1941; Culatta, 2015; Huitt & Monetti, 2008), who made an attempt "to develop a theory that would encompass psychodynamic theory, learning theory, and the influence of sociocultural factors" (Kelland, 2015). Using the Hull's stimulus-response theory of learning, Miller and Dollard (1941) postulated that motivation and need could lead people to learn particular behaviors through observations and imitations; this is positively reinforced through social interactions (Kelland, 2015). Later, Rotter stretched the behaviorist theories and studied personality as an interaction between the individual and the environment (Kelland, 2015); this was viewed as the first step to cognitive approaches to learning. Rotter's work thus hinted that learning is also dependent on cognitive factors (Willard, 2015). In addition, Chomsky (1959) believed that the stimulus-response behaviorist theories alone were not sufficient to explain language acquisition, invoking some "unknown cognitive mechanism" to help people acquire language. The works of both Rotter and Chomsky were thus the first attempts to show that behaviorist approaches were not strong enough to explain learning; they believed that cognitive factors also played a role in people's learning (Kelland, 2015; Kihlstrom, 2014; Stone, 1998; Wikipedia, 2017).

Dollard and Miller based themselves on the Hullian Theory (Kelland, 2015) and Rotter made an attempt to explain learning from "generalized expectancies of reinforcement and internal/ external locus of control (self-initiated change versus change influenced by others)" by examining cognitive social learning (Kelland, 2015; Stone,

1998). However, only Bandura was able to establish social learning as a theory stepping away from the long-acclaimed behaviorist approaches (Kihlstrom, 2014). Even though Bandura placed great focus on cognitive aspects, he was of the view that cognitive development alone could not explain behavioral changes and believed that people can learn by watching and observing others (referred to as "observational learning" or "modeling"; Huitt & Monetti, 2008; Kelland, 2015). Indeed, by analyzing the ways in which people function cognitively on their social experiences and the influences of the latter on behavior and development, Bandura put forward his Social Learning Theory. This theory was a pioneering one in that it was the first one to include "modeling" or "vicarious learning" as a form of social learning (Kelland, 2015). The origin of his theory was also based on his famous Bobo doll study which clearly highlighted the importance of modeling on behavior. This study showed that children who watched a film showing adults mistreating and aggressive toward a Bobo doll, displayed similar aggressive behavior with the Bobo doll when placed in a room with toys including the doll (Huitt & Monetti, 2008). Nevertheless, though Bandura acknowledged the importance of modeling and reinforcement in learning social skills, he also reported children's predisposition to imitate others of higher prestige or status (e.g., parents, teachers, and national figures). According to Fontana (1995), Bandura's theories are referred to as social learning theories because "they suggest that social contact in itself produces learning."

Essential Features of the Social Learning Theory—Observational Learning and Modeling

Let us now focus on the educational perspective of Bandura's Social Learning Theory and its applications. Two important aspects of the Social Learning Theory include observational learning and modeling (also called vicarious learning; Edinyang, 2016; Kelland, 2015). As far as observational learning is concerned, it does not limit itself to observing a *live model* (another person displaying or acting the behavior), but it can also involve a "verbal instructional" model (descriptions and explanations of the behavior) or a" symbolic" model (children observing characters demonstrating the behavior in books, films, television or other media; Kelland, 2015). The term modeling in the Social Learning Theory can either imply the model demonstrating the behavior for the learner or the learner observing and imitating the displayed behavior (Ormrod, 2008). Distinction has also been made between the terms "imitation" and "modeling" in the SLT (Edinyang, 2016). The ability of the learner to reproduce or replicate the behavior which has been observed again and again is referred to as imitation, while modeling is a more complex process involving four important steps to ensure effective observational learning according to SLT. The four steps in the modeling process comprise attention, retention, reproduction (also referred to as production by some authors) and motivation as illustrated in Fig. 7.1. If any one of these steps is missing, observational learning and modeling will not take place.

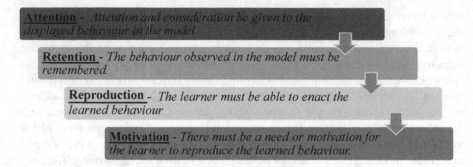

Fig. 7.1 Observational learning and the modeling process

The learner must pay attention to the model for observational learning to take place. Observing a model without any particular attention is unlikely to result in learning. Further, the information must be stored and remembered (retention). This implies that when required, the learner must be able to retrieve the information and re-enact or reproduce the observed and learned behavior (reproduction). Last but not least, to complete the modeling process the need for reproducing the observed and learned behavior must be felt by the learner. In other words, there must be a stimulus or a reason (motivation) for the learner to reproduce the observed behavior. The motivation can be in the form of reinforcement or punishment. Thus, this motivational aspect of the SLT is regarded as the most important factor that would drive the learner to perform the learned behavior. Sternberg and Williams (2009) have reported three types of reinforcement, namely:

(i)   direct reinforcement which involves rewarding the person for enacting or modeling the learned behavior.
(ii) vicarious reinforcement occurs when the learners are motivated by observing the model being rewarded on displaying the behavior.
(iii) self-reinforcement which implies the learners rewarding themselves for enacting the learned behavior.

The latter type of reinforcement is reported to encourage "self-regulation".

Implications of the Social Learning Theory on Science Education

As highlighted earlier in this chapter, Bandura's Social Learning Theory stresses a lot on cognitive concepts and is considered a bridge between behaviorist and cognitive approaches to learning. Indeed, Bandura believes that modeling will not occur without the learners engaging themselves cognitively by paying attention to the model or without an incentive. With this first leap toward cognitivism, the Social Learning Theory has important implications on science education as elaborated below.

Different definitions have been attributed to science; nevertheless, most of them lay emphasis on "observation" as an important aspect of science. For example, according to the English Oxford living dictionaries (2017), science is defined as "The intellectual and practical activity encompassing the systematic study of the structure and behavior of the physical and natural world through observation and experiment." In addition, learning science is reported to be essential to prepare twenty-first-century learners into responsible citizens who would not only be capable of understanding their world but would also function effectively in the science-driven world both at the personal and professional levels (Science Education for Responsible Citizenship, 2015; Ministry of Education, 2008). In view of the above, it can be seen that science acknowledges the importance of observation to gain knowledge and understanding and that science education has an important contribution in preparing learners for their roles as social beings. On the other hand, Bandura's Social Learning Theory puts forward that children can acquire and enact behaviors from hierarchically important individuals (models) in society through observation and modelling. Thus, in such a learning scenario as presented by Bandura, it is only natural to expect that his theory would have interesting and positive implications on science education. In view of the above, the SLT is expected to contribute positively to learners both in terms of science learning and in preparing them as twenty-first-century citizens. First, engaging learners in observation of their natural environment and its components as from the early years can be instrumental in arousing interest in the learning of science and in developing the right attitude toward the environment. This would, in turn, enhance conceptual understanding of science as an increase in interest will impact positively on student motivation and learning. Second, increased interest in science and enhanced conceptual understanding would promote awareness and understanding of the applications of science in real-life situations thereby preparing learners to perform effectively as twenty-first-century citizens.

Given that learning science involves the acquisition and development of necessary inquiry skills and processes, it is thus important for science educators to ensure that they display these skills correctly during the science lessons. Furthermore, practical work is an integral part of science and requires the proper and safe handling of various apparatus and measuring instruments by learners. It cannot be denied then that SLT can play a crucial role in science learning as it lays emphasis on learning through observation and modeling by learning. Indeed, continually observing and paying attention to how science educators and/or more abled peers display these skills correctly would enable learners to embrace (retain) and enact (reproduce) them as and when required (motivation) in line with the SLT. Science educators should thus aim at being "worthy" models for their learners by virtue of their role and also by virtue of their hierarchical position as Bandura asserts that children are more likely to observe and imbibe behaviors exhibited by individuals who are higher in status than themselves.

Problem-solving skills are considered to be essential for all citizens of the modern and increasingly complex scientifically-driven world. Thus, development of problem-solving skills among our learners is imperative to prepare them for their role as future responsible citizens (Mukhopadhyay, 2013; Wismath, Orr, & Zhong, 2014).

Though various definitions have been attributed to problem-solving, it is generally acknowledged that problem-solving is a process that involves several clearly-defined steps to be followed in the right order (Facione, 2007). It is often asked how science educators can promote the development of problem-solving skills among learners (Wismath, Orr, & Zhong, 2014), given skills cannot be taught directly. In this context, Bandura's Social Learning Theory can have positive implications for helping science educators to promote the development of problem-solving skills among their learners. Applying the Social Learning Theory, science educators need to present students with problem-solving situations—the Educators then clearly work out the steps to solve the problems in the classrooms. In so doing, Educators would be "modeling" the desired behavior for solving problems and thus helping students to learn and replicate the behavior as and when required. Furthermore, Educators can also model the correct problem-solving behavior by making use of problem-solving as an instructional method. Being regularly exposed to such problem-solving behavior as displayed by science educators would allow students to observe, retain and re-enact their roles as problem-solvers when motivated as claimed by Bandura's Social Learning Theory. As far as problem-solving is concerned, the SLT can be applied both in the cases of solving mathematical problems in science and in proposing solutions for science-related real-life problems such as global warming and the provision of pure safe water. Science educators need to clearly model out how they work the mathematical problems or how they carry out the step-wise procedures to propose solutions to science-related problems so that students can develop such problem-solving skills through observation, retention, and reproduction.

At this stage, it needs to be highlighted that we have considered science educators as *"live models"* to discuss the implications of the Social Learning Theory on science education. However, we would also like to argue here that both *"verbal instructional models"* and *"symbolic models"* are equally pertinent to science education. As highlighted above, verbal instructional models include people who explain and describe the desired behavior—they do not actually perform the behavior. Science educators therefore also represent verbal instructional models when they actually explain concepts, skills, and attitudes pertaining to science. In the same line of thought, science educators also act as verbal instructional models to help learners recognize when to invoke these concepts, skills, and attitudes, how to apply and reproduce them correctly in the event of an appropriate stimulus (motivation).

Symbolic models can also have positive implications on science education. Let us now consider some ways in which symbolic models can be applied in science education as postulated by the Social Learning Theory. Symbolic models include fictitious or real characters in textbooks, novels, movies, cartoons, television programs or other media sources displaying certain types of behaviors that can be observed and modeled. Encouraging children to read about the lives and discoveries of renowned scientists (symbolic models) can enhance their interest in science and support the acquisition of the right disposition (in terms of attitudes and skills) toward science. Attributes that can be observed and modeled from the renowned scientists (as symbolic models) include curiosity, persistence; fair testing, observation, hypothesizing, hypothesis-testing, accuracy, and precision among others. Other ways in which symbolic models can be applied in science education include relevant videos of practical work being carried out. Educators can make use of ICT to project appropriate videos with symbolic models carrying out practical work, properly handling apparatus. Symbolic models can also be in the form of resource persons sharing their science-related career experiences with students. Most interestingly, symbolic models can also involve people in different situations (from movies, cartoons, case studies, true stories, events in the newspapers among others) demonstrating the right kinds of attitudes or behaviors that are in line with the aims of science education.

In view of the above discussions, it is evident that the Social Learning Theory can support the teaching and learning of science and have interesting implications on science education. Nevertheless, to ensure that the Social Learning Theory helps in achieving the aims and objectives of science education, it is important for the Educators to expose learners to the right types of models (whether live, verbal instructional or symbolic models) and provide the correct incentive to focus their attention to the desired behaviors, skills, and attitudes. In the next two sections, we elaborate more on embedding the Social Learning Theory in science teaching and learning.

Social Learning Theory Versus Socio-Constructivist Theory in Relation to Science Education

In this section, we would like to contrast the Social Theory with the Socio-constructivist Theory as proposed by Vygotsky (Amineh & Asl, 2015). Both of these theories claim that learning can occur as a result of interactions with others, in other words as a social process. Nevertheless, there is a huge disparity between the two theories in terms of student involvement in the learning process. The Socio-constructivist Theory claims that learners construct knowledge or develop understanding when they actively work and interact with others (peers or teachers in the classroom), for example by being involved collaboratively in activities or by asking questions and sharing ideas and discussing. This allows learning to take place as students can "make better sense of information and events" (Ormrod, 2008) when they actively work with others. Thus, Socio-constructivists view knowledge-construction and learning as a social process that is based on active interactions with others. The Socio-constructivist Theory will be more elaborated and discussed in detail in Chap. 18.

On the other hand, the Social Learning Theory is sometimes criticized in that it views learning as a passive process that is based on the observation of models (Laliberte, 2005). However, it can also be argued that passive observation of models will not lead to learning unless the learner focuses "active" attention on the desired behavior of the model(s) to be able to retain and remember the behavior. Furthermore, according to the Social Learning Theory, the learner must also be able to recognize a relevant or an appropriate stimulus to be "actively" motivated to display the learned behavior. As an ending note to this section, it can also be highlighted that learners can be encouraged to discuss about the observed behavior(s) in the models (live, verbal instructional or symbolic) during science teaching and learning. This would not only promote social interactions in line with the Socio-constructivist views but also render learning of the desired behaviors more meaningful.

Embedding Social Learning Theory in Science Teaching and Learning

Knowledge in science is built upon basic science concepts learnt during early childhood. Through science activities, concepts are developed and cognitive development is supported. In that way students learn about events and things in their surrounding and daily life through performance and experience, their observation skills are improved, they become more sensitive to the environment and their problem-solving skills are boosted (Saçkes et al., 2011). It is interesting to relate concept acquisition and concept development in science to Bandura's social learning theory which includes four stages in observational learning which are described in the sections which follow.

Attention

Observers cannot learn unless they pay attention to what's happening around them. This process is influenced by characteristics of the model, such as how much one likes or identifies with the model, and by characteristics of the observer, such as the observer's expectations or level of emotional arousal.

Retention/Memory

Observers must not only recognize the observed behavior but also remember it at some later time. This process depends on the observer's ability to code or structure the information in an easily remembered form or to mentally or physically rehearse the model's actions.

Initiation/Motor

Observers must be physically and/intellectually capable of producing the act. In many cases, the observer possesses the necessary responses. But sometimes, reproducing the model's actions may involve skills which the observer has not yet acquired.

Motivation

This relates to both extrinsic and intrinsic factors. The extrinsic includes the model observed and the extent to which the model has been capturing the attention and elicit the engagement of learners. Intrinsic relates to the perception and interest of students toward the tasks or activities being put in place.

How do the above stages relate to science teaching and learning? This section will consider typical lessons in science and will make a correlation with the four stages of Bandura's social learning theory.

Science teaching and learning is a dynamic activity where teachers and pupils are engaged in a process of constructing new knowledge or concepts. However, teaching students about science means more than teaching scientific concepts. There are three dimensions of science that are all important, namely, science content, science processes, and science attitudes.

Science Content

This dimension of science includes the scientific knowledge and the scientific concepts to be learnt. It is the dimension of science that most people first think about, and it is certainly very important.

Science Processes

The science processes include skills that scientists use in the process of doing science. Thus, science processes are also referred to as "doing science". It means that science is about asking questions and finding answers to questions, these are actually the same skills that we all use in our daily lives as we try to figure out everyday questions. When we teach students to use these skills in science, we are also teaching them skills that they will use in the future in every area of their lives. One of the main science skills which we promote among learners "doing science" is to make decisions on data and evidence. This skill is very fundamental in this century since UNESCO is encouraging and supporting education systems to make provisions for developing informed decision-making skills among their school youth in their national curriculum.

Science Attitudes

The third dimension of science focuses on the characteristic attitudes and dispositions of science. These include such things as being curious and imaginative, as well as being enthusiastic about asking questions and solving problems. To sum up, it can be argued that to ensure acquisition and development of science-related concepts, the environment that the child interacts with should be enriched in a way allowing the acquisition and development of science-related concepts (Greenfield et al., 2009; Oğuz, 2007).

Characteristics of Science Teaching and Learning from a Social Learning Theory Perspective

The ability to make good observations is essential to the development of other science process skills: communicating, classifying, measuring, inferring, and predicting. The simplest observations, made using only the senses, are qualitative observations.

Qualitative observation is the driving element in Social Learning Theory. The first step in this theory is attention capture. Unless there is focused observation, there will not be attention capture. This first stage in Social Learning Theory, attention capture, is also the first step in an active learning situation. For example, in an inquiry and problem-solving-based learning, the first step is to present the learners with a relevant context whereby they have to explore and formulate ideas. These ideas are then used to engage the learners in seeking information. This search for information could be either a documentary search or an investigation by experiment within laboratory set up or investigation out of the classroom such as fieldwork or surveys.

Scientific investigations form an integral part of science education and involve a number of steps or activities such as asking questions, hypothesizing, planning and carrying out experiments, collecting data and making conclusions (Hackling, 2005). In other words, implementing scientific investigations in science lessons allow learners to work like scientists. Engaged in this type of teaching and learning, learners feel like they are wearing the hat of a typical scientist. They are made to enact the behavior and model out the work of a real scientist. They will have to explore an event, a phenomenon or an object which will lead them to ask questions and generate hypotheses. In this way, learners with the help of the teacher will conduct investigation, collect data, analyze and interpret the same to make inferences.

The importance of modeling to promote understanding of science has earlier been reported by various authors (Jonassen & Strobel, 2006). Models can be used

Fig. 7.2 Increased engagement and enhanced performance. *Source* Blunsdon, Reed, McNeil, and McEachern, (2003)

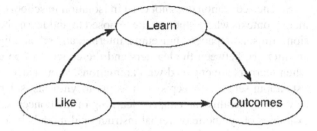

to explain a concept and are used as a tool for student interactions and they are perceived by teachers as physical representations. Such types of models are more likely to be physical objects that can help learners to better visualize concepts or phenomena. These physical models can be used to explain a concept to students, or as a way for students to explain a concept to themselves or each other. In addition, computer simulations or animations may also be used to model science concepts or phenomena. However, these models contrast with Bandura's models which display desirable behavior, skills or attitudes pertaining to science while the former is used to represent science concepts or phenomena.

It has been reported that students are more likely to "engage with a problem" if it is based on something or an issue that interests and makes sense to them (Hung & Swe Khine, 2006). They are thus more likely to focus their attention on such issues or related problems and this would ultimately lead to increased engagement in learning and enhanced performance as illustrated in Fig. 7.2. In such a context, it is important for science educators to expose learners to live or symbolic models with whom they can relate to or have some sort of affinity or interest. Such models may be national or international figures in various fields like sports, cinema, medicine, politics and technology and dealing with issues that are of interest to the science learners. This would help to "capture" the learner's attention to the desired behavior displayed by the models and lead to retention and reproduction of the behavior in the event of a stimulus.

Learners can be made to interact with each other around the models' behavior through discussions and sharing of their points of view and ideas. In this way, models can support learning and allow students to learn from each other during group or whole-class discussions about the behavior displayed by the models.

Social Learning in Science Using Digital Technologies

The section below will document some insights into Social Learning Theory in technologically-based science teaching and learning. We are living in a technology era and our youth are considered to be digital natives. They are very inclined to technology, gadgets, tablets, cell and smart phones.

Science education does not exist in isolation in schools. Outside schools there are many contexts where students are exposed to and learn about science such as television, films, newspapers, museums, internet, and so on. Digital technologies provide an interface between the learners and the concept to be understood. For example, when learners are engaged with animations, short videos or explanations by scientists about science concepts such as photosynthesis, solar system, global warming or water cycle, this can enhance learning of the concepts. Such situations represent examples of symbolic or verbal instructional models in accordance with Bandura's Social Learning Theory. Thus ICT can provide a means of exposing learners to symbolic and verbal instructional models thereby facilitating integration of the Social Learning Theory in the teaching and learning of science. This may ultimately result in increased student engagement and motivation and support their learning (UNESCO, 2012) which will help as future youth and citizens to participate fully and actively in decision-making related to any socio-scientific issue thus ensuring a scientifically literate citizenry.

Conclusion

Social learning theories emphasize changes in behavior and learning through the observation and imitation of the actions and behaviors in the environment. Social Learning Theory is still a valid theory in science education. Today science education is not solely limited to learning scientific concepts. More importantly, it englobes the science process skills and scientific attitudes. These competences are a requisite for all learners to address and face local and global challenges such as food security, energy crisis, and climate change. These issues and challenges are the very concrete contextual situations that should be embedded in science teaching and learning.

Teaching and learning in science involve knowledge acquisition through learning processes put in place by science educators and owned by the learners whereby the latter are engaged in quality or systematic observation of natural phenomena or lived models. Learners then collectively find the most appropriate means to make sense and meaning of the phenomenon and models understudy and they will be required to argue on their findings and come up to a conclusion under the facilitating processes of the educator. Though these transactions of science teaching and learning corroborate with problem-based and inquiry learning strategies, this chapter documented how these current practices of teaching and learning in science align with Bandura SLT. The chapter also elucidated some features of Bandura SLT. It also showcased, using examples, that today, the theory has still its significance in teaching and learning of science.

Further Reading

Crittenden, W. F. (2005). A social learning theory of cross-functional case education. *Journal of Business Research, 59*, 960–966.
Mesoudi, A. (2017). Pursuing Darwin's curious parallel: Prospects for a science of cultural evolution. *Proceedings of the Natural Academy of Sciences, 114*(30), 7853–7860.

References

Amineh, R. J., & Asl, H. D. (2015). Review of constructivism and social constructivism. *Journal of Social Sciences, Literature & Language, 1*(1), 9–16.
Bandura, A. (1977). Self-efficacy: Toward a unifying theory of behavioral change. *Psychological Review, 84*(2), 191–215.
Blunsdon, B., Reed, K., McNeil, N., & McEachern. (2003). Experiential learning in social science theory: An investigation of the relationship between student enjoyment and learning. *Higher Education Research & Development, 22*(1).
Boundless Psychology. (2016). Bandura and observational learning. *Boundless Psychology.* Retrieved from https://www.boundless.com/psychology/textbooks/boundless-psychology-textbook/learning-7/cognitive-approaches-to-learning-48/bandura-and-observational-learning-203-12738/.
Chen, M. (2015). *Social development theory.* University of Victoria. [E-book]. Retrieved from https://onlineacademiccommunity.uvic.ca/learningdesign/wp-content/uploads/sites/1178/2015/06/Mingli-Chen-ebook.pdf.
Chomsky, N. (1959). Review of skinner's verbal behavior. *Language, 35*(1), 26–58.
Culatta, R. (2015). Social *learning theory.* Innovative Learning. [Blog post]. Retrieved from http://www.innovativelearning.com/teaching/social_learning_theory.html.
Edinyang, S. D. (2016). The significance of social learning theories in the teaching of social studies education. *International Journal of Sociology and Anthropology Research, 2*(1), 40–45.
Facione, P. A. (2007). Critical thinking: What it is and why it counts. [Blog post]. Retrieved from http://www.telacommunications.com/nutshell/cthinking7.htm.
Fontana, D. (1995). *Psychology for teachers* (3rd ed.). Hampshire and New York: Palgrave.
Greenfield, D. B., Jirout, J., Dominguez, X., Greenberg, A., Maier, M., & Fuccillo, J. (2009). Science in the preschool classroom: A programmatic research agenda to improve science readiness. *Early Education and Development, 20*(2), 238–264.
Hackling, M. W. (2005). *Working scientifically. Implementing and assessing open investigation work in science: A resource book for teachers of primary and secondary science.* Published by Western Australia: Department of Education and Training. Perth.
Huitt, W., & Monetti, D. (2008). Social learning perspective. In W. Darity (Ed.), *International encyclopedia of the social sciences* (2nd ed., pp. 602–603). Farmington Hills, MI: Macmillan Reference USA/Thompson Gale. Retrieved from http://www.edpsycinteractive.org/papers/soclrnpers.pdf.
Hung, D., & Swe Khine, M. (2006). *Engaged learning with emerging technologies.* New York, NY 10013 USA: Springer.
Jarvis, P., Holford, J., & Griffin, C. (2003). *The theory and practice of learning* (2nd ed.). London: Kogan- Page.
John-Steiner, V., & Mahn, H. (1996). Sociocultural approaches to learning and development: A Vygotskian framework. *Educational Psychologist., 31*(3/4), 191–206.
Jonassen, D. H., & Strobel, J. (2006). Modeling for meaningful learning. In D. Hung & M. Swe Khine (Eds.), *Engaged learning with emerging technologies* (pp. 1–27). Springer.

Kelland, M. (2015). Learning theory and personality development. OpenStax-CNX module: m58073. [Blog post]. Retrieved from https://cnx.org/contents/R3cpfhGP@1/Learning-Theory-and-Personalit.

Kihlstrom, J. (2014). *The evolution of cognitive social learning theory*. [Blog post]. Retrieved from http://socrates.berkeley.edu/~kihlstrm//MemoryWeb/learning/SocialLearningTheory.html.

Laliberte, M. D. (2005). A (very) brief history of learning theory. Worcester Polytechnic Institute NERCOMP SIG. Presentation Retrieved from http://file.upi.edu/Direktori/FPIPS/JUR._PEND._SEJARAH/195704081984031-DADANG_SUPARDAN/BRIEF_HISTORY_OF_LEARNING.pdf.

Oğuz, A. (2007). A look at the theories on the formation of science concepts via samples from theory to practice. *Education, Science, Society Journal, 5*(19), 26–51.

McDevitt, T. M., & Ormrod, J. E. (2002). *Child development and education*. Upper Saddle River, NJ: Merrill Prentice Hall.

McLeod, S. A. (2016). Bandura—Social learning theory. [Blog post]. Retrieved from www.simplypsychology.org/bandura.html.

Miller, N. E., & Dollard, J. (1941). Social learning and imitation. In R. Culatta (Ed.), (2015) *Innovative learning*. New Haven: Yale University Press. Retrieved from http://www.innovativelearning.com/teaching/social_learning_theory.htm.

Ministry of Education. (2008). *The Ontario curriculum grades 11 & 12, science*. Retrieved from http://www.edu.gov.on.ca/eng/curriculum/secondary/2009science11_12.pdf.

Mukhopadhyay, R. (2013). Problem solving in science learning—Some important considerations of a teacher. *Journal of Humanities and Social Sciences, 8*(6), 21–25.

Ormrod, J. E. (2008). *Educational psychology: Developing learners*. Upper Saddle River, N.J. Pearson/Merrill/Prentice Hall. 6th Ed.

Saçkes, M., Trundle, K. C., Bell, R. L., & O'Connel, A. A. (2011). The influence of early science experience on kindergarten on children's immediate and later science achievement: Evidence from the early childhood longitudinal study. *Journal of Research in Science Teaching, 48*(2), 217–235.

Science Education for Responsible Citizenship. (2015). Report to the European Commission of the Expert Group on Science Education. Directorate-General for Research and Innovation. European Commission B-1049 Brussels. Retrieved from http://ec.europa.eu/research/swafs/pdf/pub_science_education/KI-NA-26-893-EN-N.pdf.

Sternberg, R. J., & Williams, W. M. (2009). *Educational psychology*. Merrill. Pennsylvania State University.

Stone, D. (1998). Social cognitive theory. Article Retrieved from http://mrspettyjohn.pbworks.com/f/SocialCognitiveTheory.pdf.

UNESCO. (2012). The positive impact of eLearning—2012 update, white paper. Education Transformation. Retrieved from http://www.unesco.org/fileadmin/MULTIMEDIA/HQ/ED/pdf/The%20Positive%20Impact%20of%20eLearning%202012UPDATE_2%206%20121%20(2).pdf.

Wikipedia. (2017). Social learning theory. Retrieved August 25, 2017, from https://en.wikipedia.org/wiki/Social_learning_theory.

Willard, E. (2015). Origins of social learning theory. [Blog post]. Retrieved from https://www.tutor2u.net/psychology/reference/origins-of-social-learning-theory.

Wismath, S., Orr, D., & Zhong, M. (2014). Student perception of problem solving skills. Transformative dialogues. *Teaching and Learning Journal, 7*(3), 1–17.

Anwar Bhai Rumjaun is an Associate Professor in the Science Education Department at the Mauritius Institute of Education in Mauritius. He is engaged in teaching and teacher programme, school curriculum and textbook development, and research in education. He is currently a Senior Honorary Research Associate at the UCL Institute of Education. He also supervises Master and Doctoral students registered with MIE and with University of KwaZulu-Natal (South Africa) and University of Brighton (UK). His research interests are in Science/Biology Education, Environmental Education/ESD, and Policy responses to Science Education.

Fawzia Narod is an Associate Professor in the Department of Science Education at the Mauritius Institute of Education. In addition to teaching and coordination of courses and programmes, she is actively engaged in research in education and development of curriculum and curriculum materials. Dr. Narod also supervises MA and Ph.D. research dissertations for the University of Brighton (UK) and University of Kwazulu-Natal (South Africa). Her research interests include the use of ICT as a pedagogical tool, Chemistry Education, teacher development, and educational management among others.

Chapter 8
Connectionism—Edward Thorndike

Richard Brock

Introduction and Chapter Map

Edward Thorndike's research was hugely influential in the United States for at least half a century and he is still regarded by many contemporary psychologists as a significant thinker. He produced over 500 publications across a diverse range of topics, many related to education (Mayer, 2009). In a survey of 1,725 members of the American Psychological Society, psychologists were asked to name the greatest psychologist of the twentieth century in their specialisation; Thorndike was ranked in joint ninth position with Carl Rogers (Haggbloom et al., 2002, p. 144). Thorndike's revered status may be linked to his championing of the assumption that learning could be studied and theorised, and that classroom practice should be influenced by research evidence. Though Thorndike has been classified as a behaviourist thinker (see Chapters 6 and 7), and his early writings contain descriptions of learning as the association between stimulus and response, in this chapter I will argue that he might additionally be conceptualised as a proto-constructivist (see Section IV) as his model of learning also contained themes that pre-empted the ideas of later constructivist thinkers. His work contains proposals that remain pertinent for teachers and researchers interested in learning science.

This chapter begins with a description of Thorndike's early research into learning in animals and the 'laws' he developed from this empirical research will be discussed. The experiments sparked Thorndike's interest in moments of insight which will be considered in comparison to research into conceptual change in science education. Finally, the relevance of Thorndike's ideas for contemporary science educators and researchers will be considered. It is hoped that this chapter will encourage readers to engage with the prescient and wide-ranging research of one of the founding figures of educational research.

R. Brock (✉)
King's College London, London, UK
e-mail: richard.brock@kcl.ac.uk

© Springer Nature Switzerland AG 2020
B. Akpan and T. Kennedy (eds.), *Science Education in Theory and Practice*,
Springer Texts in Education, https://doi.org/10.1007/978-3-030-43620-9_8

Thorndike's Early Research

Thorndike was born in Williamsburg, Massachusetts and studied at Wesleyan and Harvard Universities. Whilst at Harvard, he worked with the philosopher and psychologist William James and developed an interest in animal learning. Due to personal and financial difficulties, he left Harvard to continue his research at Columbia University. Having previously carried out research on learning with chickens in William James' cellar (Mayer, 2009), Thorndike switched his attention to studying how cats and dogs learn. Perhaps the most well-known of Thorndike's animal experiments are his investigations using puzzle boxes. In these experiments, a hungry cat, dog or chick was placed in a specially designed puzzle box in which the door was fitted with an unlatching device, for example, a loop of wire or a handle, that the animal would learn to manipulate in order to escape and gain access to some food. Thorndike (1898, p. 30) remarked that: 'Never will you get a better psychological subject than a hungry cat.' He recorded the amount of time it took the animal in the puzzle box to perform the required action that would release the door and allow it to escape. Typically, the animal's initial attempts took relatively long periods of time as they engaged in a series of unsuccessful behaviours before they tripped the latch, sometimes in an accidental manner. Thorndike noted that, over subsequent trials, the time taken to escape slowly and irregularly decreased and, after around 10 or 20 attempts, tended towards a stable, relatively short escape time.

His observations of the gradual decrease in escape times led Thorndike to hypothesise that animals were not capable of moments of insight, that is, sudden progressions in learning. Rather, he argued, animal learning proceeded gradually through the prioritisation of actions that were rewarded. Perhaps because of such work, Thorndike would later develop an interest in the effects of punishment and reward on learning in school and therefore may be seen as a pioneer of the study of the role of motivation in learning.

Thorndike's Model of Learning

The puzzle box experiments allowed Thorndike to formulate three principal 'laws' of learning: the laws of effect, exercise and readiness (Thorndike, 1898, pp. 244–250). The law of effect proposes that a response followed by a satisfactory outcome will be repeated whereas a response that leads to discomfort is less likely to be repeated (see Section II: Behaviourist Theories). Thorndike initially argued that punishment and reward were equally effective in modifying behaviour but, in a later text (Thorndike, 1932), argued that rewards more dependably shifted behaviours than punishments. The law of exercise states that the connection between a stimulus and response will grow as the two events repeatedly co-occur. For example, if a student is frequently rewarded for good behaviour, they are likely to develop a strong connection between the behaviour and a satisfactory outcome. In later writing, Thorndike (1905, p. 207, original emphasis removed) added that the connection may be '…with the total frame of mind in which the situation is felt' suggesting a broader view than a simple connection between stimulus and response. The law of readiness suggests that a learner has a greater likelihood of acquiring certain stimuluse-response connections than others due to their pre-existing frame of mind (Thorndike, 1913). One student may, for example, have a stronger predisposition to learn the skill of drawing than another.

Though he labelled his conclusions as 'laws', the statements appear to be conclusions that Thorndike felt had some general applicability rather than the status of absolute and universal principles (Hilgard & Bower, 1975). In addition to the three principal 'laws', Thorndike (1913, pp. 23–31) proposed five 'subordinate laws' which can be conceptualised as pre-empting some ideas in later models of learning in science education.

(a) *The law of multiple response* (Thorndike, 1913) suggests that a learner, when faced with a challenge, will attempt a number of alternative responses until they find the most appropriate behaviour for that context. Experience may lead the student to come to prefer one type of approach over others and that behaviour may then be used more frequently as a result. The law of multiple response might be seen to foreshadow the notion that learners possess multiple ways of understanding the world, for example, they may possess several coexisting models of heat (Mortimer, 1995). An important facet of learning science is the appropriate activation of one conception, from amongst multiple possible alternatives, in a given context.

(b) *The law of set or attitude.* As will be discussed in detail below, Thorndike goes beyond a conceptualisation of learning as a direct connection between stimulus and response. Thorndike argued learning is more than a 'simple equation' linking a learner with their environment and that learning is influenced by the 'mind's set' at the time of learning (Thorndike, 1913, p. 13). In his model, mental connections have a complex structure and, though they might be modelled simplistically in experiments, students' learning in the classroom requires a richer conceptualisation than a simple stimulus–response connection. Thorndike acknowledged that previous experiences may influence current learning and that: 'The things connected may be subtle relations or elusive attitudes and intentions' (Thorndike, 1932, p. 353). Thorndike's stance seems to prefigure contemporary models of conceptual structure which acknowledge the systemic nature of conceptual change (Amin, Smith, & Wiser, 2014) and the significance of tacit knowledge elements (Brock, 2015).

Thorndike's version of connectionism might be interpreted as an intermediate step between purely behaviourist models, which conceptualised learning as the development of links between stimuli and responses, and models that represent learning as changes to the relationships between abstract concepts in conceptual structure, as the extract below indicates:

> Learning is connecting, and man is the great learner primarily because he forms so
> many connections. The processes described … change the man into a wonderfully
> elaborate and intricate system of connections. There are millions of them. They include
> connections with subtle abstract elements or aspects or constituents of things and
> events, as well as with the concrete things and events themselves. (Thorndike, 1913,
> p. 54)

Thorndike (1913) called for the production of a volume showing the connections that would be developed during different classroom activities prefiguring representations such as concept maps and learning progressions that are now available for a wide range of topics in science education.

(c) *The Law of partial activity* suggests that responses may be triggered by certain elements of a situation or that one element of a context may be dominant in determining a response (Thorndike, 1913). This 'law' pre-empts science education researchers' reports that expert and novice learners may perceive different elements of problems as significant: experts tend to focus on the 'deep structure' of a problem (for example, underlying principles such as energy conservation) whilst novices may fixate on surface features that may not be relevant to solving the problem (such as the shape of a moving object) (Chi, Feltovich, & Glaser, 1981). The law highlights that students may, in some contexts, possess appropriate knowledge to answer a problem but fail to activate it in that situation.

(d) *The law of assimilation or analogy.* Thorndike was an early proponent of the value of analogical thinking for making sense of novel situations. He argued that developing scientific thinking was a 'struggle' because it required learners to treat perceptibly different entities, for example, coal dust and diamonds, as sharing some abstract properties (Thorndike, 1913, p. 29). He argued that an important element of learning is the ability to appreciate qualities, or groups of qualities, that are shared by concepts, even when other differences exist. Thorndike (1913, p. 37) advanced a model of transfer that argued learners should develop the ability to separate 'a subtle element' from the totality of a situation and thereby become able to transfer those principles to novel contexts. This notion anticipated researchers in science education who have noted that students' thinking may become grounded in particular contexts leading to an inability to apply ideas to novel situations (e.g. Brock & Taber, 2017a). Thorndike emphasised the importance of introducing students to concepts in a range of different contexts and providing explicit support to enable them to perceive underlying commonalities between different situations. He proposed that textbook authors should avoid problems set in unrealistic situations and that abstract ideas should be grounded in familiar contexts.

(e) *The law of associative shifting* suggests a response can be shifted from one stimulus to another (Thorndike, 1913). In particular, a response that is initially triggered by the totality of a context may, with appropriate interventions, become associated with a single element of the context. Recent models of learning in science education have emphasised that the appropriate contextual triggering of knowledge elements is an important element of learning (e.g. Brock & Taber, 2017a). In addition to introducing students to novel concepts, science teachers might aim to reduce the inappropriate triggering of some conceptual resources in certain contexts and promote their application in others. For example, the belief that motion requires a resultant force might be appropriately triggered in the case of accelerated motion but suppressed for the context of an object travelling at constant velocity.

Thorndike and the Progression of Learning

Thorndike had a particular interest in how learning progressed over time, which arose from his experiments with puzzle boxes. Studies of the manner in which learning or the acquisition of skills occurs over time have become common in developmental psychology and a growing number of studies in science education have sought to understand the progression of learning over time (Brock & Taber, 2017b). From his observations of animals in puzzle boxes, Thorndike concluded that their learning tended to be gradual, and that animals did not experience sudden progressions in learning.

The notion of the rate of change of learning is one that is of interest to researchers in science education. Though a number of researchers have noted that conceptual change, in general, appears to be a gradual process, a number of incidents of rapid changes to understanding, moments of insight, have been reported (Brock, 2015). Thorndike (1913) argued that moments of insight in humans are not the result of a special form of processing but that they occur through the formation of connections between pieces of information that have eluded other thinkers. He suggested that learning might plateau when certain habits are insufficiently automated to allow the learner to engage with higher order tasks. For example, Thorndike (1913) described how a chemistry learner may initially need to gain familiarity with a large number of pieces of information, a process that takes time. Once the knowledge elements are sufficiently familiar, more rapid progress may result as the various resources are fluently applied to problems. Thorndike therefore suggested that learning may involve periods of gradual and apparently limited change, punctuated by periods of more rapid learning.

Thorndike's Assumptions About the Nature of Learning

Thorndike, at least at some points in his writings, explicitly adopted assumptions associated with behaviourist models of learning (see Section II). Behaviourist models suggest that the focus of researchers' interest should be on learners' behaviours and that speculation about mental processes and entities, such as conceptual structure, is fruitless. Though Thorndike referred to 'mental functions', he argued that the term was simply a label for an *actually or possibly observable event in behavior*

(Thorndike, 1913, p. 66, italics in original). He tended to link learning to physical changes on the structural level of the brain, for example, he argued that gaining the ability to spell the word 'cat' is represented by 'actual bonds between neurones' though admitted that it would be impossible to know which 'neurones' were involved or the nature of the 'bonds' between them (Thorndike, 1913, p. 66). He reasoned that an understanding of learning was best achieved through observations of students and warned that introspection does not present a privileged route to understanding learning. Whilst the formulation of learning presented in Thorndike's laws is founded on behaviourist assumptions that link stimuli with responses, there are elements of his theories that prefigure the ideas of constructivist thinkers.

Thorndike as a Proto-Constructivist

Constructivist models of learning (see Section IV) propose that humans actively build personal understandings of the world and it is expected that learners' experiences of the world lead them to develop personal and idiosyncratic constructs of scientific concepts prior to formal teaching (Taber, 2009). These arguments are hinted at in Thorndike's writing. He reported that individuals develop unique understandings and though it is framed in the language of behaviourist psychology, the constructivist notion of an interaction between intuitive and formal concepts (see Chapter 19) is found in Thorndike's writing:

> Science, as we know it, is often a struggle to educate the neurones which compose man's brain to act similarly toward objects to which, by instinct and the ordinary training of life, they would respond quite differently, and to act diversely to objects which original nature and everyday experience assimilate. (Thorndike, 1913, pp. 29–30)

Thorndike suggested that students have an 'original fund of tendencies' (Thorndike, 1912, p. 91) and that the goal of education is to work with such capabilities to modify them into the desirable outcomes of learning. For example, he suggested that children's interest in collecting and organising objects might be nurtured into an understanding of scientific classification. There are several further hints that Thorndike's model of learning is not incompatible with several themes of constructivist writing. Thorndike believed that the process of coming to know about the world is partial and does not generate a perfect copy of the information being studied (Thorndike, 1913). He argued that students do not respond to stimuli like electrical instruments, that is, they cannot be expected to have easily predictable and reproducible responses to stimuli. Therefore, whilst the fundamental principles of associative learning might be used to model the acquisition of complex sets of knowledge, such as physics or geometry, the processes involved in learning in classrooms situations are likely to be different from his proposed laws.

Thorndike was ahead of his time in his desire to produce resources for teachers grounded in research evidence, for example, his book: *The Principles of Teaching Based on Psychology* (Thorndike, 1906). His recommendations for practitioners differ from the recommendations that might be expected of a behaviourist thinker. Though his law of association might be assumed to lead to the recommendation of drill-like practice, his position was somewhat more nuanced. He argued that: '…practice without zeal …does not make perfect …little in human behaviour can be explained by the law of habit; and by the resulting practice, unproductive or extremely wasteful forms of drill are encouraged' (Thorndike, 1913, p. 22). For Thorndike (1932, p. 64) 'mere repetition' did not, in all cases, lead to effective learning.

Thorndike critiqued the 'evils of rote-memorizing' (Thorndike, 1912, p. 166) and argued that teaching should aim for knowledge of fundamental principles in preference to mere facts or knowledge of statements derived from them (see Chapter 12). In what might be taken to be an early call for the teaching of metacognitive skills, Thorndike (1912) recommended that teaching students how to educate themselves was of equal value to teaching them knowledge or skills. Thorndike's teaching handbook advocates for a form of democratic student-centred education:

> Good teaching recognizes the variety of human nature, fits stimuli to individuals as far as possible, and, when that is not possible, chooses those stimuli which are for the greatest good of the greatest number… (Thorndike, 1906, p. 98)

As has been noted above, Thorndike was interested in developing novel forms of educational texts and published an influential series of mathematics textbooks (Mayer, 2009). The books were highly successful and provided Thorndike with a larger income than his salary from Columbia. He argued that, rather than simply presenting information, textbooks ought to support students in developing conclusions for themselves. A student is less likely to understand information, Thorndike (1912) suggested, when it is simply presented to them, and more effective texts encourage students to engage with ideas in some manner. He argued that books should provide 'just enough' support to guide a student's learning as 'economically' as possible (Thorndike, 1912, p. 164), pre-empting the notion of scaffolded teaching (see Chapter 19).

Thorndike's Connectionism for Science Teaching

In addition to his laws of learning and the general recommendations he made for teaching, Thorndike made a number of comments specific to science teaching. He recommended that assessment should, as much as possible, build on students' experiences. For example, he listed examples of good problems set in contexts that would be familiar to students:

- Rain drops are coming straight down. Will a car standing still or one moving rapidly receive in one minute the greater number of drops on its roof and sides?
- It is harder to keep your hands clean in the winter than in the summer? Why?
- Does an iron ball weigh more when it is hot than when it is cold?
- Is an incandescent lamp film on fire?

- Will a pound of popcorn gain or lose weight or stay the same after it has been popped?
(Thorndike, 1911, pp. 216–217)

Though one might expect a psychologist who claimed that educational measurements are identical to scientific measurements (Thorndike, 1918) to argue that assessment of learning is objective, instead, the problems listed above exemplify Thorndike's nuanced recommendations for effective assessment:

- Distrust the repetition of words as a test of anything more than verbal memory.
- …the power to define may exist without the knowledge of the term and knowledge of a term may exist without the power to define it.
- Distrust any one particular kind of problem as a test of appreciation of a law.
- Distrust especially problems that are familiar or of a well-known type.
- Do not take it for granted that the ability to handle certain elements when isolated implies the ability to handle the same elements in complex connections.
(Thorndike, 1906, pp. 263–264)

Thorndike (1912) made a case for the use of authentic practices in the science classroom. He argued that time should not be wasted on practical tasks that are not conducive to learning and cautioned teachers against setting learning through practical work that could be taught more efficiently through theoretical discussion. Thorndike (1912, p. 195) was critical of certain kinds of discovery learning in science education and argued that, though discovery learning might be a means of cultivating skills such as creativity and self-directed investigation, the requirement that students 'rediscover facts' was 'absurd' and hence he asserted that discovery learning was not an effective approach for the acquisition of knowledge (see Chapter 13). Thorndike (1912) was interested in the nature of representations in science textbooks and argued diagrams could, in some cases, more effectively represent information if their form differed from a faithful representation. For example, a diagram of neurons might be distorted from its anatomical geometry to more clearly represent the function of the cells. Thorndike commented on many facets of science education and contemporary science teachers might find much useful guidance for their practice in his writing.

Thorndike's Contribution to Educational Research

Thorndike has had a profound influence on manner in which educational research is conceptualised. He helped to establish the *Journal of Educational Psychology* (Smith, 2001) and had an admirable commitment to disseminating his research beyond academic psychologists. He published a popular introduction to psychology (Thorndike, 1901) and a guide to teaching based on psychological principles (Thorndike, 1906). Thorndike argued that, at the beginning of the twentieth century, writing about education had largely relied on ideas drawn from philosophy or anecdote rather than on empirical observation (Thorndike, 1906) and he became an early advocate of the notion that teaching should be influenced by empirical research, an idea that is experiencing a resurgence in contemporary calls for evidence-based teaching. Though Thorndike asserted that educational 'products' might be measured with the same 'clearness and precision' as variables in the physical sciences (Thorndike, 1918, p. 279), he nevertheless displayed a sensitivity to the challenging nature of assessing learning in a meaningful manner.

Thorndike felt that an overly simple transfer of methods from the physical sciences to the study of learning would result in crude or mistaken measurements (Thorndike, 1904). He argued that an individual's reports of perceptions, images and emotions would be subjective and that the complexity of mental functions suggested that units or scales developed to measure them would be necessarily imperfect (Thorndike, 1913). Hence, Thorndike (1913) concluded that his quantitative experimental investigations of learning were necessarily limited. Though Thorndike can be characterised as a researcher who focused on developing quantitative models of learning, for example, his learning curves, he was sensitive to the partial nature of the models he produced and accepted the complexity of the challenge of representing learning.

Critiques of Thorndike's Work

As is the case with any influential body of work, a number of critiques of Thorndike's model of learning have been put forward (Mayer, 2009). Firstly, the foundation of his work on the development of links between stimuli and responses might be seen to present too narrow a view of learning. It is difficult to imagine how the laws of learning Thorndike proposed can account for the kind of creative learning that scientists engage in when studying new phenomena. Secondly, Thorndike's conception of general laws of learning neglected the complex factors that impinge on learning. For example, it has been noted that the effects of rewards on behaviour are not as simple as Thorndike imagined and, under certain circumstances, rewards may lessen a student's motivation (Mayer, 2009). Thirdly, despite his own caveats discussed above, Thorndike's quantitative representations of progression in learning have been challenged and subsequent educational researchers have used a wide range of qualitative and mixed methods approaches to develop enriched conceptualisations of learning.

Concluding Thoughts

It is perhaps the case that, more than his learning theories, Thorndike's argument that learning is worthy of methodical investigation and that teaching ought to be informed by the outcomes of such research, will be his most enduring legacy. Thorndike was a pioneer of the idea that teaching should be based upon a body of research evidence and his desire to make psychological theories relevant and accessible to the practicing teacher has been profoundly influential. Thorndike's proposition that education could and should be improved through the application of research findings might seem a truism in contemporary society but it was a significant change to the status quo at the time he began his research.

Thorndike's desire to create an empirically supported pedagogy was underpinned by the high esteem in which he held teachers. He argued that students of education should be widely read and be aware of psychological, neurological, sociological and ethical aspects of pedagogy (Thorndike, 1912), a vision that was well ahead of its time. That he took the time to create texts that presented research in a format that was accessible to teachers and relevant to their practice is a lesson that many educational researchers would do well to follow. Though Thorndike's championing of a stimulus-response model of learning may lead some readers to dismiss his ideas on education as irrelevant to a field that has largely adopted alternative constructs, Thorndike's writings nevertheless contain much valuable insight about learning and teaching that emerged from his extended and deep engagement with the subject. His prolific writings contain a wealth of research that will spark the interest of contemporary students of education and further the practice of science teachers.

Summary

- This chapter examines Thorndike's research into learning in animals and humans and introduces his principal laws of learning: the laws of effect, exercise and readiness.
- It is argued that many themes in his writing suggest that his model of learning goes beyond a behaviourist view and prefigures some ideas found in the writings of constructivist thinkers.
- Thorndike's enduring legacy is the foundation of a programme that seeks to use research to develop more effective approaches to teaching and learning.

Recommended Resources

An excellent summary of Thorndike's work and legacy may be found in Mayer's chapter:
Mayer, R. E. (2009). E. L. Thorndike's enduring contributions to educational psychology. In B. J. Zimmerman & D. H. Schunk (Eds.), *Educational psychology: A century of contributions* (pp. 113–154). New York, NY: Routledge.
Thorndike's own writing on teaching and learning are well-written and, though they contain some ideas that are of their time, remain engaging reads for teachers and researchers. His guides to teaching are a good place to begin:
Thorndike, E. (1906). *The principles of teaching based on psychology.* New York, NY: A. G. Seiler.
Thorndike, E. (1912). *Education. A first book.* New York, NY: The Macmillan Company.
For readers interested in the details of Thorndike's theories, *The Fundamentals of learning* provides a good introduction:

Thorndike, E. (1932). *The fundamentals of learning*. New York, NY: Teachers College, Columbia University.

References

Amin, T. G., Smith, C., & Wiser, M. (2014). Student conceptions and conceptual change: Three overlapping phases of research. In N. G. Lederman & S. A. Abell (Eds.), *Handbook of research on science education* (pp. 57–81). New York, NY: Routledge.

Brock, R. (2015). Intuition and insight: Two concepts that illuminate the tacit in science education. *Studies in Science Education, 51*(2), 127–167.

Brock, R., & Taber, K. S. (2017a). Making-sense of "making-sense" in physics education: A microgenetic collective case study. In K. Hahl, K. Juuti, J. Lampiselkä, J. Lavonen, & A. Uitto (Eds.), *Cognitive and affective aspects in science education research—Selected papers from the ESERA 2015 conference* (pp. 167–178). Dordrecht: Springer.

Brock, R., & Taber, K. S. (2017b). The application of the microgenetic method to studies of learning in science education: Characteristics of published studies, methodological issues and recommendations for future research. *Studies in Science Education, 53*(1), 45–73.

Chi, M. T. H., Feltovich, P. J., & Glaser, R. (1981). Categorization and representation of physics problems by experts and novices. *Cognitive Science, 5*(2), 121–152.

Haggbloom, S. J., Warnick, R., Warnick, J. E., Jones, V. K., Yarbrough, G. L., Russell, T. M., … Monte, E. (2002). The 100 most eminent psychologists of the 20th century. *Review of General Psychology, 6*(2), 139–152.

Hilgard, E. R., & Bower, G. H. (1975). *Theories of learning*. Englewood Cliffs, NJ: Prentice-Hall Inc.

Mayer, R. E. (2009). E. L. Thorndike's enduring contributions to educational psychology. In B. J. Zimmerman & D. H. Schunk (Eds.), *Educational psychology: A century of contributions* (pp. 113–154). New York, NY: Routledge.

Mortimer, E. (1995). Conceptual change or conceptual profile change? *Science & Education, 4*(3), 267–285.

Smith, W. A. (2001). E. L. Thorndike. In P. Jarvis (Ed.), *Twentieth century thinkers in adult and continuing education* (2nd ed., pp. 77–93). London: Kogan Page.

Taber, K. S. (2009). *Progressing science education: Constructing the scientific research programme into the contingent nature of learning science*. Dordrecht: Springer.

Thorndike, E. (1898). Animal intelligence: An experimental study of the associative processes in animals. *The Psychological Review: Monograph Supplements, 2*(4), i–109.

Thorndike, E. (1901). *The human nature club*. New York, NY: Longmans, Green, and Co.

Thorndike, E. (1904). *An introduction to the theory of mental and social measurements*. New York, NY: The Science Press.

Thorndike, E. (1905). *The elements of psychology*. New York, NY: A. G. Seigler.

Thorndike, E. (1906). *The principles of teaching based on psychology*. New York, NY: A. G. Seiler.

Thorndike, E. (1911). Testing the results of the teaching of science. *The Mathematics Teacher, 3*(4), 213–218.

Thorndike, E. (1912). *Education. A first book*. New York, NY: The Macmillan Company.

Thorndike, E. (1913). *Educational psychology. Volume II. The psychology of learning*. New York, NY: Teachers College, Columbia University.

Thorndike, E. (1918). The nature, purposes, and general methods of measurement of educational products. In G. Whipple (Ed.), *The measurement of educational products* (pp. 16–24). Bloomington, IL: Public School Publishing Company.

Thorndike, E. (1932). *The fundamentals of learning*. New York, NY: Teachers College, Columbia University.

Richard Brock is a lecturer in science education at King's College London. After working as a secondary science teacher in the UK, Richard now lectures on the teacher training programme and supervises MA and doctoral students at King's College London. His research interests include: conceptual change in science education, the nature of scientific understanding, and teacher well-being.

Part III
Cognitivist Theories

Chapter 9
New Media Technologies and Information Processing Theory—George A. Miller and Others

Patricia J. Stout and Mitchell D. Klett

Introduction

Technological advancements have been a constant part of the growth of human societies. Throughout history, technology has changed virtually every aspect of our lives, from how we communicate to how we learn. Learning theories that describe basic behavioral, cognitive, and constructivist approaches should take into account the shifting cultural paradigms. According to a Pew Research Center Fact Sheet published in June 2019, 81% of adults in the United States own a smartphone. This number increases to 96% for young adults in the U.S. between the ages of 18- and 29-years-old. In addition to this, 74% of U.S. adults own a desktop or laptop computer and 52% own a tablet computer (Pew Research Center, 2019). A more extensive Pew Research Center study released in 2018 surveyed 27 different countries, 18 countries with advanced economies and 9 countries with emerging economies. From this study, Taylor and Silver (2019) concluded "Younger people in every country surveyed are much more likely to have smartphones, access the internet and use social media." According to the report, ownership of smartphones among young adults between the ages of 18- and 34-years-old in emerging economies during 2013–2018 increased drastically, from 23 to 85% in Brazil, 21 to 74% in the Philippines, 22 to 75% in Tunisia, 17 to 66% in Indonesia, 29 to 66% in Mexico, 39 to 73% in South Africa, 25 to 51% in Kenya, 23 to 48% in Nigeria, and 16 to 37% in India (Taylor & Silver, 2019). Out of the 18 advanced economies surveyed, the top five countries with the largest ownership of smartphones among adults included South Korea (95%), Israel (88%), the Netherlands (87%), Sweden (86%), and both

P. J. Stout (✉)
University of Texas at Dallas, Richardson, USA
e-mail: patricia.stout@utdallas.edu

M. D. Klett
Northern Michigan University, Marquette, USA
e-mail: mklett@nmu.edu

© Springer Nature Switzerland AG 2020
B. Akpan and T. Kennedy (eds.), *Science Education in Theory and Practice*,
Springer Texts in Education, https://doi.org/10.1007/978-3-030-43620-9_9

Australia and the U.S. (81%) (Taylor & Silver, 2019). While people in advanced economies are still more likely to utilize smartphones, have access to the Internet, and actively participate in social media, the above statistics suggest that more and more people every day are becoming connected to these new technologies. This means that students and their instructors often come to class with their smartphones and other Internet-connected devices, such as tablets or laptop computers, making these devices an essential component of 21st century learning. Today, whether one is studying for a test, conducting research, writing a paper, or preparing lecture slides, students as well as educators are constantly utilizing these new devices. With the use of these new devices comes greater accessibility to new media technologies, such as blogs, wikis, discussion boards, social networking, podcasts, video sharing, and more. So, what draws us to new media technologies and content, and what can they tell us about the learning process? Cognitive scientists have been discussing these questions long before smartphones, tablets, and computers became an extended part of our daily lives.

In the first section of this chapter, we offer an overview of the information processing theory, focusing on three fundamental works that contributed to the development of the theory, as well as highlight one recent trend in criticism of this particular learning theory. We then outline how new media technologies and their content fit into our modern society. After introducing both of these key topics, we examine the impact that new media technologies have on learning and teaching practices. As a concluding argument, we suggest that understanding the participatory nature of new media technologies, as well as their individual learning functionalities, is crucial to their successful implementation in the classroom. With this in mind, we include specific examples throughout the chapter of how science educators can integrate new media technologies into the design of their lesson plans and the development of their classroom environment to promote long-term learning.

Information Processing Theory

The information processing theory is a cognitive learning model that attempts to outline the method in which the human mind observes, stores, and retrieves information. The central premise of the theory rests on the assumption that the mind functions similar to a computer—information is input, processed, and output. In the twentieth century, many well-known cognitive psychologists influenced what became known as the information processing theory. In this section of the chapter, we would like to focus on three particular works: George A. Miller's "The Magical Number Seven" (1956), Richard C. Atkinson and Richard M. Shiffrin's "Human Memory" (1968), and Fergus I. M. Craik and Robert S. Lockhart's "Levels of Processing" (1972). These three works build upon each other, further developing the claim that the human brain applies a computer-like approach to learning. Today, the manner in which we process the information around us remains a topic of interest in academic research. This is especially true within the six disciplines that George A. Miller (2003) claims worked together to create an interdisciplinary understanding of cognition in the late 1950s and early 1960s: Psychology, Linguistics, Computer Science, Neuroscience, Anthropology, and Philosophy (p. 143). Senanan's (2016) TED-Ed video entitled *How Computer Memory Works* demonstrates the interconnectedness of cognitive science and the information processing theory to computer science. Along these same lines, Dawson (1998) claims that the information processing theory serves as a connection between disciplines, providing cognitive scientists from a variety of fields with the opportunity to communicate with each other (p. 13). Even though different understandings of the information-processing perspective have emerged, the acknowledgement of an underlying method continues to promote dialogue between disciplines.

The American psychologist George A. Miller (1920–2012) is most often associated with the development of the information processing theory. Even though Miller was educated at Harvard University in the early 1940s, when behaviorism was at its prime in the United States, he went on to play a major role in the formation of cognitive psychology in the early 1950s (Miller, 2003, p. 141). In his personal account of the cognitive revolution, Miller (2003) notes that it was only in the U.S. that experimental psychologists strongly advocated behaviorism, and that when he and others began focusing on the inner-workings of the mind, it renewed communication with psychologists abroad such as Sir Frederic Bartlett, Jean Piaget, and A.R. Luria (p. 142).

In 1956, Miller published the famous article entitled "The Magical Number Seven, Plus or Minus Two," in which he claimed that there exist specific limitations to the human capacity to process information. Prior to Miller's well-known article on the magical number 7, Ebbinghaus is reported to have made a similar discovery in 1885 (Wagoner, 2012). Miller (1956) defines the human subject as a communication system; and, as such, he hypothesizes a correlation between the amount of input and output information (p. 82). Throughout the well-known article, Miller examines three different categories of tasks in which the human capacity for immediate memory is tested: (1) tasks pertaining to absolute judgment; (2) tasks focusing on memory span; and (3) tasks centered on the act of subitizing (Cowan, 2015, p. 537). Subitizing is defined as the quick, correct, and confident identification of numbers performed for just a few items. Simply put, it is the ability to recognize a small group of objects without counting. In the examples of tasks measuring absolute judgment, researchers first asked individuals to study a range of categories, and then to place each stimulus presented into the appropriate category (Cowan, 2015, p. 537). The findings from the experiments that Miller references indicate that an individual is capable of working with roughly five to nine different categories at a time (Cowan, 2015, p. 537). Cowan (2015) points out that the most popular section of Miller's (1956) article is that which is devoted to the human memory span (p. 537). The experiments referenced in this segment test the individual's ability to remember lists of digits, letters, or words. Similar to the previous results, Miller's synthesis of findings suggests that the human ability to memorize the correct items in order within a list weakens around plus or minus seven as well. It is here that Miller (1956) introduces the concept of "chunks," or groups of recognizable letters or words; if individuals identify "chunks" within the list, they are able to memorize substantially longer lists (p. 93). In the ultimate category of tasks centered on subitizing, Miller concludes that an individual can assess, without counting, up to seven objects (Cowan, 2015, p. 537). However, in the concluding paragraph of the article, he likens the findings between all three types of tasks to coincidence. According to Cowan (2015), this mention of coincidence is one reason why Miller's 1956 article, despite its vast popularity, inspired little further research on item capacity limits (p. 538). Instead, research shifted toward the structural layout and systemic workings of the human mind.

In 1968, psychology professors Richard C. Atkinson and Richard M. Shiffrin proposed a theoretical model of how the mind processes information, in which they outlined three specific components of a linear process in the human memory system: the sensory register, the short-term store (STS), and the long-term store (LTS) (p. 92). See Fig. 9.1. The Atkinson–Shiffrin model, also referred to as the multi-store model of memory, illustrates that information is first recognized by the sensory register via sight, sound, touch, smell, or taste. It then either remains in the sensory register until

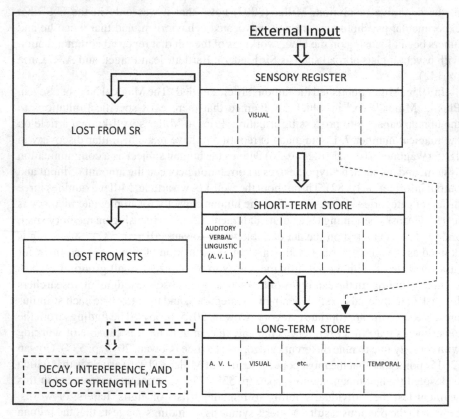

Fig. 9.1 Structure of the memory system (Atkinson & Shiffrin, 1968, p. 93). This figure has been redrawn and was originally published by Atkinson and Shiffrin to explain their three permanent components of the memory system

decaying, or is copied into the short-term store, or "working memory." Information that is ultimately copied to the long-term store becomes permanent, while information that remains in the sensory register or the short-term store deteriorates according to each store's decay rate. Thus, while it is possible for the information to transfer from one compartment to the next, the information does not simply flow between the two compartments, but rather is duplicated so that it simultaneously remains in the previous store as well until it decays (Atkinson & Shiffrin, 1968, p. 94). The dotted line from the sensory register to the long-term store in Fig. 9.1 illustrates the possibility that information is able to transfer directly into the long-term store, although Atkinson and Shiffrin (1968) noted that they do not know whether this in fact occurs (p. 94). When the subject engages in intellectual activities, information is often transferred in the reverse order (i.e., from the LTS to the STS) (Atkinson & Shiffrin, 1968, p. 94).

In addition to focusing on the three memory stores, Atkinson and Shiffrin (1968) also discuss the rehearsal process, which aids in the transferring of information from short-term memory to long-term memory. During the rehearsal process, information is repeated to initiate this movement. A very basic example of rehearsal occurs when the instructor has students repeat numerous names in a class or group intentionally in order to commit this information to their long-term memory. If repetition does not occur, then information is far more likely to be forgotten, or displaced from one's short-term memory.

Following the work of Atkinson and Shiffrin, as well as others who supported the multi-store model of memory, Fergus I. M. Craik and Robert S. Lockhart (1972) questioned the structural focus of the model and proposed a new framework of memory based on different levels of processing. Craik and Lockhart (1972) claim that the rate at which information decays, or is forgotten, depends less on the notion of memory stores, and more on the manner in which the information is encoded (p. 675). The levels of processing framework suggest that information is encoded on a continuum, and it introduces the concept of "depth of processing," in which information that is processed at a greater depth undergoes a higher level of analysis (Craik & Lockhart, 1972, p. 675). The greater the depth, the more likely one is able to remember, or learn, the information. Craik & Lockhart (1972) further explain the retention rate of information by stating, "when attention is diverted, information is lost at a rate which depends essentially on the level of analysis" (p. 677). Information processing levels in this framework are split into two types. Type I includes levels where processing of physical or sensory features occurs. Type II includes levels where pattern recognition and the extraction of meaning take place (Craik & Lockhart, 1972, pp. 675–676). Thus, the levels of processing framework place slightly more emphasis on how information is processed than that of the multi-store model.

Since we will be focusing on new media technologies in the sections that follow, we would like to point out three examples of new media content that can be used to learn about the information processing theory. The first example is a Crash Course (2014) video entitled *How We Make Memories*. This educational video is part of an online series that is dedicated to Psychology topics and is available to watch on the Crash Course YouTube channel created by John Green and Hank Green. The second example is a Khan Academy Medicine (2013) video entitled *Information processing model: Sensory, working, and long-term memory* that is available to watch on the Khan Academy YouTube channel. In addition to these two videos, a synopsis and outline of the key components of Atkinson and Shiffrin's multi-store model and Craik and Lockhart's levels of processing framework are available on the *Simply Psychology* website created by Saul McLeod.

While the information processing theory demonstrates that the manner in which material is presented to students has an effect on whether long-term learning is achievable (Gurbin, 2015, p. 2337), there are arguably some limitations to a theory that equates human cognition to computer processing. In recent years, the information processing theory has been criticized for failing to consider social and cultural influences that undoubtedly affect the manner in which one processes information. Gurbin (2015) argues that the information processing theory does not acknowledge these individualistic aspects of cognition, and that learning and teaching methods must take into account student's social and cultural influences (p. 2337). Gurbin claims, "Even before any real thinking begins, social and cultural influences affect sensory input as foundational material for cognitive processes" (2015, p. 2333). For instance, depending on the values prevalent in one's culture or social groups, an individual may decide that certain stimuli are relevant or not relevant for the task at hand, causing different sensory inputs to be discarded or retained in the sensory register (or sensory memory), which suggests that social and cultural influences affect all eight of the components associated with the information processing model—sensory input, sensory memory, attention, pattern recognition, working memory, encoding, retrieval, and long-term memory (Gurbin, 2015, pp. 2333–2337, as depicted in Fig. 9.2). With this in mind, Gurbin presents a redefined information processing model in which social and cultural influences act as an umbrella over the traditional model, potentially serving as an interim model until one defines a new model (2015, pp. 2336–2337). However, even with these criticisms, the multi-store model and the levels of processing framework cannot be discounted entirely, as they serve as a starting point for us to begin to consider how new media technologies can be used in the classroom to promote learning.

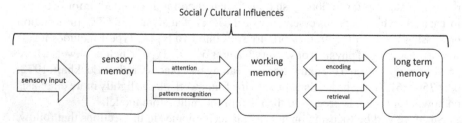

Fig. 9.2 Redefined information processing model with social/cultural influences (Gurbin, 2015, p. 2337). This figure has been redrawn and was originally published by Gurbin to outline a proposed model of information processing, which begins to take into consideration social and cultural influences

New Media Technologies in the Classroom

The Internet provides students and teachers with constant access to information, creating controversy over the role of new media and emerging technologies in the classroom. In the second edition of the book *New New Media* (2013), Paul Levinson claims that 21st century media consists of three categories: old media, new media, and "new new media." Old media, which includes print, radio, and television, employs a one-directional flow of information; it allows audience members to obtain, but not publish information. The term new media, on the other hand, commonly refers to that which exists online. However, as Levinson (2013) points out, new media does not necessarily dismantle or break away from the old media approach of communication. Many online websites continue to reinforce the "gatekeeper" mentally, in which the information published or objects sold are controlled by a select few in corporate management positions. For this reason, Levinson finds it necessary to distinguish between online platforms, such as Amazon, iTunes, and *The New York Times*, which transfer the old media approach to the web, and blogs or social networks, such as Facebook, Instagram, and Twitter, which establish a new interactive component in the dissemination process of information. The second category of blogs or social networks, or "new new media," encourages one to act as both a consumer and a producer (Levinson, 2013, p. 3). Here it is important to note that "new new media" has not replaced old media; rather, in many cases, it has become a common practice to post or tweet hyperlinks to online newspaper or magazine articles and television video clips in order to create an interconnected dialogue that flows effortlessly between medium boundaries.

It is precisely for this reason that the term new media is also commonly utilized to represent the notion of an evolving and overarching culture particular to the web. Jenkins, Purushotma, Weigel, Clinton, and Robison (2009) describe the web culture as a "participatory culture," in which members exercise the freedom to create and express themselves, under the pretext that their work will be viewed and valued by other community members with whom they share a social connection (pp. 5–6). Thus, the perception of literacy within the emerging online participatory culture moves from individual expression to community involvement (Jenkins et al., 2009, p. xiii).

Levinson's distinction between new media and "new new media" is helpful in understanding the current digital landscape of the web because it promotes the recognition that what constitutes "new" today is rapidly changing. The accelerated pace at which the understanding of new media changes into something different appears to be linked to the quick and constant development of new interactive technology associated with Web 2.0. Similar to the conceptualized stages of media, the World Wide Web has evolved over the years from the read only capabilities of Web 1.0 to the intersecting read-write availability of Web 2.0. Thus, the concept of Web 2.0 corresponds to Levinson's (2013) definition of "new new media," and Jenkins et al.'s (2009) notion of the development of a "participatory culture." While we think that it is important to recognize Levinson's concept of "new new media," for this chapter, we employ the term *new media technologies* to refer to the online apparatuses that currently enable users to participate in the ever-changing and ongoing online discussions of our time if they so choose. Likewise, we utilize the term *new media content* to refer to the information that is created and communicated through these online apparatuses. Levinson claims, "…this empowerment of everyone as producers and disseminators of information is continuing to change the ways all of us live, work, and play" (2013, p. 2). With this in mind, the next logical question is to ask ourselves whether it changes the ways in which students learn and retain information in their long-term memory.

Hew and Cheung (2013) separate the most common new media technologies available for educators to use with students into the following categories: weblog (blog), wiki, audio discussion board, social network, video sharing, podcast, social bookmarking, game virtual worlds, and social virtual worlds (p. 49). They then propose a classification system in which these technologies are cataloged by their functionality and determined to be either synchronous in nature, promoting immediate outside feedback to students, or asynchronous, encouraging students to reflect individually on their work. For example, Hew and Cheung claim that blogs tend to promote online reflection by allowing students to review previous blog posts and analyze the chronological progression of their thinking, whereas, wikis and audio discussion boards enable online collaboration and communication among students (2013, p. 49), which can be especially beneficial when working on group projects. Social networks, such as Facebook, continue to serve primarily as social spaces, while Twitter can potentially relay information from a professor to a student (Hew & Cheung, 2013, p. 49). In addition, they explain that podcasts and video sharing sites, such as YouTube, create a repository of easily accessible information that educators can use as a free source of content to engage students in class work (Hew & Cheung, 2013, p. 49). After reviewing 27 studies that examined the impact of new media technologies on K-12 and higher education student learning, Hew and Cheung found that while the use of these technologies does appear to have a positive impact on student learning, further research is needed before specific learning effects can be outlined (2013, p. 57). Unfortunately, weak methodological approaches in the establishment of experimental controls, along with the lack of longitudinal studies conducted at the time, limit the concrete analysis of results (Hew & Cheung, 2013, p. 57). Nonetheless, it is important to note that these studies revealed no negative effects or hindrances to the student learning process (Hew & Cheung, 2013, p. 57). In conclusion, Hew and Cheung suggest that it is not necessarily the technology itself that improves student learning, but rather the ways in which they are implemented in the classroom setting (2013, p. 58).

Likewise, Kalantzis and Cope (2015) claim that new media does not inherently create new learning; rather, it often serves to further extend didactic pedagogies (a.k.a. traditional teaching) (p. 376). Instead of textbooks, we now have e-books, but the content remains relatively unchanged. Similarly, it is not the testing that has changed, but rather the convenience of being able to take the test at any time or place (Kalantzis & Cope, 2015, p. 375). Gan, Menkhoff, and Smith (2015) reinforce this notion by focusing on the fact that iPad applications are allowing students to have access to information while they are outside the classroom (p. 653). In other words, it might appear that education (schools, teachers, and communities) has adopted the integration of new media technologies and content in the classroom, but the reality is that new technological devices have the overwhelming capacity to perform outdated instructional strategies. For this reason, a hands-on pedagogical approach must be implemented in combination with new media technologies.

Utilizing Technological Tools and New Media Technologies in Science Education

When analyzing the inclusion of educational technology into STEM classrooms, Connell and Abramovich (2017) highlight the fact that problems arise when educators are not adequately proficient in how to apply such technology (p. 222). The need to overcome and avoid this particular scenario often leads to the development of computer-based content as "a teacher replacement model," in which standardized content that requires little outside instruction is created and distributed to educators (Connell & Abramovich, 2017, p. 222). Computer-based content such as this generally focuses on memorization, resorting back to behaviorist views of teaching (Connell & Abramovich, 2017, p. 222). Technology, or the computer, then serves to provide instant results to online quizzes and exams, which eliminate the time necessary for students to engage in critical thinking and analysis. If we think back to Craik and Lockhart's (1972) levels of processing framework, it is this deep analysis that promotes long-term learning. Connell and Abramovich (2017) claim, "Effective STEM teaching, with its focus on critical analysis and problem solving, should not be reinforcing student beliefs that all that counts is the correct answer" (p. 223). Rather, when implemented properly, educational technology in the classroom can create an environment in which students mimic the current practices of today's scientists, mathematicians, and engineers (Connell & Abramovich, 2017, p. 223). Professionals in these three fields rely heavily on computer-based programs to assist in collecting, displaying, and analyzing data. For more information on Project and Problem-Based Teaching and Learning, see Chap. 23.

One method to promote and use Web 2.0 tools in science is to develop online learning communities for teachers and students that encourage the sharing of knowledge across cultural boundaries. An example of an organization to capitalize on this aspect is the GLOBE Program, "an international science and education program that provides students and the public worldwide with the opportunity to participate in data collection and the scientific process, and contribute meaningfully to our understanding of the Earth system and global environment" (GLOBE, 2019). In 1995, GLOBE was launched to provide opportunities for students and teachers to collaborate with scientists from around the world. The program facilitates students to input and analyze environmental data relevant to ongoing scientific research and to participate in current academic discussions within the field. An embedded video describing the program can be viewed on the GLOBE website "About" page (GLOBE, 2019). GLOBE protocols and learning activities are aligned with the U.S. national and state science education standards and the U.S. Next Generation Science Standards (NGSS). Educators around the world are easily able to fit GLOBE into their existing curriculum to promote student involvement in collaborative international research.

The incorporation of new media technologies into science education and teaching strategies should promote a "participatory culture" both within and outside the classroom. This means that rather than being solely receivers or watchers of new media content, students should also create their own podcasts, wikis, blogs, and videos as a means to demonstrate content knowledge, which promotes long-term learning. Therefore, we suggest that 21st century science learning should extend its focus beyond scientific and mathematical technological tools to include new media technologies and their content. For more information about 21st Century Learning, see Chap. 32. Considering the findings from Hew and Cheung (2013), we believe that the individual functionalities of new media technologies must be taken into consideration by educators when developing science-teaching strategies. In the science classroom and lab setting, science educators can use blogging as a means to promote individual reflection on experiments and readings conducted in the classroom. A course wiki, in which students have the ability to create and edit content collaboratively with their fellow classmates, becomes a tool that can be utilized in small groups or as an entire class in order to explore how the understanding of scientific processes and concepts change for the class throughout the semester, with the goal of developing a more thorough explanation of the topic at hand as the semester progresses. Twitter can be employed as a means for the educator to communicate with students, as well as for students and the educator to follow scientists and scientific organizations. Similarly, Facebook and Instagram can be used as tools to stay informed and to collect information about what new happenings are occurring in the science world.

An Experiment for Consideration: Popular Culture in the Science Classroom

Generally speaking, today's students are fluent in the intertextuality of Web 2.0. For this reason, we encourage science instructors to explore the use of new media technologies, such as YouTube and podcasts, as a means to make scientific connections between popular culture, including films, television shows, and current issues with which students engage, interact, and discuss on a daily basis. Since the use of these new media technologies are free upon receiving access to the Internet, and accessible to students through their smartphones, laptops, and/or tablets, utilizing such tools allows educators to incorporate visual and audio components into their lesson plans that students can re-watch or listen to on their own to review the material discussed in class. This inclusion of new media technologies has the potential to aid in the rehearsal process of information processing that is essential to learning.

Furthermore, due to the fact that such a large percentage of young adults around the world own a smartphone, watching YouTube videos and listening to podcasts becomes not only convenient, but also second nature for students whose devices are never far behind. The increased accessibility to free WiFi, which is available in most coffee shops, restaurants, and bookstores, not to mention on college campuses, makes connecting to the Internet on smart devices even easier than before.

While academics and scientists alike often dismiss popular culture, its incorporation in the classroom is currently being explored in other disciplines as a pedagogical method to increase student engagement and critical thinking. Trier (2006) posits that instructors of higher education English courses have taught with media and popular culture for eight decades (pp. 434–435). In one of his own English methods courses, Trier assigns his preservice teachers with the task of creating lesson plans that incorporate media and popular culture, which he claims they complete enthusiastically (2006, p. 436). According to a blog post by OnlineUniversities.com (2013), students in the U.S. that grew up watching educational television series, such as Sesame Street, are now accustomed to learning through entertainment. According to a recent HBO Kids video, Sesame Street is available in over 150 different countries around the world, with characters unique to the countries in which it is broadcast (HBO Kids, 2018). Since most countries around the world also have their own educational programming, this trend is most likely applicable to students internationally as well. Thus, activities similar to Trier's have the potential to be explored not only in science methods courses, but also in science labs, in which higher education students should be given the opportunity to brainstorm how popular culture texts reinforce or counter the curriculum they are teaching or learning. In this way, popular culture serves as a means to reiterate information learned in the classroom and to promote outside dialogue of science topics with family and friends.

In our search for current examples of the integration of new media technologies and popular culture in science education, we discovered *The Science Of* website and blog, created by STEM Educators Matt and Shari Brady (2018; see Recommended Sources—New Media Content). In their blog, Brady and Brady provide examples of popular culture references that they have incorporated into high school science lesson plans. In the blog post "Engagement with Pop Culture in Your STEM Classroom," they offer insight into how to begin utilizing popular culture in the classroom, noting that the key is not to show an entire film or graphic novel, but rather to make sure that the content is short and directly related to the lesson (Brady, 2016). In addition, the goal when teaching with popular culture is to use content that the students are familiar with, so that these references can serve as a quick hook into the science lesson of the day. In the humorous and blatantly true online writing style of a blogger, they remind us, "If your references are more than ten years old, they're not cool, and you suck" (Brady, 2016).

One of the highlighted stories on *The Science Of*, entitled "Getting 'Spaced' in Guardians—How Long Would You Live?," serves as an excellent example of the intertextual capabilities of new media technologies (Brady, 2017). The article utilizes a photograph of Peter Quill from the hit film *Guardians of the Galaxy* to peak interest in the science behind what really happens when the human body is exposed in space.

After a detailed explanation of the scientific process that the body goes through, a hyperlink is provided to a Facebook post in which the writer and director of *Guardians of the Galaxy*, James Gunn, addresses frequently asked questions about the film. This hyperlink to a conversation on a popular social networking site serves two purposes: (1) It reconnects the science lecture back to the initial film image, or popular culture reference; and (2) It allows students, or readers, to see that the scientific discussion ties directly into the current trending chatter on Facebook. The article ends with an embedded YouTube video clip of Astronaut Jim LeBlanc testing a pressure suit for NASA. This last video clip demonstrates how the video-sharing site YouTube is able to link new and old media by making available online a video clip from *Moon Machines*, a collection of documentaries released on the Science Channel about the Apollo program space equipment. The assimilation of new media, old media, and the traditional scientific lecture support a more flexible and organic approach to teaching, in which the use of new media technologies is dependent on the individual response from each class. As such, the popular culture content that is presented through new media technologies in the classroom can be increased or decreased with a click of the mouse. Jenkins et al. (2009) assert, "We can move in and out of informal learning communities if they fail to meet our needs; we enjoy no such mobility in our relations to formal education" (p. 11). An interview with Matt Brady, entitled "Using Pop Culture to Teach Science," was released on April 24, 2017 and is available through the *Lab Out Loud* podcast series. *Lab Out Loud* is a podcast hosted by science teachers Dale Basler and Brian Bartel, which is supported by the U.S. National Science Teaching Association (NSTA).

Student Culture in Urban Science Education

Throughout this chapter, we have emphasized the importance of thinking of students as active participants in the science classroom. New media technologies, thus, serve as tools that enable students to have access to classroom materials outside of school and to develop new ways of sharing information learned with fellow students and the outside world. This essentially provides students with an opportunity to be heard inside and outside the science classroom. By utilizing new media technologies to incorporate popular culture into science lesson plans, students begin to understand how science is relevant to their daily lives. When students and teachers share the same culture and cultural references, the challenge of utilizing new media technologies in the classroom becomes one that is centered on whether the teacher and students have access to the Internet, laptops, projectors, smartphones, tablets, and other necessary technology. However, when students and teachers do not share the same culture and cultural references, there are many considerations that must be taken into account prior to introducing and implementing new media technologies into science teaching. Here, we would like to acknowledge the recent findings and studies of Christopher Emdin, an Associate Professor of Science Education at Teachers College, Columbia University. In his book *Urban Science Education for the Hip-hop Generation* (2010),

Emdin points out that in urban schools, student culture is often very different from that of the teachers and school. He claims that today's urban science classrooms contain multiple cultures that are often at odds with one another. According to Emdin (2010), "These cultures can be grouped into four categories: the culture of science, the culture of urban teaching, the culture of urban students, and the culture of the urban teacher" (pp. xi–xii). It is the last category of culture, that of the teacher, that has the power to unite the other categories to create an environment in which learning is possible (Emdin, 2010, p. xii).

Emdin outlines hip-hop culture as a means through which science teachers can connect to urban youth. Drawing from his experiences teaching public school science in New York City, he states, "...I realized that under normal circumstances, students had to subdue parts of their hip-hop identity, or ignore their experiences that could support their learning in order to be considered a 'good science student.' I had to learn that there are parts of the students' hip-hop identities that are conducive to, and supportive of, success in science" (Emdin, 2010, p. 101). Emdin's Reality Pedagogy is based on the recognition that students' experiences with hip-hop can serve as a positive entrance into the study of science. The first step for urban science teachers is to be transparent with their students and create "cogenerative dialogues," in which students learn about the teacher, the teacher learns about the students, and they discuss the classroom environment (Emdin, 2010, 105). This autobiographical element is similar to what students would hear on a rap album; however, teachers must also convey how science became an integral part of their lives (Emdin, 2010, 108). Taking the time to listen to what is occurring outside the classroom in the students' neighborhoods provides science teachers with the opportunity to link science concepts to the experiences of their students (Emdin, 2010, p. 111). One way in which Emdin builds on the students' own ties to hip-hop is by inviting community members into the classroom to join discussions on related science topics. For instance, rappers or producers may discuss "the physics of soundproof booths in studios and the recording process," whereas graffiti artists may discuss "the chemistry of containers made of tinplated steel or aluminum used to store aerosols," or "the chemistry of the dyes and pigments used in their artwork" (Emdin, 2010, 112). The uniqueness to Emdin's (2010) approach that distinguishes him from others who have focused on the importance of understanding student culture is that he argues that there are innate characteristics in hip-hop culture, such as hip-hop lovers' passion for hip-hop and its value of creativity, that lend itself well to the learning of science. Emdin's approach to teaching science in conjunction with the acknowledgment of the power of hip-hop falls in line with the STEAM movement, in which the Arts are incorporated into STEM learning. This approach further calls for a deeper understanding of the Arts in STEAM that includes art and culture, rather than solely a highbrow understanding of art (Teachers College, Columbia University, 2016). For more information on STEAM, see Chap. 31.

Edmin has created a buzz around his teaching pedagogy through the #HipHopEd social movement, in which educators discussed connections between hip-hop and education on Twitter. As the movement gained momentum, #HipHopEd developed into a nonprofit organization that "focuses on inspiring and empowering a movement that reimagines education through the use of hip-hop as text, theory, philosophy and practice in the pursuit of emancipatory schooling" (Emdin, 2018). Emdin claims, "This focus on hip-hop as a tool for transforming science education reform has an international scope because of the visibility and accessibility of the culture to many groups of marginalized people across the globe" (2010, p. 115). Thus, this approach is useful and relevant to teachers and students internationally.

Media Literacy in the Twenty-first Century

The importance of media literacy is paramount. We live in a world immersed in technological innovations. We use smartphones, tablets, and laptop computers to access new media content on a daily basis. The information processing theory allows us to see that students must have the opportunity to think critically about the information that is presented to them in class if they are to store this information in their long-term memory. Gurbin (2015) states, "Matching new information to information that is already in existing memory is important in learning. New information must have some meaning for a person or it will not be retained" (p. 2334). Since today's students are generally fluent in the use of new media technologies, such as blogs, wikis, audio discussion boards, social networks, video sharing, podcasts, and more, incorporating their use into classroom learning is not farfetched. The reality is that most students today already use new media technologies on their own to review and learn new information. By applying the idea of new media technologies to the information processing theory, we can see that these new devices and their content serve as a means for instructors to bring meaning into lessons that apply specifically to the younger generation. As highlighted in the previous section, new media content often utilizes popular culture. Therefore, incorporating new media content into science lesson plans allows today's students to make the connections necessary for long-term memory to be established by linking it to information that they perceive to be of social value. By teaching students to also be creators of new media content, rather than just receivers, they are able to become familiar with the positive and negative aspects of new media content.

One downside to new media content is that while this information is highly accessible through the Internet, information accessed one day may not necessarily be available the next. For instance, URLs change and online content can be removed or revised without warning. Furthermore, with an abundance of new media content available, it is critical that students and instructors learn to differentiate between accurate, or credible information, and false information. Teaching with new media technologies and content requires careful planning on the part of the instructor. Nonetheless, when the instructor is able to update and adapt new media content specifically to fit with the information and interests of the class as a whole, an organic approach to teaching is possible, which we therefore suggest represents a means to incorporate social and cultural values into the information processing model.

At the end of this chapter, we have included a list of new media content that may be of use to science educators and students, and that we believe deserves further exploration. In your search for online science-related content, and upon reviewing the additional new media content at the end of this chapter, we suggest that you follow the sites of your choosing via their connected social networking pages (Facebook, Instagram, and Twitter), so that you can receive instant notification when they produce new content. By utilizing new media technologies in this manner, you will be able to gather content that may be relevant to future lesson plans or studies.

As a final note, we would like to mention the fact that a new media approach to teaching may simultaneously call for an interdisciplinary approach to learning, in which the boundaries between related fields become less defined and students question the very nature of how their areas of study relate to others, both academically and in the real world. Accessing prior knowledge and the transfer of knowledge from one learning situation to another is a way to further ideas and concepts to develop stronger processes and application in future studies. Future research should examine the benefits and disadvantages of applying an interdisciplinary approach to science learning through the incorporation of new media technologies and content in the classroom.

Summary

- The information processing theory is a learning model that equates human cognition to computer processing. It allows us to see that students must have the opportunity to think critically about the information in order to achieve long-term learning.
- The Internet provides students and teachers with constant access to information, creating controversy over the role of new media technologies and content in the classroom.
- Utilizing new media technologies and content in the classroom can promote a more flexible and organic approach to teaching that can appeal to today's students.
- Understanding the participatory nature of new media technologies, as well as their individual functionalities, is crucial to their successful implementation in the classroom.

Recommended Resources—Books

Emdin, C. (2016). *For white folks who teach in the hood…and the rest of y'all too: Reality pedagogy and urban education*. Boston, MA: Beacon Press.
Levin, I., & Tsybulsky, D. (Eds.). (2017). *Digital tools and solutions for inquiry-based STEM learning*. Hershey, PA: IGI Global.

Lin, T. B., Chen, V., & Chai, C. S. (Eds.). (2015). *New Media and Learning in the 21st Century: A socio-cultural perspective*. Singapore: Springer Singapore.

Recommended Resources—New Media Content

Bartel, B., & Basler, D. (2019). *Lab out loud: Science for the classroom and beyond* [blog and podcast]. Retrieved from http://laboutloud.com.
Brady, M., & Brady, S. (2018). *The Science Of* [website and blog]. Retrieved from https://www.discovery.com/tv-shows/mythbusters/.
Crash Course. (2014, May 5). *How we make memories—Crash course psychology #13* [video file]. Retrieved from https://www.youtube.com/watch?v=bSycdIx-C48.
Discovery Channel. (2019). *Myth Busters*. Retrieved from https://www.discovery.com/tv-shows/mythbusters/.
Emdin, C. (2012, June 18). *Urban science education for the hip-hop generation* [video file]. Retrieved from https://www.youtube.com/watch?v=P9x_IlNFvqo.
HBO Kids. (2018, February 17). *Sesame street season 48: Muppets from around the world* [video file]. Retrieved from https://www.youtube.com/watch?v=IZ4x6N1JBcg.
Hill, K. (2019). *Because science*. Retrieved from http://nerdist.com/videos/because-science/.
Khan Academy Medicine. (2013, October 24). *Information processing model: Sensory, working, and long term memory* [video file]. Retrieved from https://www.youtube.com/watch?v=pMMRE4Q2FGk.
Lab Out Loud: Science for the Classroom and Beyond. (2017, April 24). Using pop culture to teach science [podcast]. Retrieved from http://laboutloud.com/?s=using+pop+culture+to+teach+science.
McLeod, S. (2018). *Simply psychology*. Retrieved from https://www.simplypsychology.org/.
National Geographic. (2019). *StarTalk* [video files]. Retrieved from http://channel.nationalgeographic.com/startalk/.
Office Depot, Inc. (2014, April 18). *Teacher Chris Emdin finding ways to make math fun* [video file]. Retrieved from https://www.youtube.com/watch?v=Hp4wrMBZEMk.
Senanan, K. (2016, May 10). *How computer memory works* [video file]. Retrieved from https://ed.ted.com/lessons/how-computer-memory-works-kanawat-senanan#watch.
TED-Ed: Lessons Worth Sharing (Science & Technology). (2019). Retrieved from http://ed.ted.com/lessons?category=science-technology.

References

Atkinson, R. C., & Shiffrin, R. M. (1968). Human memory: A proposed system and its control processes. *Psychology of Learning and Motivation, 2*, 89–195.
Brady, M. (2016, June 23). Engagement with pop culture in your STEM classroom. Retrieved from http://www.thescienceof.org/teaching/engagement-science-classroom/.
Brady, M. (2017, May 16). Getting 'spaced' in guardians—How long would you live Retrieved from http://www.thescienceof.org/topstories/getting-spaced-guardians/.
Connell, M. L., & Abramovich, S. (2017). STEM teaching and learning via technology-enhanced inquiry. In I. Levin & D. Tsybulsky (Eds.), *Digital tools and solutions for inquiry-based STEM learning* (pp. 221–251). Hershey, PA: IGI Global.
Cowan, N. (2015). George Miller's magical number of immediate memory in retrospect: Observations on the faltering progression of science. *Psychological Review, 122*(3), 536–541.

Craik, F. I. M., & Lockhart, R. S. (1972). Levels of processing: A framework for memory research. *Journal of Verbal Learning and Verbal Behavior, 11,* 671–684.

Crash Course. (2014, May 5). How we make memories—Crash course psychology #13 [video file]. Retrieved from https://www.youtube.com/watch?v=bSycdIx-C48.

Dawson, M. R. W. (1998). *Understanding cognitive science.* Malden, Massachusetts: Blackwell Publishers.

Emdin, C. (2010). *Urban science education for the hip-hop generation.* Rotterdam: Sense Publishers.

Emdin, C. (2018). #HipHopEd. Retrieved from https://chrisemdin.com/hiphopeds/.

Gan, B., Menkhoff, T., & Smith, R. (2015). Enhancing students' learning process through interactive digital media: New opportunities for collaborative learning. *Computers in Human Behavior, 51,* 652–663.

GLOBE. (2019). About GLOBE. *The GLOBE program: Global learning and observations to benefit the environment.* Retrieved from https://www.globe.gov/about/overview.

Gurbin, T. (2015). Enlivening the machinist perspective: Humanising the information processing theory with social and cultural influences. *Procedia: Social and Behavioral Sciences, 197,* 2331–2338.

Hew, K. F., & Cheung, W. S. (2013). Use of Web 2.0 technologies in K-12 and higher education: The search for evidence-based practice. *Educational Research Review, 9,* 47–64.

Jenkins, H., Purushotma, R., Weigel, M., Clinton, K., & Robison, A. J. (2009). *Confronting the challenges of participatory culture: Media education for the 21st century.* Cambridge, MA: The MIT Press.

Kalantzis, M., & Cope, B. (2015). Learning and new media. In D. Scott & E. Hargreaves (Eds.), *The SAGE handbook of learning* (pp. 373–387). Los Angeles: SAGE Publications Ltd.

Levinson, P. (2013). *New new media* (2nd ed.). New York, NY: Pearson.

McLeod, S. (2018). *Simply psychology.* Retrieved from https://www.simplypsychology.org/.

Miller, G. A. (1956). The magical number seven, plus or minus two: Some limits on our capacity for processing information. *The Psychological Review, 63*(2), 81–97.

Miller, G. A. (2003). The cognitive revolution: A historical perspective. *Trends in Cognitive Sciences, 7*(3), 141–144.

OnlineUniversities.com. (2013, March 5). Cool teachers' guide to pop culture in the classroom. Retrieved from http://www.onlineuniversities.com/blog/2013/03/cool-teachers-guide-pop-culture-classroom/.

Pew Research Center. (2019, June 12). Mobile fact sheet. Retrieved from https://www.pewinternet.org/fact-sheet/mobile/.

Senana, K. (2016, May 10). How computer memory works. TED-Ed [video file]. Retrieved from https://ed.ted.com/lessons/how-computer-memory-works-kanawat-senanan#watch.

Taylor, K., & Silver, L. (2019, February 5). Smartphone ownership is growing rapidly around the world, but not always equally. Retrieved from https://www.pewresearch.org/global/2019/02/05/smartphone-ownership-is-growing-rapidly-around-the-world-but-not-always-equally/.

Teachers College, Columbia University. (2016, April 11). For white folks who teach in the hood…and the rest of y'all, too—A book talk [video file]. Retrieved from https://www.youtube.com/watch?v=AG7YO7nIuY4.

Trier, J. (2006). Teaching with media and popular culture. *Journal of Adolescent & Adult Literacy, 49*(5), 434–438.

Wagoner, B. (2012). Learning and Memory. In R. Harre & F. M. Moghaddam (Eds.), *Psychology for the Third Millennium: Integrating cultural and neuroscience perspectives* (pp. 116–138). London: Sage.

Patricia J. Stout is a Ph.D. candidate at the University of Texas at Dallas, where she is specializing in Cultural Studies. She holds a Masters of Art in Communication and has over 5 years of experience in Marketing and Public Relations. She has worked firsthand with the development and customization of new media technologies for both corporate clients and science education programs, as well as focused on how such technologies are changing modes of communication from an academic perspective. Her research interests lie in art as a means of communication across cultures and the use of new media technologies in the classroom.

Mitchell D. Klett is a Professor of science education and educational technology at Northern Michigan University. Dr. Klett holds a BA and MAT in geoscience and science education from the University of Texas at Dallas and a Ph.D. in education from the University of Idaho in Moscow. His teaching experience has taken him from the suburbs of Dallas, the hills of the Palouse, metropolitan Chicago, to the isolated Upper Peninsula of Michigan. He has been invited to consult with educational partners in Brazil and Colombia. His current areas of research are in the assessment and evaluation of students, faculty, and programs.

Chapter 10
Stage Theory of Cognitive Development—Jean Piaget

Brinda Oogarah-Pratap, Ajeevsing Bholoa, and Yashwantrao Ramma

Introduction

The stage theory of cognitive development is the first cognitivist theory developed by Jean Piaget almost a century ago. This chapter sets out with a brief professional profile of Jean Piaget as a cognitivist theorist. It then provides some of the historical antecedents to the Stage Theory of Cognitive Development. The key ideas underpinning the theory, such as schema, adaptation, assimilation, accommodation, stage and operations, are described subsequently. The chapter also addresses the importance of the four different stages of the cognitive development theory, namely sensorimotor, pre-operational, concrete operational and formal operational. An outline of each of the four stages is given, with a more detailed focus on the formal operational stage which we support by a specific example—the motion of a golf ball—that reflects application of selected advanced level physics concepts by physics trainee teachers enrolled in a post graduate certificate course in education (PGCE) in a teacher training institute. The findings of the study are analysed and discussed to highlight the practical applications and shortcomings of the theory, especially in relation to the appropriateness of the theory for acquisition and development of science concepts, and for fostering scientific inquiry.

B. Oogarah-Pratap (✉) · A. Bholoa · Y. Ramma
Mauritius Institute of Education, Moka, Mauritius
e-mail: b.oogarah@mie.ac.mu

A. Bholoa
e-mail: a.bholoa@mie.ac.mu

Y. Ramma
e-mail: y.ramma@mie.ac.mu

© Springer Nature Switzerland AG 2020
B. Akpan and T. Kennedy (eds.), *Science Education in Theory and Practice*,
Springer Texts in Education, https://doi.org/10.1007/978-3-030-43620-9_10

Background Information

Jean Piaget (1896–1980) was a Swiss biologist who initially started studying how molluscs adapt to their environment through experience. His interest in genetic epistemology led him to observe children, talk to them and listen to them while they were working on specific exercises he had set (Satterly, 1987). He also studied children's understanding of space, speed, time and motion which then resulted in the first major theory of cognitive development (Barouillet, 2015). The key focus of the theory was the role of biological maturation in children's capacity to understand and interact with their world. Thus, according to Piaget, intelligence is not a fixed trait; children can only undertake a task until they have reached a certain level of psychological maturity.

Piaget identified a number of key ideas underpinning his theory, namely schemas, assimilation, accommodation, adaptation, equilibrium, stage and operations. These are described in subsequent sections of this chapter.

Schemas

Schemas (also referred to as schemes) are one of the key ideas underpinning Piaget's Theory and are usually referred to as the building blocks of knowledge. They include both simple and complex actions, ideas and a set of perceptions. They help individuals to make sense of their world by organising and giving structure to their thoughts. An innate reflex of a baby is considered to be a simple schema while the ability to set up and conduct a scientific experiment constitutes a complex schema. According to Piaget, older children can perform more complex actions than younger ones since the number of schemas increases as children grow up. The ability to perform an increasing number of complex actions is the result of two key processes—assimilation and accommodation. Adaptation is the outcome of assimilation and accommodation which leads to the building of schemas (intellectual growth).

Assimilation, Accommodation and Equilibrium

Assimilation occurs when children encounter new information which they add to the existing schemas of what they are already familiar with. On the other hand, a state of disequilibrium is created when children come across new information they have never encountered before. The new information is either ignored or children will try to match it to some pre-existing schemas. Accommodation occurs when children try to do the matching, in which case the existing schemas are either modified or new ones are created to make room for the new information, thereby reinstating a state of equilibrium.

Fig. 10.1 Factors
influencing equilibrium

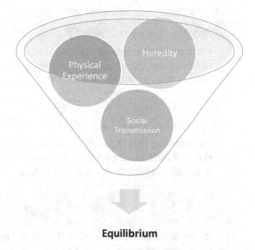

Equilibrium

Assimilation and accommodation are processes that occur automatically. They are acted upon by heredity, physical experience and social transmission. Heredity influences time and provides tools for cognitive growth. Physical experience involves concrete, hands-on experiences with objects encountered while social transmission relates to the reception of information from parents, schools and the community. The balance that is created by the forces from these three factors is referred to as equilibrium (Thomas, 1992; see Fig. 10.1).

During assimilation, equilibrium occurs as a child's schemas deal with most of the new information. When exposed to new information that does not match the existing schemas, to avoid disequilibrium, the child's schemas change to accommodate the unmatched information (accommodation). The shift from a state of disequilibrium to equilibrium is referred to as equilibration.

Stages of Cognitive Development

According to Piaget's theory of cognitive development, knowledge construction occurs through four hierarchical phases, referred to as stages. A stage refers to a period in children's development during which their thinking process allows for the understanding of a given situation. The four stages are the sensorimotor period, the pre-operational period, the concrete operational period and the formal operational period. Each stage has specific features that explain the thinking process at different periods in an individual's life (see Fig. 10.2). The assumption is that each child goes through the different stages in the same order.

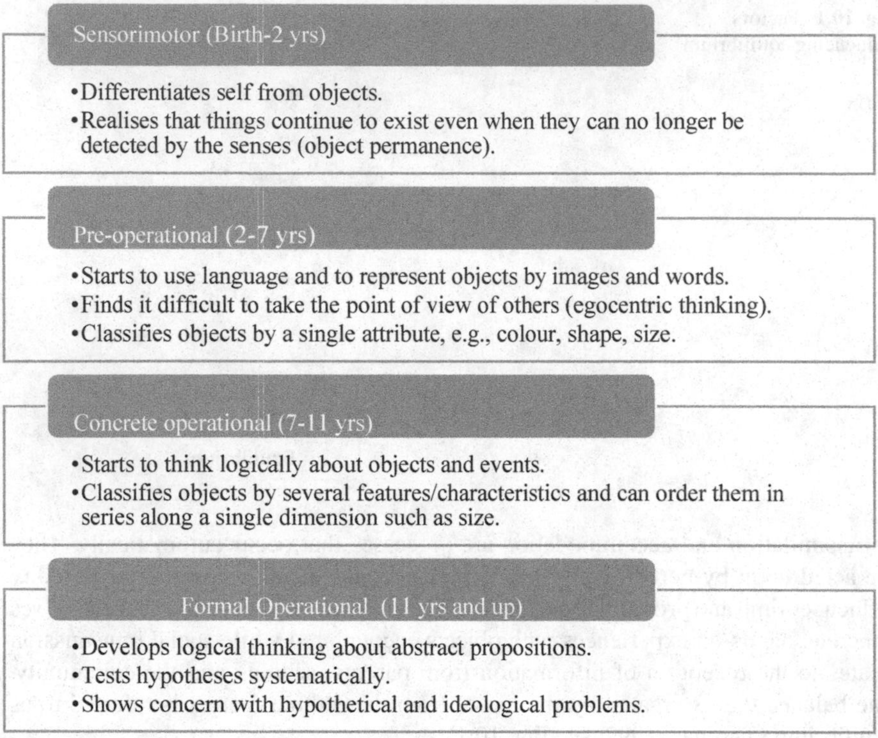

Sensorimotor (Birth-2 yrs)

• Differentiates self from objects.
• Realises that things continue to exist even when they can no longer be
  detected by the senses (object permanence).

Pre-operational (2-7 yrs)

• Starts to use language and to represent objects by images and words.
• Finds it difficult to take the point of view of others (egocentric thinking).
• Classifies objects by a single attribute, e.g., colour, shape, size.

Concrete operational (7-11 yrs)

• Starts to think logically about objects and events.
• Classifies objects by several features/characteristics and can order them in
  series along a single dimension such as size.

Formal Operational (11 yrs and up)

• Develops logical thinking about abstract propositions.
• Tests hypotheses systematically.
• Shows concern with hypothetical and ideological problems.

Fig. 10.2 Key features of Piaget's stages of cognitive development

Operations

Operations are described as organised, formal and logical mental processes (Feldman, 2001). Children develop the abilities to carry out various mental operations during the last 2 stages, that is, during the concrete operational stage (concrete operations) and the formal operational stage (formal operations).

Concrete Operations

Piaget's concrete operations stage involves a number of mental operations, namely seriation, transitivity, classification, decentering, reversibility and conservation. Seriation is the ability to order objects by a quantitative dimension such as size, weight or height. Transitivity refers to the ability to identify relationships among various items in a serial order, and perform transitive inferences. For instance, if item A is related to item B and item B is related to item C, then item A must be related to item C. Classification involves the identification and sorting of groups or a subset of another group of objects according to specific characteristics, e.g., appearance, size. Decentering is the ability to take into consideration the multiple aspects of a situation rather than focusing on only one aspect. Reversibility involves the ability of the child to recognise that numbers or objects can be changed, then returned to their original condition. Conservation is the child's ability to recognise that the quantity, length or number of an object or an item is unrelated to its arrangement or appearance of the object.

Children are able to perform all the above mentioned mental operations when they are closer to eleven years of age. As the age of children progresses, so does their ability to perform more complex operations. For instance, 7-year old children can classify or sort objects according to specific characteristics (seriation and classification), but they would be able to recognise logical relationships among elements in a serial order, and perform transitive inferences (transitivity) closer to the age of 11.

Formal Operations

According to Piaget, children have the ability to think scientifically in the formal operational stage. They can perform more complex mental operations (formal operations) such as drawing conclusions and constructing tests to evaluate hypotheses. The formal operations, also known as formal reasoning, include theoretical reasoning, combinatorial reasoning, proportional reasoning, control of variables and probabilistic and correlational reasoning (see Table 10.1). Children in the concrete operational stage are not able to perform these complex operations. According to Piaget's Theory, it is in the formal operational stage that adolescents and adults can think in an abstract manner through mental processes without dependence on concrete manipulation.

These formal reasoning patterns have been identified as essential abilities that students should develop to better perform in Science and Mathematics (Bitner, 1991; Lawson & Snitgen, 1982). Hence science activities should be designed to promote these reasoning patterns. According to Piaget, students acquire reasoning abilities through the process of equilibration rather than following direct and/or short-term teaching interventions (Valanides, 2006). Thus, the teaching and learning processes should provide opportunities for equilibration to take place. Furthermore, teachers should develop lessons that draw on their students' prior experiences and knowledge, design tasks and activities to promote higher-level thinking (Lutz & Huitt, 2004). Galotti (1989) draws our attention to the ambiguity in the usage of the term 'reasoning' which has a direct relationship with 'critical thinking' and where all information is specified in advance. This is usually the case in teaching-learning transactions. Thus, reasoning should not be confused with thinking, problem-solving and decision-making.

Exposing students to problem-based scenarios from different contexts can foster the formal operational reasoning patterns through the equilibration process, involving both assimilation and accommodation. The use of an interdisciplinary approach

Table 10.1 Reasoning patterns during the formal operational stage

Reasoning patterns	Explanation
Theoretical reasoning	Ability to apply multiple classifications, conservation logic, serial ordering, and other reasoning patterns to relationships and properties that are not directly observable
Combinatorial reasoning	Ability to consider all possible combinations of concrete or abstract items, thereby enabling the formulation of an alternative hypothesis
Proportional reasoning	Ability to state and interpret functional relationships in mathematical form Ability to construct and interpret data in tabular and graphical forms Ability to formulate a hypothesis and to interpret potential relationships between variables
Control of variables	Ability to recognise the necessity to control all relevant variables in an experimental design, except the one under investigation
Probabilistic and Correlational reasoning	Ability to interpret observations that show unpredictable variability to identify relationships among random variables

whereby the reasoning patterns are addressed in different subject areas by curriculum developers and teachers is another viable option to promote the development of the reasoning patterns. Valanides (2006) asserts that for successful implementation of the interdisciplinary approach, curriculum developers and teachers need to receive the appropriate pre-service and in-service training on how to develop an integrated curriculum and use teaching-learning approaches that facilitate the process of equilibration.

Criticisms of Piaget's Theory

Although Piaget's theory is recognised for being systematic in its approach and is considered to be one of the most influential theories of developmental psychology (Beilin & Pufall, 1992), it has nevertheless received some criticisms. The theory has been described as being too rigid and limited in the number of constructs (Barouillet, 2015). It tends to put too much emphasis on the maturation process of a child while overlooking the influence of culture, social setting and language. These factors, according to Bruner and Vygotsky, contribute substantially to the intellectual development of a child. Furthermore, interest and motivation which have a substantial role to play in the child's affective and cognitive development are ignored (Hidi, 2006; refer to Chap. 22 for an insightful perspective). Piaget's Theory also fails to recognise that children may have different intelligences as per Gardner's theory of multiple intelligence.

Moreover, it is argued that chronological ages do not always correspond to stages of development as defined by Piaget (Bastable & Dart, 2008). On the one hand, Piaget's theory is believed to underestimate the abilities of young children as it has been found that some children are able to perform concrete operations before the age of 7 years. On the other hand, the theory may be overestimating the abilities of adolescents and adults. There is evidence that formal reasoning patterns are not always demonstrated by all adults, who according to Piaget should have developed these abilities given their age (Eggen & Kauchal, 2000). The task content and instructions have been found to influence the ability of students to demonstrate formal reasoning patterns (Valanides, 2006).

Teachers' content knowledge (CK) has also been found to influence students' achievement in science (Diamond, Maeten-Rivera, Rohrer & Lee, 2014). Poor CK tends to favour expository teaching whereby the teacher relies heavily on textbooks and makes limited use of participatory approaches (Nixon, Campbell & Luft, 2016). Thus, although according to Piaget's theory, teachers should have developed the formal reasoning abilities, they may, nonetheless, not demonstrate these abilities due to limited CK. Moreover, they are more likely to use teacher-centered approaches in their class which does not favour the development of reasoning abilities among their students who are at the concrete and formal operational stages.

In the next section, we describe a study that was conducted in relation to the reasoning patterns of a group of physics trainee teachers who as per their age were expected to operate at the formal operational stage.

Methodology

Participants

The participants in this study were six physics trainee teachers who had embarked on a professional development course, namely the Post Graduate Certificate in Education (PGCE) programme. All the trainees had obtained their first degree in physics at a local university and had, after their graduation, joined the local teacher training institute on a one-and-a-half-year full-time course. The trainees were informed that research data would be used from only those who would have given their permission. They all consented to participate in the classroom-based study on a voluntary basis and showed much interest during the lesson delivery.

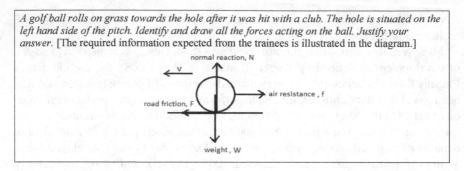

A golf ball rolls on grass towards the hole after it was hit with a club. The hole is situated on the left hand side of the pitch. Identify and draw all the forces acting on the ball. Justify your answer. [The required information expected from the trainees is illustrated in the diagram.]

Fig. 10.3 Forces acting on a rolling golf ball

The Task

For the purpose of this paper, we illustrate a task that was set to the trainees during one of the Subject Didactics modules. The participants were individually administered a problem-based scenario involving a golf ball (see Fig. 10.3) during a three-hour session of the Subject Didactics module. This module was taught by the third author and it ran over 15 weeks. The task was implemented in the 14th week and its aim was to determine to what extent the trainee teachers could demonstrate selected reasoning patterns of Piaget's formal operational stage, namely, *theoretical reasoning, combinatorial reasoning and proportional reasoning*. The concepts under consideration for the given task were weight, normal reaction, air resistance and friction (road). The trainee teachers were required to identify and draw all the forces acting on the ball and to provide the appropriate justifications of their thinking in conjunction with Piaget's formal operational stage (see Table 10.1). Our intention was to identify the extent to which the three reasoning patterns were explicitly framed while they were engaged in the given task.

Upon completion of the task, each trainee made an oral presentation (8–10 min) to communicate and justify their thinking. The presentations were video recorded. The recordings were then analysed and discussed by the three authors from the next day. The unit of analysis for the coding was any instance in the video that focused on the specificity of the reasoning elements of the formal operational stages. During the coding process, the authors met regularly to discuss, check and refine the codes until a final agreement was reached. Table 10.2 illustrates the expected responses in relation to the selected reasoning patterns.

The average scores for the reasoning items for all six participants are given in Table 10.3. The scores were computed by assigning a value of '1' to a valid response and a value '0' to either an invalid statement or a case of no response. The scores were then computed for each type of reasoning and expressed as a percentage. To illustrate our approach, the following exemplifies a valid theoretical reasoning statement involving air resistance which scored 1 mark: *'The air resistance opposes the motion of the ball'*.

Table 10.2 Piaget's theoretical, combinatorial and proportional reasonings

Factors	Formal operational elements		
	Theoretical reasoning (T)	Combinatorial reasoning (C)	Proportional reasoning (P)
Weight (W)	• The gravitational force of attraction of the Earth on the ball (mass) and it acts vertically downwards • It acts at the centre of gravity of the ball	• The magnitude of the weight vector equals to that of the normal reaction. The lengths of both vectors are the same	$W = mg$ $W = N$
Normal reaction (N)	• It acts vertically upwards • It acts from the point of contact with the surface through the centre of gravity of the ball (mass)	• Normal reaction and weight constitute the action-reaction pair in Newton's 3rd law of motion	$N = mg = N$
Air resistance (R)	• It opposes motion as a result of the collision of air molecules with the ball • This force depends on the size of the ball, nature of ball and weather condition (windy/rainy)	• At low speed, the air resistance varies with speed v • At high speed, this force varies with v^2	$f \propto v$, low speed $f \propto v^2$, high speed
Friction-road (F)	• It originates from contact between ball and surface (grass) • It opposes the motion and acts opposite to the tangential velocity	• Road friction depends on the coefficient of friction between ball and surface (grass) and normal reaction	$F = \mu N$ for motion to occur: $F > f$

Table 10.3 Reasoning (%) of participants for the golf ball problem

Reasoning	Variables			
	Weight (W)	Normal reaction (N)	Air resistance (R)	Road friction (F)
Theoretical	91.7	75.0	25.0	41.7
Combinatorial	8.3	0.0	0.0	0.0
Proportional	0.0	0.0	0.0	0.0

Findings and Discussion

The average scores obtained for the reasoning (theoretical, combinatorial and proportional) patterns of the six participants in working out the golf ball task are summarised in Table 10.3.

It can be observed from Table 10.3 that the trainees were mostly successful in providing theoretical explanation and representation of weight (92%) and normal reaction (75%). However, they were unable to provide adequate theoretical justification related to air resistance and road friction. Elements of combinatorial and proportional reasonings were comprehensively missing. For instance, none of the participants analysed the golf ball problem from the perspective of multiplicative relationships (e.g., $f = kv$ and $f = kv^2$) that characterise the structural relationships in mathematics and physics through proportional reasoning.

An in-depth analysis of the components of the theoretical reasoning revealed that the trainees performed well in identifying, representing and/or justifying *weight*, *normal reaction* and *air resistance* (see Fig. 10.4). For the convenience of analysis, we have adopted the following notations in conjunction with the information contained in Table 10.2. Three examples of such notations are illustrated:

- TW1: 1st theoretical reasoning statement related to weight.
- TW2: 2nd theoretical reasoning statement related to weight.
- TN1: 1st theoretical reasoning statement related to the normal reaction.

Analysis of the written works submitted by the trainees and oral presentations also revealed that the trainees could not justify and represent the road friction (TF2) and explain the nature of the air resistance (TR2). The findings suggest that the trainees' difficulty with reasoning involves the faulty identification of a 'forward force' (as a result of the impact of the club on the ball) to sustain the forward motion of the ball

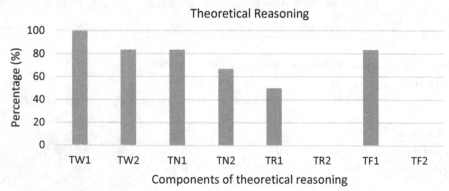

Code: T - Theoretical; W – Weight; N – Normal force; R – Air resistance

Fig. 10.4 Bar chart representing the components of theoretical reasoning for the golf ball problem ($n = 6$).

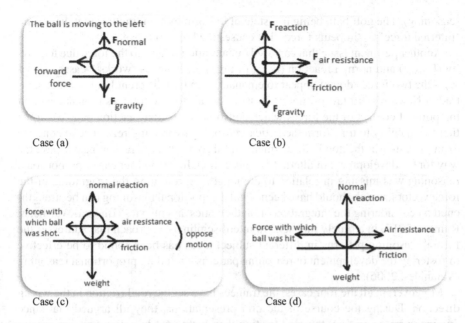

Fig. 10.5 Response from trainees (redrawn, for clarity, from their scripts) T1 (a), T2 (b), T3 (c) and T5 (d)

and a misinterpretation of the direction of the force of friction on the rolling golf ball.

In the remaining part of this section, we provide an examination of some individual attempts by the four trainees (T1, T2, T4 and T5) to solve the golf ball problem by analysing snapshots (see Fig. 10.5) of written works and excerpts from the oral presentations. The four trainees were selected on the basis of their elaborate explanations during the oral presentations.

T1 rightly identified the force due to gravity (TW1) and the normal reaction (TN1). However, the trainee erroneously considered a driving force responsible to sustain the motion. The notion of an impetus force, represented by the '*forward force*' or '*force with which the ball was pushed*' or '*propelling force*' is still rooted in the trainees' minds (Michael, 2014).

Trainee T3 did not show this 'impetus force' in her diagram. During the oral presentation, however, she explained that she had overlooked that idea but would have considered the existence of such a force. T1 and T2 justified the presence of that force by relating it to Newton's 1st law. Despite the trainees' knowledge of the underlying theory, they nevertheless could not reconcile their fragmented knowledge into a combinatorial reasoning response. In this situation, it would have been expected that the trainees would make a conceptual analysis of Newton's 1st law which stipulates that a body is either in a state of uniform motion or at rest and that the change in state results from the action of an external force (combinatorial reasoning). The golf ball, being in a state of uniform motion, implies that there is no 'internal force' or 'impetus force' that causes the body to move.

Another pertinent issue that captured our attention relates to the force due to gravity ($F_{gravity}$) and normal reaction (F_{normal}). From the trainees' works—cases (a) and (c)—the two forces do not appear to emanate from two different bodies which contradict Newton's 3rd law of motion. It is necessary for the trainees to make explicit the point of contact of the forces which constitute an action-reaction pair. We argue that the inability to transform theoretical reasoning by making reference to concrete examples, such as the golf ball, into meaningful combinatorial reasoning can pave the way for the development of alternative conception. In all the four cases, proportional reasoning was missing in relation to the length, representing the magnitude of the force vectors, which should have been equal. Proportional reasoning can be strengthened by considering the integration of mathematics and physics (Bholoa, Walshe, & Ramma, 2017) in a timely manner. An interdisciplinary approach that addresses the formal reasoning patterns in different subject areas has been found to be effective in fostering the development of reasoning patterns, including proportional reasoning (Valanides, 2006).

Moreover, in all the four cases, the trainees have set the road friction in the wrong direction. During the course of the oral presentations, they all argued that since friction opposes motion, the road friction then had to act in a direction opposite to the direction of motion of the ball. Although the trainees possess theoretical and declarative knowledge, they nonetheless lack combinatorial reasoning due to their inabilities to link the various perspectives of the inherent concepts, such as circular motion and tangential velocity at the point of contact. Upon analysing the situation, the trainees could have related the slowing down or deceleration of the golf ball to the action of the road friction which acts in a direction opposite to the tangential velocity vector of the ball.

Conclusion

If teachers were to apply Piaget's stages of cognitive development in teaching and learning, then the various mental operations which relate to concrete and formal operational should be developed along a repeated continuum rather than merely through repeating patterns of linear causality which qualifies the knowledge acquired at the formal operational stage as complete or as an end-point of a hierarchy (Seltman & Seltman, 2006). As an alternative to linear thinking, the prospective trainee teacher needs to revisit the acquired knowledge for continual growth in thinking. Knowledge acquisition being a dynamic process, the development of reasoning patterns by the trainees demand that the two stages be reviewed and refined continuously. Figure 10.6 illustrates the dynamism that should thus exist between concrete and formal operational stages.

Fig. 10.6 Cyclic interaction between concrete and formal operational stages

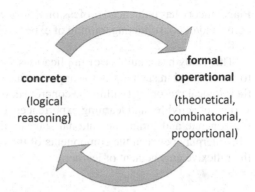

concrete

(logical reasoning)

formaL operational

(theoretical, combinatorial, proportional)

   In relation to the golf ball problem, we contend that Piaget's stages of cognitive development overestimate the reasoning ability of the trainees. We share the views of a number of researchers that cognitive development is more complex than predicted by Piaget. Moving from the concrete operational to the formal operational stage is a cyclic process whereby adult trainees can, through the manipulation of real examples, improve or review their logical understanding.

   In science, knowledge construction is largely influenced by contextual factors and its acquisition in a given domain may not be commensurate to operate in another domain be it in the same field of study or across fields. Despite the fact that the trainees had developed theoretical reasoning, they were, nevertheless, not able to apply combinatorial and propositional reasoning in a novel situation. In this particular case, a new concept—circular motion—has added to the difficulty of the trainees to integrate, by the logic of formal structures, Newton's 1st law of motion, frictional forces and change in tangential velocity to the motion of the rolling ball. Piaget's stages of cognitive development do not explicitly deal with the complex nature of logical reasoning which undergoes refinement continuously under the influence of changes happening in the areas of physical experience and social transmission (Fig. 10.1). The application of Piaget theory in teacher training has provided sufficient opportunities for the authors to identify inadequacy in the trainees' understanding and use of concepts. In fact, researchers (e.g., Tom, 1997) have reported that Piaget's theory is particularly useful for trainee teachers as it provides a framework for intertwining theory and practice of curricular and instructional significance to develop teaching and research methods that relate directly to how learning takes place. As a result of Piaget's ideas on the stages of cognitive development, new and innovative teaching methods are explored during training rather than focusing on the introduction of professional knowledge in the form of a deterministic set of procedures or generalisations hinged on the traditional beliefs consisting of teacher transmitting knowledge to learners. When teachers are allowed to experience the Piagetian theory, they facilitate learners' construction of their own knowledge. Moreover, Piaget's ideas led to the notion of 'readiness' so that concepts can be taught in a specific order. In addition,

Piaget theory has influenced professional development towards cultivating, strengthening and/or consolidating training of experimental, scientific, reasoning and logical skills.

This study has a number of implications for teacher training. Programmes have to be designed in such a way to support trainees to re-organise and further augment their logical reasoning (within the concrete developmental stage) schemas through a variety of activities and learning experiences. In this way, interconnections between the concrete and formal operational stages will be established in a cyclical mode. The interplay between the components of the two stages will pave the way towards the reflexive engagement of learners.

Summary

- The stage theory of cognitive development is the first cognitivist theory developed by Jean Piaget almost a century ago.
- The key ideas underpinning the theory include schema, adaptation, assimilation, accommodation, stage and operations.
- Assimilation and accommodation are the processes that occur automatically and are acted upon by heredity, physical experience and social transmission.
- The four different stages of the cognitive development theory are the sensorimotor, pre-operational, concrete operational and formal operational stages.
- Children develop the ability to think scientifically to perform more complex mental operations (formal operations) during the formal operational stage.
- The formal operations include theoretical reasoning, combinatorial reasoning, proportional reasoning, control of variables, probabilistic and correlational reasoning.
- Findings of the study, involving Physics trainee teachers, indicate that the development of reasoning patterns demands that the concrete and formal operation stages be reviewed and refined continuously, given that knowledge acquisition is a dynamic process.
- Piaget's theory is a prominent tool for analysing trainees' understanding and reasoning patterns of concepts in teacher training.
- The study findings support the provision of appropriate learning activities and experiences in the design of teacher training programmes to improve learners' formal reasoning abilities.

Recommended Resources

Lourenci, O., & Machado, A. (1996). In defense of Piaget's theory: A reply to 10 common criticisms. *Psychological Review, 103*(1), 143–264.

Ojose, B. (2008). Applying Piaget's theory of cognitive development to Mathematics instruction. *The Mathematics Educator, 18*(1), 26–30.

Scholnick, E. K. (1999). *Conceptual development: Piaget's legacy.* New Jersey: Lawrence Erbaulm Associates.

Young, G. (2011). *Development and causality: Neo-Piagetian perspectives.* New York: Springer.

References

Barouillet, P. N. (2015). Theories of cognitive development: From Piaget to today. *Developmental Review,* 1–12.

Bastable, S. B., & Dart, M. A. (2008). Developmental stages of the learner. In S. B. Bastable (Ed.), *Nurse as educator: Principles of teaching and learning for nursing practice* (pp. 165–216). London: Jones and Bartlett Publishers.

Beilin, H., & Pufall, P. (1992). *Piaget's theory.* New Jersey: Erbaulm.

Bholoa, A., Walshe, G., & Ramma, Y. (2017). Curriculum implications of the integration of mathematics into science. In K. S. Taber, & B. Akpan. *Science education. New directions in mathematics and science education* (pp. 211–220). Rotherham: Sense Publishers. Retrieved from https://link.springer.com/chapter/10.1007%2F978-94-6300-749-8_16.

Bitner, B. L. (1991). Formal operational reasoning modes: Predictors of critical thinking abilities and grades assigned by teachers in science and mathematics for students in grades nine to twelve. *Journal of Research in Science Teaching, 28,* 275–285.

Diamond, B. S., Maeten-Rivera, J., Rohrer, R. E., & Lee, O. (2014). Effectiveness of a curricular and professional development intervention at improving elementary teachers' science content knowledge and student achievement outcomes: Year 1 results. *Journal of Research in Science Teaching, 51*(5), 635–658.

Eggen, P. D., & Kauchal, D. P. (2000). *Educational psyhology: Windows on classroom* (5th ed.). Upper Saddle River, NJ: Prentice Hall.

Feldman, R. S. (2001). *Child development.* Upper Saddle River, NJ: Prentice Hall.

Galotti, K. M. (1989). Approaches to studying formal and everyday reasoning. *Psychological Bulletin, 105*(3), 331–351.

Hidi, S. (2006). Interest: A unique motivational variable. *Educational Research Review, 1,* 69–82.

Lawson, A. E., & Snitgen, D. A. (1982). Teaching formal reasoning in a college biology course for preservice teachers. *Journal of Research in Science Teaching, 4*(19), 233–248.

Lutz, S., & Huitt, W. (2004). Connecting cognitive development and constructivism: Implications from theory for instruction and assessment. *Constructivism in the Human Sciences, 9*(1), 67–90.

Michael, A. (2014). *Misconception in primary science.* Berkshire: Mc Graw-Hill.

Nixon, R. S., Campbell, B. K., & Luft, J. A. (2016). Effects of subject-area degree and classroom experience on new chemistry teachers' subject matter knowledge. *International Journal of Science Education, 38*(10), 1636–1654. https://doi.org/10.1080/09500693.2016.1204482.

Satterly, D. (1987). Piaget and education. In L. R. Gregory (Ed.), *The Oxford companion to the mind.* Oxford: Oxford University Press.

Seltman, M., & Seltman, P. (2006). *Piaget's logic: A critique of genetic epistemology*. London: Routledge.

Thomas, R. M. (1992). *Comparing theories of child development*. Belmont, CA: Wadsworth Publishing Company Inc.

Tom, A. R. (1997). *Redesigning teacher education*. New York: State University of New York Press.

Valanides, C. N. (2006). Formal reasoning and science teaching. *School Science and Mathematics, 96*(2), 99–107.

Dr. Brinda Oogarah-Pratap is an Associate Professor in Health and Nutrition Education at the Mauritius Institute of Education. She is involved in curriculum development at both primary and secondary levels. Her research interests include health and nutrition education, innovative practices in teacher education, including integration of online technologies. She is currently a member of a core interdisciplinary research team for a study on the content knowledge and pedagogical content knowledge of pre- and in-service trainee teachers.

Dr. Ajeevsing Bholoa is a Senior Lecturer in Mathematics Education at the Mauritius Institute of Education. He is currently the Programme Coordinator for pre-service B.Ed honours and is also involved in curriculum development at the primary and secondary levels. His research interests are related to the integration of technology as a pedagogical tool in teaching and learning of mathematics and the identification of content knowledge and pedagogical content knowledge of teachers.

Dr. Yashwantrao Ramma is a Professor of Science Education and is the Chair of Research at the Mauritius Institute of Education. As a physicist, he has worked on several research projects related to technology integration and misconceptions of both physics teachers and students. Currently, he is leading research projects on exploring teachers' content knowledge and pedagogical content knowledge across various subject areas, on indiscipline and violence in primary schools and also on students' transitions from secondary to university and teacher training.

Chapter 11
Mastery Learning—Benjamin Bloom

Ben Akpan

Introduction

Benjamin Bloom (see Box 11.1) was dissatisfied with the educational system particularly with regards to the evaluation and grading system in schools. He observed that in most cases, assessment procedures resulted in scores that approximated normal curves (Fig. 11.1).

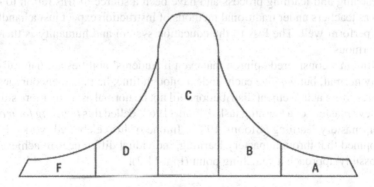

Distribution of achievement in traditional classrooms

Fig. 11.1 Normal curve of achievement using traditional methods. *Source* Akpan (2015: 21)

B. Akpan (✉)
Science Teachers Association of Nigeria, Abuja, Nigeria
e-mail: ben.b.akpan@gmail.com

© Springer Nature Switzerland AG 2020
B. Akpan and T. Kennedy (eds.), *Science Education in Theory and Practice*,
Springer Texts in Education, https://doi.org/10.1007/978-3-030-43620-9_11

Box 11.1: Benjamin Bloom Bio

Benjamin Bloom was born in 1913 in Lansford, Pennsylvania, USA. He attended Pennsylvania State University where he obtained bachelor's and master's degrees. In 1942, he received a doctorate's degree from the University of Chicago. Bloom held many positions at the University of Chicago and subsequently became a Charles H. Swift Distinguished Professor there. His most celebrated work has come to be known as Bloom's Taxonomy and is the subject of some of his writings including the *Taxonomy of Educational Objectives: Handbook 1, the Cognitive Domain*. Bloom's works also extended to mastery learning where he also had several publications including the 1971 publication: *Individual differences in school achievement: A vanishing point?* He was instrumental to the founding of the *International Association for Evaluation of Educational Achievement (IEA)* and served the *American Educational Research Association* as its president. Bloom had two sons. He died in Chicago in 1999 at the age of 86.

According to Bloom, some of the problems associated with teaching, learning, and grading in schools are due to the manner in which teachers carry out their work. When teachers believe that in a typical course of instruction about one-third of the class will perform well, another one third will perform below average while the remaining one third may just fail or obtain borderline pass marks; the corollary is that these beliefs will be translated into actual scoring practices to ascertain that the normal curve of achievement prevails. Should the achievement curve vary from the normal, there will be a strong suspicion that something went wrong.

This is irrespective of the fact that a C grade in a particular school may be as high as a B grade in another; or that an A grade in a particular year may just approximate to a C grade in another year. Bloom believes these practices have done a lot of damage to the teaching and learning process and have been a source of frustration to many students as teachers under traditional methods of instruction expect just a handful of them to perform well. The loss to the education system and humanity is therefore quite enormous.

It is Bloom's considered opinion that even if students' abilities are naturally distributably normal, but we give each student enough time, help, and encouragement, then the resulting achievement distribution will not be normal as many more students will achieve mastery of a learning task. Bloom (1968) called this *learning for mastery* and later, mastery learning (Bloom, 1971). In one of his celebrated works, Bloom (1971) opined that through mastery learning, individual differences in achievement could possibly approach a vanishing point (Fig. 11.2).

Fig. 11.2 Distribution of achievement in mastery classrooms. *Source* Akpan (2015: 21)

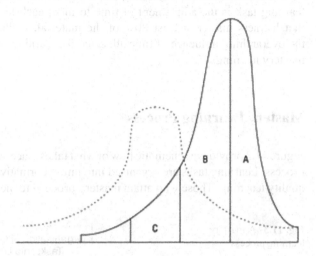

Distribution of achievement in mastery learning classrooms

Factors Affecting Mastery Learning

In any mastery learning environment, aptitude, quality of instruction, ability to comprehend instruction, perseverance, and time allocated for learning are very influential factors.

Aptitude: In everyday parlance, aptitude refers to the natural ability or skill at performing a task. However, in the context of mastery learning, aptitude is actually a measure of the amount of time that a student requires to master a particular learning task. In Bloom's opinion, the majority of learners can conceivably attain mastery of any learning task if given sufficient time such that aptitude may be seen to vary from one learning environment to another. As a consequence, it is the learning environment that has to be varied in order to attain the desired aptitude.

Quality of Instruction: The high point of mastery learning is the recognition of individual differences in learning and attainment of mastery and subsequent provision of appropriate quality of instruction to remedy the situation. The quality of instruction has therefore to relate to each learner rather than to the group of learners.

Ability to Comprehend Instruction: Learners vary in their abilities to comprehend instruction (Bloom, 1976). For example, proficiency in a particular language may affect the level and time of attainment of learning mastery in a given task. Knowing this to be the case, there is a compelling need to assure that quality instruction is varied to meet students' needs through the provision of suitable laboratory activities in science lessons, peer-tutoring, study groups, audio-visual aids, laboratory demonstrations, a variety of text books, and student workbooks.

Perseverance: The time learners willingly spend on a task (perseverance) varies from one person to another. Consequently, it is vital that a reduction in perseverance be sought through the use of measures to improve the quality of instruction such as the provision of rewards.

Time Allocated for Learning: Essential to the attainment of mastery of any learning task is the adjustment of time to meet each learner's need and to assure that learners master at least 80% of the material. It therefore bears emphasising the overarching influence of time allocated for learning on successful attainment of mastery learning.

Mastery Learning Process

Figure 11.3 provides a schematic flow of what takes place in a typical mastery learning process. Learning tasks are organised into units. Formative evaluation follows some quality teaching. Those who attain mastery proceed to the next unit while corrective

Fig. 11.3 A mastery learning process

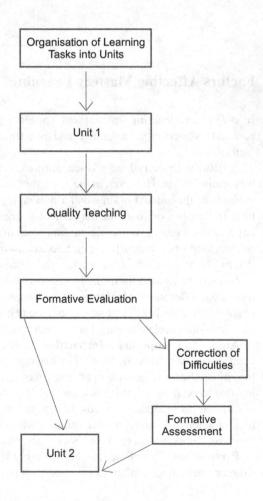

measures/formative assessment is done for the unsuccessful group until they are able to attain mastery of the task.

Mastery Learning in Practice

In practice, the following essential elements of mastery learning are important for consideration.

Feedback, Correctives, and Enrichment: For a successful mastery learning programme, there has to be a deliberate effort at providing feedback to learners. In order to do this effectively, it is inevitable to conduct formative assessments regularly. Guskey (2007) is of the opinion that such feedback must have diagnostic and prescriptive dimensions, as

> By itself…feedback does little to help students improve their learning. Significant improvement requires feedback be paired with correctives: activities that offer guidance and direction to students on how to remedy their learning problems. Because of individual differences among students, no single method of instruction works best for all. To help every student learn well…teachers must differentiate their instruction, both in the initial teaching and especially through the corrective activities…In other words, to decrease variation in results, teachers must increase variation in their teaching. (p. 16)

To enhance the effectiveness of feedback, therefore, there has to be in place a set of activities that provide systematic remedies to the learning difficulties of the students. These remedial measures are known as correctives and are usually different from the initial teaching. Guskey (2010) maintains that high-quality corrective instruction is not the same as teaching the same material again by restating the initial explanations in other ways. According to him, mastery learning teachers have to use corrective methods that take care of differences in the learning styles of students and their intelligence level. These may include, but not limited to tutoring, cooperative learning, and paraprofessional teaching aids. Such correctives may add additional time, say between 10 and 20%, to the initial estimated duration of instruction. But the good news is that whatever time is added at the early stages of mastery learning will pay off in later units as these will proceed more rapidly. According to Guskey (2007, p. 19):

> In general, teachers do not need to sacrifice content coverage to implement corrective and enrichment activities, but they must be flexible in pacing their instruction. The time used for correctives and enrichments in early units yields powerful benefits that later will make things easier. This extra time can then be recouped in later units by spending less time on reviews and increasing the instructional pace. Teachers at all levels must keep in mind what needs to be accomplished by the end of any learning sequence, but they also must see students' pathways to that end in more flexible and accommodating terms.

For the learners who do not require correctives, enrichment activities are put in place for them in the course of mastery learning. In general, these serve the purpose of widening the scope of the learning experiences. Much of the focus of enrichment programmes is towards problem-solving which according to Guskey

(2010) is not only valuable and challenging but highly rewarding. It is in the nature of mastery learning, therefore, that these enrichment activities may be drawn from outside the regular curriculum. In fact, some teachers implementing the scheme draw broadly from materials developed for the gifted and talented learners. In this direction, Sumida (2017) maintains that enrichment gives children a platform to learn about interdisciplinary issues. He gives the following example of a Japanese enrichment programme:

> …through an enrichment programme…Japanese high school students won the number one spot in a Science and Engineering Challenge and have had articles published in international academic journals. In Japan, the Ministry of Education, Culture, Sports, Science and Technology encourages student challenges in science Olympics in cooperation with science associations and universities. (p. 484)

Generally, teachers carry out feedback, corrective, and enrichment programmes using several approaches such as formative assessment involving paper-and-pencil tests, essays, projects, demonstration of skills, oral presentations, and reports. According to Guskey (2007), teachers essentially use the format of their formative assessment that aligns with the instructional goals they are pursuing.

At the end of both the corrective and enhancement activities, both groups get back to the regular class group and continue with the lessons in such a way that the sequence of instruction is not disrupted but is rather strengthened.

Aligning the Components of Instruction

Three components of instruction have been identified by Bloom as making up the teaching and learning process. These are the learning goals, instruction, and evaluation. Bloom maintains that attainment of mastery learning is easier if the three components are implemented in a consistent manner. This is what he referred to as instructional alignment. In the words of Guskey (2007):

> To ensure alignment among instructional components, teachers must make a number of crucial decisions. First, they need to decide what concepts or skills are most important for students' understanding. Teachers must determine, for example, if they want students to learn only basic skills, or if they want students to develop higher level skills and more complex cognitive processes. Second, teachers need to decide what evidence best reflects students' mastery of those basic or higher-level skills. Critics sometimes challenge teachers' abilities to make these crucial decisions… Every time they administer an assessment, grade a paper, or evaluate students' learning, teachers communicate to students what is most important to learn. Using mastery learning simply compels teachers to make these decisions more thoughtfully and purposefully. (pp. 27–28)

Formative Evaluation

According to Rossett and Sheldon (2001) evaluation is a process used in examining a programme with the aim of determining (i) what is working, and (ii) what is not working and why it is not working. In general, evaluation enables one to figure out the value of any learning or training programme thus providing the basis for decision-making, especially on ways and means of assuring improvement, if and when necessary. There are two types of evaluation—formative and summative. A summative evaluation is a means of determining the worth of a programme at the end of the programme activities. It aims to evaluate students' learning at the end of a unit of instruction through comparison with a certain benchmark. In schools, summative evaluations include, but are not limited to, terminal and end of year examinations. In this section, the focus is on formative evaluation.

Formative evaluation is a method used in determining the value of a programme while the activities for the programme are still in progress. It allows for decisions on how well the programme goals and objectives are achieved. The central attraction here is to figure out the problems, if any, and to provide adequate corrective measures where and when necessary. It allows teachers to find out the strengths and weaknesses of students and thus ensure that areas that require more work are given deserved attention. Essentially, therefore, the purpose of carrying out a formative assessment is positive in nature as it is geared towards facilitating teaching and learning; and ensures students take an active part in the learning process. Thus, within the framework of mastery learning, formative evaluation serves as a diagnostic measure and is used to determine the extent to which learners have attained mastery of a given unit. In mastery learning, units are such that could be taught within about 2 weeks and formative evaluation is carried out at the end of each unit. The result of such tests will provide the basis for use of correctives and enrichment measures earlier discussed.

Outcomes

In the context of mastery learning, outcomes are of two categories—cognitive and affective. Cognitive outcomes relate to how learners attain high grades in the learning programme—higher grades indicate higher level of mastery in the programme. On the other hand, affective outcomes relate to feelings of satisfaction and confidence by learners. Learners who attain mastery of learning tasks tend to develop more self-confidence towards the subject matter. They also are motivated to learn more.

Approaches to Mastery Learning

Two major approaches to mastery learning are Bloom's *Learning for Mastery* (LFM) and Keller's *Personalised System of Instruction* (PSI). There are also a number of instructional approaches which have been recently developed and which have their foundations on the ideals of mastery learning. These include *differentiated instruction* and *understanding by design*. It is to these four approaches that I now turn.

Learning for Mastery

The discussion so far has been based on the Learning for Mastery (LFM) approach as proposed by Benjamin Bloom. From the foregoing, it is evident that Bloom's LFM is based on a number of tenets: 1. Systematic presentation of subject matter; 2. Provision of assistance and help to remedy learning difficulties; 3. Provision of adequate time for learning mastery; and 4. Setting a clear standard of what constitutes mastery. LFM is affected by the time available for student learning—this will vary from one student to the other based on how much help is provided, how the teaching is carried out, the aptitude of the learner, and learner's verbal skills.

The Keller Plan or Personalised System of Instruction (PSI)

Under the plan: (i) learners are allowed to proceed at their own pace; (ii) learners only proceed to the next unit upon mastery of current unit; (iii) there is appropriate use of demonstrations and lectures to promote motivation; and (iv) use is made of examination supervisors (proctors) for the efficient administration of tests and tutoring (Keller, 1967, 1968).

The rationale for the use of PSI is that students vary physically as well as cognitively. Indeed, no two students will learn at the same rate. Unfortunately, the traditional mode of instruction appears to assume the contrary. PSI is an effort to address individual differences. The idea is encapsulated in this statement by Gerald Einem (Sund and Trowbridge, 1973):

> Until you break up a grade, you don't realise how futile it is to be teaching a group of kids the same material merely because they are the same age. Slow and fast learners live in a separate world. Even the teacher doesn't know how different they are until he teaches them separately. We haven't begun to realise how imaginative the fast student can be – how much challenge he needs to keep him interested – and how much specialised help the slow student needs to keep him from closing his mind and quitting. I am constantly surprised in both directions. (p. 369)

In implementing PSI, students are not held back by class average and so can advance to the next topic or unit. At the same time, slow learners have the opportunity to seek to attain higher achievement levels without hindrances. Both fast and slow learners become intrinsically motivated. Learners, therefore, proceed at their own pace and participate actively in the learning process. PSI reduces students' anxiety as they become increasingly dependent on themselves.

Differentiated Instruction

Tomlinson (2000) maintains that differentiated instruction involves tailoring instruction to meet the individual needs of learners and that four classroom elements can be differentiated by the teacher on the basis of learning profile, interest, or readiness of the learner: 1. Content—this consists of what the learner needs to study as well as how the learner has access to information; 2. Process—activities lined up for the learner in order to master the content; 3. Products—projects that enable the learner to apply what has been learnt; and 4. Learning environment—this describes the nature of the classroom such as how it works and feels.

Understanding by Design

Understanding by design provides a guide to curriculum, assessment as well as instruction by focusing on teaching and assessing for understanding and learning by transfer; and thereafter designing the curriculum 'backward' from those ends (Mctighe & Wiggins, 2012). There are seven important tenets in this approach according to them:

1. Learning is facilitated when teachers place a high premium on the planning of the curriculum.
2. Effective use of content knowledge and skill is crucial.
3. Understanding is believed to have occurred if learners autonomously make sense of their learning demonstrably by being able to, for example, explain, interpret, apply, change their perspective, or make a self-assessment.
4. Effective curriculum is planned backward based on long-term goals following a three-stage design process—desired results, evidence, and learning plan—and by so doing, the issue of treating the textbook as the curriculum rather than a teaching and learning resource is avoided.
5. Teachers serve as coaches of understanding instead of being just suppliers of content knowledge, or skill.
6. The quality and effectiveness of the curriculum are enhanced by regularly reviewing units and curriculum based on design standards.
7. The performance of learners provides information that is required to effect adjustments in the curriculum and thus assure that learning is maximised.

In their words:

> We examine our goals, examine established content standards (national, state, province, and
> district), and review curriculum expectations. Because there is typically more content than
> can reasonably be addressed within the available time, teachers are obliged to make choices
> …Learning priorities are established by long-term performance goals – what it is we want
> students, in the end, to be able to do with what they have learned. The bottom-line goal of
> education is transfer. The point of school is not to simply excel in each class, but to be able
> to use one's learning in other settings. (p. 2)

Application in Science Education

Research studies indicate the effectiveness of mastery learning approach in various
science courses such as quantitative chemistry (Mitee & Obaitan, 2015); integrated
science (Agboghorom, 2014); geometry (Sood, 2013); and physics (Achufusi &
Mgbemena, 2012; Wambugu & Changeiywo, 2008). Consequently, mastery learning
has received some attention in the area of science teaching and learning.

In applying mastery learning in science education, it is recommended that the
learning activities be broken into small units with clear objectives and assessment
procedure (Ozden, 2008); learning materials as well as instructional strategies including formative evaluation are identified; each unit begins with diagnostic tests; and the
outcomes of the diagnostic tests be used to provide both correctives and enrichment
for learners as the case may be.

The following are tips in applying the mastery learning in science classes (Akpan,
1989): 1. Students should be encouraged to have an extra credit file where they place
reports of the activities in which they have taken part such as activities done at
home, educational visits to various places, as well as information from the media;
2. Assignments in classes can be individualised such that students work at their
own rates and submit what they have been able to accomplish at the end of the
lesson; 3. Students who make above-average performance should be recognised in a
special way; 4. Use should be made of many resources including textbooks to cater
to the diverse capabilities in the class; 5. Special activities should be provided for the
academically precocious students such as having some students in the higher classes
go to the lower forms to explain some scientific phenomena; and 6. Parents should
be persuaded to make trips, along with their children, that have value for science
education.

It is noteworthy that the quality and competencies of teachers have an impact on
the implementation of mastery learning in science education. The temperaments of
teachers, their beliefs about human nature, as well as their views on the goals of
science education go a long way in determining what strategies they use in teaching.
For example, authoritarian teachers are likely to expect prompt compliance with
instructions. Behaviours of teachers that encourage reliance on authority figures are
at variance with the tenets of mastery learning. According to Wing Institute (n.d.),
there is at present a large knowledge base that reveals that in schools, teachers' role

in teaching and learning is critical as research has shown that the interaction between teachers and students as well as how instruction is carried out determine how effective schools can be. In the opinion of the Institute:

> Competencies are skills and knowledge that enable a teacher to be successful. To maximise student learning, teachers must have expertise in a wide-ranging array of competencies ...few jobs demand the integration of professional judgment and the proficient use of evidence-based competencies as does teaching...The transformational power of an effective teacher is something many of us have experienced...Research confirms this common perception of a link and reveals that of all the factors under the control of a school, teachers are (sic) the most powerful influence on student success. (p. 2)

Among the various competencies that teachers possess, the Institute lists four as yielding the greatest results: instructional delivery, classroom management, formative assessment, and personal competencies.

Indeed, over the last five decades, mastery learning has received great attention in science education (Akpan, 2015; Hussain & Suleman, 2016). Several studies as indicated earlier consistently provide support for the use of mastery learning in science teaching and learning. There are therefore some advantages to this method. Some of these are that students acquire mastery of a unit before advancing to the next unit, the teacher is better prepared for each lesson as task analysis is required, and mastery learning typically breaks the cycle of failure thus promoting affective outcomes such that more learners develop positive interests and attitudes towards science learning. This should be of very great importance for the future of science and science education (Akpan, 2017a, 2017b).

Even so, a critical aspect of mastery learning is the amount of time required to attain mastery of a given learning task. So, mastery of tasks is achieved at a great cost of time. Secondly, as not all students progress at the same time and rate, the learning process may witness a lot of interruptions. Other disadvantages are in the area of availability of a wide variety of instructional materials required for implementation and, the burden of implementing formative evaluation at some intervals that is placed on teachers.

Summary

- Mastery learning relies on the principle enunciated by Benjamin Bloom to the effect that in spite of individual differences among learners, when teachers provide sufficient time, encouragement, and help to them, most learners will attain mastery of a given learning task.
- In practice, mastery learning uses a system of feedback, correctives, and enrichment to facilitate success.
- Formative assessment is a crucial aspect of mastery learning.
- Four approaches to mastery learning have been highlighted—Learning for Mastery, the Keller Plan, Differentiated Instruction, and Understanding by Design.

- Mastery learning, especially in science education, has received great attention over the past five decades.
- The approach has numerous advantages but like any other teaching method, it also has some disadvantages.

Further Reading

Davis, D., & Sorrel, J. (1995). Mastery learning in public schools. *Educational Psychology Interactive*. Valdosta, GA: Valdosta State University. Retrieved September 11, 2017, http://www.edpsycinteractive.org/files/mastlear.html.

Education Endowment Foundation (2017). Mastery learning. Retrieved September 5, 2017, https://educationendowmentfoundation.org.uk/pdf/generate/?u=https://educationendowmentfoundation.org.uk/pdf/toolkit/?id=156&t=Teaching%20and%20Learning%20Toolkit&e=156&s=.

Edutech Wiki. (2007). Mastery learning. Retrieved September 5, 2017, http://edutechwiki.unige.ch/en/Mastery_learning.

Gagne, R. M. (1988). Chapter 4: Mastery learning and instructional design (pp. 107–124). Florida: The Learning Systems Institute, Florida State University. Retrieved March 4, 2018, http://iceskatingresources.org/chapter_4.pdf.

Martinez, J. G. R., & Martinez, N. C. (1999). Teacher effectiveness and learning for mastery. *The Journal of Educational Research*, *92*(5), 279–285. Retrieved March 3, 2018, https://faculty.weber.edu/kristinhadley/med6000/Teacher%20Effectiveness%20and%20Learning%20for%20Mastery.pdf.

Renard, L. (2017). What is mastery learning? A different approach to learning. Retrieved September 5, 2017, https://www.bookwidgets.com/blog/2017/03/what-is-mastery-learning-a-different-approach-to-learning.

Scriven, M. (1967). The methodology of evaluation. In R. W. Tyler, R. M. Gagne, & M. Scriven (Eds.), *Perspectives of curriculum evaluation* (pp. 39–83). Chicago, Illinois: Rand McNally.

References

Achufusi, N. N., & Mgbemena, C. O. (2012). The effect of using mastery learning approach on academic achievement of senior secondary school II physics students. *Elixir Educational Technology*, *51*, 10735–10737. Retrieved September 11, 2017, from http://www.elixirpublishers.com/articles/1351501686_51%20(2012)%2010735-10737.pdf.

Agboghorom, T. E. (2014). Mastery learning approach on secondary students' integrated science achievement. *British Journal of Education, 2*(7), 80–88.

Akpan, B. (2017a). Science education in a future world. In B. Akpan (Ed.), *Science education: A global perspective* (pp. 331–346). Gewerbestrasse: Springer.

Akpan, B. (2017b). Science education for sustainable development. In K. S. Taber & B. Akpan (Eds.), *Science education: An international course companion* (pp. 493–504). Rotterdam: Sense Publishers.

Akpan, B. B. (2015). The place of science education in Nigeria for global competitiveness. *Journal of the Science Teachers Association of Nigeria, 50*(1), 1–39.

Akpan, B. B. (1989). *Teaching science in primary schools: A pragmatic approach*. Calabar, Nigeria: Institute of Education, University of Calabar.

Bloom, B. S. (1968). Learning for mastery. *Evaluation Comment (UCLA—CSIEP)*, *1*(2), 1–12.
Bloom, B. S. (1971). *Individual differences in school achievement: A vanishing point?*. Blooming-ton, Indiana: Phi Delta Kappan International.
Bloom, B. S. (1976). *Human characteristics and school learning*. New York: McGraw-Hill.
Guskey, T. R. (2007). Closing the achievement gap: Revisiting Benjamin S. Bloom's "learning for mastery." *Journal of Advanced Academics*, *19*, 8–31. Retrieved September 5, 2017, http://tguskey.com/wp-content/uploads/Mastery-Learning-5-Revisiting-Blooms-Learning-for-Mastery.pdf.
Guskey, T. R. (2010). Lessons of mastery learning. Retrieved September 5, 2017, http://www.ascd.org/publications/educational-leadership/oct10/vol68/num02/Lessons-of-Mastery-Learning.aspx.
Hussain, I., & Suleman, Q. (2016). Effect of Bloom's mastery learning approach on students' academic achievement in English at secondary school level. *Journal of Literature, Languages and Linguistics*, *23*, 35–43. Retrieved September 3, 2017, http://iiste.org/Journals/index.php/JLLL/article/viewFile/31278/32116.
Keller, F. S. (1967). Engineering personalized instruction in the classroom. *Interamerican Journal of Psychology*, *1*, 189–197. Retrieved September 11, 2017, file:///C:/Users/SONY/Downloads/445-1102-1-SM.pdf.
Keller, F. S. (1968). "good-bye, teacher…". *Journal of Applied Behavior Analysis*, *1*, 79–89. Retrieved September 11, 2017, https://www.ncbi.nlm.nih.gov/pmc/articles/PMC1310979/pdf/jaba00083-0078.pdf.
Mctighe, J., & Wiggins, G. (2012). *Understanding by design framework*. Alexandria, VA, USA: ASCD. Retrieved March 11, 2018, https://www.ascd.org/ASCD/pdf/siteASCD/publications/UbD_WhitePaper0312.pdf.
Mitee, T. L., & Obaitan, G. N. (2015). Effect of mastery learning on senior secondary school students' cognitive learning outcome in quantitative chemistry. *Journal of Education and Practice*, *6*(5), 34–38.
Ozden, M. (2008). Improving science and technology education achievement using mastery learning model. *World Applied Sciences Journal*, *5*(1), 62–67.
Rossett, A., & Sheldon, K. (2001). *Beyond the podium: Delivery training and performance to a digital world*. San Francisco: Jossey–Bass/Pfeiffer.
Sumida, M. (2017). Science education for gifted learners. In K. S. Taber & B. Akpan (Eds.), *Science education: An international course companion* (pp. 479–491). Rotterdam: Sense Publishers.
Sood, V. (2013). Effect of mastery learning strategies on concept attainment in geometry among high school students. *International Journal of Behavioural Social and Movement Sciences*, *2*(2), 144–155.
Sund, R. B., & Trowbridge, L. W. (1973). *Teaching science by inquiry in the secondary school* (2nd ed.). Columbus, Ohio: Charles E. Merrill Publishing Company.
The Wing Institute. (n.d.). Teacher competencies that have the greatest impact on student achieve-ment. Retrieved March 3, 2018, https://www.winginstitute.org/quality-teachers-competencies.
Tomlinson, C. A. (2000). Differentiation of instruction in the elementary grades. ERIC Digest. ERIC Clearinghouse on Elementary and Early Childhood Education. Excerpts. Retrieved March 11, 2018, http://www.readingrockets.org/article/what-differentiated-instruction.
Wambugu, P. W., & Changeiywo, J. M. (2008). Effects of mastery learning approach on secondary school students' physics achievement. *Eurasia Journal of Mathematics, Science & Technology Education*, *4*(30), 293–302.

Ben Akpan a professor of science education is the Executive Director of the Science Teachers Association of Nigeria (STAN). He served as President of the International Council of Associa-tions for Science Education (ICASE) for 2011–2013 and currently serves on the Executive Com-mittee of ICASE as the Chair of World Conferences Standing Committee. Ben's areas of interest include chemistry, science education, environmental education, and support for science teacher associations. He is the editor of *Science Education: A Global Perspective* published by Springer and co-editor (with Keith S. Taber) of *Science Education: An International Course Companion* published by Sense Publishers. Ben is a member of the Editorial Boards of the Australian Journal of Science and Technology (AJST), Journal of Contemporary Educational Research (JCER), and Action Research and Innovation in Science Education (ARISE) Journal.

Chapter 12
Meaningful Learning—David P. Ausubel

Steven S. Sexton

Introduction

Sharon Feiman-Nemser stated, 'there is nothing as practical as a good theory' (personal communication, 19 May 2016). She noted this in her New Zealand workshop on teacher mentoring and induction. While she made this statement in relation to educative mentoring, it is applicable to David Ausubel's very practical theory of meaningful learning (Ausubel, 1968). Specifically, I will demonstrate how *The New Zealand Curriculum* (*NZC*) (Ministry of Education, 2007) provides students with education through science that is relevant, useful, and meaningful. First, I will provide a brief summary of Ausubel's (1968) *Educational Psychology: A Cognitive View* which outlines his theory of Meaningful Learning. Second, this is then followed by the New Zealand context and its national curriculum document for school-aged students in English-medium schools. Third, how New Zealand's identified effective pedagogies reflect Ausubel's theory of Meaningful Learning to include how a unit of work was designed, developed, implemented, and used to provide a demonstration to both a school's teachers and students how education through science is relevant, useful, and meaningful. Finally, the chapter concludes with a summary of the main ideas.

S. S. Sexton (✉)
University of Otago, College of Education, Dunedin, New Zealand
e-mail: steven.sexton@otago.ac.nz

© Springer Nature Switzerland AG 2020   163
B. Akpan and T. Kennedy (eds.), *Science Education in Theory and Practice*,
Springer Texts in Education, https://doi.org/10.1007/978-3-030-43620-9_12

Theory of Meaningful Learning

The following is a short summary of David Ausubel's (1968) book *Educational Psychology: A Cognitive View*. Ausubel was a cognitive learning theorist who focused on the learning of school subjects. He recognised the importance of what the student already knows as being the primary factor in what the student will learn next. For Ausubel, students seek to make sense of new material by connecting this new knowledge with what they already know. Meaning happens when new information is taken into a person's existing cognitive structure, which is the sum of all the knowledge acquired, as well as the organisation of the facts, concepts, and principles that make up that knowledge (see pp. 127–133). Ausubel explicitly distinguished between meaningful and rote learning.

Meaningful learning occurs when what students are learning relates to their pre-existing knowledge and they are able to be connected the new knowledge to this pre-existing knowledge. Rote learning, however, has no relationship to pre-existing knowledge, and therefore, quickly fades from memory. As a result, teachers must know what their students already know about the topic so that they are able to build upon this prior knowledge. Ausubel highlighted that while teachers can do everything possible, students may still not find the learning meaningful. Consequently, there is only the potential for meaningful learning. Ausubel argued that meaning does not happen until the students are able to incorporate the new knowledge into their cognitive structures. Subsumption is the process by which new material is brought into a student's existing cognitive structure. This new material is systematically compared (Ausubel referred to this as integrated reconciliation) and contrasted (or what Ausubel called progressive differentiation) with prior knowledge. More specifically, progressive differentiation is the process by which the teacher introduces new material at its highest appropriate level of abstraction. Then the teacher provides opportunities to progressively get more specific as the students contrast it with the pre-existing material in their cognitive structures. The opposing process called integrative reconciliation points out the similarities or comparisons. Through subsumption new knowledge takes on meaning and becomes anchored within the students' cognitive structures.

In meaningful learning, anchoring is the process by which the new information fits into the student's cognitive structure. This is done by comparing and contrasting it with the information that already exists. Through anchoring the student forms new relationships between the new and existing information. Learning happens when the students are able to recognise the relationships between this new knowledge and their pre-existing knowledge. Advanced organisers were the principle strategy advocated by Ausubel for teachers to support students' learning. The advance organiser activates that part of the student's cognitive structure under which the new information should fit. In particular, '*the principal function of the organizer is to bridge the gap between what the learner already knows and what he needs to know before he can successfully learn the tasks at hand*' (Ausubel, 1968, p. 148, italics in original). Just as advanced organisers set the stage for the content, practice provides the opportunities for the students to subsume the new information into their cognitive structure. Practice is important because it compares and contrasts the new information with pre-existing material already in the students' cognitive structures, thereby sufficiently anchoring the new material so it does not disappear.

The New Zealand Curriculum

NZC (Ministry of Education, 2007) is the official policy for all English-medium schools in New Zealand. As a result, this document provides the teaching and learning guidelines for approximately 95% of all children between the ages of 5 and 19 (Education Counts, 2016). While this document was required to be fully implemented in 2010, it was estimated that less than half of New Zealand's teachers were prepared for this curriculum change (Hipkins & Hodgen, 2012).

By 2010, many teachers found themselves having to implement a curriculum explicitly foregrounding effective pedagogies. Previously, teachers worked to develop essential skills and attitudes in students through a set of predetermined learning experiences in each learning area. Now teachers focus on content material that is seen as relevant, useful, and meaningful to their students (Sexton, 2017). Specifically for science, one of the reasons for teachers reporting a lack of readiness is the curriculum's emphasis on the Nature of Science and its four elements:

- Understanding about science—this requires primary students being able to ask questions about the science they are doing and accepting there may be more than one answer. Then in secondary school, students learn to collect evidence that is then interpreted through logical arguments.
- Investigating in science—means that primary students are expanding their world through activities, play, questions and/or simple models. Then secondary students work with increasingly more complex investigations and learn to evaluate the methods chosen.
- Communication in science—primary students are able to use the terminology and vocabulary appropriately as they discuss the science they are doing. Then secondary students use a wider range of vocabulary, symbols, and conventions to evaluate both popular and scientific texts.
- Participating and Contributing—first, primary students are able to relate the science they are doing to their world; and then, they are able to make decisions that impact their world based on this science. Secondary students gather relevant scientific information to draw evidence-based conclusions for appropriate actions (Ministry of Education, 2007).

Previously, teachers focused lessons on the individual content strands of the Living World (Biology), the Material World (Chemistry), the Physical World (Physics), or Planet Earth and Beyond (see, for example, Ministry of Education, 1993). Teachers should now be focussing their students' learning through the four elements of Nature of Science using whatever science content that is appropriate.

Meaningful Learning in the New Zealand Curriculum

Naku te rourou, nau te rourou, ka ora ai te iwi
(*With my basket and with your basket, the people will live*)

In New Zealand, a whakataukī, or Māori proverb, is a saying that is often shared at the beginning of a lesson or event to help foreshadow the context. The above whakataukī refers to the cooperation and combination of individual resources for the mutual benefit of all. In the context of this chapter, it means teachers, schools, and students working together for relevant, useful, and meaningful learning.

NZC is the culmination of over a century of reforms (Bell & Baker, 1997; Philips, 2000). While it is a political statement, it builds on both international and New Zealand research (Alton-lee, 2003; Nuthall, 2007). Specifically, I argue the *NZC*'s seven effective pedagogies (see pp. 34–35) directly correspond to aspects of Ausubel's theory of Meaningful Learning, see Table 12.1.

Ausubel saw learning as an active, not a passive experience, which he more thoroughly addressed in his 2000 revision of his 1963 monograph (Ausubel, 2000). Ausubel saw active learning requiring, at least, the following process; first, the cognitive analysis by the students to determine where the new material presented was relevant to their pre-existing knowledge. Then understanding the usefulness of the similarities and differences of the new material with pre-existing knowledge. Finally, the meaningful inclusion of this new material into pre-existing knowledge. New Zealand's curriculum has a similar focus on active learning. Specifically, the Ministry of Education directs schools to design and implement a curriculum grounded in relevant, useful, and meaningful learning.

NZC has eight principles that are the foundations of decision-making by both the school and its teachers: High expectations, Treaty of Waitangi, Cultural diversity, Inclusion, Learning to learn, Community engagement, Coherence, and Future focus.

Table 12.1 *The New Zealand Curriculum's* effective pedagogies as meaningful learning

Ausubel's meaningful learning	*NZC*'s effective pedagogies
Connect new material to prior knowledge	Making connections to prior learning and experience, Teaching as Inquiry
Meaningful learning not Rote learning	Enhancing the relevance of new learning, Teaching as Inquiry
Progressive differentiation	Encouraging reflective thought and action—students assimilate new learning, relate it to what they already know, adapt it for their own purposes and translate thought into action, Teaching as Inquiry
Integrated reconciliation	
Anchoring	
Practice	Providing sufficient opportunities to learn, Creating a supportive learning environment, Facilitating shared learning, Teaching as Inquiry

All planning, prioritising, and review of the school's curriculum should be underpinned by these principles as they embody what is important and unique about New Zealand identity. New Zealand is a trilingual, bicultural country in which English, Te Reo Māori (indigenous language of the Māori people), and New Zealand Sign Language are recognised as official languages of New Zealand. The 1840 Treaty of Waitangi is New Zealand's foundational document which officially acknowledges both Māori and the British Crown in terms of Partnership, Participation, and Protection in the governance of New Zealand. Of these eight principles, I argue the school and its management team should be taking the lead in seven of them. High expectations should be the focus of the classroom teacher as teachers work to empower, 'all students to learn and achieve personal excellence, regardless of their individual circumstances' (Ministry of Education, 2007, p. 9). While the eight principles guide the decision-making process in regards to what content is in the school's curriculum, the seven pedagogies reflect how this content is taught. *NZC*'s effective pedagogies (see Table 12.1) are the Ministry of Education's selection of those evidence-based approaches that have a positive impact on a student's ability to learn, achieve, and support the unique characteristics of New Zealand.

Making Connections to Prior Learning and Experience

New Zealand recognises that students learn best when they are able to integrate what they are learning to what they already know. Specifically, Graham Nuthall (2007) reported students learn what they do. Nuthall, like Ausubel, knew learning was an active process. Nuthall made his statement after compiling nearly 40 years of research into classroom practice. However, the New Zealand context also includes traditional Māori pedagogies (Bishop & Glynn, 1999; Hemara, 2000). For New Zealand, *kia piki ake i ngā raruraru o te kainga* (the learning process in school must also reflect their home life) is important. This means that part of the connections to be made include meaningful *whānau* (family) inclusion in their children's learning.

New Zealand teachers, therefore, must build upon what their students' already know and have experienced to support their ability to integrate new learning across learning areas, with prior knowledge, home practices, and the wider world.

Enhancing the Relevance of New Learning

Ausubel (2000) argued that meaningful learning depended on two factors: the first was the nature of the material to be learned and second the nature of the learner's cognitive structure. He went on to argue that material was only potentially meaningful. Meaningful learning results in this new knowledge modifying both the new material being acquired and the person's cognitive structure to which the new material is being connected. He acknowledged that in most instances the new knowledge was linking to a specific concept or proposition.

NZC positions effective teachers as those who know how to stimulate the curiosity of their students. They also know that students learn best when they know what they are learning, why they are learning this, and how this new material is relevant to their life. More importantly, effective teachers know how to challenge what their students' think they know about their world and their participation in their world (Bishop & Glynn, 1999; Sexton, 2017).

Encouraging Reflective Thought and Action

NZC encourages teachers to promote critical reflexivity in both themselves and their students. Teachers become quite adept at reflection in-action (those on the spot teaching decisions based on student engagement) and reflection on-action (taking time after teaching to evaluate what went well, what could have been done differently, and where to next in teaching). Teachers need to learn to be critical reflexive where they build upon reflection in-action and reflection on-action leading to reflection for-action (Sexton & Williamson-Leadley, 2017; Thompson & Pascal, 2012). Teachers need to consider not only their own assumptions, beliefs, values, and opinions on the content and context of learning but also their students' assumptions, beliefs, values, and opinions. As teachers learn to develop their own critical reflexivity, their students are more able to engage in the effort necessary to get actively involved in their own learning.

Ausubel (2000) argued students learn to look at the material from different angles, reconcile it, and translate it into their own frame of reference. In the New Zealand context, this becomes teachers encouraging students' *tino rangatiratanga* (autonomy and self-determination). Students learn most effectively when they develop the ability to stand back from the content and think about their own thinking.

Providing Sufficient Opportunities to Learn, Creating a Supportive Learning Environment, Facilitating Shared Learning

Ausubel devoted an entire chapter to practice (see Chap. 8, Ausubel, 1968) as it affects both learning and retention. He highlighted while practice was not a cognitive structure variable it was one of the principal factors influencing cognitive structure. In 1968, he noted that most of the knowledge regarding the effects of practice related to rote and motor learning rather than sequentially organised tasks. *NZC* acknowledges that student learns best when, 'they have time and opportunity to engage with, practice and transfer new learning' (Ministry of Education, 2007, p. 34) but just as importantly learning is inseparable from the social and cultural context in which the learning takes place (see Chaps. 7, 18, and 19 for more on the social construction of knowledge).

Nuthall (2007) summarised 40 years of classroom research into four statements:

1. Students learn what they do
2. Social relationships determine learning
3. Effective activities are built around big questions
4. Effective activities are managed by the students themselves

Learning is an active process; therefore, what students do themselves. As such, it requires students to have multiple opportunities to engage with new learning and various contexts. These contexts are situated in both social and cultural contexts. Students must feel safe in not only the learning environment but also their wider surroundings. *Ako* (reciprocal teaching and learning relationship) positions the class as a community of learners. In *ako,* both teachers and students are learners in a classroom environment that fosters positive relationships.

Teaching as Inquiry

In 1975, Lortie published his seminal work *Schoolteacher: A sociological study* in which he highlighted the concept of 'the apprenticeship of observation' in which students spend thousands of hours observing classroom teachers. In New Zealand, education is compulsory from the age of six until at least sixteen. Students experience at least ten years of the schooling system to include teachers and teaching. As such, they come into my initial teacher education Master's degree level programme with high self-efficacy and conceptions of what they believe is a teacher and teaching (Sexton, 2015). There are mixed reactions to them being told that I cannot tell them how to be a teacher. I tell them that all I can do is expose them to ideas and pedagogies that they then have to put into practice to determine what works for them. Some are relieved, as they believe they already know how they want to teach and some are upset that they are not going to be given a step-by-step guide to being a teacher. No matter what stance they start the programme with; they all learn the importance of teaching as inquiry. For many this learning to question their own teaching and learning to understand how their teaching influences students' learning comes through formative and often troubling classroom experiences demonstrating that teaching and learning are not coextensive.

Ausubel (1968) highlighted that teaching is only one of the conditions influencing learning. He then argued that while teaching and learning can be analysed independently, 'what would be the practical advantage of doing so' (Ausubel, 1968, p. 12). In the New Zealand context, this co-joint evaluation of teaching and learning is teaching as inquiry. The curriculum positions three questions teachers ask of their own teaching practice:

1. What is important given where my students are at?—focusing inquiry
2. What strategies are most likely to help my students learn this?—teaching inquiry
3. What happened as a result of the teaching and what are the implications for future teaching?—learning inquiry (see Ministry of Education, 2007, p. 35).

Most importantly for teachers is that they learn to ask themselves these questions moment by moment as the teaching takes place to be critically reflexive in how their teaching and the established classroom environment is impacting on students' learning. These three guiding questions form the basis of teaching as inquiry. In the context of education through science (or arguably any subject area), these are further broken down to facilitate meaningful learning.

Education Through Science that Is Meaningful Learning

Meaningful, useful, and relevant learning starts with teachers understanding what their students already know (Ausubel, 1968; Ministry of Education, 2007). The following is the thought process that went into planning a unit of Science around the topic of Weather as an example of how *NZC* reflects Ausubel's meaningful learning theory. This unit was then delivered to a classroom of Year 4 students (students aged 9) as an example of education through science that is meaningful, useful, and relevant to both students and the teachers observing.

Teaching as Inquiry - Focusing Inquiry

For initial planning, focus inquiry requires teachers to start questioning their practice with what they intend to do, develop, strengthen, and why this is their intent. As stated, this chapter reports on a collaborative study with a school and its teachers. Through initial discussions with the teachers, it was determined that a unit on the weather would allow the students to integrate their oral and written language (English and te reo Māori) and the measurement strand of mathematics with science. Teacher confidence and student engagement in science as barriers to learning were the priority identified by the school. Therefore, the express intent was to provide the students (and demonstrate to the teachers) with meaningful opportunities to engage in discussions about their world by making connections with what they already know and the new content.

In the *NZC*, the topic of weather falls under the content strand of Planet Earth and Beyond. However, this is only the context for which the Nature of Science is explored. In the 'Weather' unit there were three science concepts to facilitate students' expanding their understanding of the world around them:

1. Daily weather patterns and changes may be associated with rain, clouds, thunder and lightning, and wind

2. Weather features that we can observe, feel, or hear include rain, clouds, thunder and lightning, and wind
3. We can use our understanding of weather types and weather patterns and demonstrate this through activities in the classroom.

Teaching as Inquiry—Teaching Inquiry

With initial planning determined, teaching inquiry involves ensuring the unit plan has what is necessary for achieving what is intended to include strategies that are effective for this content. As this unit was an example of integrated learning, the students completed daily weather journals in groups. As the unit progressed, the journals would provide evidence of students' deepening understanding of not only the topic of weather but also integrating measurement strand of mathematics through rain fall, temperature (minimum, maximum, range), hours of sunshine (sunrise to sunset). Science requires students to be able to communicate the science they are doing and this requires using the correct terminology. As this unit is group work, it reduces the cognitive workload on each individual while allowing students to express amongst themselves how they are making sense of the content. These daily journals became exercises in progressive differentiation and integrated reconciliation.

Teaching as Inquiry—Learning Inquiry

Learning inquiry necessitates putting into action the planned unit of study. More importantly, it needs the teacher to monitor student activity, collect evidence of how students are making sense of the material, annotate their unit for what went well, and what needs to happen next. The focus of learning inquiry is on both the students' learning and the teacher's teaching. How are the students progressing to the intended learning outcomes? What does their progression tell me about the learning journey? What am I learning about my teaching from their progression? Do I need to change my teaching or strategies to be more effective? What are the next steps I need to take—more research on the topic? different activities? alternative resources? Learning inquiry compels the teacher to reflect for future teaching action by analysing all the data they have collected from both their students and about themselves as teacher.

Weather Unit—Teaching as Inquiry in Practice

Throughout this unit, oral communication and questioning provided insights into what the students were thinking and what successes they were having within the learning experiences. In addition, teacher professional development occurred after each teaching activity to explore what the students experienced, how they experienced the learning, and what would be the appropriate next learning activities. In this unit, each new learning topic (rain, clouds, thunder and lightning, and wind) modelling how to begin with exploring what the students already knew to facilitate challenging what they think they know. Each activity positioned questioning, i.e. Nature of Science's Understanding about science, as a key component of the learning to both help students understand what they are learning and what they could investigate next. Specifically, students learned to ask, 'what would happen if …?' and then what they would need to do to address this question, i.e. Nature of Science's Investigating in science.

The NZC encourages students to ask questions about the science they are exploring, investigating, or doing. It is these questions that highlight how the students are beginning to make sense of the science they are doing and if at all possible should lead to students into more investigating. This requires teachers to be more flexible and adaptable in their teaching so that they are able to respond to students' questions. In this Weather unit, the questions asked by the students in each activity helped to determine what activities the students would investigate next. As students worked in groups, they learned to ask questions, which required them to change variables and discuss scientific models with the activities. These meaningful activities provided the practice necessary to increase the 'stability and clarity, and hence the dissociability strength of the emergent new meaning in cognitive structure' (Ausubel, 1968, p. 274). As this unit was integrated into other subject areas of the NZC, it was supported by the anchoring ideas (Ausubel, 2000) from language and measurement related to the learning tasks.

Twenty years in the classroom working with students aged three to senior adults has taught me that I cannot see what is going on in their heads; however, if meaningful, useful, and relevant learning opportunities are presented I can hear how they are making sense of the content as they discuss what they are doing. In the 'Weather' unit oral conferences and group conversations about the activities they are doing provided insight into how these students were making sense of the targeted content. Using the correct and age-appropriate terminology is an important aspect of science and a formative assessment technique for how the students are understanding what they are doing, i.e. communicating in science. Students need to know how to use the terminology which means they need time to share with others and opportunities to practice using the terminology in meaningful contexts. These meaningful contexts provide students the opportunities to experience what I would argue as the most important aspect of the Nature of Science—Participating and Contributing.

Participating and Contributing requires the teacher to ensure that the learning experiences allow students the opportunities to understand how the science they are doing relates (i.e. contributes) to their understanding of their world so that they can then make informed decisions based on this understanding (i.e. participate in their world). The 'Weather' unit's journals were more than rote learning exercises. As the students progressed through the unit, the content experienced allowed them a deeper understanding of what they were doing which reflected in the content of the daily journals. What started as simple fill in the blank for weather facts became detailed explanations of not only what the weather was but also the impact of this weather. More importantly, these journals became evidence of how the students had gained an understanding of the interactions that take place between different parts of their world and the ways in which these interactions can be represented.

Final Thoughts

Much of Ausubel's Meaningful Learning theory is in mainstream educational psychology; even though he does not receive the credit, he should. Ausubel should be recognised for creating advanced organisers. He emphasised starting with what is now commonly referred to as the 'big picture' and then teaching to fill in the details. Ausubel argued that the most important thing influencing students' learning is what they already know, that is, the teacher knowing what the content of the student's cognitive structure is. Today many educators from around the world try to match instruction to the student's pre-existing knowledge so that it will be meaningful. For example in the United States, the vision for science education acknowledges that 'conceptual understanding built upon student's prior experiences is central to the new vision for science education' (Moulding, Bybee, & Paulson, 2015, p. 4).

Teachers need to engage their students. As a result in many countries science as a form of inquiry is at the core of their science curriculum. For example, in Singapore science teachers now work to ensure that their students learn through questioning, exploring, and evaluating (Tan, 2018). Teachers need to provide opportunities to help students make the connections between the new content introduced and what they already know so that they can make sense of it. Teachers are able to do this by, 'encouraging students to recognise and challenge the assumptions underlying new propositions, to distinguish between facts and hypotheses, and between warranted and unwarranted inferences' (Ausubel, 2000, p. 53).

Summary

- Students learn best when their teachers know what their students already know and then teach accordingly.
- Students learn best what they actually do (and often have to do the activities several times) to make sense of the new content.

- Students learn best when their learning provides them the opportunities to practice relevant, useful, and meaningful activities that challenge what they think they know.

References

Alton-lee, A. (2003). *Qualtiy teaching for diverse students in schooling: Best evidence synthesis.* Wellington, New Zealand: Ministry of Education.

Ausubel, D. P. (1968). *Educational psychology: A cognitive view.* New York, NY: Holt, Rinehart and Winston Inc.

Ausubel, D. P. (2000). *The acquisition and retention of knowledge: A cognitive view.* Dondrecht, The Netherlands: Springer-Science+Business Media, B.V.

Bell, B., & Baker, R. (1997). Curriculum development in science: policy-to-practice and practice-to-policy. In B. Bell & R. Baker (Eds.), *Developing the science curriculum in Aotearoa New Zealand* (pp. 1–17). Auckland, New Zealand: Longman.

Bishop, R., & Glynn, T. (1999). *Culture counts: Changing power relations in education.* Palmerston North, New Zealand: Dunmore Press.

Education Counts. (2016). *Number of schools.* Retrieved from https://www.educationcounts.govt.nz/statistics/schooling/number-of-schools.

Hemara, W. (2000). *Maori pedagogies: A view from the literature.* Wellington, New Zealand: New Zealand Council for Educational Research.

Hipkins, R., & Hodgen, E. (2012). *Curriculum support in science: Patterns in teachers' use of resources.* Wellington, New Zealand: New Zealand Council for Educational Research.

Lortie, D. C. (1975). *Schoolteacher: A sociological study.* Chicago, IL: The University of Chicago Press.

Ministry of Education. (1993). *Science in the New Zealand curriculum.* Wellington, New Zealand: Learning Media.

Ministry of Education. (2007). *The New Zealand Curriculum.* Wellington, New Zealand: Learning Media.

Moulding, B. D., Bybee, R. W., & Paulson, N. (2015). *A vision and plan for science teaching and learning: An educator's guide to a framework for K-12 science education, next generation science standards, and state science standards.* United States: Essentuial Teaching and Learning Publications.

Nuthall, G. (2007). *The hidden lives of learners.* Wellington, New Zealand: New Zealand Council for Educational Research.

Philips, D. (2000). Curriculum and assessment policy in New Zealand: Ten years of reforms. *Educational Review, 22*(2), 143–153.

Sexton, S. S. (2015). Student teacher learning to think, know, feel and act like a teacher: The impact of a master of teaching and learning programme. *Educational Alternatives, 13,* 72–85.

Sexton, S. S. (2017). In *The New Zealand curriculum* is it science education or education through science? One educator's argument. In B. Akpan (Ed.), *Science education: A global perspective* (pp. 219-234). Switzerland: Springer International Publishing.

Sexton, S. S., & Williamson-Leadley, S. (2017). Promoting reflexive thinking and adaptive expertise through video capturing to challenge postgraduate primary student teachers to think, know, feel and act like a teacher. *Science Education International, 28*(2), 172–179.

Tan, A.-L. (2018). Journey of science teacher education in Singapore: past, present and future. *Asia-Pacific Science Education, 4*(1), 1–16. https://doi.org/10.1186/s41029-017-0018-8.

Thompson, N., & Pascal, J. (2012). Developing critically reflective practice. *Reflective Practice, 13*(2), 311–325.

Steven S. Sexton is a senior lecturer at the University of Otago, College of Education. He obtained his Ph.D. from the University of Sydney in 2007. He has been a classroom teacher in Japan, Thailand, Saudi Arabia, Australia, and New Zealand. Currently, he delivers science education papers in both the undergraduate initial teacher education primary programme and the Master of Teaching and Learning programme. His research interest areas are relevant, useful, and meaningful learning in science education, teacher cognition, and heteronormativity in schools.

Chapter 13
Discovery Learning—Jerome Bruner

Yasemin Ozdem-Yilmaz and Kader Bilican

Introduction

Jerome S. Bruner (1915–2016) was a psychologist who has been influential in education mainly by his work in 1963, entitled "The process of education", and long before that with his work on psychology. This chapter will focus on his educational ideas as a foundation of cognitive constructionism and their impact on educational practice. The chapter is about the cognitive theory of education introduced from the perspective of Bruner and Discovery Learning as an instructional method in science courses.

In the chapter, first, Bruner's early theory of education is described as in his 1960 book *The Process of Education*. Then, cognitive constructivist theory of education with implications on science classrooms is described. Next, discovery learning is emphasized as a cognitive process that shapes the mind of a student in a functional way to learn culture. An argument for effective discovery learning strategies for science education is provided and followed by a part in which evidence about the significance of discovery learning for successful science teaching and learning is discussed. Mainly, the aim is to provide the theoretical framework that embraces and adopts discovery learning as a method, which permits students to discover their own learning in science.

In the last part, we make references to connection between major science curricula and discovery learning to demonstrate the implications of cognitive theory from

Y. Ozdem Yilmaz (✉)
Faculty of Education, Mugla Sitki Kocman University, Mugla, Turkey
e-mail: yasemin.ozdem@hotmail.com; yaseminozdem@mu.edu.tr

K. Bilican
Faculty of Education, Kırıkkale University, Kırıkkale, Turkey
e-mail: kaderbilican@kku.edu.tr; kader.bilican@gmail.com

© Springer Nature Switzerland AG 2020
B. Akpan and T. Kennedy (eds.), *Science Education in Theory and Practice*,
Springer Texts in Education, https://doi.org/10.1007/978-3-030-43620-9_13

Bruner's perspective in practice. Finally, the chapter ends with a discussion on the critiques to discovery learning and suggestions for both science teachers and science educators.

Cognitive Constructivism: The Historical and Theoretical Background

For some philosophers, the roots of cognitive movement can be dated back to the time of ancient philosophers, such as Aristotle and Plato, who first looked at mind as the source of knowledge. However, cognitive science as an intellectual movement emerged in the late 1950s. In that period psychological scientists began to crack the walls of behaviorism. Psychologists such as George Miller, James McClelland, Philip Johnson-Laird, and Steven Pinker; linguistic scientists, such as Noam Chomsky and George Lakoff, and computational scientists, such as Marvin Minsky, Alan Turing, John von Neumann have been influential in the development of interdisciplinary nature of cognitive science. The movement was in search for the internal processes in mind that could not be explained only through stimulus and response theory of behaviorism. Cognitive science is, therefore, described as an interdisciplinary approach, in which psychology, anthropology, and linguistics come together in the study of the mind so as to understand the nature of human learning and behavior (Miller, 2003).

In the study of the mind, learning has been a fundamental question. According to cognitive theorists, the learning from the cognitive perspective involves the acquisition or reorganization of the cognitive structures in mind (Good & Brophy, 1990, p. 187). Internal coding and structuring are the mental processes through which the knowledge is acquired and actively organized into existing or new cognitive structures. Thus, cognitive scientists are interested more on how learners come to acquire knowledge and focus on helping the learners make meaning of it, organize, and connect it to what they already know (Yilmaz, 2011).

Constructivism, on the other side, has been a concern throughout the twentieth century especially in the fields of development psychology and cognitive psychology, and is mainly shaped by Bruner, Kelly, Piaget, von Glaserfeld, and Vygotsky. Constructivism has also been used as a major concept in educational studies, which focus on learning and teaching processes. There is an agreement among constructivists on the idea that knowledge is a human construction. Yet, there are different approaches to the constructs such as knowledge, reality, and learning within constructivism. Therefore, there are different constructivist approaches to learning that can be grouped as cognitive and developmental approach, social constructive approach (Chap. 18 in this book), and radical constructivism (Chap. 24 in this book). Among these, cognitive constructivism has its roots in cognitive learning psychology and in the works of cognitive learning psychologist, Jean Piaget.

The main argument of cognitive constructivism is that children construct their own knowledge as they interact with the world around them. Steiner (2014) describes these interactions that "enable students to create schemas or mental models; the models are changed, enlarged, and made more complex as children continue to learn." (pp. 319–320) The information that the individual possesses to that end, and the cognitive structures that this information creates are the starting points for this approach. In other words, the individual realizes new learning by associating it with existing knowledge. Thus, learning in cognitive constructivism is described as a mental process in which knowledge is structured internally through experiences, which are interpreted, analyzed, and synthesized. Based on the work of Piaget, as provided by Oogarah-Pratap, Bholoa and Ramma in this book (Chap. 10), for the children to assimilate new information into their existing mental constructs, direct and repeated experience is suggested. What if the experience does not fit into existing mental constructs? Then, the existing constructs will be accommodated or modified to reach cognitive stability or equilibrium (Steiner, 2014).

Jerome Seymour Bruner, beginning from the early 1950s, was an influential psychologist in cognitive and constructivist studies especially after the foundation of the "Cognition Project" set up at Harvard in 1952. The following is how Jerome S. Bruner's ideas about cognitive science evolved through his career.

Jerome Bruner's Cognitive Constructivism

Jerome Bruner was an important researcher in "Cognitive Revolution" of the late 1950s. He and his colleagues, through a series of research and publications dating back to the 1940s and early 1950s, revolted against the behaviorist theories. *The Process of Education* (1960) by Bruner, translated into many languages, has been recognized as an effective study of the reorganization of curricula in many countries. Their movement was "broader and deeper" in terms of how it changed the understanding of knowledge from *gathering* as the correlation of sensory input and behavioral output to *construction* as an active selection and culturally situated meaning-making of experience (Bruner, 1983, p. 103).

In his early works, Bruner shares the same perspective with Piaget in cognitive science, but later departs from this perspective (Takaya, 2013). Bruner adopts a different view from the one Piaget asserted about the developmental stages. He believes that every subject could be taught to every child, provided that it was presented in an appropriate format whatever the developmental stage the child is in. In his words, "We begin with the hypothesis that any subject can be taught effectively in some intellectually honest form to any child at any stage of development" (Bruner, 1960, p. 33). According to Bruner, there are many versatile and diverse perspectives in real life, and this phenomenon is acquired at a very early age.

In the developmental sense, children make sense of their experiences in three ways: using their actions (enactive representation), visual aids (iconic representation), and language (symbolic representation). Bruner believes that cognitive structures are mostly formed by the child's experiences and impressions, and symbolic representation is very important for cognitive development. Since language is the primary means of symbolizing the world, Bruner attaches great importance to language in determining cognitive development (McLeod, 2008).

According to Bruner, the main objective of cognitive development is to provide individuals with a model of the world and the truth. In his later works, Bruner adds learning with social and cultural content. Like Vygostky, as described by Taber in this book (Chap. 19), Bruner emphasized the role of the social environment in the development of the child. Through the process called "enculturation", Bruner suggests, the individual forms a complex thinking structure that interacts with the environment. The model of the world is constructed by an individual's interaction with objects, people, words, and ideas. The resulting information is stored in memory (Woolfolk, 1993). During their interactions, individuals create an appropriate framework shaped by the cultural traditions, including how to interpret and accept certain experiences and meanings. Eventually, this framework influences the subsequent learning.

Educational Implications

For Bruner (1961), the purpose of the education is to facilitate the thinking and problem-solving skills of a child so that these skills can be transferred to various situations. In Bruner's 1960 book, *The Process of Education*, the main premise was that the students are active learners of their own knowledge. The student chooses knowledge, hypothesizes, and makes decisions in order to integrate new experiences into existing mental structures. In this definition, learning is considered as a cognitive construct that provides meanings, constructs experiences, and allows the information to cross the boundaries of information.

Bruner believes that all children have a natural curiosity and desire to become acculturated in various subjects; but if the subject is very difficult, they will get bored. Because of this, the lessons in the school should be processed in a way that is appropriate for the developmental stage of the child. Bruner, like Vygotsky, emphasized the social nature of the learner. He stated that adults should help a child develop skills by means of 'scaffolding'. Bruner describes scaffolding as "the steps taken to reduce the degrees of freedom in carrying out some task so that the child can concentrate on the difficult skill (which) she is in the process of acquiring" (Bruner, 1978, p. 19). By scaffolding, adult and children interact such that adult assists the children to achieve their goals. Therefore, in education, the student must discover the principles of learning in an effective conversation with the teacher during the teaching process.

Bruner asserts that an instructional theory should have four characteristics: (1) motivation; stimulating interest and curiosity in learning; (2) structure; a knowledge structure and level that learners can best assimilate knowledge; (3) organization; find the best possible ways to present the material; and (4) consolidation; to make the best use of rewards and punishments for motivation. Besides, according to Bruner, when students are encouraged to learn new principles on their own, the feeling of independence, which is the essential result of effective teaching, is developed in the students. Therefore, the teaching should minimize the feeling of failure. For this, Bruner suggests the curricula should be organized in a spiral structure that allows students to build on what they have learned before. This way, information is structured so that complex ideas can first be taught at a simplified level and then revisited at more complex levels. In other words, topics should be taught at increasingly difficult levels. By this way, theoretically, students should be able to solve problems by themselves at increasingly complex levels.

In short, the principles that shape Bruner's theory can be summarized like:

- Education should support the experiences that make the student willing and open to learning.
- Education should be structured in such a way that the student can easily understand (spiral configuration).
- Education should be designed to facilitate the use of acquired knowledge in different situations.

In summary, according to Bruner (1960), learning is an active process in which learner is actively engaging in objects around his world to construct knowledge based on his previous experiences. Thinking is the major outcome of cognitive development which results in drawing conclusions from experience. Bruner (1957) explained that process as "generic coding systems that permit one to go beyond the data to new and possibly fruitful predictions" (p. 234). Based on that notion, Bruner asserted that to improve that coding system for better thinking, students should be provided learning environments to discover (Bruner, 1960).

Discovery Learning

Bruner (1961) stated that the act of making sense of the learning experiences relied on an internal cognitive structure. Accordingly, he defined discovery learning as an inquiry, that takes place in problem-solving situations. Other researchers made similar definitions of discovery learning. For example, the definition made by Ormrod (1995) as discovery learning is "an approach to instruction through which students interact with their environment by exploring and manipulating objects, wrestling with questions and controversies, or performing experiments" (442). Another definition provided by van Joolingen (1999) is that:

> Discovery learning is a type of learning where learners construct their own knowledge by experimenting with a domain and inferring rules from results of these experiments. The basic idea of this kind of learning is that because learners can design their own experiments in the domain and infer the rules of the domain themselves, they are actually *constructing* their knowledge (p. 386).

The major focus in these definitions is that the act of "discovery" should not be regarded as "the act of finding out something that before was unknown to mankind, but rather [included] all forms of obtaining knowledge for oneself by the use of one's own mind" (Bruner, 1961, p. 22). For the actualization of discovery learning, learners are required to find out the targeted information as a result of exploring the objects or material provided for them on their own. Correspondingly, through discovery learning, learners go through a process in which they take the responsibility of their learning culminating in not only learning vast amount of knowledge but also gaining higher order thinking skills.

Discovery learning, when it is first revealed by Bruner in 1961, was not only considered to be a teaching method. Bruner stated that discovery is used for "all forms of obtaining knowledge for oneself by the use of one's own mind" (p. 22). Accordingly, Bruner was asserting that the students should be encouraged to make their own discoveries. However, this approach is also mostly critiqued in such that students cannot make *scientific discovery* by just following the propositions made by discovery learning (Ausubel, 1961). Prior to the nineteenthcentury, scientific discovery is defined as both the search for scientific knowledge (scientific inquiry) and the result of this search. However, now scientific discovery is still at the epicenter of philosophical discussions. Schickore (2014) explains the discussions in brief:

> Most philosophical discussions of scientific discoveries focus on the generation of new hypotheses that fit or explain given data sets or allow for the derivation of testable consequences. Philosophical discussions of scientific discovery have been intricate and complex because the term "discovery" has been used in many different ways, both to refer to the outcome and to the procedure of inquiry.

Distinct from the scientific discovery, discovery learning is considered as a means of teaching and learning by Bruner. In this method, students are encouraged to discover the scientific phenomenon usually with the aid of teacher and through inductive logic. However, he also conceded that this does not mean waiting forever for students to discover or "leave the curriculum completely open". (1960, p. 613) Keeping in mind the fact that "children are naturally scientists", long-lasting learning takes place if only children's curiosity continues.

Considering the discovery learning as making sense of one's own environment by exploring and interacting with the objects around himself, it is not surprising that discovery learning has a considerable influence on science education. Regardless of level, discovering the knowledge about the world based on ones' experiences could be stated as overall goal of science education for students. As one of the major obstacles for science learning is the lack of motivation of students towards science, discovery learning can best support students' curiosity, hands-on, minds-on activity, and learning. Efforts to reform science education heavily stressed the abilities such as conducting investigations, making observations, and drawing explanations and in using what is in mind. These abilities were also covered under the framework of "abilities necessary to do scientific inquiry" in the National Science Education Standards (National Research Council [NRC], 1996). Implementation of discovery learning entails engaging learners with identifying a problem, posing questions, formulating hypothesis, conducting and designing experiments to investigate phenomena and drawing conclusions for solving a problem (Gillani, 2003). In general, in discovery learning, students confront a problematic situation and go through a process in which they explore the subject on their own. It is the students to decide what is to be solved and how it is to be solved. In other words, it is the students' role to recognize the problem, decide possible solution strategies, develop their own procedures to test the ideas, formulate explanations and make conclusions. Teacher's role is to be a facilitator, acting as a guide for students and organizing required resources for students while they are setting up on their own. They are urged to provide minimal guidance for students. Students are active inquirers designing their own unique investigations to explore the problem that they have chosen. The main characteristics of discovery as opposed to traditional modes of instruction are summarized as follows (Castronova, 2002):

- Students are active learners participating in hands-on, problem-solving activities.
- Focusing on *process* rather than *product* enhances mastery and application of skills.
- Chance for learning from failure increases motivation for developing solutions to the problem.
- Discovery learning supports natural human curiosity and interest for learning.
- Feedback, collaborations, and discussions are vital parts of discovery learning thus promoting deeper understanding.

Despite the convincing arguments in favor of the discovery approach, it is not simply the hands-on activities that enable students to have scientific understanding. In a meta-analysis conducted by Alfieri, Brooks, Aldrich, and Tenenbaum (2011), the task of discovery learning can range from implicit pattern detection to constructing explanations and from working through manuals to creating simulations. The level of teacher involvement specifies the types of discovery learning. Based on teachers' extent of guidance, the following three modes of discovery learning within a continuum have been proposed (Moore, 2009):

Guided discovery: Teacher decides on the content and the directions for the problem-solving. Teacher involvement occurs most in that level compared to other two modes of discovery.

Modified discovery: The task provided by the teacher but the procedure for problem-solving is determined and designed by students.

Open/Unassisted/Pure discovery: Students decide on the content to be learned and the procedure to solve the problem. Students create their own unique investigations to explore the problem and draw conclusions based on their investigations. In this mode of discovery, teacher involvement is less, and student autonomy is the most related to learning of students. In pure discovery learning, the student is required to discover new content with little or no assistance.

Bruner suggested that "Practice in discovering for oneself teaches one to acquire information in a way that makes that information more readily viable in problem-solving" (Bruner, 1961, p. 26). However, Bruner also said that the students should be able to benefit from their prior knowledge and past experiences. In other words,

Bruner emphasized that pure discovery-based learning could enhance the entire learning experience, yet it requires the students to have a priori or at least some base of knowledge of the topic under investigation. Otherwise, the student will have difficulty in discovering new content, with little or no help from the teacher.

Indeed, Mayer (2004) argued that pure, unassisted discovery learning practices should be questioned because of the insufficient evidence that concludes such applications really lead to the achievement of learning outcomes. His analysis of the literature demonstrated that unassisted or pure discovery learning does not help students to discover problem-solving rules, conservation strategies, or programming concepts. Mayer emphasized that although constructivist approaches are useful for learning under certain conditions, unassisted discovery learning does not appear to be advantageous because of its structural deficiencies. On the other hand, Alfieri et al. (2011) found that guided or modified (enhanced) discovery learning is more effective, because these instructions assist learners to interact with materials, manipulate variables, discover phenomena, and apply their prior knowledge or learning principles.

Bruner (1961) hypothesized four benefits of discovery learning: increased intellectual potency, intrinsic motivation, the learning of the heuristics of discovery, and enhanced use of memory. Bruner noted (p. 31) that "Once the heuristics of discovery have been mastered, they constitute a state of problem-solving or inquiry that serves for any kind of task one may encounter." The importance of active student involvement is reflected in statements like "… the schoolboy learning physics is a physicist and it is easier for him to learn physics behaving like a physicist than doing something else." (Bruner, 1960, p. 21). Proponents of discovery learning also claimed that successful discovery learning environments support students' learning in various dimensions such that:

- Enhances active engagement of students in learning process for higher achievement.
- Foster students' curiosity to learn and investigate.
- Enable students' autonomy in developing their own inquiry procedures.
- Enable learners to take the responsibility of their own learning.
- Increase one's use of creativity and higher order thinking skills.
- Encourage learners to master problem-solving skills.
- Fosters life-long learning.
- Provides individualized learning experience based on the learner's pace.
- Enriches retention of knowledge.
- Enhances the transfer of knowledge in a variety of situations.

On the other hand, the fact that Bruner's description of the nature of teaching is entirely different from theorists such as Ausubel and Skinner has led to much discussion of discovery learning. The arguments against discovery learning are based on the following issues: discovery learning

- Requires more amount of time and effort for teachers to prepare and manage discovery learning activities.

- May allow possible alternative conceptions.
- May be ineffective in large classes.
- May overwhelm students who need more direction and feedback for their learning.
- May demotivate students due to cognitive load.
- May hinder teachers from providing meaningful discussion environment.
- May prevent teachers to detect 'weak students' having difficulties in engaging in tasks.
- May stress out teachers due to pressure to cover content.

Research on the Effectiveness of Discovery Learning

Research on discovery learning provided evidence-based claims mostly in favor of discovery learning. One of the studies was on related influence of discovery learning on 7th-grade students' academic achievement, retention, and perception of inquiry skills (Balim, 2009). The quasi-experimental design was applied in the study. Comparing the control group with the experimental one, students in discovery learning group significantly performed better in the scores of academic achievement and retention. Regarding perceptions of inquiry skills, the experimental group performed better than the control group. That is, students taught through discovery-based learning were better in the understanding importance of inquiry skills. In another study, students' misconceptions regarding water cycle were aimed to be eliminated by using discovery learning. The participants of the study were 150 8th-grade students (14–15 years old). The results of the study revealed that the treatment group which was exposed to discovery learning-based instruction had better academic achievement compared to the control group. Moreover, the study supported the superiority of discovery learning in terms of gaining meaningful knowledge and deep understanding of the concept.

Besides the studies investigating the influence of discovery learning on content knowledge gain, there are more focusing on the skills. For example, a study by Wartono, Hudha, and Batlolona (2017) aimed to report the influence of inquiry-discovery learning on critical thinking skills. The inquiry-discovery learning was described as modified discovery mode where the task is provided by teacher but the procedure for problem-solving is created by students. The participants of the study were 67 senior high school students. Students in experimental group were taught physics by inquiry-discovery learning and students in control group were taught physics by traditional teaching method. The study reported a significant difference in the critical thinking skills in favor of inquiry-discovery class.

As summarized in the studies above, discovery teaching fosters the critical thinking skills, concept understanding, and understanding of practice of inquiry skills which meets the goal of scientific literacy proposed to be a major goal by many science curricula, science education scholars, and policy documents (Driver, Leach, Millar, & Scott, 1996; Kolstø, 2001). Science education is not only drive for an

economic capital but also development of human, to increase the democratic participation of the one related to policy-making. Actively participating in policy-making
process on science-based societal issues at both local and global level (e.g., climate
change) depends on the level of scientific literacy. The definition of scientific literacy involves "*the knowledge and understanding of scientific concepts and processes
required for personal decision making, participation in civic and cultural affairs, and
economic productivity*" (NRC, 1996). Based on the definition, it could be inferred
that, engaging in informed decision-making is closely related to how one understands
the scientific concepts and appreciates scientific thinking to solve the problems one
encounter with at both local and global level. Therefore, students should be provided
with learning experiences of decision-making on science-based societal issues to be
able to make more informed choices. Consequently, it is reported that understanding
of scientific inquiry would enhance students' competences to deal with societal-based
issues (Lee, 2007). That is, instructional approaches such as discovery learning might
serve a context to enhance informed decision-making skills. In that sense, engaging
in discovery learning, would assist students to become more actively play role in
policy-making on science-based social issues.

Despite several benefits of the discovery learning reported in many studies, some
debates continue regarding the effectiveness of discovery learning. It is argued that
pure discovery learning does not guarantee deep learning due to the unstructured
content, overload cognition of students and lack of feedback and guidance based
on the pace of students (Mayer, 2004). Additionally, other studies revealed some
mixed results regarding relative effect of discovery learning. For instance, the study
comparing the effect of direct instruction versus discovery learning reported results
challenging the superiority of discovery learning (Klahr & Nigam, 2004).

Suggestions for Science Teaching

In general, discovery learning is still undeniably an effective approach to provide
learning environment to students in which they can challenge their misconceptions,
approach problems from their angles, and propose solutions on their own. In teaching
by discovery, the teacher presents the examples and works with examples until the
students have discovered the structure of the topic; the basic relationships, principles,
and features of ideas. For this reason, Bruner advocates that the learning is formed
through induction. By this way, general principles are formulated using specific
examples. For example, a teacher can provide the students with examples of animal
and plant cells and ask students for their discovery of the characteristics of animal
and plant cells.

Inductive approach requires the student to think intuitively. Bruner suggests that
students should try to make predictions based on incomplete evidence to feed their
intuitive thinking and then systematically research these predictions by inquiry. For
example, teacher may demonstrate some materials' weights when they are in air and
when in water and ask what causes the differences. Students predict that water exerts

a force opposite to the gravitational force. Then the teacher may ask them to calculate the difference.

To get an optimal effect of discovery learning, teachers could make several arrangements regarding the feedback and guidance given. The following suggestions are made in the research while planning discovery-based activities for teachers:

- Plan extra time for the discovery-based activities.
- Revise activities to give scaffold to learners needing additional assistance.
- Plan extra time for feedback.
- Follow up learners to ensure they are on track.
- Record each learner's process.
- Ensure learners discuss and review their outcomes.

According to Bruner, the individual acquires knowledge (or constructs models) in three different ways during cognitive development: action, imagination, and symbols. For this reason, information on the organization of teaching activities should be presented in accordance with the characteristics of the development period. In the operational period, the information is gained by establishing a direct relationship with the object. In this period, the child learns by using sensory organs. In the imaginary period, the models in the memory of the individual are formed with visual images. For this reason, pictures and photographs can be used in teaching. In the symbolic period, language and symbols become important. Individuals can use symbols to develop new models without having concrete experience. In this period, new information can be given to the students by written and verbal symbols.

Bruner suggested that the best way to gain the ability to learn independently is to allow the students to orient themselves to the activities in their own interest, to make inventions, and to satisfy their curiosity. Instead of giving answers to the students, it is necessary to encourage them to solve problems by themselves or in small groups and find answers by themselves. This will require adequate time to be given to the students to solve problems, tools be provided when necessary, students be guided by questions and tips, and that they are given the opportunity to solve the problems on their own.

An important aspect of teaching discovery for teachers is the students' attitude towards learning. According to Bruner, in order to develop a positive attitude towards learning, it is necessary to motivate and to raise the curiosity about the subject that will be learned. Therefore, one of the most effective ways will be to put curiosity into action to create a certain degree of uncertainty in students' minds. In applying the teaching approach based on discovery learning, the lesson with a certain level of ambiguity will give students a sense of curiosity and a sense of learning.

In this process, the teacher's support and guidance is important to the students; it is imperative to reduce the risk of failure and motivate the students. The teacher must always act with a set of questions, exercises, examples, and lesson plans. However, one of the important points to be taken into consideration is that when asking student questions, they should be organized according to difficulty beginning with the easy ones, from concrete to abstract, from simple to complex and adjusted for students' readiness.

In sum, discovery learning can be considered as the background for the notion of 'learning by doing'. The learners actively seek information and construct knowledge through the meaning-making process of their experiences. In education, discovery learning has implications such as the use of spiral curriculum, teachers' scaffolding to assist learners, presentation of information from concrete and simple to abstract and complex structures, and encouraging students have interactions with the material and social environment. Despite pure discovery is criticized in terms of the time required and the lack of evidence in student gains, the enacted discovery is found to be more effective since it allows learners to interact with materials, manipulate variables, discover phenomena, and apply their prior knowledge or learning principles.

Summary

- Cognitive development theory is based on the process of information processing. This understanding implies that the cognitive processes of teaching should be realized in accordance with the developmental stages.
- Learning is a cumulative process. New learning is based on previously learned information. For this reason, it is necessary that the new experiences that the child will have to match the old ones.
- Learning is a discovery process. For this reason, the motivation of students to explore and the creation of a sense of curiosity are important activities in learning. It is necessary to provide the necessary conditions in order to activate the desire for curiosity and learning in the student.
- In discovery learning, the teacher presents the examples and the structure to the student; students work with examples until they discover the basic relationships, principles, and properties between ideas. For this reason, Bruner advocates that the learning is formed through induction in discovery learning.
- Discovery learning takes time to implement because all learners cannot be expected to learn at the same pace in a certain period of time. However, the learning is expected to be more permanent as the student is actively involved in the class.

Further Reading List

Bruner, J. (1960). *The process of education.* Cambridge, MA: Harvard University Press.
Bruner, J. S. (1961). *The act of discovery.* Harvard educational review.
Bruner, J. (1966). *Toward a theory of instruction.* Cambridge, MA: Harvard University Press.
Bruner, J. (1973). *Going beyond the information given.* New York: Norton.
Bruner, J. S. (1978). The role of dialogue in language acquisition. In A. Sinclair, R. J. Jarvelle, & W. J. M. Levelt (Eds.), *The child's concept of language.* New York: Springer.
Bruner, J. (1983). *Child's talk: Learning to use language.* New York: Norton.
Bruner, J. (1986). *Actual minds, possible worlds.* Cambridge, MA: Harvard University Press.
Bruner, J. (1990). *Acts of meaning.* Cambridge, MA: Harvard University Press.

References

Alfieri, L., Brooks, P. J., Aldrich, N. J., & Tenenbaum, H. R. (2011). Does discovery-based instruction enhance learning? *Journal of Educational Psychology, 103*(1), 1–18. https://doi.org/10.1037/a0021017.
Ausubel, D. P. (1961). Learning by discovery: Rationale and mystique. *The bulletin of the National Association of Secondary School Principals, 45*(269), 18–58.
Balim, A. G. (2009). The effects of discovery learning on students' success and inquiry learning skills. *Eurasian Journal of Educational Research, 35,* 1–20.
Bruner, J. S. (1957). *Going beyond the information given.* New York: Norton.
Castronova, J. (2002). Discovery learning for the 21st century: What is it and how does it compare to traditional learning in effectiveness in the 21st century? *Action Research Exchange, 1*(1), 1–12.
Driver, R., Leach, J., & Millar, R. (1996). *Young people's images of science.* Bristol: Open University Press. Available at https://files.eric.ed.gov/fulltext/ED393679.pdf
Gillani, B. B. (2003). *Learning theories and the design of e-learning environments.* Lanham, MD: University Press of America.
Good, T. L., & Brophy, J. E. (1990). *Educational psychology: A realistic approach* (4th ed.). White Plains, NY: Longman.
Klahr, D., & Nigam, M. (2004). The equivalence of learning paths in early science instruction: Effects of direct instruction and discovery learning. *Psychological Science, 15*(10), 661–667.
Kolstø, S. D. (2001). Scientific literacy for citizenship: Tools for dealing with the science dimension of controversial socioscientific issues. *Science Education, 85*(3), 291–310.
Lee, Y. C. (2007). Developing decision-making skills for socio-scientific issues. *Journal of Biological Education, 41*(4), 170–177.
Mayer, R. E. (2004). Should there be a three-strikes rule against pure discovery learning? The case for guided methods of instruction. *American Psychologist, 59,* 14–19.
McLeod, S. A. (2008). *Bruner.* Retrieved from www.simplypsychology.org/bruner.html.
Miller, G. A. (2003). The cognitive revolution: A historical perspective. *Trends in Cognitive Sciences, 7*(3), 141–144.
Moore, K. (2009). *Effective instructional strategies.* Thousand Oaks, CA: Sage Publications Inc.
National Research Council. (1996). *National science education standards.* National Academies Press.
Ormrod, J. (1995). *Educational psychology: Principles and applications.* Englewood Cliffs, NJ: Prentice-Hall.
Schickore, J. (2014). Scientific Discovery. In E. N. Zalta (Ed.), *The Stanford encyclopedia of philosophy.* Retrieved from https://plato.stanford.edu/entries/scientific-discovery/.
Steiner, D. M. (2014). Learning, constructivist theories of. *Value inquiry book series: Global studies encyclopedic dictionary, 276,* 319–320.

Takaya, K. (2013). *Jerome Bruner: Developing a sense of the possible.* New York, USA: Springer.

van Joolingen, W. (1999). Cognitive tools for discovery learning. *International Journal of Artificial Intelligence in Education, 10,* 385–397.

Wartono, W., Hudha, M. N., & Batlolona, J. R. (2017). How are the physics critical thinking skills of the students taught by using inquiry-discovery through empirical and theoretical overview? *Eurasia Journal of Mathematics, Science and Technology Education, 14*(2), 691–697.

Woolfolk, E. A. (1993). *Educational psychology.* Boston: Allyn and Bacon.

Yilmaz, K. (2011). The cognitive perspective on learning: Its theoretical underpinnings and implications for classroom practices. *The Clearing House: A Journal of Educational Strategies, Issues and Ideas, 84*(5), 204–212.

Dr. Ozdem-Yilmaz is a faculty at Mugla Sitki Kocman University in Turkey. She was graduated from Middle East Technical University (METU) with a Bachelor of Science degree in Elementary Science Education program in 2003. She worked as a teacher for 5 years in private and public schools in Turkey and in the USA. She completed her Ph.D. in the field of Science Teacher Education in 2014 at METU. She was awarded by ESERA with travel award for young researchers and supported by the Academic Training Program to conduct research at the University of Bristol, UK. She completed her post-doctoral research on Science Centres at Great Lakes Science Center in 2016. Her research interests are Argumentation, Inquiry-based learning, Science Teacher Education, andScience Centres.

Dr. Kader Bilican is a faculty in the department of primary education at Kırıkkale University in Turkey. She received her Ph.D. degree in science education at Middle East Technical University in Turkey. She had been as a visiting scholar at Indiana University in IN, USA for a year and joined international projects in collaboration with the USA. She also led national funded projects in addition to being as an associate partner of projects. Her research areas of interest are professional development of science teachers, pre-service and in-service science teachers' scientific epistemic views and practice, and teaching science to young students. She has several national and international publications. She also attended several important international science education conferences such as NARST and ESERA.

Chapter 14
Guided Discovery—Robert Gagné

Yashwantrao Ramma, Ajeevsing Bholoa, and Mike Watts

Introduction

Research increasingly supports the idea that the adoption of research-based practices (Wieman & Perkins, 2005) in classroom transactions allows learners to increase retention dramatically and thereby improve test and examination scores. Recourse to various teaching-learning theories has the specific objective of enabling learners in their construction of subject-based knowledge. However, success in this respect rests upon teachers' ability to select and carefully use appropriate theory for planning and teaching a specific concept. There is a growing realisation of the importance of attending to students' needs to maximise academic growth, and meaningful learning may require a combination of theories. So, a teacher may well use a range of classroom strategies, including traditional exposition and/or a guided discovery approach, depending on the context and the process by which learning is expected to take place.

In this chapter, we discuss Robert M. Gagné's theory on guided discovery and we target our exploration of his theoretical ideas principally at teachers—with learners securely in mind. Gagné initiated his instructional theory during World War II for the process of training pilots in the Air Force. He later developed a sequence of requirements, clearly defined and he codified what principles educators should use as they developed instruction. The purpose behind Gagné's seminal theory (Gagné, Wager, Golas & Keller, 2005) was to provide teachers and instructional designers with a sense of direction in preparing lessons, the overall objectives being to foster

Y. Ramma (✉) · A. Bholoa
Mauritius Institute of Education, Reduit, Mauritius
e-mail: y.ramma@mie.ac.mu

A. Bholoa
e-mail: a.bholoa@mie.ac.mu

M. Watts
Brunel University London, Uxbridge, UK
e-mail: Mike.Watts@brunel.ac.uk

© Springer Nature Switzerland AG 2020   191
B. Akpan and T. Kennedy (eds.), *Science Education in Theory and Practice*,
Springer Texts in Education, https://doi.org/10.1007/978-3-030-43620-9_14

and enhance students' thinking and achievement. For example, his nine events of instruction are intended to help build a framework which teachers can use to prepare and deliver instructional content. Gagné's theory resulted essentially from a fusion of behaviourist concepts into a complex—and more complete—theory of instruction (Romiszowski, 2016). In his writing, he has distinguished his theory from purely behaviourist ones because he includes in it an appreciation of a variety of different types of learners and modes of learning. His work has placed considerable emphasis on the individuality of learners in the instructional process, and has acknowledged the importance of mental processes in learning, teaching and training. After gradual refinement, the current theory (Gagné & Driscoll, 1988) comprises three central components: (i) a taxonomy of learning outcomes, (ii) the conditions required for learning and (iii) the nine events of instruction.

To begin, Gagné's theory rests upon five taxonomies of learning (Driscoll, 2004; Gagné & Driscoll, 1988; Petry, Mouton & Reigeluth, 1987):

- Verbal information—through which connection is made with the learner's prior knowledge and understanding. This prior knowledge can be derived from learners' everyday life experience and/or from the previous lessons. To enable meaning-making, new information must always be related in some way to what learners already know.
- Intellectual skills—these relate to the ability to discriminate among items, concepts and facts, and to the selection of appropriate rules, principles and laws. Similar to verbal information, intellectual skills build upon already acquired skills. The set of new skills to be learned during the lesson should be presented gradually within a Vygotskian-style 'zone of proximal development' of the learner (van den Broek, 2012).
- Cognitive strategies—these relate to the ability to make use of acquired knowledge and skills to solve a given conceptual task. To operate at this level, learners need to have internalised specific concepts as well as relevant skills. In addition, the adoption of informative feedback is an important stepping stone towards helping learners to situate whether their strategic efforts are effective and innovative (Driscoll, 2004).
- Motor skills—these entail making use of correct practice in a coordinated manner to perform particular tasks. During the learning process, learners develop motor skills and this process should be followed by appropriate feedback to guide the learners to display the acquired skills in a variety of contexts (Gagné, Briggs & Wager, 1992). At the same time, learners should also be encouraged to make use of mental practice (critical thinking).
- Attitudes—the adoption of positive dispositions (could be in group) to perform a given task. To learn and express attitude, learners must already possess a set of information (preliminary data) and qualities (for example, confidence, optimism, resilience and kindness).

Furthermore, the theory introduces nine instructional events and corresponding cognitive processes (Driscoll, 2004; Petry, Mouton & Reigeluth, 1987) for conducting lessons. These events may vary to a certain extent depending on the strategies adopted by the teacher. For instance, if the teacher chooses to use group work, then the proceedings of the events will change as compared to individual work. The events are as follows:

- *Gaining attention*—the teacher arouses interest in the subject matter by relating the lesson with lived experiences of the learners. This might be through stimulating the students with novelty, uncertainty and surprise; posing thought-provoking questions or having students ask questions to be answered by other students.
- *Informing learners of the lesson objective*—the teacher articulates the learning outcomes to the class and ensures that learners are well informed about what is to be expected by, for instance, describing criteria for standard performance.
- *Stimulating recall of prior learning*—the teacher facilitates the connection between the prior experiences and knowledge of learners (for example, by means of a mind map) with the concept that learners will study in the current lesson. A discussion on the issues will ensure synchronisation among learners and the teacher (Zhang & Lu, 2011), thus ensuring that learners' attitudes are catered for.
- *Presenting the stimulus*—step-by-step organisation and explanation of concepts is developed. Two-way communication is established between teacher and learners. This might be through the provision of examples, by presenting multiple versions of the same content, e.g. through video, demonstration, lecture, podcast, group work and so on.
- *Providing learning guidance*—learners are actively engaged in the tasks with the continuous support of the teacher. This supportive guidance might be through 'scaffolding' (giving cues, hints, prompts, mnemonics); through organising varied learning strategies (concept mapping, role playing, visualising); using examples and non-examples to help students see what not to do or the opposite of examples, or providing case studies, analogies, visual images and metaphors.
- *Eliciting performance*—learners are provided with new tasks that serve as evidence of internalisation of learning. Engaging learners in performing authentic tasks (Mueller, 2017) is a useful way to situate whether—and what—learners have internalised while learning. During the process, the teacher can ask deep-learning questions, making reference to what students already know, and thereby spot shortcomings and immediately provide remediation support.
- *Providing feedback*—teachers offer feedback based on the interactions and evidence of learning. This event is not a stand-alone one but cuts across any of the events. The teacher helps students integrate new knowledge by providing real-world examples, and fosters autonomy among learners by means of additional independent practices.
- *Assessing performance*—in order to evaluate the effectiveness of the instructional events, the teacher must test to see if the expected learning outcomes have been achieved—performance being based on the previously stated objectives. At each stage of the step-by-step process, learners' understanding is challenged in the form of diagnostic and formative assessments. The recourse to diagnostic assessment enables learners to 'identify the core principles, issues or concepts associated with

the task in the early stages of a course [and which] could promote an attitude of self-regulation' (Crisp, 2012, p. 40) in the learners.

- *Enhancing retention and transfer*—students are invited to apply newly constructed knowledge to real-life situations through, for example, paraphrasing content, using metaphors, generating examples, or creating concept maps or outlines.

Influence of Gagné's Theory in Education

Throughout his work, Gagné has argued that teachers should use a variety of instructional methods to meet the needs of their students. The teacher's role is an attempt to reduce the cognitive load of the learners by limiting the amount of material presented to them and to engage them in the organisation of concepts into a suitable ordered structure (Wieman & Perkins, 2005). The conditions are both external to the learner (such as objects, pictures and verbal communication) and internal (such as prior knowledge, positive attitude and interest), and should be taken into consideration while designing the instruction to take place. The nine instructional events, in turn, are derived from his own, and others' experiments in cognitive psychology. This also means that a particular lesson has to be constructed within a differentiated perspective to cater for the individual needs of learners.

Methods

To explore Gagné's theory, we describe and discuss a case study of a mathematics in-service trainee teacher involved in the peer/microteaching module of a Post Graduate Certificate in Education (PGCE) course. One particular trainee teacher at the heart of the case study, a young woman we pseudonymise as Nita (holder of a B.Sc. Mathematics), has 5 years teaching experience at secondary school level and was part of a group of 13 trainee teachers following a 45-hour module. In that module, the group of trainee teachers were required to design, plan and implement lessons based on Gagné's theory. They were also expected to offer their reflections, including self-reflection on the episodes of their teaching session. All thirteen peer/microteaching sessions were video-recorded, and Nita was selected for the case study on the basis of her profound critical reflection carried out on the lesson she taught. The topic of the lesson was 'connected particles': an applied mathematics/mechanics (physics) problem. She presented her peer/microteaching, based on Gagné's theory with the following objectives:

The diagram shows two particles A and B of masses 1 kg and 4 kg respectively, attached at the ends of a light inextensible string which passes over a fixed smooth pulley. Particle B hangs in the air while particle A lies on a smooth horizontal plane. The system is released from rest. Find

    (a) the acceleration of the particles; and
    (b) the tension in the string.

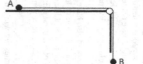

Fig. 14.1 Question on 'connected particles'

– To represent the forces acting on each particle, connected by a light inextensible string and passing over a smooth pulley;
– To apply Newton's second law of motion to each particle to form two linear simultaneous equations in terms of the tension in the strings and the acceleration of the two particles.

The 'connected particles' mathematics problem was set to the class as shown in Fig. 14.1.

The diagram shows two particles A and B of masses 1 kg and 4 kg, respectively, attached at the ends of a light inextensible string which passes over a fixed smooth pulley. Particle B hangs in the air while particle A lies on a smooth horizontal plane. The system is released from rest. Find

(a) the acceleration of the particles; and
(b) the tension in the string.

Findings

See Table 14.1.

Discussion

Our discussion is focussed primarily on Nita's implementation of the lesson on 'connected particles' whilst being guided by Gagné's nine events of instruction. The five domains of knowledge are also considered as they inform the manner in which the teaching strategies and methods have been employed to accomplish the goals of the lesson.

Table 14.1 Nita's interpretation of Gagné's nine events of instructions on 'connected particles'

Events	Observation of the peer/microteaching session	Components not observed
1. Gaining attention	Nita, the trainee teacher, missed this part	Peers should have been welcomed—greeted and asked for their well-being to establish a rapport of mutual respect To arouse their interest, Nita might have related everyday life experiences, such as the towing of vehicles, raising of construction materials on a building site by means of pictures or short videos A pre-test (or diagnostic test) could also have been administered to promote the participation of learners (Cheung, 2016). This diagnostic test could have served to create a situational awareness of learners' existing knowledge about real life events that have some relationship with the lesson
2. Informing the learner of the lesson objective	Two learning objectives, mentioned earlier, were brought to the attention of the class	Nita stated the learning outcomes but without making any reference to the previous instructional events. Furthermore, Nita did not fully engage her peers in clarifying the meaning of 'fixed smooth pulley', 'light inextensible string', and 'Newton's second law'
3. Stimulating recall of prior learning	Nita recapped on some real-life application of 'connected particles' and some of the basic concepts and associated assumptions through quiz sessions in a group setting Quiz 1: State one real-life application of 'connected particles' Quiz 2: The weight of the string is negligible as the string is……. (inextensible, light). Quiz 3: Since the string is inextensible, the……….. of the 2 particles is equal. (tension, acceleration, weight) Quiz 4: The smooth pulley ensures that the………. in the string on either side of the pulley is equal. (tension, weight)	**Intellectual skills:** In the case of this lesson, Nita used four class quizzes In Quiz 1, discussion should have been held around the terminology 'particle' in relation to mass and size to maintain consistency with a physics lesson In Quiz 2, the statement is misleading as comparison of the weight of the string ought to have been made with the weight of the other connected bodies In Quiz 3, both tension and acceleration are correct answers whereas the trainee considered only acceleration to be the correct one. She should have mentioned that the string is taut and that a frictionless situation is being considered In Quiz 4, discussion should have been held over the significance of the term 'smooth'. Furthermore, the choice of the term 'weight' as an option does not fit in the sentence structure

(continued)

Table 14.1 (continued)

Events	Observation of the peer/microteaching session	Components not observed
4. Presenting the stimulus material with distinctive features	Nita described the 'connected particles' problem via instrumental understanding using a numerical exercise From the start to presenting the material there was very little participation by the learners. For example, in one instance, a peer was requested to draw the forces acting on particles A and B. However, there was no instruction from Nita to provide justifications to support learning and identifying areas of difficulties	**Intellectual skill:** The link between the previous stages to this one was hardly made It was expected that Nita would introduce the 'connected particles' case as two disjoint systems and she would scaffold learners in identifying, with justifications, the various forces acting on each particle **Cognitive strategy:** Once learners would have developed understanding of the forces acting on the separate systems, they would have been guided to relate the acquired knowledge to this new situation of 'connected particles'
5. Providing learning guidance	Nita made some attempt to elaborate on the direction of tension for each particle using a book attached to a string as a concrete example. Instances of application of frictional forces pertaining to the given examples were provided. However, some of the verbal statements made by Nita could conflict with terminologies used in physics. For instance, relating 'g' to 'gravitational force' is conceptually problematic. In another instance, Nita mentioned that 'the tension acts away from the load (i.e. point mass)'; this could mislead the learners due to ambiguity in relating tension to a point mass However, no attempt was made to introduce the concepts related to the 'connected particles' problem. Reference was made to the following terminologies: 'taut', 'inextensible', 'a fixed pulley', 'a reference point'	**Verbal information, intellectual skill, motor skill and cognitive strategy:** It was expected that Nita would start the lesson by guiding learners, through scaffolding, to select the appropriate rules for identifying the forces in the 'connected particles' problem and then to decide on the course of action in order to develop the simultaneous equations

(continued)

Table 14.1 (continued)

Events	Observation of the peer/microteaching session	Components not observed
6. Eliciting performance	Nita's peers were requested to draw the forces acting on each particle on the board Once the forces have been drawn by the group, Nita acknowledged the correctness of the answers without engaging the peers into critical thinking with reference to other avenues	**Verbal information, intellectual skill, motor skill** and **cognitive strategy**: The peers should have been requested to justify their course of action with the support of the appropriate rules (e.g. why does the weight act downward? Why is tension equal along the string?) Furthermore, insightful information should have been obtained from her peers about associated concepts from mathematics and physics

(continued)

Table 14.1 (continued)

Events	Observation of the peer/microteaching session	Components not observed
7. Providing informative feedback	Feedback to the whole class was given after a particular peer had attempted the task on the board	**Verbal information, intellectual skill, motor skill and cognitive strategy**: Most of verbal communication emanated from Nita. Individual feedback should have been offered to that particular peer in the very first instance. For example, students peers should have been asked to explain the choice of directions of tension for each mass as separate systems before considering them as a single system When the two systems are put together, tensions at the pulley should also be considered
8. Assessing performance	Assessment of the 'connected particles' problem occurred when student peers were called upon to attempt specific task on the board, such as drawing forces and calculating the weight, drawing the tensions, forming and solving the simultaneous equations	**Verbal information, Intellectual skill, motor skills and cognitive strategy**: Another similar situation of 'connected particles' but in the opposite direction could have been considered. In addition, concrete objects could have been used to mimic real-life situations
9. Enhancing retention and learning transfer	Worksheets with more problem-solving tasks have been provided, with one of the problems involving two particles attached to a string passing over fixed pulley	**Verbal information and intellectual skills**: Two 'connected particles' tasks could have been given with different magnitude of the masses before giving the problem involving two particles attached to a string passing over a fixed pulley

Event 1—Gaining Attention

Nita, the trainee teacher, started the lesson, as illustrated by Episode 1.

> Episode 1: "Good afternoon everyone. Before we start the lesson on 'connected particles', we will start with our learning objectives for today's lesson."

Right at the beginning of the lesson, opportunities for declarative knowledge were somehow scanty. By directly opening the lesson with an explanation on the learning objectives, Nita missed the opportunity to gain the attention of the class by creating links between everyday life experiences of her learners (in this case, with her peers) and the subject matter of her lesson. Gaining attention is a first step to direct learners' attention towards concepts that are already familiar to them from the environment. Thus, the link between known (life experiences) with the unknown (new or abstract concepts) has been missed.

This first event, which was overlooked by the trainee teacher, is meant to direct learner's attention towards the teaching and learning process. This event is regarded as difficult (Belikuşakh-Çardak, 2016) since the activities gaining the students' attention need to arouse the interest of the learners for the underlying concepts of the lesson. Such activities could have included a short video, an interesting object or picture related to 'connected particles'.

Event 2—Informing the Learner of the Lesson Objective

Nita informed the class of the lesson objectives so as to offer a sense of direction to her peers as illustrated by Episode 2.

> Episode 2: "By the end of our lesson, each student should be able to represent the forces….be acquainted with light, inextensible string passing over a fixed pulley…. Second, we will have to use Newton's Second Law of Motion, i.e. *resultant force* $= m \times a$ to each particle to form 2 linear simultaneous equations in terms of the tension T in the string and the acceleration a of the two particles."

According to Gagné, Nita's lesson objectives excluded verbal information and focused more on intellectual skills, especially concepts and procedures. Elements of verbal information would have been apparent if the objectives would have included statements such as

(a) Define a system of 'connected particles';
(b) State the modelling assumptions of the system of 'connected particles'.

The elements of verbal communication were mostly observed during the 'Stimulating Recall of Prior Knowledge' event but were restricted to transmission of teacher's knowledge of facts rather than engaging the peers in relating prior knowledge with the objectives of the current lesson on 'connected particles'.

While informing the class about the learning objectives, it would have been appropriate for Nita to probe into the pre-existing knowledge of learners about the meaning of 'light, inextensible string' and 'fixed pulley'. These concepts relate to notions that form the basis for constructing new schemas on existing ones. Nothing should be taken for granted in the memory reorganisation process (Derry, 1996).

Event 3—Stimulating Recall of Prior Learning

For this learning event, Nita envisioned a quiz in groups of four by means of a PowerPoint presentation. Though it was a commendable initiative, she did not, however, promote any group dynamics for attempting the quiz. It would have been appropriate, for example, to raise the interest and curiosity of her class and also identify areas for knowledge consolidation. She simply accepted individual responses and gave the explanation herself as illustrated below:

> Episode 3: "Do you agree with her [another group member's] answer because she was the one to give the example of a car pulling a trailer. Yes, because they are 2 connected bodies as they are connected by a string."

In Quiz 1, Nita could have made the integration of related concepts, such as 'particle' and 'mass', between mathematics and physics to ensure that there is no misinterpretation or misunderstanding of the concepts when they are dealt with in the distinctive subject areas. Misunderstandings in this part of physics are commonly legion, and deserve acknowledgement, with a view to challenging them (see Warren, 1979) for a classical treatise on understanding Newtonian forces).

Quiz 2, involves the notion of comparison and, as such, further discussion should have been held in the form of a diagnostic exercise and which could then have been taken up during the course of the lesson.

In Quiz 3, conditions such as the string being taut and the pulley being frictionless were missed in the discussion. Thought-provoking situations could have led the learners to situate how these conditions influence both the tension and acceleration. The cause-effect relationship was not apparent as declarative knowledge was given prominence.

Event 4: Presenting the Stimulus Material with Distinctive Features

It is a recurrent feature in mathematics teaching in using a numerical exercise to drive the lesson development. For conceptual understanding, it is envisaged not to focus on numerical values but at identification and representation of the forces acting on the various components constituting the system. Once the distinct parts in the system have been analysed, the whole system can then be studied.

Event 5: Providing Learning Guidance

> Episode 4: "For example, if I am trying to pull this book towards me, where will the force tension act? So, if I pull it towards me, it will be away from the load, that is tension is a pulling force."

Though it is commendable that Nita attempted to illustrate the concept of tension in the 'connected particles' problem with a demonstration, she has nevertheless added a new element—which she had neither explored herself, nor allowed her peers to do. In this case, her hand and the string (in contact with her hand) constitute one system, while the string in contact with the book constitutes another system. The tension constituting the pulling force relates to the sting-book system and not to the hand-string system.

Event 6: Eliciting Performance

There was evidence of formative assessment when Nita invited her student peers to proceed to the whiteboard and to attempt individually a specific part of the question. However, Nita should have infused some elements of critical thinking in the formative assessment tasks to allow her class the opportunity to express their understanding of the concepts through reflection (Ramma & Bholoa, 2018). Once, when one of her class had represented the forces on particles A and B, Nita proceeded with the justification of her student's work by answering herself all the questions that she raised as illustrated below by episode 5.

> Episode 5: "Particle A has mass 1 kg, how is the force calculated? It will be *mass × gravitational force,* which is 10. Thus, it will be 4 multiplied by 10, which is 40 N. The system is now released from rest. If it is released from rest, you will be having some kind of acceleration. Here we know the acceleration of the 2 particles will be the same and what about the tension? It will be the same along the string."

Nita could have set some thought-provoking questions to drive learners into states of cognitive conflicts, and could have developed some discourse as follows: How are the 10 N and 40 N forces represented vectorially? How would you establish the occurrence of an acceleration of the particles? What are the conditions necessary for acceleration of both particles to be the same? What are the conditions necessary for tension along the string to be the same?

From Episode 5, we surmise that Nita herself holds misconception in that, for example, she related 'g' to 'gravitational force' rather than 'acceleration due to gravity' or 'acceleration of free fall'. She maintained this on three occasions during the course of the lesson.

Event 7: Providing Informative Feedback

Episode 6: "Since particle A is resting on the horizontal plane, is there any other force that acts on the particle in the upward direction? [All peers gave 'normal reaction' as the answer] Can one of you come and show the normal reaction of particle A [one of her peers proceeded to draw the normal reaction] ... So, here R equals to the normal reaction of the particle acting vertically upward."

Nita simply acknowledged the normal reaction as an upward force acting on the particle but did not offer feedback, through questioning, to guide her peers to situate that the normal reaction acts on the particle from the surface where contact is made. This results from Newton's Third Law of Motion, and that the length of the weight and normal reaction vectors are equal. Additional feedback as to why there was no motion on the vertical direction could have been explored. The feedback could have enabled her peers to identify any deficiencies.

Event 8: Assessing Performance

Nita adopted a unilateral approach to enable her peers to solve the numerical problem. It is important for the trainee to examine learners' understanding of the concept of 'connected particles' and their abilities to transfer acquired knowledge to a completely new situation, like particle B is now connected directly with another particle C as illustrated in Fig. 14.2.

Fig. 14.2 A new situation on 'connected particles'

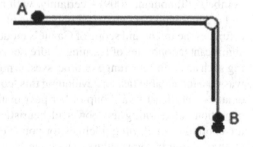

Event 9: Enhancing Retention and Learning Transfer

Nita used worksheets on 'connected particles' as a form of summative assessment. She made a brief recap of what the lesson was about and then gave a brief overview of the questions to be attempted. The work was to be submitted during the next lesson.

Conclusion

According to Jerome Bruner (1966), an instructional theory should deal with four major elements: (i) the learners' predispositions; (ii) the design of the concepts to be presented and a structure for ease of understanding; (iii) the most successful progression of ideas in which to present a body of knowledge; and (iv) the administration of rewards and punishments. Therefore, an instructional theory that focuses on the overall structure of learning would serve to provide the most successful learning experience. At one level, Gagné's model is certainly a successful approach to learning design, and it is certainly not without merit. Despite some of the limitations we discuss below, Gagné's work has made an enormous impact in the field of instruction; aspects of his system have become the foundation, for example, of the computer-assisted instructional design approaches to e-learning that continue to be influential in the field today.

Gagné recognised that learners bring 'conditions' with them to a learning activity (previous experiences, attitudes and prior knowledge, for example) that have a significant influence on the learning process itself. Gagné has argued that an instructor needs to understand these conditions of learning in order to optimise the learning interaction accordingly. It is clear from our discussion and illustrative episodes above from Nita's teaching that she was only partially successful in achieving this. Nita's ability to implement Gagné's nine events of instruction was superficial as she was largely influenced by the prevailing examination-oriented system (Ramma, Samy & Gopee, 2015). It is true that Gagné's work has been dismissed as 'teaching by numbers' (Taubman, 2009)—certainly, with nine steps, the approach can feel long and arduous, and Nita managed only some of them. In our view, though, this underestimates the intent and scope of Gagné's model—designing a lesson that covers five significant taxonomies of learning, addresses conditions for learning and incorporating such an ambitious range of processes, is no simple matter. In our case study, Nita was set a formidable task in developing this 'connected particles' lesson within mathematics—not least to a group of her peer mathematicians. In this context, Gagné's conditions of learning are powerful heuristics, and the nine events of instruction are enormously helpful guidelines for young educational designers—they provide a strong starting framework upon which teachers can base their lessons.

Rowlands, Turner, Thwaites & Huckstep (2009) discuss their development of a 'knowledge quartet' related to the structuring of teaching—in their case—of primary mathematics. Gagné's Nine Events fit well with what these authors call *foundation*, *transformation* and *connective* knowledge—the knowledge and skills required of a teacher to plan, illustrate, explain and connect the subject matter of the lesson to their students and any previous knowledge and learning. To use Gagné's terminology, the learner's *internal conditions* must then be paired with the correct *external conditions* as manifested in the form of the *instructional events*, the type of instruction given (Gagné, 1985). In other words, the programme should be adjusted to the learner's needs. Our trainee teacher, Nita, did not always manage this: her unsteady start of the lesson failed to help her audience to relate the topic of the lesson to something they already knew. There were few, if any, deep-level questions that related to her learners' prior knowledge, and—while she certainly embedded quizzes in her lesson—it is not clear that these served to internalise new knowledge and expertise.

It is clear, too, that Nita's planning sometimes skips between the Nine Events, which is perhaps a recognition that the pre-structured framework of the learning design can be overly prescriptive and leave little room for pedagogical elements that we now recognise as important for successful learning experiences. Ravenscroft (2003), for example, has pointed out that when teachers follow the Nine Learning events rigidly, learners have little or no control over pace, timing and sequence. In Nita's case, there was no evidence that she encouraged her learners to engage in higher level reasoning, reflection, exploration or creativity. This may, of course, have been a consequence of her own tentative understanding of the applied mathematics behind the 'connected particles' problem, and the fact that she was, after all, working with a class of her trainee peers. These 'case study' caveats aside, however, we do see Gagné's approach as failing to cope with much ambiguity, failing to teach the learner how to use their judgment in situations that are new or in problem-solving effectively. This fourth part of Rowland et al. (2009)'s quartet is called *contingent* knowledge, understandings that allow the teacher to manage unplanned situation where, for example, a student asks an awkward question, where the teacher's own subject matter falters, when confronted by an unexpected result from a calculation, or when an example fails to deliver the understanding needed. No amount of planning can predict all of the possible occurrences within a lively and active-learning lesson, and we would want to work with teachers to cater for such circumstances.

Gagné's theory is generally designed to enhance concept acquisition by learners during face-to-face teaching-learning transactions given that there is a logical sequencing in the nine instructional events and that each one is contingent on the subsequent level. It rests on the teacher to organise the instructional events so that there is direct relationship between one event and the next.

Our own assessments of Gagné's instructional-guided discovery based upon our own research-based experiences (Ramma, Bholoa, Watts & Nadal, 2017), do concern the growing datedness of his work. So, for example, we would argue that the nine steps would have to take account of new information technologies that speak to the needs of mobile twenty-first century learners. Another shortcoming is the linearity in the sequencing of the instructional events and the eventual implementation . Fortunately, rapid developments in this 'blended' area of teaching and learning are bringing about major changes both in the relationship between teachers and the taught (Thaufeega, Watts & Crowe, 2016) as well as in the structure of educational encounters between them.

Summary

- Gagné's theory on guided discovery rests on five taxonomies of learning: verbal information, intellectual skills, cognitive strategies, motor skills and attitudes.
- Nine instructional events are captured within the theory, namely, gaining attention, informing learners of the lesson objective, stimulating recall of prior learning, presenting the stimulus, providing learning guidance, eliciting performance, providing feedback, assessing performance, and enhancing retention and transfer.
- Gagné's theory is most suitably used during classroom transactions during which the teacher can monitor progress of learners when the nine instructional events are being enacted. Concept acquisition by learners can then be achieved sequentially and logically.
- Teachers are largely influenced by the traditional mode of teaching which inhibits their ability to successfully enact the nine instructional events. Reflection on taught lessons should therefore help teachers to improve their practices.
- Learners' prior knowledge and experiences have to be taken into consideration during the course of the lesson delivery as they have significant bearing on the teaching-learning transactions.
- Teachers may hold misconceptions about certain concepts, even in their areas of specialisation and they should try to dispel them, by engaging in constructive discourse with their peers.

Recommended Resources

Gagné, R. M., & Brown, L. T. (1961). Some factors in the programming of conceptual learning. *Journal of Experimental Psychology, 62,* 313–321.

Gagné, R. M., Wager, W. W., Golas, K. G., & Keller, J. (2005). *Principles of instructional design*. Toronto: Thomson Wadsworth.

Martínez-Plumed, F., Ferri, C., Hernández-Orallo, J., & Ramírez-Quintana, M. J. (2015). Forgetting and consolidation for incremental and cumulative knowledge acquisition systems. https://arxiv.org/abs/1502.05615.

Warren, J. W. (1979). *Understanding force*. London, UK: Murray

References

Belikuşakh-Çardak, Ç. S. (2016). Instructional process and concepts in theory and practice. In C. Akdeniz (Ed.), *Models of teaching*. Singapore: Springer Science+Business Media.

Bruner, J. (1966). *Towards a theory of instruction*. Cambridge: Harvard University Press.

Cheung, L. (2016). Using an instructional design model to teach medical procedures. *Medical Science Education, 26*, 175–180. https://doi.org/10.1007/s40670-016-0228-9.

Crisp, G. T. (2012). Integrative assessment: Reframing assessement practice for current and future learning. *Assessement and Evaluation in Higher Education, 37*(1), 33–43.

Derry, S. J. (1996). Cognitive schema theory in the constructivist debate. *Educational Phycologist, 31*(3/4), 163–174.

Driscoll, M. P. (2004). *Psychology of learning for instruction* (3rd ed.). New York: Pearson.

Gagné, R. M. (1985). *The conditions of learning and theory of instruction*. Rinehart and Winston: Holt.

Gagné, R. M., & Driscoll, M. P. (1988). *Essentials of learning for instruction* (2nd ed.). Boston, MA: Allyn and Bacon.

Gagné, R. M., Briggs, L. J., & Wager, W. W. (1992). *Principles of instructional design* (4th ed.). New York: Harcourt Brace College Publishers.

Gagné, R. M., Wager, W. W., Golas, K. G., & Keller, J. (2005). *Principles of instructional design*. Toronto, ON: Thomson Wadsworth.

Mueller, J. (2017). *Authentic assessment toolbox*. Retrieved from http://jfmueller.faculty.noctrl.edu/toolbox/.

Petry, B., Mouton, H., & Reigeluth, C. M. (1987). A lesson based on the Gagne-Briaggs theory of instruction. In C. M. Reigeluth (Ed.), *Instructional theories in action: Lessons illustrating selected theories and models* (pp. 10–43). Hillsdale, NJ: Lawrence Eribaum Associates.

Ramma, Y., & Bholoa, A. (2018). A critical evaluation of a teacher professional development model—A case study of a pre-service physics teacher. In S. Ladage & S. Narvekar (Ed.), *epiSTEME 7* (pp. 285–293). Mumbai: Cinnamon Teal. Retrieved from https://episteme7.hbcse.tifr.res.in/wp-content/uploads/2018/01/epiSTEME-7-pages-1-474-without-header.pdf.

Ramma, Y., Bholoa, A., Watts, M., & Nadal, P. (2017). Teaching and learning physics using technology: Making a case for the affective domain. *Education Inquiry*, 1–27. Retrieved from https://www.tandfonline.com/doi/full/10.1080/20004508.2017.1343606.

Ramma, Y., Samy, M., & Gopee, A. (2015). Creativity and innovation in science and technology—Bridging the gap between secondary and tertiary levels of education. *International Journal of Educational Management, 29*(1), 1–17. Retrieved from http://www.emeraldinsight.com/doi/full/10.1108/IJEM-05-2013-0076.

Ravenscroft, A. (2003). From conditioning to learning communities: Implications of fifty years of research in e-learning interaction design. *Association for Learning Technology Journal, 11*(3), 4–18.

Romiszowski, A. J. (2016). *Designing instructional systems—Decision making in course planning and curriculum design*. London: Routledge Taylor and Francis Group.

Rowland, T., Turner, F., Thwaites, A., & Huckstep, P. (2009). *Developing primary mathematics teaching. Reflecting on practice with the knowledge quartet*. London: SAGE Publications Ltd.

Taubman, P. M. (2009). *Teaching by the numbers: Deconstructing the discourse of standards and accountability*. New York: Routledge.

Thaufeega, F., Watts, D. M., & Crowe, N. (2016). Are institutes and learners ready for e-learning in the Maldives? In *International Technology, Education and Development (INTED) Conference*. Valencia, Spain.

van den Broek, G. S. (2012). *Innovative research-based approaches to learning and teaching*. Nijmegen: OECD. https://doi.org/10.1787/5k97f6x1kn0w-en.

Warren, J. W. (1979). *Understanding force*. London, UK: Murray.

Wieman, C., & Perkins, K. (2005). By using the tools of physics in their teaching, instructors can move students from mindless memorization to understanding and appreciation. *Physics Today, 58*(11), 36–50.

Zhang, W., & Lu, J. (2011). Dynamic synchronisation of teacher—Students affection in affective instruction. *International Education Studies, 4*(1), 238–241.

Dr. Yashwantrao Ramma is a Professor of Science Education and Chairperson of Research Unit at the Mauritius Institute of Education. As a physicist, he has worked on several research projects related to technology integration and physics misconceptions of both physics teachers and students. Currently, he is leading research projects on exploring teachers' content knowledge and pedagogical content knowledge across various subject areas, on indiscipline and school violence in primary schools and also on students' transitions from secondary to university and teacher training.

Dr. Ajeevsing Bholoa is a Senior Lecturer in Mathematics Education at the Mauritius Institute of Education. He is currently the Programme Coordinator for pre-service B.Ed honours and is also involved in curriculum development at the primary and secondary levels. His research interests are related to the integration of technology as a pedagogical tool in teaching and learning of mathematics and the identification of content knowledge and pedagogical content knowledge of teachers.

Dr. Mike Watts is a Professor of Education at Brunel University, London, conducting 'naturalistic' people-orientated research principally in science education and in scholarship in higher education. He has conducted major studies in both formal and informal educational settings in the UK and abroad, and has published widely on his research through books, journal articles and numerous conference papers. His work is international and relates to 'Academic development in universities', 'Public understanding of science' and 'Identity and science education', alongside many other issues. He teaches at all levels within Brunel's Department of Education and is currently supervising 16 Ph.D. students.

Chapter 15
Developing Intellectual Sophistication and Scientific Thinking—The Schemes of William G. Perry and Deanna Kuhn

Keith S. Taber

Introduction

William G. Perry proposed a theory of the stages of intellectual and ethical development that he identified from work with undergraduate college students. At the time when his work was proposed, it seemed to be most relevant to young adults who would be expected to have successfully passed through the stages of cognitive development that had been identified by Jean Piaget in his work with children and adolescents. However, it is now clear that the stages of development discussed by Perry are very relevant to the school science curriculum, and so to the types of thinking often now expected from school students when studying science.

Deanna Kuhn has worked with children exploring the development of scientific thinking and developed a model of the development of critical thinking that has strong links to the scheme proposed by Perry. One interpretation suggested by comparing their work is that school science now routinely challenges pupils to demonstrate a level of epistemological sophistication that was often still being formed in many undergraduate students in the mid-twentieth century.

Cognitive and Moral Development

It is widely recognised, indeed it is commonplace experience, that development from a neonate through childhood and adolescence into adulthood is not simply a matter of physical growth. A young child does not access all the kinds of thinking available to a mature adult. Part of development is acquiring new modes of thinking about the world, as, for example, when language is internalised (see Chap. 19). There have

K. S. Taber (✉)
Faculty of Education, University of Cambridge, Cambridge, England
e-mail: kst24@cam.ac.uk

© Springer Nature Switzerland AG 2020   209
B. Akpan and T. Kennedy (eds.), *Science Education in Theory and Practice*,
Springer Texts in Education, https://doi.org/10.1007/978-3-030-43620-9_15

been a number of key theorists who have studied and sought to understand the nature of how such development occurs.

Jean Piaget (see Chap. 10) focused on the development of cognition. He posited a complex stage theory that had four main stages characterised by increasingly sophisticated levels of thinking (Piaget, 1970/1972). In Piaget's model, the fourth stage was called formal operations. This implied that a person was capable of highly abstract thinking and able to undertake mental operations on internal mental representations. This was very relevant to learning science as many science topics taught in school involve theoretical abstractions that students are expected to engage with, and indeed apply, in the absence of the natural phenomena from which those ideas were initially abstracted. Examples might be clades in biology which concern the evolutionary relationships between organisms (which do not necessarily all exist at the same time or place); notions of flux density in magnetic fields (which are not visible but may be represented by visualising imaginary field lines); or oxidation states used to represent redox processes (understood in terms of shifts in electron density that are conceptualised as partial electron relocations in molecules, that is, subtle modifications in particles theorised to exist at a scale many orders of magnitude removed from direct observation). Given that many secondary school learners are not considered to have fully developed formal operational thinking, it was argued that learning difficulties students face in school science may often result from a mismatch between the demands of the curriculum and the level of cognitive development of many of the students (Shayer & Adey, 1981).

Another key thinker, Lev Vygotsky (see Chap. 19), considered that adult ways of thinking could be understood as a culturally developed resource (that is, a resource that had been developed historically within a cultural group), into which young people could be inducted by mediation from more advanced members of the community, supported by such tools as language and other shared forms of symbolic representation. Even in a scientifically literate society, children will not develop conceptions that closely match canonical scientific concepts without formal instruction or other mediation (e.g. through books, websites, documentaries, etc.).

Other theorists considered a different aspect of development, related to moral growth (Kohlberg, 1973) (cf. Chap. 5). Cognitive development related to the ability to think in more sophisticated (and abstract) ways, whereas moral development related to the development of a system of values. This was more concerned with making 'good' or 'wise' choices when taking practical action, rather than being able to solve logical puzzles or apply technical concepts. When Benjamin Bloom (see Chap. 11) set out taxonomies of educational objectives to guide pedagogy, he developed distinct taxonomies for the cognitive domain (Bloom, 1968) and the affective domain (Krathwohl, Bloom, & Masia, 1968). To be characterised at the highest level of the affective domain required "an internal consistency to the system of attitudes and values at any particular moment" that gave a 'predisposition' or "basic orientation which enables the individual to reduce and order the complex world… and to act consistently and effectively in it" (Krathwohl et al. 1968, p. 48). Such an individual would develop a worldview that offered a coherent philosophy of life that guided judgements across all domains (Taber, 2015).

Piaget's work was strongly linked to the development of the kinds of concepts met in school science and mathematics, and its relevance to science teaching was clear. As the affective domain concerns values, rather than conceptual understanding, it can appear to be more relevant to learning about areas of the curriculum traditionally associated with values—the arts and humanities—yet an authentic science education must introduce learners to the values inherent in science (open-mindedness, seeking evidence and so forth) and teaching about the applications of science in relation to public policy engages value judgements as well as knowledge.

Considerations of moral development are less about evaluation of the specific moral decisions a person makes (i.e. whether one might agree with a person's decisions or consider they have behaved in a good way), but more about the sophistication of the thinking, and the coherence of the value system that underpins this. Arguably, fundamentally, the thinking skills being applied are not distinct from those that pertain when evaluating cognitive development. Perry (1985) proposed a theory of the development of student thinking that encompassed intellectual and ethical development within the same scheme.

Development Beyond Piaget's Formal Operations and Scientific Thinking

Piaget's scheme considered cognitive development to be complete with the acquisition of formal operations. However, there were suggestions that there might be further progression beyond the Piagetian scheme. For example, Arlin (1975) explored the idea that whilst formal operations provided the ability to engage in successful problem-solving, further development was needed to be an effective problemfinder—development that might be considered a fifth stage. This skill is clearly important in scientific work: a key feature of research is in identifying, and conceptualising, potentially productive questions. In science, logical thought works in coordination with creative thinking (Taber, 2011), and this becomes especially salient when school science is expected to engage students in enquiry activities (see Chap. 23).

School science had traditionally taught a model of 'the scientific method', that is, the use of control of variables to design experiments, and formal operations provided the means to use logic to apply hypothetico-deductive thinking in such 'fair tests'. However, it is increasingly thought that an effective science education (at school level, as well as in higher education) must have a strong focus on enquiry, where the earlier phases of investigations—such as recognising suitable research topics, refining research questions and then designing studies to address those questions— is as important as later applying logic to make deductions from experimental results (Riga, Winterbottom, Harris, & Newby, 2017). This could be considered to require the kind of 'fifth stage' that Arlin investigated.

It was also asked whether acquisition of formal operations was sufficient to treat knowledge as non-absolute, or to cope with contradictions (Kramer, 1983). This is especially relevant to school science in contexts where it is considered important that students not only learn some science, but also learn about the nature of science (Taber, 2017). Formal operations work when logic is sufficient to reach a conclusion—for example, in mathematical systems where the notion of proof applies. A modern understanding of science suggests that a naive positivism is misguided, and that all scientific findings should be seen as potentially provisional and open to reconsideration in the light of either new evidence or a new perspective to reconceptualise evidence. That is, scientific knowledge is not absolute and is theoretical (and so reliant on some commitments that have to be assumed a priori and cannot be demonstrated).

In much scientific research it is not even possible to draw absolute conclusions when working within a particular theoretical framework: scientific results are seldom unequivocal, as they are subject to both limitations of measurement and observation, and sometimes human error, and, moreover, nature is often more subtle and complex than the models being used to conceptualise and design studies. Scientists often have to deal with contradiction, and fuzzy data, and be able to make judgements about the extent to which robust conclusions can reasonably be drawn in the face of imperfect (in the sense of not entirely matching the predictions of any particular hypothesis) datasets.

The kinds of understanding of the processes of science that are set out as target knowledge in many national school systems rely then on learners exhibiting thinking that has been considered characteristic of a fifth stage *beyond* the formal operational level—when that stage itself is not thought to be fully acquired by all secondary school-age learners. Piagetian theory assumes a constructivist process where each stage is slowly built through experiences deriving from the regular application of the operations that have been acquired in the preceding stage (see Chap. 10): so (from this perspective) only students having fully acquired formal operations would be ready to start constructing a 'fifth stage' of post-formal operations. Development of such thinking skills is therefore a topic of great importance for curriculum development and pedagogy in school science.

Perry's Study of Undergraduate Thinking

Perry carried out his work in the mid-twentieth century with college students in the United States, that is, undergraduate students studying for degrees. Moreover, he worked with students at the elite Harvard and (to a lesser extent) Radcliffe Colleges, exploring their experience of engaging with the study of a range of subjects. (At the time of the work, Harvard College only accepted male students and Radcliffe College only accepted female students—the institutions later merged). So, Perry was working with young adults who had successfully completed schooling and had been admitted to prestigious degree courses. It should also be noted that undergraduate education in the United States is somewhat different to that in some other parts of the world, in that a first (bachelor's) degree course often comprises a wide curriculum, rather than being specialised within a single discipline such as anthropology, chemistry or zoology. Perry's team talked to students over the 4 years of their undergraduate degree. Perry characterised the data collection as 'open' interviews that sought to elicit the participants' ways of making sense of their experiences.

Perry (p. 48) reported finding a developmental pattern in the data "in the special sense originally derived from biology in that it consists of an orderly progress in which more complex forms are created by the differentiation and reintegration of earlier simple forms". He described this development as an ability to make sense of increasingly nuanced information or situations:

> In its full range the scheme begins with those simplistic forms in which a person construes his [sic] world in unqualified polar terms of absolute right-wrong, good-bad; it ends with those complex forms through which he undertakes to affirm his own commitments in a world of contingent knowledge and relative values. The intervening forms and transitions in the scheme outline the major steps through which the person, as evidenced in our students' reports, appears to extend his power to make meaning in successive confrontations with diversity (p. 3).

Perry's model differed from the kind of scheme offered by Piaget in that, although it represented a course of development, Perry noted that individual students could 'retrogress' at any point. That is, even when a student had demonstrated thinking characteristic of a higher position in the scheme, they might later offer thinking linked to an earlier position. In Piaget's scheme such 'décelage', where a student reverts to thinking typical of an earlier stage, might be explained as a lack of familiarity with a novel context or topic area. Perry's scheme by contrast was linked to developing a personal value system, and retrogression might reflect broader considerations (e.g. times of personal stress or contexts related to existential issues that may seem to threaten existing beliefs).

Perry characterised his scheme in terms of nine steps, and he offered two overviews of the sequence: either viewed from the midpoint or in terms of three major divisions (pp. 64–65). This is represented in Fig. 15.1. Point 5 represents a perception of knowledge and values as relative, contingent and contextual—representing the outcome of a slow shift from an earlier position where it is considered all knowledge claims or value positions can be simply judged true or false. From this central position of a generalised relativism, the individual develops personal commitments that are no longer considered absolute, but which are a suitable basis for making meaningful evaluations.

In the first part of development (positions 1–3), the individual slowly modified an absolutistic right-wrong outlook to begin to admit a degree of pluralism. In the second part (positions 4–6), there is a deepening appreciation of the problematic nature of laissez-faire relativism. In the final part (positions 7–9), the individual draws upon their experience to develop their own personal system of commitments. The reader is referred to Perry's (1970a, 1970b) own account for details of the nine positions.

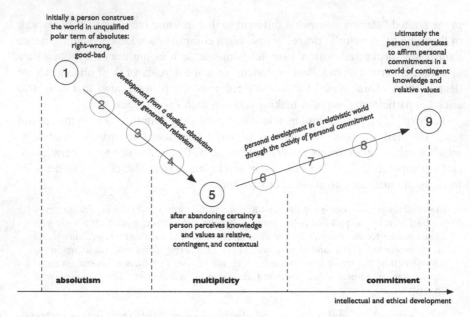

Fig. 15.1 A representation of Perry's developmental scheme (Adapted from Taber, 2013, Fig. 14.3, p. 265)

The Challenge of Becoming a Scholar

Perry found that even intelligent, highly motivated undergraduates struggled with the kinds of work they were set in some classes. These students expected their teachers to set out a particular perspective of a topic that needed to be understood, and which the student might later apply and be tested on. Yet, in many humanities classes, teachers did not offer this. When they were set alternative readings offering contrary viewpoints, these students assumed they were expected to identify with one of the approaches and they also expected their teachers to later confirm which was the superior position. Instead, they were often exposed to diverse perspectives, asked to appreciate them all, but not told which account they should believe or which standpoint they should adopt.

In simple terms, Perry found students were looking for a 'right' answer that could clearly be distinguished from the alternatives, and so often assumed their teachers were expecting them to work out which of the set readings they were meant to agree with. They were often then frustrated when their teachers refused to cooperate through indicating that a particular take on a topic was to be preferred. The teachers, however, recognised that there were multiple valid views supporting ongoing debates in many fields and saw their job as introducing perspectives and encouraging the students to think their way through to their own positions.

The realisation that they were not meant to find right answers could lead students to come to the view that there were not any right or wrong answers, because it was all a matter of personal opinion—so that anyone's take on a situation was as good as anyone else's. This still fell short of what was expected, which was that students could recognise the strengths and weaknesses of different positions; appreciate that judgements were informed by values and come to their own evaluations based on personal sets of values that could be articulated and so recruited to argue for a position. Over time, many students, but not all, would manage this.

For students studying a modular degree, these challenges to their developing thinking were not necessarily the same in all areas of the curriculum. History might offer alternative explanations of events; there might be different interpretations of texts in literature and different aesthetic judgments of the relative merits of different authors and their works; there might be different ideological political positions deriving from the perspectives of different interest groups: but in the natural sciences, these challenges were less extreme.

Science teaching tended to offer canonical understandings, and (at undergraduate level, at least) the basis for scientific knowledge was often presented in terms of clear-cut critical experiments. Science is not only written by the 'victors' (cf. history), but it is the 'victors' who come to be heavily cited, and then featured in the textbooks. Scientific reports deal with the context of justification and generally hide the messy aspects of the context of discovery (Medawar, 1963/1990): the cul-de-sacs, the human mistakes and the role of serendipity. Scientific accounts privilege the logical thinking underpinning the deductive nature of reaching conclusions in studies, rather than the creative thinking required to imagine those possibilities to be considered and tested (Taber, 2011).

The logical argument from evidence can be audited by the scientific community, whereas the creative insights that made a study possible are not open to any objective validation. That many scientific discoveries emerge from messy research programmes that only slowly lead to a consensus position is usually ignored in textbook accounts reduced to a rhetoric of conclusions (Niaz & Rodriguez, 2000). When science teaching follows this pattern, it may not seem to require students to have developed far along Perry's progression.

The Relevance of Perry's Scheme to Socio-Scientific Thinking

The science curriculum now often requires students to appreciate more of the nature of science and the complexities around actual scientific work. Moreover, increasingly school science encompasses socio-scientific issues (Zeidler, 2014), where science interacts with the wider society. There are many important matters of public policy, of global, international or just local concern, where scientific knowledge is needed to inform decision-making, but where, of itself, science is insufficient to reach a judgement. Often different groups in society take different views in debates about such matters: perhaps because they have different interests (perhaps the wider community will benefit from the new airport, power station or chemical refinery: but those living in the immediate vicinity may have good reason to oppose the development) or different ideological and value positions (there is no objective view on how to balance economic wealth against environmental protection) or different perceptions of risk (as when the best advice is that there is a possibility of a serious disaster, but with a very small chance of it occurring).

For students to engage in these areas of learning they have to not only understand the science but also appreciate and empathise with different standpoints and value positions, and then apply their own values to reach a recommendation. This requires schoolchildren to engage in just the kinds of thinking that Perry found many undergraduates at elite institutions were still developing. This potentially presents something of an enigma. In the 1980s, the school science curriculum was criticised because it expected students of around 14–16 years of age to master abstract scientific concepts when many were still in the process of fully acquiring the requisite formal operational thinking skills (Shayer & Adey, 1981). Yet in the twenty-first century, the school curriculum in many countries has been reformed to ask students to appreciate a more nuanced understanding of scientific enquiry that forms provisional knowledge from messy datasets, and to engage in debate over socio-scientific issues drawing upon diverse value-based standpoints, that is, activities requiring what has been characterised 'post-formal' thinking.

Perry's Model Informing Science Pedagogy

Perry's model can be seen as descriptive, rather than prescriptive. That is, Perry undertook detailed and careful enquiry at a particular time. His scheme describes what he found among undergraduate students who experienced a particular college curriculum, and more importantly had previously passed through a particular school curriculum. It might be argued that a school curriculum that largely presents canonical accounts to be understood, learnt and applied, does not give learners the necessary experiences to fully develop from expecting right and wrong answers, through a form of contextual relativism, towards a position of personal commitment based on a system of coherent values (i.e. the kind of value system Bloom and his colleagues saw as the highest level of their taxonomy of educational objectives in the affective domain).

If it is accepted that the forms of thinking developed depend upon the educative experiences provided in a culture (Luria, 1976), then the levels of intellectual development supported depend upon educational aims and their enactment in what learners are expected to do and achieve. After all, if IQ tests are considered to offer useful measures of human intelligence, then measured human intelligence increased substantially in many countries during the twentieth century (Flynn, 1987)—presumably reflecting greater levels and standards of education (as there was negligible physiological evolution over that period). Perry (1985) reported that "a study of examination questions given to freshman at Harvard at the turn of the [Twentieth] Century reveals them all to ... ask for memorised facts and operations in a single assumed framework of Absolute Truth" (p. 5) and suggested that over a period of 25 years he had seen that the "position [on his scheme] of the modal entering freshman at Harvard has advanced from around Three to nearly Five" (p. 12).

The educational theorist Jerome Bruner (see Chap. 18) claimed that it was possible to teach any subject, in some intellectually honest manner, to a learner of any age (Bruner, 1960). This attitude suggests that it should be possible to teach school students richer accounts of the nature of science, and to engage them in debate over socio-scientific issues, as long as they are suitably supported by teachers structuring appropriately engaging and accessible learning activities (cf. Chap. 19). If the message to take from Perry's work is that higher levels of intellectual and ethical development do not occur automatically (that is, purely under biological control) but require suitable educational experiences (cf. Chap. 19), then appropriate pedagogy needs to be developed.

Kuhn on the Development of Critical Thinking

Deanna Kuhn is an educational psychologist who has taken great interest in the development of thinking skills, such as scientific reasoning. Her work explores a range of themes important to science teaching and indeed to education more widely. This includes aspects of informal reasoning and argumentation, and approaches to pedagogy. One particular theme in her work is critical thinking, and how this develops. She is also interested in metacognition, which she considers as strongly linked to critical thinking. The treatment here is necessarily limited to offering a flavour of some of her most important work.

Kuhn sees the origins of what might be called 'scientific thinking' in developing epistemological understanding—understandings relating to the nature and sources of knowledge (Kuhn & Pearsall, 2000). This links to the appearance of what is sometimes known as a theory of mind (Wellman, 2011). Usually by the age of 5 children recognise that statements people make about the world are actually statements about the claimants' beliefs about the world. So young children will come to appreciate that an actual state of affairs may not be the same as a person's construal of the state of affairs: people may have false beliefs. This is a starting point for developing the ability to coordinate theory and evidence, which Kuhn considers the essence of scientific thinking.

Metacognition is cognition about cognition—so could be considered to encompass judgement about others having false beliefs. However, usually the term refers to thinking about one's own cognition. Kuhn (1999, p. 18) argues that "thinking about one's thought—in contrast to simply engaging in it—opens up a whole new plane of cognitive operations that do not exist at a simple first-order level of cognition". Students may be said to show different levels of metacognitive awareness and can be encouraged to develop metacognitive skills. This links to themes such as being a reflective learner and developing what are sometimes called 'study skills'. An

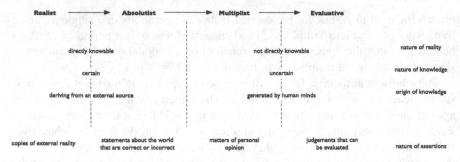

Fig. 15.2 Kuhn's model of levels of epistemological understanding

effective learner needs to have knowledge of their own current knowledge level (i.e.
meta-knowledge), and whether it matches educational goals (that may either be set
by the learner or provided by a teacher or other external agent); to appreciate which
activities are likely to help them learn and to be able to monitor their own learning so
that they can know when (and to what extent) they have been successful—and can
judge when a learning activity is proving unproductive and some change in activity is indicated (a different approach, taking a break, seeking additional support).
Metacognition is important to effective learning, in science as in other curriculum
areas.

Kuhn (1999) proposed a four-stage model of levels of epistemological understanding (see Fig. 15.2). Young children consider reality to be directly knowable, so
that assertions can be considered unmediated accounts of reality, but later they come
to develop greater epistemological sophistication and appreciate that such direct
access to the way things are is not possible. That is, they start to appreciate that
knowledge is something generated within human minds, rather than taking the naïve
view that reality imposes itself on mind. This can be considered as moving to a
constructivist position (see the contributions in Sect. 4 of this volume), appreciating
that knowledge takes the form of conjectures, ideas, theories and so forth—constructions put upon perceptions—rather than perfect impressions of an actual state
of events. This reflects a contemporary understanding of the nature of science that
sees science as a reliable—but not infallible—means of generating and evaluating
theoretical knowledge.

Kuhn's model comprised of four stages labelled as realist, absolutist, multiplist
and evaluative—a model that has strong parallels with Perry's scheme for intellectual
and ethical development:

> The absolutist sees knowledge in largely objective terms, as located in the external world
> and knowable with certainty. The multiplist becomes aware of the subjective component
> of knowing, but to such an extent that it overpowers and obliterates any objective standard
> that would provide a basis for comparison or evaluation of opinions. Only the evaluativist
> is successful in integrating and coordinating the two, by acknowledging uncertainty without
> forsaking evaluation. (Kuhn, 1999, pp. 22–23).

In the realist stage, the child simply accepts that assertions made by others report
the world as it is, but when they come to appreciate there can be false beliefs they shift

to an absolutist position that some assertions are indeed statements reflecting reality, and others are not. This allows a role for critical thinking in making judgements about which assertions are true, and which are false. This absolutism is similar to the starting point of Perry's scheme (see Fig. 15.1)—Perry had not included children in his study and did not find any undergraduates holding a realist position.

However, the child later moves to a multiplist position where it comes to appreciate that absolute and certain knowledge of the world is not possible, as knowledge is generated within minds, which admits scope for subjectivity in all human knowledge. (Science may be seen as a system to minimise the subjective aspect of human knowledge.) Given that assertions cannot simply be considered true or false—as how things seem often depends upon one's viewpoint, or the perspective adopted—there is then considered to be no sense in seeking to apply critical thinking to evaluate various assertions. This position may be more productive in some contexts than others. We live in pluralist societies, where democracy requires respecting and valuing the views of those we disagree with. However, science depends upon critical evaluation of ideas and is not generally considered consistent with a multiplist position. While some philosophers of science have argued that some degree of pluralism within science is valuable when exploring complex phenomena (Mitchell, 2003), this would generally be considered an epistemological stance rather than an ontological commitment. That is, reality is seen as having a unitary nature, but when our models and conceptions are imperfect accounts of that nature, then working with several complementary partial accounts can sometimes be valuable. Pluralism is then adopted pragmatically (see Chap. 16), rather than on principle as a commitment to the nature of reality.

School science, and arguably especially chemistry, commonly presents students with pluralism in terms of the models and representations used in teaching. So electrons may be located in shells or in orbitals—or even outside those orbitals when they are understood as probability envelopes—or as being diffuse clouds; solids may be hard and incompressible because they are composed of particles in contact—but those same solids may be subject to thermal expansion and contraction due to the variable amount of space between the particles from which they are composed. It is assumed that students will have the sophistication to appreciate that this pluralism of models and representations sometimes reflects limitations of knowledge, and more often the challenges of expressing nature in ways we can easily comprehend and visualise, rather than being a realistic account of nature itself. Yet, this is something that needs to be taught and is unlikely to simply be intuited (Taber, 2010).

A young person who moves beyond multiplism comes to appreciate that even if there cannot be absolute certainty, it is still possible to critically evaluate ideas and make choices between alternatives. Good scientific practice includes being self-critical, always looking for alternative explanations, never prejudging results, identifying weaknesses in positions adopted, being open to revisit conclusions in the light of new evidence or conceptualisations, and so forth: but also, ultimately, in making judgements about the extent to which the best available interpretation of the evidence supports mooted hypotheses. This allows the positing of provisional knowledge that is seen as the best currently available way of making sense of some aspect of nature—and evaluating how robust and refined it seems to be.

If school-age students are working at different levels of epistemological understanding then this has consequences for how they make sense of the science they are taught. One interview study of 13–14-year olds suggested most of those participating had a naive view of the epistemic basis of scientific knowledge—often little more than someone having a hunch that could be tested and shown to either be true or false (i.e. an absolutist stance). So, theories were not considered substantially different in nature from hypotheses and were seen as uncertain simply because the necessary determination had not yet been made:

> there was limited evidence that these students saw scientific knowledge as existing on a continuum that allowed continuous variation (and change) in the extent to which ideas might be considered as reliable scientific knowledge as, over time, different evidence is collected, critiqued, checked, compared etc. Rather, these secondary students tended to think scientists carried out experiments that prove a theory to be correct...or obviously wrong...The general impression was that theories were largely seen as yet-to-be-supported products of imagination, and that testing them was largely straightforward. (Taber, Billingsley, Riga, & Newdick, 2015, p. 390).

However, whilst these students were best understood as at the 'absolutist' stage, often the same students would adopt a multiplist position when asked about what they were taught in religious studies lessons—where different positions were seen as a matter of personal opinion or choice, and it was considered as inappropriate to critique someone else's convictions about religious or ethical issues. This suggests that individual learners may appear to be at different positions on schemes such as those of Perry (see Fig. 15.1) and Kuhn (see Fig. 15.2) when asked about different domains of knowledge.

Conclusion

Models necessarily simplify reality, but the general pattern identified by Perry, and reinforced in the work of Kuhn and others, seems to be robust. Perry acknowledged that individuals can regress, and (as Piaget found in his work on cognitive development) setting tasks in different domains of experience may lead to individuals appearing to operate at different levels. It is important to acknowledge that Perry's work has been subject to critique, in particular, that females were underrepresented in his sample—an issue later explored in the programme to elicit women's ways of knowing (Finster, 1989)—although later work at Wellesley College (an elite U.S. institution educating women) supported Perry's general findings (Ashton-Jones & Thomas, 1990).

Regardless of such caveats, this chapter has discussed a general pattern in the development of thinking that has great significance for science education. That is, there is a form of intellectual maturation which allows individuals to move from assuming statements can unproblematically be shown to be right or wrong, to accepting that the evaluation of an assertion may differ according to perspective—but where all arguable positions are considered of similar merit—to appreciating that, although knowledge is a human construction with a subjective element, it is often still possible to identify criteria that allow one to evaluate alternatives and make a rational and justifiable (if potentially fallible) choice between them.

At one level, this assertion about the development of intellectual sophistication could be seen as a potential restriction on science education, highlighting aspects of the curriculum that students may struggle to engage with. Alternatively, such a scheme may be seen as the basis for organising educational experiences (e.g. Finster, 1991) to support—and perhaps even accelerate—progression. For example, from a sociocultural perspective (e.g. see Chap. 19), awareness of this pattern of progression may suggest an important dimension for diagnostic assessment, which can then inform the extent to which teachers need to offer mediation to support learners in appreciating and adopting, and so slowly internalising, more mature epistemological stances.

Further Reading

Perry, W. G. (1970). *Forms of intellectual and ethical development in the college years: A scheme.* New York: Holt, Rinehart & Winston.
Taber, K. S. (2017). Beliefs and science education. In K. S. Taber & B. Akpan (Eds.), *Science education: An international course companion* (pp. 53–67). Rotterdam: Sense Publishers.

References

Arlin, P. K. (1975). Cognitive development in adulthood: A fifth stage? *Developmental Psychology, 11*(5), 602–606.
Ashton-Jones, E., & Thomas, D. K. (1990). Composition, collaboration, and women's ways of knowing: A conversation with Mary Belen. *Journal of Advanced Composition, 10*(2), 275–292.
Bloom, B. S. (1968). The cognitive domain. In L. H. Clark (Ed.), *Strategies and tactics in secondary school teaching: A book of readings* (pp. 49–55). London: MacMillan.
Bruner, J. S. (1960). *The process of education.* New York: Vintage Books.
Finster, D. C. (1989). Developmental instruction: Part 1. Perry's model of intellectual development. *Journal of Chemical Education, 66*(8), 659–661.
Finster, D. C. (1991). Developmental instruction: Part 2. Application of Perry's model to general chemistry. *Journal of Chemical Education, 68*(9), 752–756.
Flynn, J. R. (1987). Massive IQ gains in 14 nations: What IQ tests really measure. *Psychological Bulletin, 101*(2), 171.
Kohlberg, L. (1973). Stages and aging in moral development—Some speculations. *The Gerontologist, 13*(4), 497–502. https://doi.org/10.1093/geront/13.4.497.
Kramer, D. A. (1983). Post-formal operations? A need for further conceptualization. *Human Development, 26,* 91–105.

Krathwohl, D. R., Bloom, B. S., & Masia, B. B. (1968). The affective domain. In L. H. Clark (Ed.), *Strategies and tactics in secondary school teaching: A book of readings* (pp. 41–49). New York: The Macmillan Company.

Kuhn, D. (1999). A developmental model of critical thinking. *Educational Researcher, 28*(2), 16–46.

Kuhn, D., & Pearsall, S. (2000). Developmental origins of scientific thinking. *Journal of Cognition and Development, 1*(1), 113–129. https://doi.org/10.1207/S15327647JCD0101N_11.

Luria, A. R. (1976). *Cognitive development: Its cultural and social foundations.* Cambridge, MA: Harvard University Press.

Medawar, P. B. (1963/1990). Is the scientific paper a fraud? In P. B. Medawar (Ed.), *The threat and the glory* (pp. 228–233). New York: Harper Collins, 1990 (Reprinted from: The Listener, Volume 70: 12th September, 1963).

Mitchell, S. D. (2003). *Biological complexity and integrative pluralism.* Cambridge: Cambridge University Press.

Niaz, M., & Rodriguez, M. A. (2000). Teaching chemistry as a rhetoric of conclusions or heuristic principles—A history and philosophy of science perspective. *Chemistry Education: Research and Practice in Europe, 1*(3), 315–322.

Perry, W. G. (1970). *Forms of intellectual and ethical development in the college years: A scheme.* New York: Holt, Rinehart & Winston.

Perry, W. G. (1985). Different worlds in the same classroom: Students' evolution in their vision of knowledge and their expectations of teachers. *On Teaching and Learning, 1*(1), 1–17.

Piaget, J. (1970/1972). *The principles of genetic epistemology* (W. Mays, Trans.). London: Routledge & Kegan Paul.

Riga, F., Winterbottom, M., Harris, E., & Newby, L. (2017). Inquiry-based science education. In K. S. Taber & B. Akpan (Eds.), *Science education: An international course companion* (pp. 247–261). Springer.

Shayer, M., & Adey, P. (1981). *Towards a science of science teaching: Cognitive development and curriculum demand.* Oxford: Heinemann Educational Books.

Taber, K. S. (2010). Straw men and false dichotomies: Overcoming philosophical confusion in chemical education. *Journal of Chemical Education, 87*(5), 552–558.

Taber, K. S. (2011). The natures of scientific thinking: Creativity as the handmaiden to logic in the development of public and personal knowledge. In M. S. Khine (Ed.), *Advances in the nature of science research—Concepts and methodologies* (pp. 51–74). Dordrecht: Springer.

Taber, K. S. (2013). *Modelling learners and learning in science education: Developing representations of concepts, conceptual structure and conceptual change to inform teaching and research.* Dordrecht: Springer.

Taber, K. S. (2015). Affect and meeting the needs of the gifted chemistry learner: Providing intellectual challenge to engage students in enjoyable learning. In M. Kahveci & M. Orgill (Eds.), *Affective dimensions in chemistry education* (pp. 133–158). Berlin Heidelberg: Springer.

Taber, K. S. (2017). Reflecting the nature of science in science education. In K. S. Taber & B. Akpan (Eds.), *Science education: An international course companion* (pp. 23–37). Rotterdam: Sense Publishers.

Taber, K. S., Billingsley, B., Riga, F., & Newdick, H. (2015). English secondary students' thinking about the status of scientific theories: Consistent, comprehensive, coherent and extensively evidenced explanations of aspects of the natural world—or just 'an idea someone has'. *The Curriculum Journal, 26*(3), 370–403. https://doi.org/10.1080/09585176.2015.1043926.

Wellman, H. M. (2011). Developing a theory of mind. In U. Goswami (Ed.), *The Wiley-Blackwell handbook of childhood cognitive development* (2nd ed., pp. 258–284). Chichester, West Sussex: Wiley-Blackwell.

Zeidler, D. L. (2014). Socioscientific issues as a curriculum emphasis: Theory, research, and practice. In N. G. Lederman & S. K. Abell (Eds.), *Handbook of research on science education* (Vol. 2, pp. 697–726). New York: Routledge.

Keith S. Taber is the Professor of Science Education at the University of Cambridge. Keith trained as a graduate teacher of chemistry and physics, and taught sciences in comprehensive secondary schools and a further education college in England. He joined the Faculty of Education at Cambridge in 1999 to work in initial teacher education. Since 2010, he has mostly worked with research students, teaching educational research methods and supervising student projects. He was until recently the Lead Editor of the Royal Society of Chemistry journal 'Chemistry Education Research and Practice', and is Editor-in-Chief of the book series 'RSC Advances in Chemistry Education'. His main research interests relate to conceptual learning in the sciences, including conceptual development and integration. He is interested in how students understand both scientific concepts and scientific values and processes.

Part IV
Constructivist Theories

Chapter 16
Pragmatism—John Dewey

Fran Riga

Introduction

John Dewey is the world-renowned American philosopher, psychologist, and social reformer who had a profound influence on educational practice and research throughout the twentieth century—and continues to do so to this day. This chapter focuses on Dewey's ideas relating to pragmatism, or as he referred to it, *instrumentalism*. The chapter begins by examining the origins of pragmatism and particularly the meaning that Dewey attached to the term. Dewey's theory of pragmatism (with its emphasis on *inquiry*) is then set out within the contexts of philosophy and education, with special emphasis on its relevance to science education.

Dewey and the Origins of Pragmatism

Pragmatism is a term that has perplexed educators and philosophers alike for some time and will, in all likelihood, continue to do so in the foreseeable future. Owing to its relatively recent appearance on the philosophical stage, a clear-cut, all-encompassing definition of the term is proving elusive, although Talisse & Aikin (2008) suggest that this need not necessarily be viewed in a negative light, saying that pragmatism is 'a *living philosophy* rather than a historical relic' (p. 3). That is to say, it is still evolving.

It is now widely accepted that pragmatism is a school of thought whose origins may be found in the works of Charles Sanders Peirce in the late 1800s. Peirce's ideas were then taken up, interpreted in various ways, extended, and popularized, in the first half of the twentieth century by its earliest protagonists—William James and

F. Riga (✉)
Faculty of Education, University of Cambridge, Cambridge, UK
e-mail: fr223@cam.ac.uk

© Springer Nature Switzerland AG 2020

B. Akpan and T. Kennedy (eds.), *Science Education in Theory and Practice*,
Springer Texts in Education, https://doi.org/10.1007/978-3-030-43620-9_16

John Dewey. It has therefore been hailed as 'the first truly *American* philosophical movement' (Biesta & Burbules, 2003, p. 4). Although Dewey would probably not dispute this, he argues that American thought is really a continuation of European thought, contending that European ideas (like the American language, laws, institutions, morals, and religion) were 'imported' from Europe, *but* then re-adapted to fit American life and conditions. He describes the pragmatic movement as an attempt at 're-adaptation' (to American life) and suggests that 'the practical element ... found in all phases of American life' is the reason why pragmatism (or, 'instrumentalism', as he calls it) places great importance on 'the teleological phase of thought and knowledge' (Dewey, 1925, in Hickman & Alexander, 1998a, pp. 11–12). Dewey insists on action being intelligent and reflective, with thought being the cornerstone of life. Consequently, pragmatism moves the *individual* into centre stage—however, this is 'an individual who evolves and develops in a natural and human environment' (Dewey, 1925, in Hickman & Alexander, 1998a, p. 12).

> It is beyond doubt that the progressive and unstable character of American life and civilization has facilitated the birth of a philosophy which regards the world as being in continuous formation, where there is still place for indeterminism, for the new and for a real future (Dewey, 1925, in Hickman & Alexander, 1998a, p. 12).

Hence, although pragmatism is chronicled as being 'rooted in the Western philosophical tradition', it is unique in insisting that 'philosophy should take the methods and insights of modern science into account'. Dewey, in particular, emphasized that modern science's experimental method should serve as 'a model for human problem-solving and the acquisition of knowledge' (Biesta & Burbules, 2003, p. 5). Dewey describes Peirce as 'an empiricist ... an experimentalist ... a man whose intelligence is formed in the laboratory' (Dewey, 1925, in Hickman & Alexander, 1998a, pp. 3–4), where *experimentalists* may be thought of as individuals who believe that all their knowledge can be arrived at through scientific experiments. This led to the coining of yet another name by which pragmatism is known, i.e. *experimentalism*. To avoid confusion, the term pragmatism will be used throughout this chapter rather than *experimentalism* or *instrumentalism*.

Returning now to the origin of the term, Peirce's conceptualization of pragmatism may be found in the following extract:

> To develop [a thought's] meaning, we have simply to determine what habits it produces, for what a thing means is simply what habits it involves (Peirce, quoted in Talisse & Aikin, 2008, p. 9).

Talisse & Aikin (2008, p. 9) go on to explain:

> By 'habit' Peirce means a standard course of action undertaken in response to specific conditions. For any thought, then, one may extract its complete meaning by drawing out the proposals for action that it suggests.

This thesis has become identified as the *pragmatic maxim*—or in Peirce's words:

> Consider what effects, that might conceivably have practical bearings, we conceive the object of our conception to have. Then our conception of these effects is the whole of our conception of the object (Peirce, quoted in Talisse & Aikin, 2008, pp. 9–10).

That is to say, to be able to understand the meaning of a concept, one must be able to *apply* the concept to fulfil some purpose or action (in real life). One needs to see how the concept becomes enacted in real life, because it is through the modification of one's behaviour—in response to its application for a purpose—that the true meaning of the concept may be known. Dewey concurs, saying that '[I]n order to be able to attribute a meaning to concepts, one must be able to apply them to existence [i.e. human conduct]' (Dewey, 1925, in Hickman & Alexander, 1998a, p. 4).

In addition to conceiving the idea for a new philosophical movement, Peirce was also keen to ally pragmatism with the concept of *inquiry* (or what both he and Dewey referred to as 'logic'). Dewey, too, perceived inquiry as being crucial, although their interpretations of the term were not identical. Peirce viewed inquiry as the 'struggle' that ensues when changing from a position of 'doubt' to a position of 'belief', where 'belief' (for Peirce) was "a state which 'guide[s] our desires and shape[s] our actions'" (Peirce, quoted in Talisse & Aikin, 2008, p. 17). True to the *pragmatic maxim*, he saw *inquiry* as establishing in one's nature a habit that would determine one's actions. Doubt, on the other hand, represented a disturbance of the harmony of one's actions. He viewed doubt was an 'uneasy state from which we struggle to free ourselves' (Peirce, quoted in Talisse & Aikin, 2008, p. 18). In fact, Peirce saw the two as going together—he perceived inquiry as the *reliever* of doubt, and whose purpose was to remove doubt and so arrive at belief. For Peirce, any *process* which resulted in change from a state of doubt to a state of belief *was inquiry*, and he argued for the method of science as being the only way doubt could be transformed into belief. For him, the *only* purpose for undertaking inquiry was for 'the settlement of opinion', in other words, to eliminate doubt and so arrive at belief (Talisse & Aikin, 2008, p. 18).

> Inquiry, when properly conducted, is the process of attempting to arrive at a belief that would never occasion doubt, a belief that would not give rise to recalcitrant experiences (Talisse & Aikin, 2008, p. 20).

Dewey perceived inquiry slightly differently. Dewey defines inquiry as follows:

> Inquiry is the controlled or directed transformation of an indeterminate situation into one that is so determinate in its constituent distinctions and relations as to convert the elements of the original situation into a unified 'Whole' (Dewey, 1938b, pp. 104–105. Italics are Original).

Dewey replaces Peirce's keyword 'opinion'—which is associated with *people*—with 'situation'.

> What is designated by the word "situation" is *not* a single object or event or set of events. For we never experience nor form judgments about objects and events in isolation, but only in connection with a contextual whole. This latter is what is called a "situation". ... In actual experience, there is never any such isolated singular object or event; *an* object or event is always a special part, phase, or aspect, of an environing experienced world—a situation (Dewey, 1938b, pp. 66 and 67).

Like Peirce, Dewey sees inquiry in terms of a reaction to a particular experience, which both he and Peirce call 'doubt'. However, according to Dewey, it is not we (as persons) that are doubtful, but *situations*—*situations* have traits such as 'disturbed, troubled, ambiguous, confused, full of conflicting tendencies, obscure, etc.', *not individuals*. He writes '[W]e are doubtful because the situation is inherently doubtful' (Dewey, 1938b, pp. 105–106). So, Dewey's conception of inquiry incorporates the notion of a 'situation' (which could be either 'determinate' or 'indeterminate'). Hence, inquiry is a response to one's experience of doubt, and doubt is attributed to situations (not people). Doubt is an *experience*, but Dewey's perception of experience went beyond the traditional notion of this concept—which had viewed experience as 'receiving impressions of external objects through the sense organs' (Talisse & Aikin, 2008, p. 22) and being an antithetical term to thought (Dewey, 1917, in Hickman & Alexander, 1998a, p. 48). To Dewey, experience was essentially 'a process of undergoing; a process of standing something; of suffering and passion, of affection, in the literal sense of these words' (Dewey, 1917, in Hickman & Alexander, 1998a, p. 49). And, the process of 'undergoing' is not passive. It involves action. Moreover, experience is experimental—endeavouring to change what is given. Hence experience is 'a future implicated in the present!'. In other words, experience involves an adjustment of the organism to the environment where 'every step in the process is conditioned by reference to further changes which it effects' (Dewey, 1917, in Hickman & Alexander, 1998a, p. 49). Dewey argued that experience is 'assuredly … an affair of the intercourse of a living being with its physical and social environment' (Dewey, 1917, in Hickman & Alexander, 1998a, p. 47).

According to Dewey, an indeterminate or 'problematic situation' (such as doubt) would be one in which the constituent elements of the situation are not in harmony, and a determinate situation (such as belief) would be one in which the constituent elements are in harmony—keeping in mind that a situation involves an interacting blend of biological and social environments. The whole purpose of inquiry for Dewey is to construct a new 'situation' that did not exist prior to engaging in the inquiry.

Dewey's Pragmatism—Action, Meaning and Knowledge

Much has been made of pragmatism's preoccupation with action. In keeping with Peirce's ideas, pragmatism has been described as *a theory of action* (Miettinen, 2006), primarily concerned with the ability to understand and/or clarify concepts (Dewey, 1925, in Hickman & Alexander, 1998a). Understanding, which precedes knowledge, involves the capacity to attribute meaning to things/concepts. However, to do so one needs to 'be able to apply them [concepts] to existence', and according to Dewey, 'it is by means of action that this application is made possible' (Dewey, 1925, in Hickman & Alexander, 1998a, p. 4). This means that to fully understand a particular concept, one would need to be able to apply it (the concept) to fulfil some purpose or action or human conduct—where 'action' may be understood to be 'the means by which a problematic situation is resolved' (Dewey, 1929a, p. 244).

To Dewey, knowledge, too, comes about as a result of the processes which 'transform a problematic situation into a resolved one' (Dewey, 1929a, p. 242). He perceives knowledge as an action. Dewey (1929a) proposes the idea that interaction is 'a universal trait of natural existence' (p. 244), and action is that mode of the interaction which emanates from the organism. Knowing is not something which is imposed from *outside* (which he refers to as 'spectator theory of knowing'), but is something that occurs *within* nature (from the organism)—'an act which modifies what previously existed' (p. 245).

Hence, for Dewey, pragmatism tries to make sense of a concept by seeing how the concept is enacted in real life. Such enactment (or action) will, in due course, produce some modification in the conduct of one's life, and such modification—resulting from this application—is what constitutes the true meaning of the concept (i.e. knowledge of the concept is continually being appropriated by the organism).

In short, pragmatism insists on the 'necessity of human conduct and the fulfilment of some aim in order to clarify thought' (Dewey, 1925, in Hickman & Alexander, 1998a, p. 4). Moreover, the greater the range of applications of the concept to human conduct/life, the more the meaning of the concept/term can be generalized.

> [T]he rational purport of a word or other expression, lies exclusively in its conceivable bearing upon the conduct of life; so that, since obviously nothing that might not result from experiment can have any direct bearing upon conduct, if one can define accurately all the conceivable experimental phenomena which the affirmation or denial of a concept could imply, one will have therein a complete definition of the concept (Peirce, quoted in Dewey, 1925, in Hickman & Alexander, 1998a, p. 4).

By continuously interacting with our environment and trying to achieve a state of harmony with it (our environment), we acquire certain patterns of probable behaviour, called *habits*. These habits are unique to each individual, and cannot develop through repetition—they form the basis of meaning (which is a precursor to knowledge). Dewey perceived meaning as 'the way in which the organism responds to the environment', i.e. it is 'a property of behaviour' (Biesta & Burbules, 2003, p. 36; Dewey, 1929b, p. 179)—habits can be thought of as individual configurations of objects/events that make meaning possible.

Dewey believed that the experimental method of modern science served as a model both for problem solving and for the acquirement of meaning/knowledge (Biesta & Burbules, 2003, p. 5). Dewey gave much thought to knowledge and how it might be acquired through the philosophical 'lens' of *action*. He rejected Descartes' assumption of the dualism of mind and matter, i.e. that reality consists of two separate types of 'stuff'—mind and matter (Biesta & Burbules, 2003, p. 5). In attempting to address the dilemma of how an immaterial thing (such as the mind) can apprehend knowledge of a material world (composed of matter), Dewey's starting point, quite different from Descartes' dualism of mind and matter, looked to the interactions (or 'transactions') occurring in nature, where nature is perceived as 'a moving whole of interacting parts' (Dewey, 1929a, p. 291). Given the constantly changing events/interactions in nature, he could not view the structure of knowledge as a static and unchanging system. Hence, Dewey believed that knowledge could only be apprehended and progressed through *action*—'knowing goes forward by means of doing' (Dewey,1929a, p. 290).

For Dewey, it was through the process of conducting scientific inquiry that the conflict between knowing and doing, and, theory and practice, could be discarded.

The old notion that one comes to 'know' things inside oneself and where one assumes 'a definite separation between the world in which man thinks and knows, and, the world in which he lives and acts'—is abandoned in favour of Pierce's new idea, *pragmatism*, which sees 'indefinite interactions taking place within a course of nature which is not fixed and complete, but which is capable of direction to new and different results through the mediation of intentional operations' (Dewey,1929a, p. 291). Dewey described this 'change in the method of knowing' as a revolution in people's attitude towards 'natural occurrences and their interactions' (Dewey,1929a, p. 85). He saw this transformation in attitude towards the traditional relationship between knowledge and action as representing a Copernican-type *revolution*. He referred to it as a 'shift from knowing which makes a difference to the knower but none in the world, to knowing which is a directed change within the world' (Dewey, 1929a, p. 291).

> There is no practical point gained in asserting that a thing is what it is experienced to be apart from knowledge … Knowledge is instrumental (Dewey, 1929a, p. 294, 298).

For Dewey, there appear to be three ways or 'actions' through which one arrives at knowledge:

- experience/action (especially physical),
- thought,
- communication.

Experience. 'Experience' once meant the results accumulated in memory of a variety of past doings and undergoings that were had without control by insight, when the net accumulation was found to be practically available in dealing with present situations (Dewey, 1929a, p. 81).

Experience can be thought of as an action through which knowledge may be acquired. It does not occur 'in a vacuum', but is an action which is social in nature as it involves both 'contact and communication' (Dewey, 1938a, p. 40, 38).

Dewey sets out two principles that guide the interpretation of experience—*continuity* and *interaction*—asserting that these principles are not mutually exclusive but 'intercept and unite' (Dewey, 1938a, p. 44). *Continuity* of experience 'means that every experience both takes up something from those which have gone before and modifies in some way the quality of those which come after' (p. 35)—an underlying idea that permeates pragmatism. In large measure, the world in which we live *is* the way it is because of things, events and experiences that have gone before (p. 39). *Interaction* involves the interplay between what goes on inside a person's body and mind on the one hand, and, the external conditions which affect experiences on the other. As already indicated, Dewey calls the interaction between these two a 'situation' (p. 42), and living in the world implies that individuals live in a 'series of situations' (p. 43).

> An experience is always what it is because of a transaction taking place between an individual and what, at the time, constitutes his environment … the environment … is whatever conditions interact with personal needs, desires, purposes, and capacities to create the experience which is had (Dewey, 1938a, p. 43–44).

Thought. Understanding concepts (the precursor to thinking) involves thinking, and thinking is the mechanism/action through which ideas develop. In *How We Think* (1910, 1933), Dewey describes thinking as the 'procession of mental states' through the mind (1933, p. 4), and suggests that thinking processes can be interpreted in terms of four 'senses' or meanings:

(1) Thinking, 'in its loosest sense', is that 'uncontrolled coursing of ideas through our heads' (1933, p. 2, 4), i.e. the kind of thinking that we engage in all the time in our conscious and our sleeping (or dreaming) hours.

(2) In its second sense, thinking is restricted to the things we cannot perceive directly through our senses—those sorts of narratives that children frequently blurt out, possibly on a wave of emotion, and whose main objective is a well-structured (possibly exciting!) plot. However, the aim of such narratives is *not* 'belief about facts or... truths', but rather 'successions of imaginative incidents and episodes which, having a certain coherence, hanging together on a continuous thread, lie between kaleidoscopic flights of fancy and considerations deliberately employed to establish a conclusion' (1933, p. 3).

(3) In a third sense, thinking may be characterized by beliefs that are easily accepted, but *without* the individual having properly considered any grounds/evidence for their acceptance—moreover, very often individuals are not even consciously aware of how or where these thoughts were picked up.

(4) In its final sense, thinking is ascribed to those thought processes which Dewey refers to as 'reflective thinking'—in his view, this is the highest category of thought process. Accordingly, Dewey defines reflective thinking as follows.

Reflection involves not simply a sequence of ideas, but a consequence—a consecutive ordering in such a way that each determines the next as its proper outcome, while each outcome in turn leans back on, or refers to, its predecessors. The successive portions of a reflective thought grow out of one another and support one another; they do not come and go in a medley. Each phase is a step from something to something—technically speaking, it is a term of thought. Each term leaves a deposit that is utilized in the next term. The stream or flow becomes a train or chain. There are in any reflective thought definite units that are linked together so that there is a sustained movement to a common end (Dewey, 1933, p. 4–5).

Dewey's description of reflective thinking resonates with his account of the *continuity* principle of experience (discussed earlier)—both rely on what went before to influence change in what follows—again, an idea that permeates pragmatism.

Dewey considers thinking as a kind of 'inner experimentation' (Dewey, 1929b, p 166), with thinking being another way in which knowledge may be acquired. In other words, he perceives thinking as a way of enacting different scenarios without having to (physically) 'suffer' the consequences of these scenarios—ultimately however, to check the applicability of a particular scenario, one would have to en*act* it in 'real life' conditions (Biesta & Burbules, 2003).

Communication. Dewey maintained that discourse was both instrumental and consummatory. For him, communication was: 'an exchange which procures something wanted; it involves a claim, appeal, order, direction or request, which realizes want at less cost than personal labour exacts, since it procures the cooperative assistance of others' (Dewey, 1929b, p. 183). When individuals communicate with one another, 'all natural events are subject to reconsideration and revision'—they are adjusted according to the needs of the conversation, whether the consequence of the conversation is further conversation (i.e. 'public discourse'), or whether the consequence of the conversation is internal dialogue (i.e. 'thinking') (Dewey, 1929b, p. 166). Resonating with both *experience* and *thinking*, *communication* involves considering something which went before (e.g. in the form of an idea from another person), and then possibly adding to or modifying this, according to the requirements of the conversation—for the purpose of sustaining 'movement to a common end', i.e. to progress knowledge (Dewey, 1933, p. 4–5). Hence, when individuals communicate with one another to achieve a common end, not only are their individual views, approaches and habits modified to produce (as a consequence) an integrated and harmonized response, but, through this process, their individual worlds are also transformed (Biesta & Burbules, 2003, p. 12).

> Where communication exists, things in acquiring meaning, thereby acquire representatives, surrogates, signs and implicates, which are infinitely more amenable to management, more permanent and more accommodating, than events in their first estate (Dewey, 1929b, p. 167).

> The heart of language is … is communication; the establishment of cooperation in an activity in which there are partners, and in which the activity of each is modified and regulated by partnership (Dewey, 1929b, p. 179).

To sum up. Experience, thought and communication are seen as ways in which meaning is activated. Through the processes of experiencing, thinking and communicating, we continuously interact with our environment—we change it, are changed by it, and come to know things (either for the first time or more deeply).

> In the act of knowing—and hence in research—both the knower and what is to be known are changed by the transaction between them (Biesta & Burbules, 2003, p. 12).

Dewey's Pragmatism and Science Education

Dewey was one of the earliest leading figures who advocated for activity-based, hands-on approaches to teaching and learning—what is now encapsulated in an approach to teaching often referred to as *inquiry-based education*. Viewing experience as an action through which concepts/things may become known, Dewey observed 'no experience is educative that does not tend both to knowledge of more facts and entertaining of more ideas and to a better, a more orderly, arrangement of them' (Dewey, 1938a, p. 82). Essentially, activity-based approaches try to engage students in situations that are both appealing and tap into their curiosity (Prawat, 2000). Although Dewey believed in *teachers* taking the lead in creating experiences that would be 'educative', he did not view the teacher as simply the 'designer of problematic environments' (with complete control of what they wanted students to learn), but rather as "an intellectual leader—a person who can get the class, as a 'social unity', interested and excited about ideas" (Prawat, 2000, p. 810). The teacher is seen as the instigator of situations and experiences that inspire students to immerse themselves in a process that leads to constructing knowledge.

For Dewey, inquiry is a particular type of experience. It is the process through which a belief that has become problematic is scrutinized and resolved by taking action—'It is a process of making choices by asking and answering questions, in which those questions concern the likely outcomes of applying current beliefs to future action' (Morgan, 2014, p. 1047). Morgan describes Dewey's approach to inquiry as involving the following five steps:

1. Recognizing a situation as problematic;
2. Considering the difference it makes to define the problem one way rather than another;
3. Developing a possible line of action as a response to the problem;
4. Evaluating potential actions in terms of their likely consequences;
5. Taking actions that are felt to be likely to address the problematic situation; Morgan, (2014), p. 1047.

Moreover, Dewey's notion of inquiry as an experience is inextricably linked with knowledge. 'Through the inquiry process *knowledge* becomes a verb'—i.e. the process of knowing (rather than the noun 'knowledge'). So, knowledge assumes a different meaning from the traditional view where it is seen as an objective *thing*—a 'bank' of information to be 'acquired, transmitted, and maintained' (Breault, 2014, p. 190). Knowledge, for Dewey, is a far more fluid, dynamic process—knowing is a process that is *continuous*, and *contingent upon* experience.

In particular, Dewey argued that *science* should be presented to students in ways that would kindle their curiosity and stimulate their thinking, rather than as the transmission of unchanging facts from teacher to student (with students having to commit these facts to memory). Over the last fifty years or so, a resurgence of Dewey's ideas has been taking place—advocating for science to be taught as an 'effective method of inquiry' (Dewey, 1910, p. 124), which incorporates his pragmatist view of science as a subject with laws and concepts that are constantly open to inspection, challenge and revision (Riga, Winterbottom, Harris, & Newby, 2016). This has coincided (not coincidentally!) with the advent of constructivist theories of education which challenged behaviourist ideas that, up to that time, had viewed learning as something imposed externally by an educator (and which used *conditioning* in order to promote favourable behaviours and discourage unfavourable ones). Dewey asserted that individuals *themselves* constructed their own knowledge through experience, thinking and communication. Hence, nowadays, Dewey would be described as having a social constructivist approach to teaching and learning.

In more recent years, Dewey's ideas have had a considerable impact on the development of what could be termed the 'inquiry-based science education movement', which has seen the development of curricula and guides to teaching and learning science—such as the *National Science Education Standards* (1996) and *Inquiry and the National Science Education Standards* (2000). In these National Research Council

(NRC) publications, inquiry is portrayed 'as an approach to teaching science which engages students in the same sorts of activities, practices, and thinking processes that scientists use in their work, i.e. in their pursuit of scientific inquiry' (Riga et al, 2016, p. 250). Dewey's notion of inquiring into 'authentic questions' which arise from students' prior experiences is presented as the key strategy for teaching science (NRC, 1996, p. 31)—a standpoint founded on constructivist principles.

> A key ingredient in inquiry-based approaches in education has been getting students to think for themselves—both while working together collaboratively in groups and when working alone. How proficient a student becomes in the practice of thinking independently is what will ultimately determine whether s/he will be a 'disciple' or an 'inquirer' (Dewey, 1980) (Riga et al, 2016, p. 259).

The NRC presents the core components of *inquiry* as follows:

> Inquiry is a multifaceted activity that involves making observations; posing questions; examining books and other sources of information to see what is already known; planning investigations; reviewing what is already known in light of experimental evidence; using tools to gather, analyse, and interpret data; proposing answers, explanations, and predictions; and communicating the results (National Research Council, 1996, p. 23).

The extent of autonomy and direction which teachers might give their students when pursuing inquiry-based work may vary considerably from doing highly structured 'recipe-style' activities/tasks at one end of the spectrum, to undertaking open-ended project-work at the other end (where students would enjoy complete autonomy) (For more detailed discussion of inquiry-based approaches to science teaching and learning, refer to Riga et al, 2016).

Concluding Thoughts

In closing, it is important to draw attention to a common misconception associated with Dewey's beliefs regarding pragmatism. Dewey's strong emphasis on practical action, i.e. what he called the 'method of science' as a way of resolving problematic situations, has led some authors to claim that Dewey believed that all knowledge must be firmly rooted in reality and that the natural sciences are the only route to apprehending this reality (and hence knowledge). Biesta and Burbules strongly argue against this supposition recalling that Dewey thought it one of the greatest mistakes of modern philosophy to equate what we *know* to what is *real* (Biesta & Burbules, 2003). In Dewey's own words,

> Physical inquiry has been taken as typical of the nature of knowing. The selection is justified because the operations of physical knowledge are so perfected and its scheme of symbols so well devised. But it would be misinterpreted if it were taken to mean that science is the only valid kind of knowledge" (Dewey, 1929a, p. 250–251).

Dewey's 'pragmatic' ideas have frequently been characterized as being 'extremely challenging to implement', possibly because of the influence that a number of factors (such as social conditions, educational research, the experiences and interests of teachers, and standardized assessment procedures) have had on education—or, as Dewey calls it, the *educative process*. For Dewey, it is *not* the practitioner's values and practices that will determine the *educative process*, but the other way around—it is the *educative process* that will determine one's educational values and practices (Hickman, 2014).

> The norms of education cannot be conveyed in "cookbook" fashion... They emerge, instead, as teachers interact with students in the process of learning. Education—the educative process—is autonomous, Dewey reminds us. In a robust democracy, it must be free to determine its own ends (Hickman, 2014, p. 208).

Further Reading

Biesta, G. J. (1994). Education as practical intersubjectivity: Towards a critical-pragmatic understanding of education. *Educational Theory, 44*(3), 299–317.

Biesta, G. J. J. (1995). Pragmatism as a pedagogy of communicative action. In J. Garrison (Ed.), *The new scholarship on John Dewey*. Dordrecht: Kluwer Academic.

Committee on a Conceptual Framework for New K-12 Science Education Standards. (2012). *A framework for K-12 science education: Practices, crosscutting concepts, and core ideas*. Washington: The National Academies Press. http://nap.edu/13165.

Dewey, J. (1908). What does pragmatism mean by practical? *The Journal of Philosophy, Psychology and Scientific Methods, 5*(4), 85–99.

Eames, S. M. (2003). *Experience and value: Essays on John Dewey and pragmatic naturalism*. In E. R. Eames & R. W. Field (Eds.). Carbondale: Southern Illinois University Press.

Khasawneh, O. M., Miqdadi, R. M., & Hijazi, A. Y. (2014). Implementing pragmatism and John Dewey's educational philosophy in Jordanian public schools. *Journal of International Education Research, 10*(1), 37.

Next Generation Science Standards. (2013). https://www.nextgenscience.org/.

Sleeper, R. W. (2001). *The necessity of pragmatism: John Dewey's conception of philosophy*. Urbana and Chicago: University of Illinois Press.

Vanderstraeten, R., & Biesta, G. (2006). How is education possible? Pragmatism, communication and the social organisation of education. *British Journal of Educational Studies, 54*(2), 160–174.

References

Biesta, G., & Burbules, N. C. (2003). *Pragmatism and educational research*. Lanham: Rowman & Littlefield.

Breault, D. A. (2014). Inquiry and education: a way of seeing the world. In D. A. Breault & R. Breault (Eds.), *Experiencing Dewey. Insights for today's classroom*. (2nd ed., pp. 189–191). New York: Routledge.

Dewey, J. (1910). *How we think*. Mineola, New York: Dover Publications Inc.

Dewey, J. (1917). The need for a recovery of philosophy. In L. A. Hickman & T. M. Alexander (Eds.), *The essential Dewey. Volume 1. Pragmatism, education, democracy*. Bloomington and Indianapolis: Indiana University Press.

Dewey, J. (1925). The development of American pragmatism. In L. A. Hickman & T. M. Alexander (Eds.), *The essential Dewey. Volume 1. Pragmatism, education, democracy*. Bloomington and Indianapolis: Indiana University Press.

Dewey, J. (1929a). *The quest for certainty: A study of the relation of knowledge and action*. New York: Minton, Balch and Company.

Dewey, J. (1929b). *Experience and nature*. London: George Allen and Unwin LTD.

Dewey, J. (1933). *How we think: A restatement of the relation of reflective thinking to the educative process*. Boston: D.C. Heath & Co.

Dewey, J. (1938a). *Experience and education* (The Kappa Delta Pi Lecture Series). New York: Collier Macmillan Publishers.

Dewey, J. (1938b). *Logic. The theory of inquiry*. New York. Henry Holt and Company.

Dewey, J. (1980). Democracy and education. In J. A. Boydston (Ed.), *John Dewey: The middle works, 1899–1924* (vol 9, 1916, pp. 1–370). Carbondale: Southern Illinois University Press.

Hickman, L. A., & Alexander, T. M. (Eds.). (1998a). *The essential Dewey. Volume 1. Pragmatism, education, democracy*. Bloomington and Indianapolis: Indiana University Press.

Hickman, L. A., & Alexander, T. M. (Eds.). (1998b). *The essential Dewey. Volume 2. Ethics, logic, psychology*. Bloomington and Indianapolis: Indiana University Press.

Hickman, L. (2014). Autonomous education. Free to determine its own ends. In D. A. Breault & R. Breault (Eds.), *Experiencing Dewey. Insights for today's classroom* (pp. 207–208). New York: Routledge.

Miettinen, R. (2006). Epistemology of transformative material activity: John Dewey's pragmatism and cultural-historical activity theory. *Journal for the Theory of Social Behaviour, 36*(4), 389–408.

Morgan, D. L. (2014). Pragmatism as a paradigm for social research. *Qualitative Inquiry, 20*(8), 1045–1053.

National Research Council. (1996). *National science education standards*. Washington, DC: National Academy Press. Retrieved from http://www.nap.edu/catalog/4962/national-science-education-standards.

National Research Council. (2000). *Inquiry and the national science education standards: A guide for teaching and learning*. Washington, DC: National Academy Press.

NGSS Lead States. (2013). *Next generation science standards: For states, by states*. Washington, DC: The National Academies Press. Retrieved from: https://www.nextgenscience.org/.

Prawat, R. S. (2000). The two faces of deweyan pragmatism: Inductionism versus social constructivism. *Teachers College Record, 102*(4), 805–840.

Riga, F., Winterbottom, M., Harris, E., & Newby, L. (2016). Inquiry-based science education. In K. S. Taber & B. Akpan (Eds.), *Science education: An international course companion*. Rotterdam: Sense Publishers.

Talisse, R. B., & Aikin, S. F. (2008). *Pragmatism: A guide for the perplexed*. London, UK: Continuum.

Dr. Fran Riga is a teaching and research associate in the Faculty of Education at the University of Cambridge, where she has worked on a number of research projects in the following areas: science education for the gifted, trainee teachers' conceptualization of assessment, inquiry-based science education, dialogic approaches in secondary education, computer-based adaptive learning, and promoting science to 'science-disadvantaged' communities through science centres. She comes from a background of teaching science and mathematics in secondary schools, and in her Ph.D. research she investigated secondary students' conceptual development in astronomy topics.

Chapter 17
Experiential Learning—David A. Kolb

Louise Lehane

Introduction

This chapter presents Kolb's Experiential Learning Theory (ELT) with particular emphasis on how the theory can translate into practice in the science classroom. It will look at connecting Kolb's ELT with, among others, the Piagetian theory of learning through lived experiences in order to understand the cognitive processes which in turn allow for meaningful learning.

It is firstly important to note that experience is subjective which suggests the individual nature of how students learn. It can be argued that this subjectivity is often forgotten in how we, as teachers, teacher educators, etc., plan our lessons. Therefore, in a sense constructing a lesson plan results in almost uniform planning, regardless of the students' prior experience and their subjective reality. A teacher's pedagogical content knowledge (PCK) is therefore a critical consideration in the planning and delivery of lessons, in recognising and being able to adapt to the individual learner within a particular context. PCK is effectively the knowledge of content and how to teach that content within a particular context (Loughran, Mulhall, & Berry, 2006). Among others, two components of PCK are a teacher's knowledge of the students' understanding of science and knowledge of instructional strategies (Magnusson, Krajcik, & Borko, 1999). It is within these components that understanding of the unique learning experience, that is experiential learning, needs to be considered.

L. Lehane (✉)
St. Angela's College, Sligo, Ireland
e-mail: llehane@stangelas.nuigalway.ie

© Springer Nature Switzerland AG 2020
B. Akpan and T. Kennedy (eds.), *Science Education in Theory and Practice*,
Springer Texts in Education, https://doi.org/10.1007/978-3-030-43620-9_17

Kolb: Learning Through Experience

It is firstly important to note that the Experiential Learning Theory (ELT) is in effect a holistic model of the learning process (Kolb, Boyatzis, & Mainemelis, 2000). As the name suggests, the theory is focused on learning through experience which in effect distinguishes it from other learning theories (Kolb et al., 2000). When looking at others, they very much reflect two other theories of learning, the cognitive and behaviourist theories of learning. Experiential learning takes a further integrative perspective to consider learning that combines experience, perception, cognition and behaviour (Kolb, 1984). It looks at engaging students in an experience that will ultimately have real consequences. Instead of hearing from or reading the experiences of others, students make their own discoveries themselves, e.g. through online research, collaborative learning, etc. In science, this can also be done through experiencing experimental investigations, mirroring the practice of a scientist, which will be further discussed over the course of this chapter.

Origins of ELT

ELT draws on the work of many prominent scholars from the twentieth century in terms of how experience influences human learning and development. Kolb drew his ideas about the nature of experiential learning mainly from the works of Dewey, Lewin and Piaget. Collectively, through Dewey's pragmatic approach, Lewin's theories on social psychology and Piaget's theory of cognitive development, a unique perspective on teaching and learning is considered through the ELT (Kolb, 1984). All three theorists emphasise the traditions of experiential learning towards a life of purpose and self-directed learning as the guiding principle for education (Kolb, 1984). Kolb uses Lewin's tradition of action research and the laboratory method and Dewey's work on educational research to develop his model (Kolb, 1984). In relation to Piaget's theory of learning, it consisted of two major aspects, the process of coming to know and the stages humans move through as we acquire this ability. Piaget considered that learning should involve a process of constructing understanding through lived experiences (Kolb, 1984). In other words, humans can understand and process information more effectively if they have constructed that knowledge themselves. Piaget appreciated the importance of building constructs and internalising knowledge rather than accepting information as presented through rote learning. Piaget's theory of how we should construct knowledge therefore intimately links with Kolb's ELT.

Model of ELT

In effect, experiential learning is made up of three key stages: the planning stage, doing stage and reviewing stage. In its totality, experiential learning is governed by two processes; experience and reflection. For learning to be effective a person must have four different types of abilities which are depicted in the ELT model. The ELT model shows two dialectical perspectives on **grasping** and **transforming** experience (Kolb et al., 2000). There are two modes of **grasping** experience—Concrete Experience (CE) and Abstract Conceptualisation (AC) and two modes of **transforming** experience—Reflective Observation (RO) and Active Experimentation (AE) (Ibid). Learners must have all four of these abilities and must be able to choose which of these learning abilities bring to bear an understanding of the concept being learned (Ibid).

A distinctive variation exists between the different modes and can be surmised as follows:

- CE is about doing and having an experience
- AC is about concluding and learning from experience
- RO is reviewing/reflecting on the experience
- AE is planning and trying out what is learned (Kolb, 1984).

Learners must be able to involve themselves in new experiences (CE), reflect on their experiences from different perspectives (RO), create concepts that integrate their observations into sound theories (AC) and use these theories to make decisions and solve problems (AE) (Kolb et al., 2000). Therefore, as mentioned above experiential learning is focused on experience and reflection.

It must be argued that this process is not easy. For example, how can one act and reflect at the same time? It can be suggested that this takes experience and needs to be scaffolded initially to make this transition more meaningful. This scaffolding will be described in due course with reference to the practice of scientific inquiry.

Focus on Experiential Learning in Different Contexts

Experiential learning is so dominant in international contexts that there are even job opportunities for experiential learning officers in some countries! The Chinese post-primary science curriculum is heavily focused on integrated experiential learning for the development of scientifically literate citizens. The Irish post-primary science curriculum is directed towards creating an autonomous learner who should experience science using nature of science as the holistic model guiding their practice. In the Netherlands, experiential learning is a key focus as it is in Canada, America, New Zealand and Australia, for example. Indeed, Finland, who is often seen as the 'poster child' of education, places a big emphasis on experiential learning.

The model of experiential learning put forward by Kolb has been made into more structured steps for curriculum implementation in some countries. For example, in Ireland experiential learning can be considered in terms of four stages: experiencing, processing, generalising and applying (NCCA, 2001). This perspective has been used in the development of curricula for subjects such as social, political and health education (NCCA, 2001). *Experiencing* is the activity stage, while *processing* is a stage where students reflect on their experience. This is followed by the *generalising* stage where comparisons between students' answers are made and finally the *applying* stage where students apply the learning to new contexts (NCCA, 2001). The following is an example in science where this could be done.

A jigsaw methodology is used to get students to read different pieces of text to comprehend the key pieces of information from their texts. The jigsaw methodology is a co-operative learning strategy which not only helps students to construct knowledge on something but seeks to develop additional life skills such as the ability to work in groups, communicate effectively and engage in critical thinking.

The jigsaw methodology uses home groups and expert groups. Members of the home group go to an assigned expert group where they are each given a task to complete (**experiencing stage**). They will return to their home groups with the key pieces of information discussed from their expert groups. They are asked to bring together the information by the creation of a product, e.g. a poster. Each group then presents its poster. The whole class then engages in group processing where they discuss (**processing stage**) on the experience, comparisons (**generalising stage**) are made between groups/students' experiences and the key ideas they learned. Finally, the students are asked to apply (**applying stage**) their new knowledge to a real-life context. They are also asked to consider the new life skills that they have learned from engaging in the process.

The jigsaw methodology can be used for a paper and pencil inquiry activity or indeed it can be used for students to engage in mini-experiments within their expert groups.

Other Examples of Experiential Learning in Science Education

It is important to note here that variations often exist between the theoretical/research perspective and the practical application of a theory in the classroom context. Indeed, this is an important variation to help teachers apply a theoretical perspective to their own learning context. Often research literature is focused on a fixed definition of a construct; therefore the purpose of the upcoming section is to illustrate the flexible nature of experiential learning. This section will look at how the theory influences different aspects of science education, namely concept acquisition, the discussion of which started in the previous section by outlining the jigsaw methodology as a vehicle for experiencing learning and will be presented towards the end of this section. It will also look at how ELT relates to scientific practices both inside and outside of the classroom through consideration for its relationship to scientific inquiry. The section will also focus on how ELT can be used to guide classroom practices focused on developing students understanding of socio-scientific issues, the goal of which is to make them active, informed citizens. Finally it will look at the different learning abilities of students and how ELT activities can be used for supporting the learning of different students.

As well as the applications of ELT, the limitations of this theory will also be briefly discussed throughout.

Firstly, it is important to say that experiential learning is not a repertoire of methodologies but it is a statement of fact, people learn from experience (Kolb, 1984). Therefore, the following presentation of ideas look at methodologies aligned with concept acquisition through experience. It must also be noted that some of the theories presented in this book look at similar methodologies, but the key to the discussion in this chapter is the focus on experience and reflection as part of the learning process. It is crucial however to understand that the methodologies discussed do not solely allow for concept acquisition but allow students to experience how science is typically practiced in the real world and the need to create a socially responsive science citizen who understands socio-scientific issues and the need to be informed through research.

As the discussion of experiential learning in respect to science begins, it is important to consider that learning science should be culturally constructed. Therefore, it can be argued that students should experience science similar to how a scientist engages in scientific practices. In a way, all of the learning abilities mentioned above mirror the practice of a scientist as they actively engage in doing, reflecting, concluding and connecting with existing theories. Thus experiential learning is intimately linked to the holistic model of the nature of science. Nature of science reflects how science works and specifically how we carry out investigations, how we communicate in science and how we develop an appreciation for the contribution of science and scientists to society (NCCA, 2016). All of these features are inherently focused on experience as a key enabler in the act of mirroring the practice of a scientist. This is a critical consideration in the teaching of science as it is necessary in authentic scientific practices to mirror how science is practiced outside of the classroom walls. Students need to understand the often messy nature of how scientists practice science in the real world, separate to the rigorous following of a 'cookbook style', scientific method approach to investigating that is often part of the students learning experience. This overly structured approach does not allow an opportunity for authentic experiential learning to take place as students are not given the opportunity to reflect on what they have done and what they would do differently next time, as what scientists would do in the real world context. The act of mirroring the practice of a scientist often considered in terms of scientific inquiry which refers to the diverse ways that a scientist studies the world, therefore is intimately linked with the experiential learning theory of which will be further explored in due course. Scientific inquiry contains many features such as designing investigations, observing, collecting and analysing data and drawing conclusions by connecting with existing theory and is in itself an idealised form of how students should experience science. When one looks at the features of experiential learning it can be considered that scientists engage in similar practices of reflection, synonymous with experiential learning and indeed other theories discussed in this book such as constructivism. Trying to bring this into classroom practice is a goal for many involved in science education, be it policymakers, teachers, teacher educators, industry, etc.

It can also be suggested that learning is socially constructed, as discussed in the chapter looking at social constructivism, and this notion is true in the real world scientific landscape as scientists rarely work in isolation; thus students should experience science by socially constructing their understanding of the concept under consideration.

To extend the discussion, Fig. 17.1 provides an extension to Kolb's ELT model but with a particular focus on scientific practices through the inquiry process.

Figure 17.1 presents one way for learners to experience science through scientific inquiry. Of course, it is not always possible, or realistic, to have every lesson formatted using the above processes. A limitation of the ELT and the practice of scientific inquiry is that many teachers feel constrained by a heavy curriculum with limited

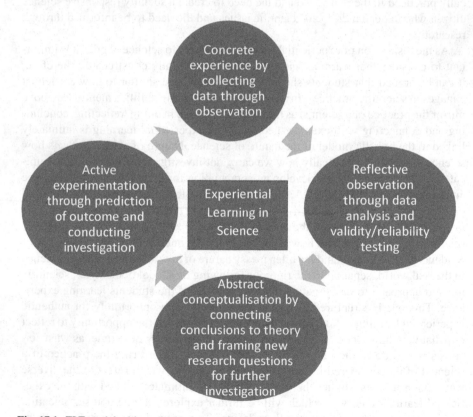

Fig. 17.1 ELT model with particular focus on scientific practices

time to complete science in the way it should be practiced. Additionally, the teachers' own orientations towards science teaching, in other words their beliefs about how science should be learned, can be the ultimate determinate of how students experience science. The act of reflection can also be difficult for both the teacher and the students as examination of one's own work can often be a 'raw' experience as some people struggle to appraise something they have produced or experienced. Another limitation is that while it is necessary to recognise that learning science should be more than just learning content; learning scientific practices and skills (mirroring the practice of a scientist) should be equally important but yet this is not often the case in the typical classroom setting. To that end, it is important to get the balance right, not every lesson should use the same methodologies but the students should experience a repertoire of active learning strategies and nature of science foci during their concept acquisition. Often the methodology/scientific practice takes precedence when trying to engage in constructivist learning in the classroom, the content to be learned is equally important as the experience of how students learn in novel ways.

Additionally, experiential learning allows for the opportunity for students to engage in other aspects of the nature of science such as science as a socio-institutional system (Erduran & Dagher, 2014). Looking at the act of CE, AC, RO and AE, discussed previously in the chapter, students can actively engage in debate related to particular socio-scientific issues (using secondary scientific research data as a tool for engaging in active dialogue), form group conclusions based on the evidence presented, reflect on the experience of engaging in critical, informed debate and develop a research question for further exploration. The goal ultimately from this is to create informed, active citizens who are tentative in their conclusions and who search for and use evidence in informing their opinion. This act of reflection is intimately related to the ELT.

While the development of informed, active citizens is crucial, again limitations such as curricular constraints and the teachers' own self-efficacy and their belief system in how and what students learn can act as a disabling factor towards this. Teachers are the change agents towards promoting experiential learning and their voice as well as the students' voice must not be lost in the discussion of bringing experiential learning into the classroom.

How and what we learn is also specific to different contexts such that the learners must decide which learning ability, i.e. CE or AC that they will use in a given situation. The students' learning ability can be both a constraint and an opportunity to learning using an experiential learning approach. Learning is sometimes best done through experiencing the concrete and in other times, best served through symbolic, abstract representations. However, it is important to reflect on the stages of cognitive development outlined by Piaget which suggests that cognitively, humans do not develop the ability to think abstractly until they reach the formal operational stage of cognitive development (Woolfolk, Hughes, & Walkup, 2008). This suggests therefore that experiential learning is very much linked to a person's cognitive development. This must be considered in the teaching of young people as the formal operational stage of cognitive development does not typically begin until age 12, even as research suggests that some adolescents still cannot think in an abstract way suggesting their cognitive development is not yet at the formal operational stage (Bliss, 1995).

While the above provides context in terms of grasping experience, processing experience is also specific to the person. Some learners prefer to watch others engaged in an experience and reflect on it while others prefer to engage in active participation. It can be argued that in terms of intelligence, the latter would very much reflect a bodily kinaesthetic intelligence while the former would reflect a visual learner's preference (Gardner, 2011). A person who is a reflective observer would actively learn from teacher demonstration followed by a discussion (reflection) on the key foci of the demonstration. A person who is an active participant, prefers to experience learning through active engagement in scientific practices. From a curricular policy viewpoint, the focus is on active experimentation through scientific inquiry (NCCA, 2016; NGSS, 2013). Are current curricular policies targeting one type of experiential learner and thereby limiting the concept acquisition and development of some? In a sense, different types of learners need to be catered for in the overall context of the diverse classroom and while experiential learning is crucial, it is necessary to understand that it has different facets.

Kolb further groups the learners into different categories:

1. Diverging learner—their dominant learning abilities are CE and RO.
2. Assimilating learner—their dominant learning abilities are AC and RO.
3. Converging learners—their dominant learning abilities are AC and AE.
4. Accommodating learners—their dominant learning abilities are CE and AE (Kolb et al., 2000).

The assimilating learning style is important for science careers while people with the diverging learning style, for example, like working with people, are emotional and tend to work in the arts (Kolb et al., 2000). The reason why the assimilating learning style is considered advantageous to those pursuing a career in science is due to the fact that these learners are typically interested in ideas and abstract concepts. Creation of models would serve as a pedagogical strategy which caters for these learners as well as the active pursuit of new ideas through research. Learners with a diverging learning style benefit more than other types of learners from formative assessment strategies such as personalised feedback. They tend to be more open-minded and therefore respond more positively to constructive criticism.

Learners with a converging learning style prefer to solve problems and would be best served in specialist and technology careers. An accommodating learner works best from problems solved by others and is most effective in careers in marketing and sales. The accommodating learners would benefit from co-operative learning strategies where action is a key part of the learning experience.

A number of factors shape and influence learning styles such as personality, adaptive competencies and educational experience (Kolb et al., 2000). It is not within the remit of this chapter to focus heavily on these factors with the exception of discussion around educational experience as this is very much directly influenced by the teacher.

Other Instructional Strategies that Focus on Experiential Learning

The content above provided a general understanding of how ELT links with different practices in the classroom, the remaining content in the section will look at specific methodologies to primarily allow for concept acquisition and development.

General group work is an activity central to experiential learning. But for group work to be effective, a number of considerations need to be made such as the following:

1. *Assignment of task.* Clear instructions need to be given, including timing considerations and providing students with a purpose to the activity.
2. *Group selection.* When selecting groups, this can be done through random selection or through mixed ability groupings. The latter selection serves best for a more meaningful learning experience for all.
3. *Roles within a group.* It is important to consider assigning different roles in a group, e.g. a chairperson, a scribe, a presenter and a timekeeper. The teacher can ask for volunteers or purposefully select the roles of the individual group members.

The following are some other examples of strategies which can be part of experiential learning: role-play, simulation, brainstorming, generation of creative products (e.g. a poster), project work, having a visitor or role model talk to the students, case studies, walking debate or engaging with a piece of text (NCCA, 2001). I have been engaged in projects where role models, i.e. practicing scientists have visited students and the response from students generally is that their stereotypical image of a scientist is altered based on the experience. Indeed, that is the very essence of experiential learning which describes the end product of learning to allow students to apply knowledge, skills and feelings in a real-life context. Meeting with practicing scientists is a critical learning experience for the students as it can make them aware that science in the real-world context is vast, differentiated and responsible for several real-life experiences that students have. For example, in an age of the smartphone it is important for students to understand that scientists have a part to play in making the gadgets that they use every day. This discussion is a critical part of the nature of science. This experience also helps make science more relevant to students and helps to increase interest and motivation in science. From an occupational perspective, it can help students to consider pursuing a career in science by having a broad understanding of scientific practices outside the four walls of the classroom. One limitation is obvious in the organising of such a visit but as it is not a typical, frequent methodology that a teacher will use, it is something that can be organised well in advance.

Another example of an experiential learning activity is the construction of a concept map (Clark, Threeton, & Ewing, 2010). A concept mapping activity can be used as an orientation activity, assessment activity or as a way of combining groups' ideas. It can also be an assessment tool, a practical learning tool and an instructional tool

(Broggy, 2010). With scientific inquiry practices it can be used to bring together the data collected by the groups to form an overall consensus. A concept map begins with a focus question and this focus question can be the research question on which the investigation was based. A 'focus question' clearly specifies the problem or issue that the concept map should help to resolve (Novak & Cañas, 2006). In a concept map, two or more concepts (nodes) are linked by words that describe their relationship. They allow a large amount of information to be represented visually and the act of constructing a concept map allows students to experience and reflect, thereby engaging in experiential learning. I would use concept maps often in my teaching and one of the limitations of using concept mapping is that students struggle with the initial design of the concept map, in particular the hierarchical placing of concepts and the linking phrases between concepts and they can often become frustrated at the same time. But looking at learning from an ELT lens, reflecting on the experience can allow students to consider what they would need to do differently to enhance their concept maps in future.

It is here I think it is important to note that concept acquisition and development should not be a linear, straightforward process when applying the ELT to student learning. Authentic and real learning takes place when students engage in critical thinking tasks that often result in frustration but ultimately lead to a fruitful learning experience; students remember these uncertainties towards the final product much more than the straightforward, surface thinking that can be found in a more traditional classroom.

Engaging with a piece of text can be considered experiential learning if (1) the students are aware of the purpose of the text; and (2) it is the main resource for the processing phase of experiential learning (NCCA, 2001) and again students reflect on the experience. However while engaging in text can also help to develop students' literacy skills, for some students they may struggle to maintain focus and understand the content if their comprehension skills are weak. Using a reciprocal reading technique can be useful in this case (see Petty, 2009 for more information).

Learning logs can be used as an effective learning aid during experiential learning which allows students to track their learning and reflect on the experience. For example, taking part in a field trip can be considered a valuable learning experience both from the perspective of learning new content and learning about how science works in practice (nature of science). A learning log can be a purposeful tool to document new learning and allow students to reflect on the meaning behind their experience. A limitation of using learning logs however is the need for consistent monitoring of students' work to ensure students do not hold misconceptions related to the content learned. While it takes time, this monitoring and feedback serves as a necessary tool to understanding students' progression of learning.

To conclude this section, experiential learning as a theoretical construct can not only be applied to methodologies which allow for concept acquisition but allow for students to experience and reflect on scientific practice similar to that of a scientist, using scientific inquiry as a vehicle to engage in this experience. Experiential learning allows opportunities for the creation of active, informed citizens to engage in critical debate and discussion on socio-scientific issues that allow students to see beyond the four walls of the classroom and an assessment-driven curriculum.

However in order for authentic experiential learning to take place, the system needs to allow it to happen. Recent changes in curricula, such as that in Ireland, are removing the rigid science curriculum to allow students more time to engage in authentic scientific practices which model an experiential learning approach. However change takes time and both the teacher and students' voice needs to be at the forefront of the change process.

Assessing Experiential Learning

A key concern among teachers when they use alternative approaches, such as experiential learning, is how/what to assess. Unfortunately, assessment is the tail that wags the curriculum dog (Hargreaves, 1989) and unless the assessment values experiential learning as a vehicle to enhance students' knowledge, classrooms will remain a rote learning, product-focused environment where the teacher is the dominant voice and actor.

Ultimately, both the curricular and the classroom focus should be on assessing more than students' ability to recall information, it should be about assessing students developing knowledge and skills through experience. For example, a technology teacher should be concerned with understanding students' reaction to particular experiences in the technology classroom and science is no different. Assessment for learning opportunities is crucial in this regard. Indeed, the feedback processes are critical to experiential learning and constitute a particular focus of the Lewinian model (Kolb, 1984). The opportunity needs to be present for teacher feedback as well as peer and self-assessment strategies. For example, the product of an inquiry activity could be the generation of a poster. One possible way of assessing through feedback is the use of a criterion-referenced rubric. Table 17.1 presents a rubric I use to assess students' posters following an inquiry activity.

The scales in the table (exceptional, above expectations, etc.) are taken from the new Junior Cycle Curriculum in Ireland (NCCA, 2016).

A teacher could also allow students to define how their work will be judged. In a sense they could choose what criteria will be used to assess their work or indeed help create a grading rubric. This should be done after students have experienced assessment through an already designed rubric. This activity helps to create the autonomous learner which is a critical part of a learner's education experience.

Other learning products which could potentially be assessed include the following:

- Creating a reflective journal or portfolio.
- Reflection on critical events that took place during the experience through group discussion.
- Essay, report or oral presentation.

Table 17.1 Assessment Rubric

Criteria	Exceptional	Above expectations	In line with expectations	Yet to meet expectations
Awareness of the scientific inquiry process				
Quality of content of poster with reference to the chosen theme				
Presentation of the poster				
Level of engagement in the experience				
Clarity of student explanation				
Originality of the poster				

The latter also serves to develop general life skills including the development of students' verbal and non-verbal communication skills and ability to articulate information. This is an explicit focus of curricular policies internationally in progressive education countries such as Ireland (NCCA, 2016).

- Self-awareness tools and exercises (e.g. questionnaires about learning patterns). These would be examples of self-assessment strategies where students are asked to consciously reflect on what they learned and what they still have difficulty in mastering.
- Short answers to questions of a 'why' or 'explain' nature (e.g. 'What did you learn through this assignment? What did you not learn that you would like to?'). In other words, higher-order questions are very important in assessing experiential learning outcomes. Short answers help to develop students' ability to be precise in explanation.
- One-on-one oral assessments with the teacher. Such teacher/learner interactions are not only critical to student learning but imperative in the creation of an open, positive classroom environment.
- A project that develops ideas further (individually or in small groups). This would be considered an extended activity which is often the result of an inquiry-based activity. In real life, scientists generate new questions from the investigative experience, therefore students should be able to enact further extension works (Teaching and Learning Services, 2014).

All of these assessment strategies should place some emphasis on how the experience facilitated the students in developing skills and not just on the content that they learned from the experience. This is in line with Dewey's model which sees learning as a developmental process (Kolb, 1984). The reason for this is that experiential learning reflects learning in real life situations and how authentic knowledge development should take place.

Effect of Experiential Learning on the Teacher

While the learner is a key player, the effect of having learners engaged in experiential learning from the teacher's perspective needs to be considered. I would argue that teachers need to remain motivated and interested in their domain in order to translate this in their classroom. Recent research showed that teachers who engaged in experiential learning practices were more motivated to continue to try out best practices in their classroom (Zhang & Campbell, 2012). Indeed, I can relay my experiences of using an experiential learning approach in teaching, both in post-primary and third-level contexts. I continue to see the impact of this type of learning on the students, regardless of their stage of cognitive and social development. I appreciate that experiential learning does take more time in planning for but I am of the view that the effect on students both from a learning and skills development perspective is worth the additional planning required. It is important to note that planning is more than just about resources; it is also about revising the key content needed in the delivery of the lesson. Often teachers feel restricted to rote learning because they believe that they do not have the necessary pedagogical content knowledge (PCK) to allow learners to engage in experiential learning. In other words, they often have low self-efficacy. I admit that this was an issue for me, my own self-efficacy, but it is a necessary action if students are to fully become active and informed learners and for the teacher to be an effective facilitator of learning. This planning has become less over the years as my repertoire of resources and PCK have developed.

Research Against Experiential Learning

Some view experiential learning to be a fad where the focus is on the process rather than the content to be learned and it must also be noted that some scholars do not consider Kolb's interpretation of ELT to provide an accurate presentation and extension of the original theory. Miettinen (2000), for example, implies that Kolb does not refer to the Lewinian model but instead uses a secondary source to develop his model. Secondly, the same author suggests that Kolb selectively uses existing literature to develop his model, therefore reducing the authenticity of his model. Despite these assertions, the extent to which Kolb's model has been used in curriculum design and as part of theoretical frameworks for research purposes, suggests that generally speaking, educationalists view Kolb's model with respect.

In terms of research, while the majority of the literature suggests that students should learn through active experience, there are those who think that student-led

Table 17.2 Levels of Scientific Inquiry (Bell, Smetana, & Binns, 2005)

Low level inquiry-very teacher focused	Becoming less teacher focused	Becoming more learner focused	High level inquiry-very learner focused
Confirmation inquiry	*Structured inquiry*	*Guided inquiry*	*Open inquiry*
Question given by teacher	Question given by teacher	Question given by teacher	Question derived by learner
Procedure given by teacher	Procedure given by teacher	Procedure developed by learner	Procedure derived by learner
Outcome known in advance	Outcome not known in advance	Outcome derived by learner	Outcome derived by learner

practices such as experiential learning, inquiry-based learning and discovery learning are not effective at facilitating students developing knowledge and understanding about scientific concepts. This can be seen in the research conducted by Kirschner, Sweller and Clark (2006) who proposed that such practices do not consider the characteristics of the working memory, long-term memory and the intricate relationship between the two (Kirschner et al., 2006). They note that research trying to validate the use of experiential learning has not, up to now, been successful. What the authors talk about is the use of minimal guidance through experiential learning, etc. I suggest that experiential learning practices should begin at a more structured level and then move to a more open/student-directed level, once they are comfortable and consistent in classroom routines focused on learning through experience. Table 17.2 can be used to describe the different levels of scientific inquiry which, as mentioned previously, is a related practice to experiential learning.

Table 17.2 shows the changing focus from teacher-directed learning to student-directed learning where autonomy is central to the experience. This is critical in providing students with true learning experiences. This progressive spectrum is not considered in the work reported by Kirschner et al. (2006).

Conclusion

I have spent years trying to instil these ideas into pre-service science teachers but often their apprenticeship of observation (Lortie, 1975) leads to resistance to alternative ways of learning. Effectively they are products of a success system where one form of learning worked for them, often through rote learning, due to one dimensional assessment structures. It is critical to assure pre-service teachers understand the existing assumptions as a way to try to alter their beliefs/values towards how science should be taught.

Within initial teacher education (ITE), it is advised that pre-service teachers engage in experiential learning as a vehicle to develop as a teacher while concurrently positioning themselves as learners engaged in experiential learning. This can be done through the pre-service teachers actively understanding the importance and engaging critically in school placement opportunities and reflecting on experiences of such. As teacher educators we must apply the theories we advocate if they are to have any real meaning and relevance for future educators.

My final message serves as a direct appeal to all teachers, both practicing and trainees. Learners, generally speaking, remember the teachers they had during their schooling. We all remember teachers who impacted both positively and negatively on us as learners. My question is, do you want to be remembered for the right reason, i.e. that you facilitated your students in positively experiencing science the way it is meant to be practiced?

Summary

- This chapter has provided some practical insight into how a particular learning theory can be translated into the science classroom. It has looked at developing a model of experiential learning centred on practice in the science classroom. With this, it has focused on the different learning styles of students and the pedagogical practices, in line with experiential learning, which can facilitate their construction of new knowledge.
- It has provided insight into how experiential learning links with philosophical/holistic underpinnings of science education, such as nature of science and scientific inquiry. The most important message that hopefully comes from this chapter is the idea that learning is an individual endeavour and a one-size-fits-all mentality does not work in a productive learning environment. It is critical to vary your pedagogical approaches to challenge and motivate your learners.

Further Reading

Darling-Hammond, L. (2008). Teaching and learning for understanding. In L. Darling-Hammond, B, Barron, P. D, Pearson, A. H, Schoenfield, E. K, Stage, T. D, Zimmerman, G. K, Cervetti, & J. L, Tilson (Eds.), *Powerful learning: What we know about teaching for understanding*. San-Francisco: Jossey-Bass.

Kolb, D. A. (1984). *Experiential learning: Experience as the source of learning and development*. New Jersey: Prentice-Hall.

Kolb, D. A., Boyatzis, R. E., & Mainemelis, C. (2000). Experiential learning theory: Previous research and new directions. In R. J. Sternberg & L. F. Zhang (Eds.), *Perspectives on cognitive, learning, and thinking styles*. NJ: Lawrence Erlbaum.

Petty, G. (2009). *Evidence based teaching: A practical approach* (2nd ed.). Cheltenham: Nelson Thornes Ltd.

References

Bell, R., Smetana, L., & Binns, I. (2005). Simplifying inquiry instruction. *The Science Teacher, 72*(7), 30–34.

Bliss, J. (1995). Piaget and after: The case of learning science. *Studies in Science Education, 25,* 139–172.

Broggy, J. (2010). *Concept mapping in physics in an Irish university: An investigation into the application of the tool with particular reference to its relevance to problem solving and the use of scientific language.* Unpublished thesis (Ph.D.), University of Limerick.

Clark, R. W., Threeton, M. D., & Ewing, J. C. (2010). The potential of experiential learning models and practices in career and technical education & career and technical teacher education. *Journal of Career and Technical Education, 25*(2), 46–62.

Erduran, S., & Dagher, Z. R. (2014). Regaining focus in Irish junior cycle science: Potential new directions for curriculum and assessment on nature of science. *Irish Education Studies, 33*(4), 335–350.

Gardner, H. (2011). *Frames of mind: The theory of multiple intelligences.* Basic books.

Hargreaves, A. (1989). *Curriculum and assessment reform.* Milton Keynes: Open University Press.

Kirschner, P., Sweller, J., & Clark, R. (2006). Why minimal guidance during instruction does not work: An analysis of the failure of constructivist, discovery, problem-based, experiential, and inquiry-based teaching. *Educational Psychologist, 41*(2), 75–86.

Kolb, D. A. (1984). *Experiential learning: Experience as the source of learning and development.* New Jersey: Prentice-Hall.

Kolb, D. A., Boyatzis, R. E., & Mainemelis, C. (2000). Experiential learning theory: Previous research and new directions. In R. J. Sternberg & L. F. Zhang (Eds.), *Perspectives on cognitive, learning, and thinking styles.* NJ: Lawrence Erlbaum.

Lortie, D. (1975). *Schoolteacher: A sociological study.* Chicago, IL: The University of Chicago Press.

Loughran, J., Mulhall, P., & Berry, A. (2006). *Understanding and developing science teachers' pedagogical content knowledge* (1st ed.). Rotterdam: Sense Publishers.

Magnusson, S., Krajcik, J., & Borko, H. (1999). Nature, sources, and development of pedagogical content knowledge for science teaching. In J. Gess-Newsome & N. G. Lederman (Eds.), *Examining pedagogical content knowledge: The construct and its implications for science teaching* (pp. 95–132). Boston: Kluwer.

Miettinen, R. (2000). The concept of experiential learning and John Dewey's theory of reflective thought and action. *International Journal of Lifelong Education, 19*(1), 54–72.

NCCA. (2001). *Social, political and health education: Junior cycle guidelines for teachers.* Dublin: Department of Education and Science.

NCCA. (2016). *Specification for junior cycle science.* Dublin: NCCA.

NGSS (2013). *DCI arrangements of the next generation science standards.* https://www.nextgenscience.org/sites/default/files/NGSS%20DCI%20Combined%2011.6.13.pdf.

Novak, J., & Cañas, A. (2006). *The theory underlying concept maps and how to construct them* (Vol. 01). Florida Institute for Human and Machine Cognition: Technical report IHMC CmapTools.

Petty, G. (2009). *Evidence based teaching: A practical approach* (2nd ed.). Cheltenham: Nelson Thornes Ltd.

Teaching & Learning Services. (2014). *Guidelines for assessment of experiential learning.* Montreal: Teaching and Learning Services. McGill University.

Woolfolk, A., Hughes, M., & Walkup, V. (2008). *Psychology in education.* Harlow: Pearson Prentice Hall.

Zhang, D., & Campbell, T. (2012). An exploration of the potential impact of the integrated experiential learning curriculum in Beijing, China. *International Journal of Science Education, 34*(7), 1093–1123.

Dr. Louise Lehane is a lecturer in Education at St Angela's College, a college of the National University of Ireland, Galway. She lectures in the areas of general and science pedagogics, sociology of education and history and policy of education. She is a qualified science teacher and, following the completion of her initial teacher education programme, embarked on a Ph.D. Her thesis was focused on the use of a pedagogical content knowledge (PCK) lens to capture preservice science teachers' scientific inquiry orientations within a professional learning community. Her main research interests include PCK, scientific inquiry, nature of science, curriculum policy and the continuum of professional development.

Chapter 18
Social Constructivism—Jerome Bruner

Miia Rannikmäe, Jack Holbrook, and Regina Soobard

Overview

In this chapter, rather than giving an overview of all branches of constructivism, the authors focus on constructivism but in particular on the value of social constructivism in education and in particular the input of Jerome Bruner's ideas. Jerome Bruner has, arguably, given the latest and most updated influence into widening social constructivism and in highlighting its value in modern societies. This chapter starts with brief introduction of Bruner's work, followed by a comparison of constructivism, Bruner's work and a comparison of constructivism and social constructivism. In reflecting on important issues in contemporary education, the authors use an example from a European Commission-funded project, MultiCo, to show the role of social constructivism in science education to meet the needs of society, especially in increasing the awareness of young people about science-related careers.

Introduction

In this chapter, a model of a social constructivist way of teaching and learning in Estonia is introduced, developed within the framework of an EC-funded project called 'MultiCo' (http://www.multico-project.eu/). Estonia is a country in which students have demonstrated high achievement in PISA studies (OECD, 2016), and which

M. Rannikmäe · J. Holbrook (✉) · R. Soobard
Centre for Science Education, University of Tartu, Tartu, Estonia
e-mail: jack@ut.ee; jack.holbrook@ut.ee

M. Rannikmäe
e-mail: miia@ut.ee

R. Soobard
e-mail: regina.soobard@ut.ee

has adopted a competence-based curriculum. Emphasising the focus on science as a social endeavour, science education is strongly encouraged to follow an 'education through science' philosophy (Holbrook & Rannikmäe, 2007), promoting a more social view of science learning.

General Overview of Constructivism

Constructivism is a well-known educational theory, strongly advocated over the last 50 years. During this period, rapid changes in the society have taken place, educational reforms have occurred and new emphases for the goals of education have appeared. In these societal settings, the emphases placed on constructivism have changed.

Constructivism is seen as being associated with a way of teaching through which learners play an active role in their own learning. Learners construct knowledge based on personal experiences. Being members of society, learners' interpretations of their experiences depend on the social environment surrounding them. This environment is challenging learners to look beyond current experiences and predict new events in the future, considering cultural factors associated with the environment. In a constructivist classroom, educational developments are geared to the students constructing, rather than the teacher. Unlike the Skinner idea of behaviourism, where the teacher ('as the knowledgeable expert') pours information into passive students, the constructivist model views the students as actively engaged in their own process of learning (Cooper, 1993).

Constructivism can be discussed from two aspects: the nature of learning and the conception of knowledge, both relevant at the student level, or at the teacher level.

In constructivism, learning is (Taber, 2011):

(a) An active process, which is constantly open to change to construct new ideas, or concepts;
(b) An interactive process of adjusting currently held mental models to make sense of the physical, cognitive, emotional and social experiences, by interpreting, representing and restructuring pre-existing knowledge;
(c) A social process in which one's learning is intimately associated with that of other human beings, including teachers, peers and other community members. It is recognized that the social aspect of learning uses conversation, direct interaction with others and the application of knowledge as integral aspects of learning;
(d) A contextual process, in that we do not learn isolated facts and abstract theories separated from learning associated with the rest of our lives.

The key component in constructivism is the conception of knowledge acquisition seen as:

(a) Constructed, not transmitted or reproduced;
(b) Subjective, as each person creates personal meaning out of experiences and integrates new ideas into existing knowledge structures;
(c) Adding to previous knowledge constructions, beliefs and attitudes, which are considered and which impact on the knowledge construction process.

The role of the learner is associated with applying their existing knowledge and real-world experiences, learning to hypothesise, testing their theories, trying things that may not work, asking questions, sharing with each other and reflecting on their experiences. In so doing, students construct their own understanding of the world in which they live, accommodating and assimilating new information with their current understanding. In this way, students play central roles in mediating and controlling learning to set their own goals, regulate their own learning processes and even undertake self-assessment. Although students may feel the need to maintain their established ideas and thus may reject new information that challenges their prior conceptions, this may be because their views are strongly held and there is a need for strong group pressure to instigate a level of doubt and to consider a change of view. Nevertheless, students need to know how to learn, or change their thinking/learning style. And learners need to use and test ideas, skills and information through relevant activities.

In the constructivist classroom, both teacher and students think of knowledge as dynamic, or ever-changing, with students needing to develop the ability to successfully stretch and explore their view—certainly not simply seeing knowledge as inert fragments to be memorised. Key assumptions associated with this perspective include (Jordan, Carlile, & Stack, 2008):

1. The current beliefs of students, whether correct or incorrect, are important.
2. Despite having the same learning experience, students base their learning on understanding and meaning, which is personal to them.
3. Understanding or constructing meaning is an active and continuous process.
4. When students construct a new meaning, they may give it provisional acceptance, or they may even reject it.
5. Learning is an active, not a passive, process and depends on students taking responsibility for their learning.
6. Learning may involve conceptual changes.

Also, it is worth noting that constructivism supports the need for the teacher to play an active role. The role of the teacher, as a holder of expert knowledge and as a facilitator guiding the student learning, is still crucial, although the teacher's role may need modification or adaptation, so that students have opportunities to self-construct their own knowledge. The teacher thus serves as a mentor—guiding, monitoring, tutoring—or facilitator of the students' learning. In this respect, the teacher (Jordan et al., 2008):

(a) Uses approaches recognising students' prior knowledge, instead of following a textbook presentation or curriculum content;
(b) Needs to ensure flexibility in the development of student inquiry;
(c) Provides a motivational and inspiring environment for student interactions;
(d) Relies heavily on open-ended questioning, student hands-on problem-solving though promoting inquiry-based learning, creating situations where the students are motivated, to ask questions and reflect on their learning;
(e) Scaffolds students' development (within groups or with teacher support) so as to encourage them to seek to perform just beyond the limits of their ability when working alone;
(f) Triggers extensive dialogue and collaboration to expose the learner to alternative viewpoints and multiple perspectives among their fellow students, supporting collaboration in constructing knowledge, rather than in competition;
(g) Nurtures students' natural curiosity and seeks to promote their motivation, autonomy and self-regulation;
(h) Utilises formative and embedded assessment.

Bruner on a Cognitive Revolution Leading to Constructivism

Besides Vygotsky, Piaget and Dewey, the most recent contribution to conceptualising and researching constructivism, over the last three decades, comes from Bruner (Table 19.1). **Jerome S. Bruner** (1915–2016) is seen as a key figure in the so called 'cognitive revolution' within the field of education. He has published several books, for example, *Towards a Theory of Instruction in 1966, and The Process of Education*, the first version published in 1960, the latest revised version in 2009 (Bruner, 2009), both of which are widely referenced as classics. His view of children as active problem-solvers, ready to explore 'difficult' subjects, while being out of step with the dominant view in education at that time, stimulated many to consider a different point of view.

Towards the end of the 1990s, Bruner became critical of the 'cognitive revolution' and looked to the building of a cultural psychology that took proper account of the historical and social context of participants. In his 1996 book, *The Culture of Education,* these arguments were developed with respect to schooling. In this book, Bruner highlighted four views of education, all of which were seen as applicable in today's scientific and technological world:

– Students are imitative learners whose focus is on demonstrative (everyday) activity, where their knowledge and skills may appear;
– Students learn from didactic approaches, where concepts, facts, theories are presented and first form part of compulsory learning, followed by applications;
– Students are thinkers, who make sense of their world, the essential part being on discussion and collaboration;

- Students are knowledgeable; in teaching, it is essential to help students to distinguish between personal knowledge and that taken to be known by the culture (Bruner, 1997; Smith, 2002).

In the 1960s, Jerome Bruner developed a theory of cognitive growth, which in contrast to Piaget, looked to focus on environmental and experiential factors. Bruner suggested that intellectual ability developed in stages through step-by-step changes, based on how the mind was utilised. Bruner's thinking became increasingly influenced by Lev Vygotsky (see Chap. 19) and he began to be critical of the intrapersonal focus he had taken, and the lack of attention paid to a social and political context (Smith, 2002).

Social Constructivism and Bruner

Social constructivism is seen as a sub-set of constructivism. The major difference is with the greater emphasis placed on learning through social interaction (Kukla, 2000). Vygotsky indicates that culture provides students the cognitive environment needed for development, with adults in the students' environment providing the channel for culture to play a constructivist role, be it language, electronic forms of information access and processing, cultural history or social context (Vygotsky, 1978).

A learner's prior knowledge is individual; it depends on personal attitudes and methods used towards processing prior knowledge (Capel, Leask, & Turner, 2000). Therefore, it is important to give students possibilities to construct new knowledge in a situation similar to real life, thereby ensuring that it is created in an authentic context. Social constructivism explains that learners actively construct their own knowledge through experiences and interactions with others (Bruner, 1966; Vygotsky, 1962).

Bruner's views are particularly valuable in conceptualising social constructivism. He suggests (Capel et al., 2000):

- The basis for learning and the development of thinking within social constructivism is activity. Therefore, actively engaging students in the learning process and in social interactions with others allows them to construct their own meaning from self-regulated new knowledge. Based on their views, the relationship between students' own talking and thinking is important. That means that students need opportunities to express their own views to aid and consolidate the thinking process. Therefore, social interactions with others are seen as facilitating learning;
- Learners use different thinking strategies, depending on their previous knowledge, situation and type of learning materials used. Therefore, the important role in using social constructivism in classrooms lies in the learning approach, the situation created and the learning materials used, which actively support students' engagement and social interactions in the learning process;
- Learning involves searching patterns, regularity and predictability. In this process, the role of the teacher needs to support and assist students to find and

formulate patterns and regularities and through this, expanding the learners' prior knowledge.

This led Bruner to define three stages of learning related to thinking about the world, each facilitated by social constructivism (Bruner, 1960):

(a) Enactive representation: implication for teaching is providing opportunities for 'learning by doing'.
(b) Iconic representation: thinking about something through concrete images for understanding it.
(c) Symbolic representation: thinking abstractly about things.
    Scott, Asoko and Leach (2007), building on Bruner's three stages of learning, expand the component of symbolic implication and thus a fourth stage can be considered:
(d) Learning science involves learning the social language of the scientific community.

Social Constructivism in the Classroom

In social constructivist classrooms, collaborative learning is a process of peer interaction that is, at a minimum, permitted, but more importantly, mediated and structured by the teacher. Discussion can be promoted by the problems or scenarios, and is guided by means of effectively directed questions, the introduction and the clarification of concepts and information, (Jordan et al., 2008). Social constructivism focuses on a social nature of cognition and suggests approaches that facilitate a community of learners to engage in activity, discourse and reflection, encouraging students to take on more ownership in the putting forward of ideas to share with others and to pursue autonomy with a view to interacting in mutual reciprocity in dealing with social relations.

Summarising, social constructivism related to teaching and learning can be considered from four aspects: the nature of knowledge, the nature of learning, the nature of the reality and the nature of motivation:

- Knowledge is a human product, which is socially and culturally constructed in an active manner and not something which can be looked up from books (Gredler, 1997). Accordingly, to psychology, knowledge is neither tied to the external world, nor wholly to the working of the mind, but exists as outcomes of mental contradictions that result from interactions with other people and the environment (Schunk, 2012);
- Learning is based on real life, including problem-solving, which takes place in a social manner through shared experiences and discussion with others. And it takes place such that new ideas are matched against existing knowledge and the learner adapts rules to make sense of the world. Social constructivism focuses on the role of a social group, and sees learning as something that emerges from group

interaction processes, not as something which takes place solely within the individual. Learning is seen as an active, socially engaged process, not one of passive development in response to external commands. Therefore, social constructivism acknowledges the uniqueness and complexity of the individual learner and values, utilizes and rewards it as an integral part of the learning process (Wertsch, 1997);

- Social constructivists believe that reality is constructed, not discovered through human activity, so that societies together invent the world (Kukla, 2000). Social constructivism maintains that while it is possible for people to have shared meanings, which are negotiated through discussion, it also acknowledges that no two people can have exactly the same discussions with exactly the same people. To this extent, social constructivism allows multiple realities to exist;
- The motivation of the learner is regarded as having both intrinsic and extrinsic roots. The intrinsic motivation is created through curiosity about the world, while the extrinsic motivation can be provided by the rewards, which can be accessed through the fruits of the interaction.

Table 18.1 is a comparison of constructivism with social constructivism (seen as a subset) related to the work of such educationalists. While constructivism relates to the general behaviour by the teacher, or by the student as individuals, social constructivism sees the behaviour stemming from operating within, or promoting by, collaborative efforts. Such examples of collaborative efforts are engaging a student with a teacher, or other students, or with student/teacher interactions. While constructivism is seen as acting at a personal level, not imposed by others, in a social constructivist environment, the personal level learning is collaborative; at a simple level, by classmates or the teacher and for a wider interpretation, by society.

A Science Teaching–Learning Practice Enabling Social Constructivism

The overall goal of science education is seen as promoting scientific literacy (Estonian Government, 2011; Roberts, 2007). Scientific literacy can be taken to mean developing an ability to creatively utilise appropriate evidence-based scientific knowledge and skills, particularly with relevance for everyday life and a career, in solving personally challenging yet meaningful scientific problems, as well as making responsible socio-scientific decisions (Holbrook & Rannikmäe, 2009).

To enhance SL or STL based on a social constructivist approach, Holbrook and Rannikmäe (2010) put forward a teaching-learning model involving 3-stages. The first stage of the model is based on self-determination theory (Ryan & Deci, 2002) and seeks to highlight the importance of intrinsic motivation in driving human behaviour (students' learning). This viewpoint is seen as being in line with an 'education through science' approach, as opposed to a curriculum content approach, described as 'science through education' (Holbrook & Rannikmäe, 2007). Science learning

Table 18.1 Comparing constructivism and social constructivism based on the work of Vygotsky, Piaget, Dewey and Bruner

Aspect	Constructivism viewpoint	Social constructivism viewpoint
Contributors	*Vygotsky*: zone of proximal development (ZPD): see Chap. 20 **Piaget**: Stage Theory of Cognitive Development (sensorimotor, pre-operational, concrete, formal): see Chaps. 11 and 18 **Dewey**: Learning by doing: see Chap. 17 **Bruner**: Discovery Learning Theory (guided discovery, problem-based learning, simulation-based learning, case-based learning and incidental learning): see Chap. 14 **Bruner**: Cognitive revolution (the role of structure in learning, readiness for learning, intuitive and analytical thinking, motives for learning)	*Vygotsky*: (a) Social development theory relating to social interaction; the more knowledgeable other) (b) Constructivist theory (the role of culture in providing cognitive tools) **Dewey**: Both teachers and students need to learn together how best to enhance student learning **Bruner**: Theory of cognitive growth (focus on environmental and experiential factors) **Bruner**: Critical of Vygotsky's intrapersonal focus and the lack of attention paid to a social and political context
Key features	Emphasis on knowledge construction, which is based on personal experiences Teacher as the facilitator of the learning process	Emphasis on the collaborative nature of learning The importance of a cultural and social context Learning as a process by which learners support and encourage one another
Appearance in the classroom	Self-regulated learning, e.g. metacognition, problem-based learning, critical thinking, concept mapping, inquiry-based learning, web-based scientific inquiry	Collaborative learning, e.g. situated learning, collaborative inquiry, problem-solving as anchored instruction, stimulating interest of students (example in this case, a scenario) Informal learning and forming learning communities
Misconceptions appearing during implementation	Teachers believing they can construct concepts for students; (not appreciating students need to construct everything themselves) Teachers' belief that a stimulating environment is sufficient for students to actually construct knowledge (a confusion between the teaching providing a stimulating learning environment and the need for students to self-regulate their learning)	The curriculum is based solely on social interactions (an emphasis on subject conceptualisation is no longer essential but on guiding students to do the constructing)

is initiated by a *familiar contextual frame* of reference, intended to link to a per-ceived need in the eyes of students. In this initial stage, the aspect of relevance is seen as a major focus. The learning, stemming from relevant aspects for students, within a society, pays careful attention to the development of personal and social attributes, seen to be part of the overall goals of education. Anchoring the relevance of the instruction, by utilising approaches promoting the interest of students, is seen as a further need to create social constructivist, motivational learning. In this stage within the 3-stage model, an essential component of the relevance-anchored, interest-enhanced instruction is a student motivational scenario. Within the learning process, the scenario is *contextualised* by students and, with help of the teacher, is used to stimulate science ideas, initiating the science learning from a known to an unknown situation, through a student-perceived, collaborative learning need.

Stage 2 is driven by a 'need to know' science frame, which provides a scien-tific bearing on the social concern/issue, expressed in the scenario. The ensuing, *de-contextualised* learning focuses on the scientific ideas, solving inquiry-based sci-entific problems and the seeking and evaluating of relevant scientific information. This stage builds on students' prior learning (as identified by the teacher during stage 1) and, with appropriate scaffolding (guidance, support and extrinsic motivation) by the teacher and enhanced by student collaboration, seeks to promote the development of students' intellectual self-actualisation and self-efficacy. The conceptual science learning that evolves needs to be seen by the teacher as being at a level commiserate with the students' learning potential, based on the creation of a zone of proximal development (Vygotsky, 1978). In this *de-contextualised* mode, the teaching is no longer context-based learning, but engaging in an inquiry-based, science education approach.

Stage 3 is commensurate with consolidation of the scientific learning through transference to the earlier contextual frame and promoting socio-scientific decision-making. This is a phase of *re-contextualisation*, heavily driven by a social construc-tivist viewpoint. This is the stage where students are encouraged to feel themselves as members of a social environment, interact with stakeholders (for example, as in a role play, or debate) and make reasoned decision, which consider value (ethical, moral, social, environmental, economic) aspects.

Illustrating a Social Constructivist Approach Within Stages 1 and 2

The approach is taken from research within an EC project called, MultiCo, designed to attract more students towards studying science. The project recognises that an evidence-based, attractive science education provision can enable all citizens to play a more active role in the science, technology and engineering processes, to make informed choices and to more fully engage in a knowledge-based society. Within MultiCo, the stage 1 is initiated via a scenario which further plays a central role in promoting competences and awareness about science-related careers among students. Such scenarios are created through stakeholder co-operation between scientists in education, experts from science fields and industry and involving also civil society organisations, non-formal science educators and students. The following principles from Bruner's work were taken into account in developing the scenario:

- Instruction needs to be concerned with the experiences and contexts that make the student willing and able to learn (readiness).

  - In MultiCo, the first stage involves creating a scenario for students to be able to relate to a real-life issue and seeking to trigger motivation to learn.

- Instruction needs to be structured so that it can be easily followed by students (spiral organisation, Bruner, 1966).

  - In MultiCo, this is ensured by allowing students to ask questions during the scenario presentation, guided by the teacher and therefore the scenario presentation is a collaboration between teacher and students.

- Instruction needs to be designed to facilitate extrapolation and/or fill in the gaps (going beyond the information given).

  - In MultiCo, at the end of scenarios, students are faced with concepts or ideas, which are new to them and therefore this encourages them to prepare for learning new science content in the next phase.

The scenario creation is approached from two different aspects:

(a) The need to see the scenario as a situation, involving a concern or issue and presented in a student-relevant context (students construct their initial ideas).

(b) The need to ensure progress from the scenario into actual classroom science teaching (considerations need to focus on including a teaching element, recommendations on how best to get students involved and how to move into curriculum-related science learning e.g. a 'scientific' question to be investigated—for this, educators/teachers construct the teaching component of the scenario).

Parameters which are considered in developing a scenario:

1. The scenario needs to be 'relevant in the eyes of the students' (not relevant as perceived by the teacher).

   The scenario context is thus most likely connected with:

   - Students' personal life, either now or in the future (*personal relevance*);
   - A social problem/issue or problems/issues, which may have a (hidden) science component (*social relevance*), and/or connected;
   - Updated global, or local problems/issues (*media relevance*).

2. The scenario is interesting for students (this is intended to mean interesting to students in general and hence the scenario is not gender specific).

With this in mind, the orientation needs to be towards an attractive problem or issue, or an unexpected or extraordinary situation, with the possibility to involve students in an unusual scientific, hands-on activity.

3. The scenario (for this project) includes career parameter(s).

   This derives from a focus on industries related to science themes, which form the challenge: e.g. (in the project) energy, water, waste, climate change, food, health, transportation.

4. The scenario is expected to be an initiator, leading to learning that is related to the intended science curriculum, both in terms of subject matter and general (cross-curricular) competences.

5. The introductory scenario is expected to provide the rationale for gaining new knowledge and competences, as outlined in the curriculum, and is thus anticipated to have a positive impact on students becoming intrinsically motivated.

Examples of possible scenarios, incorporating a career awareness focus:

1. An industry visit (purpose of the visit can be descriptive, or problem-oriented).
2. Virtual scenario (e.g. a video showing work in industry, or a video of a visit, pointing out different aspects).
3. A career story (given as a text, cartoon, or possibly as a role play, such as involving interviews).
4. An issue (socio-scientific), or a problem (science related), which includes career-related aspects.
5. A recognised problem (industry linked, science-related).

Exemplary Scenario 1 (Created by Students, Modified by Teachers for Actual Teaching)

This scenario is created by 24 (grade 7, 13–14-year old) students, working in groups. The given task is to develop a scenario, seen as motivating for other students (in this case, social constructivism learning within science for teaching; SCT). To facilitate this, the students are taken on a visit to a famous international beer and soft drinks factory and introduced to the process by which lemonade and beer is produced, starting from the raw materials and involving the development of economical technologies, quality control plus advertising new products to the public. During the visit students are encouraged to ask any questions they wished and make personal contacts with employees of the factory during the visit and, if necessary, afterwards. After the curriculum-related visit, students are asked to work in groups of two or three to draft a learning scenario, which they are expected to introduce, during the next science class, to other students (social constructivism for learning, SCL). The best scenarios, from those presented, are chosen although students are allowed to make modifications based on comments received (social constructivism for learning, SCL).

One of the best scenarios, created as powerpoint slides, is about applying a sugar tax to soft drinks (a political consideration under review at that time) and how this

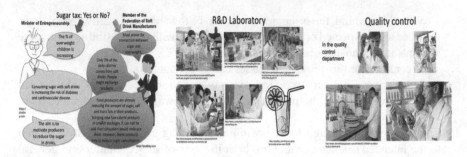

Fig. 18.1 A Set of PowerPoint slides from the scenario 'A sugar tax for soft drinks. Yes or No?'

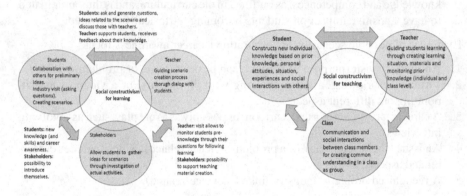

Fig. 18.2 Social constructivist use of scenarios in learning (cycle 1) and teaching (cycle 2)

may influence lemonade production. The context of this scenario is seen by the teachers as relevant for the science content students are expected to learn within stage 2. It is also seen as creating a readiness and willingness for students to learn new science knowledge and skills, through the included attitudinal and value-related aspects (e.g. new taxes, lemonade consumption, sugar consumption, health, different professions in one factory, qualification needed for professions, science-related career awareness). Illustrated in Fig. 18.1 is the set of powerpoint slides, created by students, initiating the motivational student-relevant learning (stage 1).

The social constructivist cycles which take place during the scenario (cycle 1) and the subsequent, stage 2, teaching module for the development and implementation (cycle 2) are illustrated in Fig. 18.2.

The Overall Teaching Module (Created by Teachers)

The following (Table 18.2) describes the teaching approach, indicating activities undertaken and how the module incorporates a social constructivist approach.

Table 18.2 A teaching module showing aspects included which relate to a social constructivist classroom

The Lemonade 3-stage module	Aspects of the (Social) constructivist classroom
Stage 1(contextualisation) Presentation of a perceived student relevant, motivational, socio-scientific scenario, on which is built student-relevant science content, on a 'need to know' basis **Activities**: Discussion to solicit students' prior ideas about a sugar tax The lemonade factory scenario introduces all departments in the factory, plus the science content (knowledge and skills) needed to work in those departments, and possible impact on workers in the context of a potential introduction of a sugar tax	Curriculum emphasis linked to the various factory operations, beginning with considering the factory operations as a whole
Stage 1 Students ask and generate questions/ideas related to the scenario **Activities**: A commitment to ask questions related to the factory slides (e.g. What kind of knowledge and skills are needed to work there? What are these workers doing?)	The pursuit of student questions and interests are valued
Stage 1 Discussion around the scenario helps the teacher to be aware of students' pre-knowledge **Activities**: group work to seek more information about a sugar tax (for or against groups were formed who then use smart technologies, computers, school books, etc., to seek information). Later, in a class collaborative discussion, all students share their findings and put forward their opinions as to whether they agree with a sugar tax or not	Learning is interactive, building on students' prior knowledge
Stage 2 (de-contextualisation) Visit to a Lemonade factory **Activities**: students given an opportunity to ask questions from employees working in the lemonade factory and to discuss with them their own findings and related to knowledge and skills needed and the necessity of a sugar tax	Learning is interactive, building on students' prior knowledge
Stage 2 Inquiry-based science learning **Activities**: experimentation, modelling the lemonade production	Materials and apparatus used are based on the ideas discussed with factory staff members

(continued)

Table 18.2 (continued)

The Lemonade 3-stage module	Aspects of the (Social) constructivist classroom
Stage 2 Inquiry-based science learning **Activities**: the teacher is the facilitator assisting student groups in designing experiments (e.g. pH measurement with indicator strips; determination of the mass of CO_2 by weighing)	Teacher discusses with students, helping students construct their own knowledge
Stage 2 Construction of students' interdisciplinary knowledge **Activities**: students draw concept and consequence maps. (concept maps—interlinking perceived concepts with interlinking descriptors; consequence maps putting concept map perceptions into a learning sequence)	Knowledge is seen as dynamic, changing with gained experiences
Stage 3 Summary of the learning **Activities**: role play (deciding for or against the sugar tax); degustation (food tasting); industrial planning to design new drinks	The teacher's role is interactive, rooted in negotiation. (recognising the student's zone of proximal development to encourage ideas/interactions from all; poor ideas challenged to stimulate further thought)
Stage 1–3 Teacher formatively assesses students on • collaboration and group work • inquiry-based practical work (undertaken as a group) • creativity (as a group) • design skills (as a group)	Assessment is both formative and summative (includes student work, observations and points of view, as well as tests). The process is seen as important, as well as the product
Stage 1–3 The whole module is based on group work	Students work primarily in groups

Evaluation of the Module

Students from five different European countries evaluated the scenario. Students liked the format of the scenario and considered this type of teaching to be unique. This indicated that creating scenarios, following ideas based on social constructivism (e.g. involving students, giving them an opportunity to work with stakeholders), was seen as useful in classrooms for motivating students to learn science. Most students declared having better knowledge about possible science-related professions in industry and the working life skills needed. Interviews with practicing teachers highlighted aspects fitting with social constructivism in the process of teaching. However, teachers seldom used educational terminology related to this, but they showed a readiness to seek deeper insights into associated teaching processes. This suggested that in order to support teachers' thoughtful actions in the classroom, more in-service in the area of educational theories was needed. The paradigm shift, from constructivism towards social constructivism in science education, was needed to ensure the teaching moved away from a surface concept formation emphasis towards socio-scientific decision-making and student competence development.

Summary

General Overview of Constructivism

Bruner on a cognitive revolution leading to constructivism.
Social constructivism and Bruner.
Social constructivism in the classroom.

A Science Teaching–Learning Practice Enabling Social Constructivism

Illustrating a social constructivist approach within stages 1 and 2.

Exemplary Scenario 1 (Created by Students, Modified by Teachers for Actual Teaching)

The Overall Teaching Module (created by teachers).
    Evaluation of the module.

References

Bruner, J. S. (1960). *The process of education*. Harvard University Press.
Bruner, J. S. (1966). *The process of education*. New York: Vintage.
Bruner, J. S. (1997). A narrative model of self-construction. *Annals of the New York Academy of Sciences, 818*(1), 145–161.
Bruner, J. S. (2009). *The process of education, Revised Edition*. Harvard University Press.
Capel, S., Leask, M., & Turner, T. (2000). *Learning to teach in the secondary school. A companion to school experience* (2nd ed.). Great Britain: TJ International Ltd.
Cooper, P. A. (1993). Paradigm shifts in designed instruction: From behaviorism to cognitivism to constructivism. *Educational technology, 33*(5), 12–19.
Estonian Government. (2011). Gümnaasiumi riiklik õppekava (National curriculum for gymnasium). *Regulation of the Government of the Republic of Estonia, No. 2*. Estonia: Tallinn.

Gredler, M. E. (1997). *Learning and instruction: Theory into practice* (3rd ed.). Upper Saddle River, NJ: Prentice-Hall.

Holbrook, J., & Rannikmae, M. (2007). Nature of science education for enhancing scientific literacy. *International Journal of Science Education, 29*(11), 1347–1362.

Holbrook, J., & Rannikmäe, M. (2009). The meaning of scientific literacy. *International Journal of Environmental & Science Education, 4*(3), 275–288.

Holbrook, J. & Rannikmäe, M. (2010). Contextualisation, de-contextualisation, re-contextualisation—A science teaching approach to enhance meaningful learning for scientific literacy. In: I. Eilks, & B. Ralle (Eds.), *Contemporary science education* (pp. 69–82). Shaker Verlag.

Jordan, A., Carlile, O., & Stack, A. (2008). *Approaches to learning: A guide for teachers.* Berkshire: McGraw-Hill, Open University Press.

Kukla, A. (2000). *Social constructivism and the philosophy of science.* New York: Routledge.

MultiCo. (2015). Promoting youth scientific career awareness and its attractiveness through multi-stakeholder cooperation. Retrieved from http://www.multico-project.eu/.

OECD (2016). *PISA 2015 results (Vol. 1): Excellence and equity in education.* Paris: OECD Publishing.

Roberts, D. A. (2007). Scientific literacy/science literacy. In S. K. Abell & N. G. Lederman (Eds.), *Handbook of research on science education* (pp. 729–780). Mahwah: Lawrence Erlbaum Associates.

Ryan, R. M., & Deci, E. L. (2002). An overview of self-determination theory. In E. L. Deci & R. M. Ryan (Eds.), *Handbook of self-determination research* (pp. 3–33). Rochester, NY: University of Rochester Press.

Schunk, D. (2012). *Learning theories: An educational Perspective* (6th ed.). Boston, MA: Pearson Education.

Scott, P., Asoko, H., & Leach, J. (2007). Student conceptions and conceptual learning in science. In S. K. Abell & N. G. Lederman (Eds.), *Handbook of research on science education* (pp. 31–56). United States of America: Lawrence Erlbaum Associates.

Smith, M. K. (2002). Jerome S. Bruner and the process of education. *The encyclopedia of informal education.* Retrieved from http://infed.org/mobi/jerome-bruner-and-the-process-of-education/.

Taber, K. S. (2011). Constructivism as educational theory: Contingency in learning, and optimally guided instruction. In J. Hassaskhah (Ed.), *educational theory* (pp. 39–61). New York: Nova.

Vygotsky, L. S. (1962). *Thought and language.* Cambridge, MA: MIT Press.

Vygotsky, L. S. (1978). Problems of method. *Mind in society* (M. Cole, Trans.). Cambridge, MA: Harvard University Press.

Wertsch, J. V. (1997). *Vygotsky and the formation of the mind.* MA: Cambridge Press.

Miia Rannikmäe is Professor and Head of the Centre for Science Education, University of Tartu, Estonia. She has considerable experience in science education within Estonia, Europe and worldwide (Fulbright fellow—University of Iowa, USA). She is an honorary doctor in the Eastern University of Finland. She has a strong school teaching background, considerable experience in pre- and in-service teacher education and has strong links with science teacher associations worldwide. She has been a member of a EC high level group publishing a report on 'Europe needs more Scientists'. She has been running a number of EC-funded projects and Estonian research grants. Her Ph.D. students are involved in areas such as scientific literacy, relevance, creativity/reasoning, inquiry teaching/learning and the nature of science.

Jack Holbrook is a visiting professor at the Centre for Science Education, University of Tartu, Estonia. Initially trained as a chemistry/maths teacher in the UK (University of London), Jack spent five years as a secondary school teacher before moving into teacher training, first in the UK followed by Tanzania, Hong Kong and Estonia. Currently, Jack is involved in guiding science education Ph.D. students, European science education projects and being an International Consultant in Curriculum, Teacher Education and Assessment. Jack's qualifications include a Ph.D. in Chemistry (University of London), FRSC from the Royal Society of Chemistry (UK) and Past President and Distinguished Award Holder for ICASE (International Council of Associations for Science Education). Jack has written a number of articles in international journals and as a co-editor a book entitled 'The Need for a Paradigm Shift in Science Education in Post-Soviet Countries.'

Regina Soobard is a research fellow in the Centre for Science education, University of Tartu, Estonia. She earned her Ph.D. in science education at the University of Tartu (2015) on gymnasium students' scientific literacy development based on determinants of cognitive learning outcomes and self-perception. She is teaching at the M.Sc. level and holding the position of director of the gymnasium science teacher programme, as well as co-supervising Ph.D. students in science education and educational sciences. She has been awarded BAFF a scholarship for research in Michigan State University, USA.

Chapter 19
Mediated Learning Leading Development—The Social Development Theory of Lev Vygotsky

Keith S. Taber

Introduction

Lev Vygotsky worked in the Soviet Union (CCCP: Со́юз Сов́етских Социалист́ических Респ́ублик) in the first third of the twentieth century, before dying of tuberculosis at 37 years of age. Considering his early death, and considerable political censure (at one point some of his work could only be read by those to whom the KGB, the CCCP 'secret police', issued a special library pass), Vygotsky's influence on education internationally today is noteworthy. He was very interested in cognitive development, and his work is relevant to education in general (e.g. in terms of pedagogy and assessment) as well as having particular value in supporting learners with specific developmental or learning difficulties and gifted learners. Vygotsky was also very interested in literature and the arts more generally.

Vygotsky wrote in Russian, and most of his writing is in the form of discrete papers. He is best known in the English-speaking world through two books: 'Thought and Language' (1934/1986) and 'Mind in Society' (1978), the latter edited together from a number of his discrete works. An English publication of 'Thought and Language' (it is sometimes considered that it might have been better translated as 'Thinking and Speech' and appears under that title in other editions) included an introduction by Jerome Bruner (see Chap. 13) who recognised the potential importance of Vygotsky's work and sought to publicise it the West.

Vygotsky worked with a number of collaborators (perhaps the best known in the West is Alexander Luria), and his ideas have been adopted, adapted and developed by a range of thinkers working in different national contexts. This chapter introduces Vygotsky's work in terms of some of his best-known ideas with relevance to research and practice in education. In particular, the chapter considers his emphasis on language and the use of symbols as tools, the sociocultural aspect of education

K. S. Taber (✉)
Faculty of Education, University of Cambridge, Cambridge, England, UK
e-mail: kst24@cam.ac.uk

© Springer Nature Switzerland AG 2020
B. Akpan and T. Kennedy (eds.), *Science Education in Theory and Practice*,
Springer Texts in Education, https://doi.org/10.1007/978-3-030-43620-9_19

and development, the zone of proximal development, and his model of cognitive development. These themes are interlinked, and the treatment here will reflect that.

Vygotsky's ideas are complex and have been much discussed and developed. As with all texts, his writings are open to interpretations, something perhaps especially significant when reading in translation. Vygotsky's early death prevented him from fully developing and refining many of his ideas. For example, Vygotsky is said to have dictated the final chapter of *Thought and Language* on his deathbed, giving him no opportunity to review the overall text once the draft was finished. If we see writing as potentially a tool for thinking (a notion that fits well with Vygotsky's perspective), we would expect an author's ideas to develop through the process of writing a book, and authors often review their manuscripts after drafting to ensure consistency. This luxury was not afforded to Vygotsky. This chapter focuses on introducing some of the areas where the legacy of Vygotsky's writings influences current thinking and practice in relation to teaching and learning, and the nature of schooling.

The Importance of the Social in Learning and Development

Vygotsky was interested in human development, and he thought that a full understanding of this topic needed to consider four quite distinct levels or scales. One had to understand the development of the human species as a biological entity, the history of human peoples as they developed culture, the general course of the development of an individual and the development of particular psychological processes as they appear in an individual. The latter required microgenetic studies (Brock & Taber, 2016) that intensely investigated an individual during the time when new processes developed. Vygotsky noted that when such opportunities occurred during psychological experiments (exploring children's responses to tasks under controlled conditions) his contemporaries were usually interested in looking at stable patterns and so ignored the 'training' phase, whilst those patterns were being established. It was that stage of cognition in flux that Vygotsky thought offered most interest.

A key focus of Vygotsky's work was the social nature of learning and development (cf. Chap. 7). He considered that the ability to teach others, and to learn from others, was a characteristic quality of human beings (Moll, 1990). Indeed, Vygotsky went as far as suggesting that it was generally the case that the learning of an individual always involved a process of internalising (to an intra-personal or intra-mental plane) what is first experienced in interaction with others (i.e. experienced on an inter-personal or inter-mental plane) who had already previously internalised that learning. This then is an emphasis on the role of culture (and therefore less directly, history) in the development of the individual. That which affords one to develop as an adult mind operating in some particular society at some point in its history would not be available to a lone epistemic subject learning directly from interactions with the physical/natural (non-social) environment.

This is perhaps obvious in the context of formal education such as in science lessons—children are taught, with varying degrees of success, about Newtonian physics, the circulatory system, atomic structure, and so much more: knowledge they would have negligible chance of acquiring simply through lone direct interrogation of nature. However, Vygotsky was thinking more widely—so even before school the young child learns about the world supported by parents and others. For Vygotsky, development was not purely related to the child being supported to transition into an adult through social mediation. Rather, the nature of human society is that we continue throughout our lives to learn, and develop, through the mediation provided by the culture, that is, through interactions (directly or mediated through various media) with others. Taking this view seriously should have implications for what we see education to be preparation for, and how we consider it is best organised, as well as how we view new forms of media that can mediate enculturation (see Chap. 9).

People then, by the nature of what it is to be human, exist within some specific culture (Geertz, 1973). Such cultures have developed historically, such that they represent the combined development of many generations. Enculturation depends upon mediation by others who already share in aspects of the culture being acquired. However, it is also important to note that Vygotsky's theories were dialectical in nature (he was working in a Marxist state, in more than one sense)—so he did not conceive of a one-way process of the individual absorbing a static culture (cf. Collins, 2010), but rather he thought that the changes the learner goes through can change the context itself. Cultures are themselves in flux (thus history) and subject to diverse influences—so they are always in a kind of unstable equilibrium that may be readily shifted. Vygotsky himself lived in revolutionary times.

Social Constructivism

One area where this social focus is important is the manner in which Vygotsky may be considered a constructivist—in the sense of someone who believes that knowledge is actively constructed (rather than being already innately present in some sense, and being revealed by contemplation or experience; or being acquired by sense impressions that impress fully formed knowledge directly onto mind). Vygotsky was contemporaneous with (the early) Piaget and read and commented on his work. Piaget (see Chap. 10) assumed that the learner was an active constructor of knowledge, and his perspective focused on the learner's actions in and on the environment (Piaget, 1970/1972). Piaget certainly acknowledged the role of social interaction in some learning, but he largely wrote about his epistemic subject as if the social was secondary—and considered young children as too egocentric to effectively learn through social interaction. For Piaget, when young children play together, they are really each playing alone within the same social space, and the ability to genuinely share in authentic collective activity only develops over time (Piaget, 1932/1977).

Vygotsky, however, considered social interaction to be a central part of all human learning. Whereas Piaget's research programme was one of genetic epistemology (finding the common cognitive development sequence that each individual person would be expected to pass through), Vygotsky's programme was sociohistorical: that is, it took the perspective that human psychological developments are mediated by culture and so ultimately contingent on history (Cole, 1990, p. 91). Vygotsky believed that from the age of about two years, development is closely influenced by the young learner's interactions with other minds (Crain, 1992). Vygotsky's perspective, unlike Piaget's, did not suggest a single pattern of development as inevitable for all humans, regardless of their cultural context.

For Piaget, action on the environment supported by existing cognitive structures allowed the development of more advanced structures: which in turn allowed more advanced learning. The nature of science (as primarily a body of theoretical knowledge that develops through the interplay between theory and empirical observation and hypothesis testing) suggests that understanding much school science depends on learners having already acquired the stage of formal operations (Shayer & Adey, 1981). So, for Piaget, "development explains learning" (Piaget, 1997, p. 20).

In contrast to this, Vygotsky considered that learning should lead development. He suggested at one point that "the only 'good learning' is that which is in advance of development" (Vygotsky, 1978, p. 89). At first sight, this seems problematic— if the learning of certain material requires a particular level of development, then without that degree of development the learning should not be possible. However, for Vygotsky 'good learning' is initiated on the inter-mental plane, mediated by others who are further ahead in their own development, so that the learner vicariously experiences what is to be learnt. At this point, the learner is (to borrow a term) a legitimate but *peripheral* participant in the activity (i.e. one who would no longer be able to continue the activity successfully without the support of others—see Chap. 20). Yet, by engaging in the interaction, the learner can begin to internalise and take ownership of the knowledge—and so is able to eventually become a full participant (Lave & Wenger, 1991). Once this process is complete, the individual will be able to demonstrate the learning without the support of the interaction with others. This process is possible because of tools such as symbolic systems that support both (a) communication with others and (b) thinking for oneself.

Tools and Mediation

Vygotsky saw an extensive use of tools as something specifically human. Although he was aware some other animals used tools, he considered human tool use as different in extent and kind. In particular, humans can *use tools to make and improve other tools*, and Vygotsky thought that this second-level use of tools was important to our development. There is a parallel here with Piaget's notion of formal operations, the most developed of his four main stages of cognitive development, where a person can not only undertake mental operations to model aspects of the world, but is able to *mentally operate on those mental operations themselves*. Tools include artifacts such as a stepladder or hammer, but could also be tokens and other signs and symbols.

Another key term in Vygotsky's thinking is mediation. Mediation allows what would otherwise not be possible. Others can mediate for us; and we can use tools (in the external world, or intra-mentally) to mediate activities. This is seen as essentially social in nature, even when a child is solving a problem alone, because the tools they use (be that physical objects or physical tokens of other objects or symbolic tools used in thought) are provided by the culture. The child who has internalised symbolic tools (such as number systems, or, say, chemical formulae) and can now apply them unaided only does so following previous mediated access to such systems in interaction with others. Teaching is the process by which such mediation of learning is deliberately carried out.

One area of work that has developed from that of Vygotsky and his colleagues is that of activity theory (see Chap. 21), or cultural–historical activity theory (CHAT). Vygotsky's work is considered to be the first generation in this tradition and is associated with the mediation triangle which sets out graphically the subject (acting person), object (to be acted upon to some effect) and mediating tool, as the apices of a triangle. This is seen as a dialectical system with each component influencing the rest.

This simple image (Fig. 19.1) is itself of course an example of a symbolic tool. It has been pointed out (Taber, 2014) that it has a strong parallel with the idea of the experiential gestalt of causation, which has been suggested to be a common way in which people understand action in the world, and which influences much learning about natural mechanisms in science classes. Andersson (1986) has suggested that a wide range of reported alternative conceptions in science may be understood in terms of this pattern of thought. Leontiev and others developed a 'second generation' of CHAT which extended the mediation triangle to collective activity by including rules, community and division of labour. CHAT offers an important theoretical perspective for understanding and analysing education (Smardon, 2009).

Fig. 19.1 The general form of the semiotic triangle

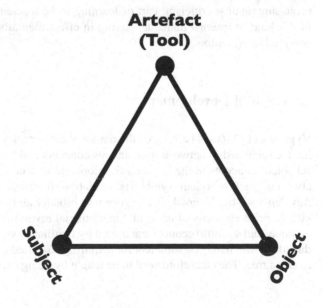

Language in Development

Vygotsky put a strong focus on the role of language in human learning and development. For example, he looked at the role of private speech, talking to oneself, that is, common among young children. Piaget was also interested in this feature, and for him it linked to the egocentric nature of the child: the difficulty young children have decentering from their own perspective and seeing the world from a different viewpoint (Piaget, 1959/2002). Most adults sometimes talk out loud to themselves, but most of their internal dialogue is undertaken as verbal thought without being spoken. Children, however, often accompany an activity with a commentary that is spoken out loud even though only intended for themselves.

Talking to ourselves, whether out loud or internally, invites an explanation. Language is not necessary for thought (not all our conscious thinking is verbal) but is needed for communicating *with others*. Vygotsky suggested that private talk actually had a strong social element, as language had its origins in the need for people to communicate with each other. Vygotsky considered that the child adopted the tools of communication with another as a means to help plan and carry out actions—even when no one else was present. Later, such talk would be internalised, but in its origin private talk is social in nature (Vygotsky & Luria, 1994). This reflects the general principle that in development what is acquired on the intra-personal plane (within the mental life of the individual) follows what is acquired on the inter-personal plane (in interaction with others). Once this tool becomes available, it could be used not only to communicate with others, but to support the individual's thinking, and so aid planning, problem-solving, reviewing experience, etc. A key skill for a scientist is to be able to critique their own ideas, considering the potential objections and challenges others may suggest, and so looking to weed out weak ideas, and strengthen the more promising against criticism. Part of learning to be a scientist is to learn to engage in this kind of internal dialogue, having in effect mentally modelled (internalised) potential interlocutors.

Conceptual Development

Vygotsky (1934/1994) discussed the nature of concept development and, in particular, the relationship between spontaneous concepts and 'scientific' (or academic, or schooled) concepts in the learner's development of a conceptual system. Scientific concepts (such as xylem, symbiosis, oxidation, transition metal, photon, magnetic flux density—but Vygotsky's category was broader and would also include gestalt switch, price elasticity of demand, the industrial revolution, distributive justice, the baroque, and so forth) cannot be acquired by familiarity with instances met in everyday life—which may be sufficient for acquiring so-called natural kind concepts such as cat or tree. They therefore need to be taught by being explained through language.

In particular, Vygotsky considered words to be key tools, acquired through mediation, which were essential to developing high-level thinking and mature concepts. Children may learn new words from conversations without initially having a sophisticated understanding of their intended meaning—clearly limiting their communicative affordances (Fodor, 1972). Vygotsky thought that personal word meanings evolve—a process that can be mediated by talk with others and through internal processes of conceptual development. Once a word is acquired, initial impressions of what it could mean can be tested and developed in conversations with others, and indeed in internal dialogue.

Vygotsky saw conceptual development as an interaction between spontaneous and scientific concepts. Spontaneous concepts, with their experiential grounding, allow scientific concepts to be understood as more than just formal definitions—so, in effect, the student can develop a 'feel' for what is meant by technical notions such as momentum or density or combustion or excretion, or indeed (by building up layers of concepts ultimately grounded in spontaneous concepts) what is meant by atomic orbital, electromagnetic induction or cellular respiration. Scientific concepts provide sophisticated tools for thinking and communicating about spontaneous concepts. So, spontaneous concepts abstracted from perceived regularities in experience can come to be understood in terms of, for example, friction or viscosity or thermoregulation.

The notion of a dialectic is operating here as both types of concept are themselves changed in the interaction—Vygotsky used the image of the spontaneous and scientific concepts moving or growing towards each other. In effect, the resulting system of concepts is neither spontaneous nor scientific (nor just a collection of these two types) but some kind of hybrid that is the synthesis of the thesis–antithesis of spontaneous and scientific concepts. Our mature concepts are actually melded concepts that draw upon both sources (Taber, 2013).

These ideas are reflected in more recent influential work exploring the metaphorical nature of human concepts (Lakoff & Johnson, 1980). This suggests that our abstract conceptions are built upon direct perception in terms of metaphors that allow us to extend the use of terms that originally had direct experiential referents. So, we know what a *big* mistake is, and why the time to the holiday is described as *long*, and so forth. We refer to the element carbon being 'above' that of silicon in a reproduction of a periodic table laid flat on a desk, and to nucleophiles being 'hard' or 'soft'. Darwin (1871/2006) wrote of the 'Descent of Man...', which was an enquiry into whether Homo sapiens had 'descended [sic] from some pre-existing form' (p. 778), rather than an account of his 'fall' from grace.

This is again consistent with the general principle of constructivism: human cognition builds up complex abstractions incrementally from what can be directly perceived in the world (Taber, 2014). Language is a core resource for these processes. A child who understands what *big* means in relation to a big dog, a big chair, a big bed and a big box (i.e. examples where big is something perceived as large in relation to others of its kind) is through mediation via dialogue able to appreciate how within the culture an idea can be said to be big even though an idea is not perceivable and does not have a physical size.

Even with the tools of language, communication between minds is inevitably fallible, and the teaching of concept abstractions is clearly challenging. The teacher is charged with introducing the learner to the cultural tools of the subject being taught (e.g. concepts such as oxidation, transition element, alkali metal, halogen) and helping the learner to engage with these tools with support till the learner can internalise them so that they become part of the available repertoire of interpretive resources for making sense of, and communicating, experience. The skilled teacher will use models, stories, gestures, images, analogies, similes and various other mediating tools (Lemke, 1990; Ogborn, Kress, Martins, & McGillicuddy, 1996): Vygotsky's perspective would suggest that these devices support the process of understanding the abstract concepts in terms of the learners' existing interpretive repertoire of spontaneous concepts (or existing melded concepts deriving partially from them).

Given the importance of spontaneous concepts in concept development, it can be valuable to spend time eliciting student ideas at the start of a topic—a very common constructivist technique in science teaching (Driver & Oldham, 1986). The effective teacher does not just present the academic ideas in the abstract but tries to work with the students' own thinking and shift it towards the target knowledge (Scott, 1998). Dialogue between pupils to share, explain and challenge ideas has been found to have much potential to support this process (Tudge, 1990). Mortimer and Scott (2003) highlight the importance of dialogue in science teaching, and the role of the teacher in eliciting students' ideas and supporting the process of engaging students in active dialogue as they move towards understanding and adopting authoritative science concepts.

The Zone of Proximal Development and Assessment

One of Vygotsky's best-known ideas is the so-called 'zone of proximal development' (ZPD). Vygotsky considered that the usual approach to assessing students by giving them a test they should complete unaided and without reference materials was often inappropriate. He discussed the kind of intelligence testing that calculated students' mental ages. Binet had introduced such tests to identify pupils who were retarded in their development compared with their physical age and who would not benefit from being in class with their same-age peers (Gould, 1992). This was progressive at the time (certainly an improvement on the previous method of measuring the size of a pupil's head). Vygotsky's insight was that several pupils of the same mental age may have very different potentials for further learning in the near future.

Vygotsky imagined a kind of 'phase space' relating to the potential competencies of a learner. At any moment in time, a learner's current level of development would encompass a wide range of competencies, a zone of actual development (ZAD), outside of which lie all those things they cannot yet do (techniques they have not mastered, problems they could not solve, etc.). In effect, traditional educational assessment looked to identify the extent of the ZAD in relation to some particular domain—such as perhaps what the student already knows and understands about acids or the extent to which the student can find solutions to exercises requiring the use of the equations of motion. Vygotsky, however, considered it was much more useful to know about the extent of the zone around the ZAD which reflected what the learner could not yet do autonomously but was ready to do with suitable support (i.e. the ZPD). This zone of next, or proximal, development would (like the ZAD) vary from student to student, and indicated what the student was ready to learn.

If we want to assess people purely in terms of what they can do unaided without support, the traditional test or examination makes sense. If, however, education is about preparing people for their roles in society—where their work will be mediated by others and a wide range of cultural tools—then it would seem to make more sense to assess people in situations that better reflect how people actually work, and learn, in the workplace, in organisations and in other social contexts (see Chap. 20). So, contexts such as project-based learning, working in teams, open-book exams, assessment by interactive interview, etc., would seem much more useful foci for assessment (cf. Chap. 23). In recent decades, there has been a strong emphasis in many countries on a shift from summative assessment to diagnostic and formative assessment—assessment to support learning—at least *during* educational courses if not at their conclusion. Vygotsky was arguing for diagnostic assessment—assessment in, and of, the ZPD—in the 1930s. In science education, there has been an ongoing programme of work to develop diagnostic tools to support diagnostic assessment in teaching (Treagust, 1995).

Scaffolding and Pedagogy

A key notion developed from Vygotsky' ideas is that of 'scaffolding' learning (Wood, Bruner, & Ross, 1976). If one accepts Vygotsky's principle that learning precedes development, then teachers should be looking to get their students working in their ZPD. Students can be very busy (and successful) working in their ZAD, but this does not support further development. Drill and practice might increase efficiency (accuracy, speed) but does not help a student move on to a new level of skill or understanding (cf. Chap. 11). However, by definition, a student given a task considered beyond their ZAD, in their ZPD, will fail: *unless* they are given suitable support (see Fig. 19.2). So, learning activities need to be both beyond the ZAD, and yet mediated to allow success with suitable support. Scaffolding is structure put in place to enable the learners to succeed in such a way that they will learn new competencies.

Scaffolding has entered the educational lexicon, and the term is sometimes used very loosely. Designing educational scaffolding is a challenging task because it has to be matched to the ZPD (Taber & Brock, 2018). Insufficient support leads to frustration and failure. Yet, support that takes over too much of the task will not encourage learning. The scaffolding therefore has to be dynamic, so it moves the learner in manageable stages from legitimate peripheral participation (sometimes starting as just an observer) to taking over full central participation (with the teacher now being purely an observer), giving the learner full agency and allowing the learner to internalise

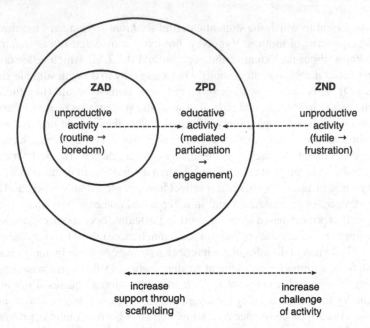

Fig. 19.2 Educative learning activities are those that balance task challenge and the support provided

the new competency. This model is used, for example, in the professional preparation of new school teachers. A new science teacher-in-preparation initially observes the regular teacher at work, before assisting them (perhaps by supporting students during deskwork), before taking responsibility for leading on particular tasks (e.g. introducing a laboratory activity) and so on until they are eventually preparing, teaching and assessing sequences of lessons monitored by the experienced class teacher. Such preparation may include regular shifts between studying in the university and teaching on school placement (Taber, 2017), potentially supporting the development of personal concepts melding classroom experience and taught pedagogic theory.

As one example from school science learning: at the end of secondary chemistry education, a student might be expected to identify an unknown (a cation, an oxidation state, the concentration or purity of a reagent, etc.) through a series of measurements involving back titration. As well as carrying out the laboratory actions, the student will need to access and manipulate chemical equations, and undertake a series of calculations—having first mentally mapped out the activity to conceptualise how a series of processes can lead to the solution to the task. It is expected that a successful student in advanced chemistry can undertake and solve such a problem. Very few students studying at senior school level are initially able to complete such a task even when the appropriate mediational tools (laboratory apparatus, reference works, the relevant symbolic systems, etc.) are available. This is so, even when the components of the process are individually within their ZAD (they know the chemical equations, have the required mathematics, etc.).

The teacher's role here is to set up the learning so that the scaffolding is initially rich enough to take the learner through the task, but is then gradually reduced (the term used is 'faded') as the student internalises more of the individual components. The teacher could begin by reducing the whole activity to a recipe to be followed, but that would likely support limited learning (cf. Fig. 19.2: a key competency here is understanding, and being able to plan according to, the overall logic of which measurements and symbolic manipulations are needed).

The teacher might then instead decide to provide a briefing sheet asking basic questions about relevant prerequisite knowledge that will be needed (perhaps about titration, redox, balancing equations, half-equations, etc.) and a flow chart with all the required stages (to help the student appreciate the logic and see where the steps fit into a larger picture), and a list of the relevant chemical equations. The teacher may also encourage students to work in pairs as this will require them to talk through and explain their thinking to each other. Later in the term when a similar activity is undertaken, the flow chart provided may omit some information that the students have to identify and input, and no chemical equations will be provided. At some later stage, the students will be expected to build the flow chart themselves when planning their activity. Eventually, students would be expected to design and undertake the activity alone, drawing only upon reference materials.

There are different types of possible scaffolding tools that can be introduced. Elsewhere, I have suggested scaffolding planks (platforms for new knowledge) and poles (provided outlines lending support, or provided outlines lending epistemological support) as two examples (Taber, 2002). The 'planks' help the learner identify and organise existing prerequisite knowledge and the 'poles' help set out a framework for carrying out the new activity. Both may be considered to help limit the 'degrees of freedom' among which choices might be made (Taber, 2018). In the back titration example, the titration practical briefing sheet (a scaffolding plank) sets out which previous learning is going to be called upon, and the flow chart (a scaffolding pole) directs the purpose and nature of each stage of the laboratory work and subsequent analysis.

Much teacher scaffolding uses speech. The typical nature of the language game in the classroom, where, for example, teachers ask series of questions to which they already know answers (Edwards & Mercer, 1987), can be seen as functioning as part of the scaffolding process by breaking ideas down into manageable learning quanta, limiting the degrees of freedom within the talk—reducing memory load by highlighting what is to be considered now (a kind of scaffolding plank)—and managing the sequencing of ideas being presented and considered (a kind of scaffolding pole).

Special Needs and Gifted Pedagogy

One area of Vygotsky's work was 'defectology' (a term which seems ugly and incorrect in modern English usage), the study of children for whom development was impeded by some defect. Regardless of the term, Vygotsky's perspective was progressive. Vygotsky felt that too much emphasis was placed on measuring the level of defect, rather than looking to compensate for it. Vygotsky's theoretical perspective implied that for learning activities to be educative they needed to challenge the students in their ZPD but provide support to allow the student to achieve. This suggested that if a child was visually impaired or deaf, for example, this would exclude them from some of the usual cultural mediation supporting the acquisition of the symbolic tools that were the basis of higher cognitive functioning. A child with some disability would fail to develop normally in terms of cognitive development not because of lack of potential of the cognitive apparatus, but rather because normal development would not be mediated in the usual ways. For Vygotsky then the aim was to find compensatory means to provide the tools needed for development. Students need to be provided with support in their ZPD, and if the usual means of mediation were not accessible, alternatives needed to be found or developed. (An example would be braille as an alternative to print—an alternative tool for accessing texts.)

One area sometimes classed under special needs or inclusion is the issue of those students who are considered as 'gifted' (or in different educational contexts, 'talented' or of 'high ability'). Conceptions and definitions of giftedness vary, but in many educational contexts there will be some students who have developed further than their peer group such that learning activities which are appropriate for most of the group have little value for the gifted learner (Taber, 2007).

Whilst some traditional approaches treat the gifted child as a discrete category from others, it is also possible to see the label of giftedness as dynamic and contextual—that is, specified in relation to a particular lesson and activity—so that who is considered as gifted might vary over time and according to curriculum subject or even topic. Some students will have extensive experience of part-time work or hobbies or cultural traditions which put them at a very different starting point for learning particular material. An obvious example would be a child from a bilingual home in a class being introduced to a 'foreign' language that is effectively L1 (first language) for *that* child. In science, some students bring to class extensive experience of building mechanisms or circuits, or collecting natural history specimens, or amateur astronomy.

Vygotsky's theory suggests that 'good learning' takes place in the ZPD, and therefore educative experiences are those experiences that are both challenging and suitably supported (see Fig. 19.2). Activities that are within the ZPD of most students in a class may well fall within the ZAD for gifted students and so have little educative value for them. In principle (if not in practice), the solution is simple: the teacher needs to shift the balance between challenge and support for the different students in a class. Gifted learners require more challenging activities, or less scaffolding, than others in the class (Taber, 2016).

Whilst the need for more challenge for these students is widely recognised, Vygotsky's theory offers a novel perspective suggesting starting planning teaching so lesson activities are challenging for the most able in the class, and then designing differentiated scaffolding to provide the optimal balance of challenge and support for all the different students in the class. If teachers are able to plan differentiated teaching in this way, there ceases to be any value in labelling particular students in a class as gifted or having special needs.

Further Reading

Daniels, H. (2001). *Vygotsky & Pedagogy*. London: RoutledgeFalmer.
Newman, F., & Holzman, L. (1993). *Lev Vygotsky: Revolutionary scientist*. London: Routledge.

References

Andersson, B. (1986). The experiential gestalt of causation: A common core to pupils' preconceptions in science. *European Journal of Science Education, 8*(2), 155–171.

Brock, R., & Taber, K. S. (2016). The application of the microgenetic method to studies of learning in science education: characteristics of published studies, methodological issues and recommendations for future research. *Studies in Science Education*, 1–29. https://doi.org/10.1080/03057267. 2016.1262046.

Cole, M. (1990). Cognitive development and formal schooling: The evidence from cross-cultural research. In L. C. Moll (Ed.), *Vygotsky and education: Instructional implications and applications of sociohistorical psychology* (pp. 89–110). Cambridge: Cambridge University Press.

Collins, H. (2010). *Tacit and explicit knowledge*. Chicago: The University of Chicago Press.

Crain, W. (1992). *Theories of development: Concepts and applications* (3rd ed.). London: Prentice-Hall International.

Darwin, C. (1871/2006). The descent of man, and selection in relation to sex. In E. O. Wilson (Ed.), *From so simple a beginning: The four great books of Charles Darwin* (pp. 767–1248). New York: W W Norton & Company.

Driver, R., & Oldham, V. (1986). A constructivist approach to curriculum development in science. *Studies in Science Education, 13*, 105–122.

Edwards, D., & Mercer, N. (1987). *Common knowledge: The development of understanding in the classroom*. London: Routledge.

Fodor, J. A. (1972). Some reflections on L. S. Vygotsky's Thought and Language. *Cognition, 1*(1), 83–95.

Geertz, C. (1973). *The interpretation of cultures: Selected essays*. New York: Basic Books.

Gould, S. J. (1992). *The mismeasure of man*. London: Penguin.

Lakoff, G., & Johnson, M. (1980). The metaphorical structure of the human conceptual system. *Cognitive Science, 4*(2), 195–208.

Lave, J., & Wenger, E. (1991). *Situated cognition: Legitimate peripheral participation*. Cambridge: Cambridge University Press.

Lemke, J. L. (1990). *Talking science: Language, learning, and values*. Norwood, New Jersey: Ablex Publishing Corporation.

Moll, L. C. (1990). Introduction. In L. C. Moll (Ed.), *Vygotsky and education: Instructional implications and applications of sociohistorical psychology* (pp. 1–27). Cambridge: Cambridge University Press.

Mortimer, E. F., & Scott, P. H. (2003). *Meaning making in secondary science classrooms*. Maidenhead: Open University Press.

Ogborn, J., Kress, G., Martins, I., & McGillicuddy, K. (1996). *Explaining science in the classroom*. Buckingham: Open University Press.

Piaget, J. (1932/1977). *The moral judgement of the child*. Harmondsworth: Penguin Books.

Piaget, J. (1959/2002). *The language and thought of the child* (3rd ed.). London: Routledge.

Piaget, J. (1970/1972). *The principles of genetic epistemology* (W. Mays, Trans.). London: Routledge & Kegan Paul.

Piaget, J. (1997). Development and learning. In M. Cole, S. R. Cole, & M. Gauvain (Eds.). *Readings on the development of children* (2nd ed., pp. 19–28). New York: W. H. Freeman.

Scott, P. H. (1998). Teacher talk and meaning making in science classrooms: A review of studies from a Vygotskian perspective. *Studies in Science Education, 32*, 45–80.

Shayer, M., & Adey, P. (1981). *Towards a science of science teaching: Cognitive development and curriculum demand*. Oxford: Heinemann Educational Books.

Smardon, R. (2009). Sociocultural and cultural-historical frameworks for science education. In W.-M. Roth & K. Tobin (Eds.), *The world of science education: Handbook of research in North America* (pp. 15–25). Rotterdam, The Netherlands: Sense Publishers.

Taber, K. S. (2002). *Chemical misconceptions—Prevention, diagnosis and cure: Theoretical background* (Vol. 1). London: Royal Society of Chemistry.

Taber, K. S. (2007). Science education for gifted learners? In K. S. Taber (Ed.), *Science education for gifted learners* (pp. 1–14). London: Routledge.

Taber, K. S. (2013). *Modelling learners and learning in science education: Developing representations of concepts, conceptual structure and conceptual change to inform teaching and research*. Dordrecht: Springer.

Taber, K. S. (2014). *Student thinking and learning in science: Perspectives on the nature and development of learners' ideas*. New York: Routledge.

Taber, K. S. (2016). Giftedness, intelligence, creativity and the construction of knowledge in the science classroom. In K. S. Taber & M. Sumida (Eds.), *International perspectives on science education for the gifted: Key issues and challenges* (pp. 1–12). Abingdon, Oxon: Routledge.

Taber, K. S. (2017). Working to meet the needs of school pupils who are gifted in science through school-university initial teacher education partnerships. In M. Sumida & K. S. Taber (Eds.), *Policy and practice in science education for the gifted: Approaches from diverse national contexts* (pp. 1–14). Abingdon, Oxon.: Routledge.

Taber, K. S. (2018). Scaffolding learning: Principles for effective teaching and the design of classroom resources. In M. Abend (Ed.), *Effective teaching and learning: Perspectives, strategies and implementation* (pp. 1–43). New York: Nova Science Publishers.

Taber, K. S., & Brock, R. (2018). A study to explore the potential of designing teaching activities to scaffold learning: understanding circular motion. In M. Abend (Ed.), *Effective teaching and learning: Perspectives, strategies and implementation* (pp. 45–85). New York: Nova Science Publishers.

Treagust, D. F. (1995). Diagnostic assessment of students' science knowledge. In S. M. Glynn & R. Duit (Eds.), *Learning science in the schools: Research reforming practice* (pp. 327–346). Mahwah, New Jersey: Lawrence Erlbaum Associates.

Tudge, J. (1990). Vygotswky, the zone of proximal development, and peer collaboration: Implications for classroom practice. In L. C. Moll (Ed.), *Vygotsky and education: Instructional implications and applications of sociohistorical psychology* (pp. 155–172). Cambridge: Cambridge University Press.

Vygotsky, L. S. (1934/1986). *Thought and language*. London: MIT Press.

Vygotsky, L. S. (1934/1994). The development of academic concepts in school aged children. In R. van der Veer & J. Valsiner (Eds.). *The Vygotsky reader* (pp. 355–370). Oxford: Blackwell.

Vygotsky, L. S. (1978). *Mind in society: The development of higher psychological processes*. Cambridge, Massachusetts: Harvard University Press.

Vygotsky, L. S., & Luria, A. (1994). Tool and symbol in child development. In R. van der Veer & J. Valsiner (Eds.). *The Vygotsky reader* (pp. 99–174).

Wood, D., Bruner, J. S., & Ross, G. (1976). The role of tutoring in problem solving. *Journal of Child Psychology and Psychiatry, 17*(2), 89–100. https://doi.org/10.1111/j.1469-7610.1976.tb00381.x.

Keith S. Taber is the Professor of Science Education at the University of Cambridge. Keith trained as a graduate teacher of chemistry and physics, and taught sciences in comprehensive secondary schools and a further education college in England. He joined the Faculty of Education at Cambridge in 1999 to work in initial teacher education. Since 2010, he has mostly worked with research students, teaching educational research methods and supervising student projects. Keith was until recently the lead Editor of the Royal Society of Chemistry journal 'Chemistry Education Research and Practice', and is Editor-in-Chief of the book series 'RSC Advances in Chemistry Education'. Keith's main research interests relate to conceptual learning in the sciences, including conceptual development and integration. He is interested in how students understand both scientific concepts and scientific values and processes.

Chapter 20
Situated Cognition and Cognitive Apprenticeship Learning

Gultekin Cakmakci, Mehmet Aydeniz, Amelia Brown, and Joseph M. Makokha

Introduction

Throughout the educational literature, there has been a shift from the behaviorist to constructivist theories of learning (Aikenhead, 1996). Besides, within constructivist theories of learning, there has been a substantial shift from radical to social constructivism theories of learning. Accordingly, these theories have been influential in the design of a number of curricula. For instance, in some countries such as Germany, France, Switzerland, and South Korea, with dual education systems, people engage in many apprenticeship occupations (e.g., carpenter, dentist's assistant, electrician) in collaboration between companies/industries and schools. The dual education system is seen as an effective system in particular in vocational schools for creating a fourth industrial revolution (often called industry 4.0) ecosystem (Leopold, Ratcheva, & Zahidi, 2016) and also for promoting participants' social and emotional skills (OECD, 2017). In the U.S. constructivist views of learning has dominated the discussion around curriculum development efforts in science and mathematics education as well. With the publication of the National Science Education Standards in 1996 and the Next Generation Science Standards (Achieve Inc., 2013) in 2013,

G. Cakmakci (✉)
Hacettepe University, Ankara, Turkey
e-mail: cakmakci@hacettepe.edu.tr; gultekincakmakci@gmail.com

M. Aydeniz · A. Brown
University of Tennessee, Knoxville, USA
e-mail: maydeniz@utk.edu

A. Brown
e-mail: aabrown@vols.utk.edu

J. M. Makokha
Stanford University, Stanford, USA
e-mail: makokha@stanford.edu

© Springer Nature Switzerland AG 2020
B. Akpan and T. Kennedy (eds.), *Science Education in Theory and Practice*,
Springer Texts in Education, https://doi.org/10.1007/978-3-030-43620-9_20

a greater emphasis has been placed on students' participation in authentic scientific practices such as inquiry, modeling, and argumentation. Student's effective and meaningful participation in such practices can best be guided and interpreted through situated cognition and cognitive apprenticeship theories of learning.

This chapter discusses situated cognition and cognitive apprenticeship learning, which are situated within social constructivist approaches to instruction. More specifically, we focus on the contributions of Brown, Collins, and Duguid (1989) and Collins, Brown, and Holum (1991) to the establishment and evolution of these theories and the relevance of these theories for reform efforts in science education.

According to situated learning theory, learning is a social activity that takes place when someone is doing something in a social context; accordingly, learning environment has social, cultural, and physical contexts (Brown et al., 1989; Lave & Wenger, 1991; Vygotsky, 1978). Cognitive apprenticeship learning reflects situated learning theory (Collins et al., 1991; Rogoff, 1990). The notion of apprenticeship has been influential in teaching and learning throughout history. Children learn their first languages from their families; novices learn how to grow crops, make houses, do farming, cooking; employees learn job skills; and scientists learn how to carry out research by working with seniors. Cognitive apprenticeship focuses on enculturing learners into adopting the cognitive processes and skills of those who are legitimate participants of a particular community through scaffolding. Therefore, cognitive apprenticeship suggests that the learning environment should be designed to make targeted cognitive processes explicit and visible so that students can observe, enact and practice them in contexts that make sense to them and can enhance their domain specific as well as domain-general knowledge and skills.

Situated Cognition

Educational scientists have used learning theories to understand how learning takes place, how knowledge and skills are acquired, and how these knowledge and skills are used in different contexts. Empirical and theoretical developments in learning sciences have led to the emergence of the situated cognition (Brown et al., 1989; Collins & Greeno, 2010; Smith & Semin, 2004), whose main argument assumes that cognition is fundamentally a social activity, and is distributed across members of a learning community and that knowledge is situated in the contexts, cultures, and activities, in which it is produced and used (Clancey, 1997; Robbing & Aydede, 2009; Roth & Jornet, 2013; Wilson & Myers, 2000). Two other assumptions of situated cognition theory are that: (i) "cognition arises in, and for the purpose of, action, thus cognition is enacted" and "cognition is distributed across material and social settings because of features" (Roth & Jornet, 2013, p. 2); and (ii) *cognition becomes distributed when* a team of people engages in solving a problem through talk, questioning, and coordination of cultural and representational tools (Hutchins, 1995). These assumptions of situated cognition are rooted primarily in *Cultural Historical Activity Theory*, which considers "thinking, acting (praxis), and environment as interacting and dependent parts of the same analytic unit" (Roth & Jornet, 2013, p. 2).

In the 1980s Brown et al. (1989) published one of their most influential seminal works *Situated Cognition and the Culture of Learning*. Since this publication, the impact of the situated cognition perspective on learning has influenced many fields ranging from education, social psychology, communication, and computation. Yet, the social cognition perspective has had a tremendous impact on the field of education. According to situated learning theory, learning arises from the dynamic interaction between the learner and the environment in which the learning takes place (Roth & Jornet, 2013). Therefore, scholars who have conducted research in this domain, have focused on learner's actions in connection with their cognition in a specific social, cultural, and physical context as opposed to taking learners' mental processing of information as the sole unit of analysis (Roth & Jornet, 2013). One of the main contributions of situated cognition to the field of learning is that "perceiving, remembering, or reasoning are not independent phenomena—to be explored as operations of the brain alone—but are integral to agents-in-their-context-acting-for-a-purpose-and-with-tools." (Roth & Jornet, 2013, p. 473). In science education, these assumptions have become very instrumental in our understandings of what and how students learn science in authentic scientific contexts. For instance, situated cognition can help us explain how students may be able to appropriate the goals, epistemologies and practices of scientists as they learn science.

Lave (1991) located cognition in practices, rituals (patterned actions) that are specific to certain cultural communities, and learning as a process of legitimate peripheral participation (Lave & Wenger, 1991) in these patterned actions and of cognitive apprenticeship (Lave, 1988). This is a radical shift from traditional views of cognition where learning is viewed as the acquisition of knowledge through information processing and construction of mental representation of the external world. This new perspective views learning "in terms of expanding the learner's action possibilities in larger systems of activity" (Roth & Jornet, 2013, p. 4) rather than limiting learning to cognitive phenomena solely to encoding, retrieval, or processing of information.

Language and other cultural tools of practice play a crucial role in this new perspective on learning. According to situated cognition, "language is not a system of correspondences between symbols and elements in the world, but a means for humans to coordinate their situated actions, with others and for agents to stimulate their own minds" (Roth & Jornet, 2013, p. 468). From a situated cognition perspective, "cognitive phenomena are not restricted to what happens inside the brain, but refer to the interactions within the person-in-situation unit" (Roth & Jornet, 2013, p. 468), often via language.

If teaching practices and methods were viewed as an evolutionary timeline, most of the timeline would be dominated by what are commonly referred to as conventional teaching methods. These are the methods that many of us experienced in school, such as lectures, presentation, note-taking, memorization practices and techniques, worksheets, and many more. These conventional teaching methods take on multiple manifestations in the classroom, but share the common characteristics of teachers somehow being in charge of transferring required knowledge to students.

This experience is counter to how science is practiced and how scientific models are constructed, evaluated, and critiqued in authentic scientific contexts.

These conventional teaching practices are more recently referred to as the "Banking" model of education, based on the writings of Paulo Freire. Freire (2005) used the term "Banking" to intentionally show that teachers were in control of depositing information into students, and students were thus passive (and thus in a power-negative and oppressive situation) in the learning process. Freire argued for liberating educational practices; namely, educational practices that empower instead of oppress students. Freire proposed multiple methods of achieving liberating education, including allowing students to construct their own learning by recognizing the cultural capital of students and the context in which the learning takes place as essential to the learning process.

Freire's concepts are often combined with the works of Dewey and Piaget to form the basis for a modern constructivist model of education. Piaget, often called the father of constructivism, tirelessly promoted the importance of human experiences and the learning process. Dewey echoed these calls, especially in the realm of science, by encouraging laboratory experiences in the sciences to encourage real-world learning experiences and problem-solving skills.

Research on educational methodology based on the theoretical frameworks of Freire, Piaget, and Dewey is now commonplace. The past 50–75 years on our educational timeline shows a clear shift away from the conventional banking model of education toward the various methods that a constructivist and/or liberating construct of education can manifest in a teaching and learning environment. An examination of this research shows two related yet distinct veins of investigations: research into the social interactions and contexts of the educational process (largely related to Freire's concepts of liberating education) and research into the cognitive and conceptual processes and procedures of knowledge acquisition (largely related to constructivist theories of education).

Situated cognition (also referred to as situated learning) recognizes the importance of overlapping these two research veins and theoretical frameworks. As defined by Collins and Greeno (2008), situated cognition is "the view that knowing and learning by individuals are inextricably situated in the physical and social contexts of their acquisition and use" (p. 335). Vosniadou, Loannides, Dimitrakopoulou, and Papademetriou (2001) explain the situationality of knowledge by stating "students do not come to school as empty vessels but have representations, beliefs, and presuppositions about the way the physical world operates" (p. 392). Brown et al. (1989) further elaborate that all knowledge is situated, not just in the teaching and learning process but also in the "context and culture in which it is developed and used" (p. 32). This has significant implications for science education. Scientific practice, its goals, epistemologies, the knowledge it produces and the process that lead to the production of that knowledge is not only context driven but also influenced by sociocultural practices of the community in which it is being practiced.

If students were empty vessels no construction of knowledge would be needed; we could simply fill the empty vessel with knowledge. Instead, effective science instruction must recognize that culture and society frame both the knowledge a student possesses upon entering school, as well as the knowledge and skills the student is expected to obtain once in the classroom setting.

In order to understand the role of situated cognition in education and research, we need to clarify what is meant by the terms knowledge, the role of context in learning, social context, cultural context, physical context, and activity.

Knowledge

Situated cognition recognizes that "knowledge is social, and no other knowledge is more social than any other" (Khan, Mitchell, Brown, & Leitch, 1998, p. 772). Examples of this viewpoint of knowledge abound, including Brown et al.'s (1989) description of language acquisition. Brown et al. (1989) point out that while dictionaries are useful resources, we do not teach children to read, write, and speak by sitting them down in front of a dictionary. Language acquisition cannot happen by an individual alone, even with useful resources; acquiring the knowledge of using and understanding a language is a social event that requires multiple interactions between several individuals in the social system.

Given the social nature of knowledge, we also can see that knowledge is contextual. If learning is social, that means that all learning has a social context, and thus all learning is contextual. Brown et al. (1989) explain the contextual nature of knowledge by simply pointing out that the jargon, slang, accent, and even the language that a child learns to use depends directly on the cultural context where their socially dependent learning takes place.

Finally, with a recognition that knowledge is both social and contextual, one can naturally ask the question of how to take this social and contextual knowledge and transition into the more specialized body of knowledge required of many scientists. Children may learn to read and write and communicate from the social interactions driven by the rich cultural tapestry where they spend their formative years; but how do these children socially and culturally learn the knowledge and skills necessary to perform surgery or conduct research or engineer technological improvements? For this explanation, we look to the concept of cognitive apprenticeship (Collins, Brown, & Newman, 1987), which will be discussed later in this chapter.

Role of Context in Learning

Now that we understand how knowledge is defined, we will look closely at the essential component of situated cognition, which is that knowledge, is "inextricably situated" in context (Collins & Greeno, 2008). Situated cognition recognizes several contexts that are closely linked to knowledge acquisition and use, including social context, cultural context, and physical context. While we want to emphasize that these contexts are all interrelated, we now look at them individually in order to examine the unique applications to science education of each individual context type.

Social Context

In addition to recognizing the role of social interactions in the learning process, situated cognition recognizes that the social constructs and identities of the members of a community of learners impact the learning process (Gee, 1997). Of particular interest to science educators is the role of social identity and how that contributes to the social context of learning science. Social identities have been shown to directly impact both achievement in science as well as motivation to pursue science higher education and careers. One example is that students with female gender identity are often less likely to pursue science fields in higher education and/or careers in the sciences (Aydeniz & Hodge, 2011; Carlone & Johnson, 2007). Identity development as females in a social context that promotes males as the dominant learning group of scientists perpetuates this participation gap. Tan, Calabrese-Barton, Kang, and O'Neil (2013) observed that when the school classroom environment is not supportive of identity-based learning, female students who had previously expressed interest in science actually lost interest or distanced themselves from pursuing higher science education. Riedinger (2015) found that youth derive their sense of self and identity from perceived membership and belonging in a learning group. Thus, negotiating and developing one's identity as a member of the learning group, such as female science students needing to navigate social roles and power dynamics unique to female science students, are essential to science learning. The importance of female science students needing to navigate identity development in the science classroom is only one example of social context and its impact on science learning. In a social context where females were not statistically shown to participate in science careers at much lower rates than their male counterparts, or in a social context where textbooks and other learning media did not over-represent males as practitioners of science, the role of identity development in science learning for female students would not be of much concern. It is the role of a practitioner of situated cognition to identify the social contextual factors unique to their learning environment and recognize these as a part of daily practice.

Cultural Context

Brown et al. (1989) place such importance on the cultural context of learning that they create a term for this: "enculturation" (p. 33). While it is easy to understand how a child's language acquisition (to refer to our earlier example) is dependent on the cultural context in which learning takes place, many struggle to see how this concept applies to science learning. Science taught in schools often minimizes or leaves out entirely the cultural context of the scientific understanding in favor of the scientific facts as they are currently understood and explained. Thus, when scientific understanding involves as a product of new technology or new research, many students of science are left behind, clinging to their notions of science as they learned them in school based on the misguided misunderstandings that science is universally above cultural influence. The recognition of the cultural context of learning as provided by the situated cognition framework is especially helpful to science teachers as a method to prevent these common misconceptions regarding the nature of science.

Physical Context

The physical context of where learning takes place is often seen as troublesome from the situated cognition standpoint. While Dewey (1938) was successful in implementing more experiential learning in the sciences through additions of laboratory activities, more recently we have begun to question the authenticity of these science learning experiences. The idea that students must engage in practices common to their subject area, as well as learning experiences that are meaningful to the social and cultural world outside of school is often referred to as authenticity.

Brown et al. (1989) point out that school activities are inherently inauthentic for several reasons: (1) often school activities do not incorporate the social and cultural aspects of learning, as discussed above, making them inauthentic learning experiences, (2) the practices taught and expected in school are not the practices expected by experts or practitioners in the field, and (3) even if a school or teacher attempts to address either or both #1 and #2, the culture of the school and the classroom context often overshadow these attempts, creating at best a "hybrid" learning activity rather than an authentic learning activity (p. 34). In addition to promoting the benefits of authentic learning, Brown et al. (1989) caution that inauthentic school activities and assignments lead to ineffective learning, stating that these inauthentic environments "create a culture" of "phobia" for the subject area being presented. (p. 34). Echoing Brown et al.'s sentiments, Bricker and Bell (2014) state that school can be disruptive to science learning, specifically that the formality of the classroom setting is not conducive to a learning pathway that considers culture and identity as an aspect of science learning (Aikenhead, 1996). As creating a phobia or lack of motivation toward science is not the goal of any conscientious science teacher, special attention needs to be allocated toward the contextual authenticity of learning experiences in the science classroom.

We must also recognize that the physical context of learning—where the learning takes place—is largely dependent on the social and cultural context of learning. School quality, both in teacher quality and availability of resources, varies widely based on the socioeconomic level and cultural respect for education in the area in which the school is located. This has led to the proliferation of alternate learning environments, often referred to as place-based learning or out-of-school learning, in science education. From the situated cognition standpoint, there is certainly potential for place-based learning and out-of-school learning to provide more authentic learning experiences than can be provided in a school classroom. However, eventually science educators will need to correct and adjust the classroom climate to provide more authentic, socially, and culturally contextual science education experiences. Relying on experiences out of school to correct for the lack of situated cognition in school is shortsighted at best, and at worst discriminatory toward those who cannot attend the out-of-school experiences.

Activity

From the situated cognition standpoint, we have discussed the nature of knowledge, and the contexts in which this knowledge occurs (or does not occur, as the case may be). There is one more component of situated cognition to discuss, and that is the activity of learning. All learning or attempted learning is activity. Brown et al. (1989) forcefully attest that "the activity in which knowledge is deployed... is not separable from or ancillary to learning and cognition. Rather, it is an integral part of what is learned" (p. 32).

Fortunately learning activities are best suited for science education and situated cognition abound. In recognition of the need for meaningful, practitioner-based activities, science education offers problem and project-based learning, modeling, visualization, argumentation, collaborative learning, questioning, forecasting, labs, and experiments, etc. The role of the teacher in the science classroom is often creating, selecting, preparing, and delivering these activities for their students. Many resources are available to teachers in the quest to select activities that will lead to knowledge. However, science educators must remember, "different ideas of what is appropriate learning activity produce very different results" (Brown et al., 1989, p. 32). This means that the activity you acquire from a science educational supplier might work one year and not the next. Or an activity you received from a colleague in a school across town might have been magical for their classes but a total failure for your class. Or, that list of labs that all science teachers in your district are supposed to complete with fidelity to the instructions—well, probably not all of them will be successful in your classroom. Why? According to situated cognition, activity is integral with learning, and learning is dependent on context, therefore the learning successes of classroom activities vary according to the classroom social, cultural, and physical context. The role of the effective science teacher is not just selecting authentic activities as good learning experiences, but tailoring and executing these activities based on their professional knowledge of the unique contexts within and surrounding their classroom and the goals of their curriculum.

Given the complexities that are now apparently involved with becoming a science teacher practitioner of situated cognition, there is no list of lesson plans or labs that we can distribute as examples of situated cognition in the science classroom. Examples do exist, yet these examples are often discussed in the context of the features they contain rather than a step-by-step implementation plan for use in the classroom. This lack of demonstrability certainly leads to the rift between the theory of situated cognition and implementation of the tenets of situated cognition in the classroom. In order to help bridge this rift, we offer the following reminders for those looking to promote situated cognition in the science classroom:

- The traditional banking model of education offers limited opportunity for situated learning to occur;
- Knowledge and learning are socially, culturally, and contextually situated;
- Promotion of identity development alongside of science learning is key to addressing the social context of science learning;
- Ignoring the cultural impact on science will not promote an accurate conception of the nature of science;
- School settings have the potential to be detrimental to authentic science learning activities;
- While no activity is fail-safe in all educational contexts, the activity chosen must allow for students to construct their own knowledge; and
- The individual responsible for tailoring instruction to meet the needs of all learners by selecting appropriate learning activities and recognizing the social and cultural components of science learning within those activities is ultimately the science teacher.

Cognitive Apprenticeship

Cognitive apprenticeship is an important construct in describing cognitive and social growth in children. Rogoff (1990) argues that children's development is an apprenticeship in nature. Children are guided to participate in social activity within the social community who supports their understanding of the cultural norms of the social group, and development of skill in using the tools of the culture they belong. Teaching and learning have been based on apprenticeship throughout history with a different emphasize. Nonetheless, in education, there has been a move from traditional apprenticeship to cognitive apprenticeship. A focus on cognitive skills and process rather than only physical skills development, the use of skills in varied contexts rather than only the context of their use, and the use of structured rather than entirely naturalistic opportunities for skill development differentiate *cognitive apprenticeship* from *traditional apprenticeship* (Collins et al., 1991). When we teach science, we are enculturating students into the community of scientists and expect them to acquire epistemology, knowledge skills, ways of thinking, and tools of the scientific or engineering community.

Collins et al. (1991) suggested four dimensions that should be considered while designing learning environment based on cognitive apprenticeship learning: *content*, *method*, *sequencing*, and *sociology*. In addition, they also suggested a pedagogical

Fig. 20.1 On the first picture two learners carry out a task from a real-world context. On the second picture, the teacher facilitates their learning by explicitly discussing key scientific concepts and practices in the task (*Photograph* © Gultekin Cakmakci)

framework that included six processes teachers would use to promote student learning: *modeling, coaching, scaffolding, articulation, reflection*, and *exploration*. In this chapter, we framed design thinking methodology from a cognitive apprenticeship perspective with these four dimensions and six processes of cognitive apprenticeship learning (Brown, 2009; Cross, 2011). We believe that as represented in Fig. 20.1, pedagogical practices of cognitive apprenticeship and strategies like design thinking (Cross, 2011) would help teachers to make key aspects of thinking visible to students (Cakmakci, 2012; Collins et al., 1991).

Design Thinking from a Cognitive Apprenticeship Perspective

Humans have been designing since antiquity. Design thinking is a method of solving problems in a practical, creative, iterative way that can be applied in different domains (Cross, 2011). In this method, one begins by identifying the need or problem, then proceeds with understanding the context, within which a solution is implemented and tested, then refined using feedback from users. This exemplifies the cognitive apprenticeship theory given that learners encounter authentic tasks and real-life situations; interact with skilled instructors and coaches to learn domain-specific and domain-general skills; focus on cognitive rather than only physical skill development through deliberately planned activities; and the use of methods that scaffold learning. The result from applying design thinking, a hands-on learning method, is that students are likely to better understand, internalize and apply learned concepts. The hands-on nature also lends itself to science teaching as well as many other domains.

This approach allows learners to encounter concepts within real-world settings where they observe from, and enact solutions with the help of their instructors—who scaffold the learners as they practice their skills. These concepts may span different disciplines (Brown, 2009) from physics, material science, anthropology, biology, psychology, and others, which together form the basis of solutions to problems ranging from tasks such as creating a better electronic device to designing a modern patient care facility that takes advantage of cutting-edge technology. Using this method, students have the opportunity to come up with different ways of applying the knowledge and skills gained from learning activities while interacting with their instructors, then crafting solutions and at times coming up with novel ideas. They make sense of their scientific knowledge within the given contexts from interacting and understanding their users in a real-world setting—and solving real-world problems with help from experts—which fosters a higher level of learning and mitigates issues around authenticity, context, and thought processes (Brown et al., 1989) common to other styles. A briefcase study using a project-based college class illustrates this learning method.

In one such year-long class that employs design thinking—in the engineering department of a leading university, (ME310 Design Methodology, www.goo.gl/W9AS8C)—novice product design students are presented with complex problems from different industries, and of varying specifications from open to very narrow prompts, and asked to craft solutions. For instance, the tasks may vary from designing the next generation space shuttle for a leading aeronautics company to coming up with a single detailed feature of a smart building to be constructed overseas. In order to successfully solve such tasks, students not only have to understand the context—physical, social, and conceptual—within which they are working, but also the relevant tools and technologies available within these contexts. These examples demonstrate the in-context nature of the learning environment, where the students explore the problem space to understand their intended users and their corresponding needs, followed by idea generation based on what they have learned from interaction with these users, and finally applying their science skills to create tangible solutions. The process is iterative since new insights from users often lead to a point of view which might inspire ideas that once prototyped, point to other new ideas and may even require a new round of observations in order to understand new aspects—which may have been previously disregarded or were deemed insignificant.

As for the cognitive process, these ME310 students engage with multiple stakeholders under the guidance of their instructors and coaches, learning by doing actual design work despite their limited experience in the industries they engage with; while creating knowledge, exploring new concepts and immediately applying new knowledge to their designs. This apprentice model therefore presents both the *maker* aspects as well as the cognitive apprenticeship characteristics through engaging with design tasks under the supervision of experienced faculty members and industry professionals. Another important feature of this program is the industry partnership from which one or more corporate personnel are provided to actively engage with each ME310 team that is working on their task. They often bring extensive knowledge and skills, as well as connections with the corporate entities that might be interested in the outcome, adding yet another resource for the students. These corporate liaisons act as sounding boards for the students' ideas and also provide guidance for the students as they get out to visit actual industries, users and spaces, allowing them to investigate every important aspect related to their task.

Students in the ME310 class come in as novices and transform within a year to accomplished engineering designers with a tangible product developed under the guidance of specialists from whom they learn along the way, while also creating new knowledge by combining different aspects of their experience. This learning experience can be simplified into three general, distinct steps: understanding the process, practicing the process, and delivering a target solution. In each of these steps, students are guided as they explore, discover, and apply new knowledge in solving complex problems under the supervision of their instructors and coaches. We can therefore view design thinking within three broad aspects under this framework: understand; practice the process; and delivering solutions.

Understanding the Process

This first stage involves getting the students to understand the design thinking methodology and equipping them with the basic skills required to effectively conduct user-research to understand the context within which their problem and solution lies. It takes advantage of the curiosity that learners have toward science, people, and their interactions with their surroundings—which effectively provides the contextual setting.

Practicing the Process

Once the students get the general ideas around design thinking, they are presented with fast-paced tasks to get them familiar with the concepts. They may be asked to identify a problem (discover a need) within a specific space, propose solutions and then test them to find out if they fit. This process is often fast paced in order to give the students a chance to explore multiple possibilities instead of concentrating on perfecting a single idea. One common introductory task is building of "paper bikes"—something that few, if any of incoming students have ever done before, allowing them to explore their creative imagination and to employ the many science skills and knowledge they already possess. Beyond this, they engage with designated industries to begin exploring their long-term project such that subsequent prototypes reflect identified problems/needs within their space.

Delivering Solutions

The student teams are each sponsored by a corporate entity, and while they are composed of students from two to three universities from around the globe, they work with and learn from all instructors and eventually deliver a finished product to their sponsor. Given that different schools offer different areas of specialization, the instructors, coaches, and partners ensure that each team leverages their differences— for example, industrial design, mechatronics, and manufacturing in one team. They use their knowledge and skills to design, manufacture parts, and assemble their prototype, then test and improve it using feedback from their intended users. Once testing is complete and modifications have been made to reflect feedback, the final product is manufactured. Some researchers argue that in some cases entrepreneurship or impact aspect could be added or explicitly addressed in the design thinking model.

This ME310 example presents a brief overview of the design thinking method in practice, including a summary of the activities that highlight different aspects and processes, to demonstrate how it is implemented in one university course. While there are many unique aspects that make the course a great fit for this method, educators in other settings may find their own ways of implementing this model of cognitive apprenticeship within their specific situations. Let us now consider the above process in terms of the dimensions for designing a learning environment (Collins et al., 1991) as well as ways in which instructors promoted student learning in ME310.

Dimensions of Cognitive Apprenticeship for Designing a Learning Environment

Collins et al. (1991) suggested four dimensions to be considered when designing learning environment based on cognitive apprenticeship learning: content, method, sequencing, and sociology. Using the ME310 class example, we see these dimensions embodied in the different components that constitute the learning space. Let us explore each of them, followed by processes that were applied to support learning.

- *Content*: The content incorporated real-life examples and scenarios that were used to model skills, as well as to generate the tasks that were assigned to student teams.
- *Method*: Learning was hands-on, iterative problem solving, which was scaffolded by instructors and coaches, allowing learners to gain and practice new skills with support from experts.
- *Sequencing*: The learning activities and tasks were deliberately planned to advance mastery by presenting just the right level of difficulty on subsequent tasks.
- *Sociology*: Learning in this class was inherently co-operative, with students learning from and interacting freely with each other, instructors, coaches, as well as potential users of their products.

Six Processes Used in Promoting Student Learning

Modeling

The instructors and coaches in ME310 begin with learning activities that allow them to model the skills as they invite students to participate through assigned tasks, such as making observations, asking questions, annotating, and others.

Coaching

Once students begin working on assigned tasks, the instructors and coaches monitor and provide directions as necessary, pointing out opportunities for best performance and successful completion of tasks. This could be in as simple a task as assigning responsibilities within a team, or setting up a shared planner/timeline.

Scaffolding

The instructors continue to monitor learning while providing specific help, directions, and opportunities to perform advanced tasks once students demonstrate mastery, or revisiting previously covered skills if necessary. An example is asking students to create multiple variations of a prototype for extra score.

Articulation

Students learn from the instructors and coaches who verbalize their thought process and describe the interconnectedness of different aspects needed to complete tasks. The instructor may explain what constitutes a great user testing, for instance.

Reflection

Once students have completed a specific task such as interviewing a user, or assembling a prototype, they reflect on the process verbally or in writing. This is shared with their team as well as with instructors.

Exploration

The students are encouraged to go beyond the examples presented by imagining new scenarios as they seek to understand and resolve problems. This is where novel ideas emerge from—such as designing a manufacturing platform as a way of inventing the future space shuttle—something that would have seemed far removed from the originally assigned task.

Conclusion

In this chapter, we discussed two fundamental learning theories, namely, the situated cognition and cognitive apprenticeship learning, which are situated within social constructivist approaches to instruction. We also supported our discussion with a case study in which engineering design was looked at and implemented through a cognitive apprenticeship perspective. While situated cognition and cognitive apprenticeship both have contributed to our understanding of learning, the characteristics of emerging learning contexts and tools have made the use of these two theories more relevant than ever. According to situation learning theory, learning arises from the dynamic interaction between the learner and the environment in which the learning takes place (Roth & Jornet, 2013). Thus, any interpretation of learning should acknowledge the social, cultural, and historical context in which learning takes place. What these two theories suggest is that learning is not only about memorizing and retaining knowledge, it is also about becoming someone, belonging to a culture, learning how to become a legitimate, competent and productive member of a group. This requires learning how to use the rules, tools, and norms of the specific culture in which one is trying to achieve legitimate membership (Lave & Wenger, 1991). Accordingly, learners' social and emotional skills are also central to this process (OECD, 2017). Science education colleagues have studied how students learn when the learning tasks are designed based on cognitive apprenticeship and the learning contexts emulate the authentic scientific contexts (Barab & Hay, 2001; Charney et al., 2007). The findings suggest that students develop more robust and meaningful understandings and acquire a deeper understanding of the nature of science (Bell, Blair, Crawford & Lederman, 2003).

Collins et al.'s (1991) emphasis on four dimensions such as *content, method, sequencing,* and *sociology* that needs to be considered while designing learning environment must be taken very seriously by educators as they design learning environments in and outside of classrooms. Applying these dimensions in design of learning environments will result in more productive student engagement. However, making learning relevant to students' lives and taking context and culture into account will make learning more authentic. This implies that the goals of our learning activities should focus on epistemologies of science, engage students in deep questions related to the nature of science, and the activities we design should engage students in such practices as modeling, argumentation, and questioning, the types of practices that are used to construct, justify, evaluate, critique, and validate the scientific knowledge. When it comes to practical applications and limitations of this theory, the blended and online learning platforms as well as online instructional videos—where learners engage with a trainer in isolation (mostly), rather than within a direct, personal, social setting—contrasts the theory and suggests a different approach. The online learning videos scenario thus limits the application of this theory, as some of the component parts that make up the theory are missing. Thus, online learning platforms need to improve their approaches in that sense.

Acknowledgements We would like to acknowledge Prof. Larry Leifer and Prof. Mark Cutkosky of Stanford University School of Engineering, whose project-based design class ME310 is sited in this article, and for their generosity in allowing us to share the experience.

Further Readings

Barab, S. A., & Hay, K. E. (2001). Doing science at the elbows of experts: Issues related to the science apprenticeship camp. *Journal of Research in Science Teaching, 38*(1), 70–102.

Bell, R., Blair, M., Crawford, B., & Lederman, N. (2003). Just do it? Impact of a science apprenticeship program on high school students' understandings of the nature of science and scientific inquiry. *Journal of Research in Science Teaching, 40,* 487–509.

Brown, J. S., Collins, A., & Duguid, P. (1989). Situated cognition and the culture of learning. *Educational Researcher, 18*(1), 32–42.

Brown, T. (2009). *Change by design: How design thinking transforms organizations and inspires innovation.* New York: Harper Collins.

Charney, J., Hmelo-Silver, C. E., Sofer, W., Neigeborn, L., Coletta, S., & Nemeroff, M. (2007). Cognitive apprenticeship in science through immersion in laboratory practices. *International Journal of Science Education, 29*(2), 195–213.

Lave, J., & Wenger, E. (1991). *Situated learning: Legitimate peripheral participation.* Cambridge: Cambridge University Press.

ME310: https://web.stanford.edu/group/me310/me310_2016/index.html

Rogoff, B. (1990). *Apprenticeship in thinking: Cognitive development and social context.* London: Oxford University Press.

References

Aikenhead, G. S. (1996). Science education: Border crossing into the subculture of science. *Studies in Science Education, 27,* 1–51.

Aydeniz, M., & Hodge, L. (2011). Identity: A complex structure for researching students' academic behavior in science and mathematics. *Cultural Studies of Science Education, 6*(2), 509–523.

Bricker, L. A., & Bell, P. (2014). What comes to mind when you think of science? The perfumery!: Documenting science-related cultural learning pathways across contexts and timescales. *Journal of Research in Science Teaching, 51,* 260–285.

Brown, T. (2009). *Change by design: How design thinking transforms organizations and inspires innovation.* New York: Harper Collins.

Brown, J. S., Collins, A., & Duguid, P. (1989). Situated cognition and the culture of learning. *Educational Researcher, 18*(1), 32–42.

Cakmakci, G. (2012). Promoting pre-service teachers' ideas about nature of science through educational research apprenticeship. *Australian Journal of Teacher Education, 37*(2), 114–135.

Carlone, H. B., & Johnson, A. (2007). Understanding the science experiences of successful women of color: Science identity as an analytic lens. *Journal of Research in Science Teaching, 44*(8), 1187–1218.

Clancey, W. J. (1997). *Situated cognition: On human knowledge and computer representations.* Cambridge: Cambridge University Press.

Collins, A., Brown, J. S., & Holum, A. (1991). Cognitive apprenticeship: Making thinking visible. *American Educator, 15*(3), 6–11, 38–46.

Collins, A., Brown, J. S., & Newman, S. E. (1987). Cognitive apprenticeship: Teaching the craft of reading, writing, and mathematics. Technical Report No. 403. Center for the Study of Reading. ERIC Document 284181.

Collins, A., & Greeno, J. G. (2008). Situated cognition. In E. M. Anderman & L. H. Anderman (Eds.), *Psychology of classroom learning: An encyclopedia.* Gale: Farmington Hills MI.

Collins, A. & Greeno, J. G. (2010). A situative view of learning. In E. Baker, P. Peterson, & B. McGaw (Eds.), *International encyclopedia of education.* London: Elsevier. Reprinted in V. G. Aukrust (Ed.). (2011). *Learning and cognition in education* (pp. 64–70). London: Elsevier.

Cross, N. (2011). *Design thinking: Understanding how designers think and work.* Oxford: Berg.

Dewey, J. (1938). *Experience and education.* New York: MacMillan.

Freire, P. (2005). *Pedagogy of the oppressed (30th Anniversary Edition: Translated by Myra Bergman Ramos with an introduction by Donaldo Macedo).* New York: The Continuum International Publishing Group Inc.

Gee, J. P. (1997). Thinking, learning, and reading: The situated sociocultural mind. In D. Kirshner & J. A. Whitson (Eds.), *Situated cognition.* Mahwah: Lawrence Erlbaum Associates.

Hutchins, E. (1995). *Cognition in the wild.* Cambridge, UK: MIT Press.

Khan, T. M., Mitchell, J. E. M., Brown, K. E., & Leitch, R. R. (1998). Situated learning using descriptive models. *International Journal of Human-Computer Studies, 49*(6), 771–796.

Lave, J. (1988). *Cognition in practice.* Cambridge, England: Cambridge University Press.

Lave, J. (1991). Situated learning in communities of practice. In L. B. Resnick, J. M. Levine, & S. D. Teasley (Eds.), *Perspectives on socially shared cognition* (pp. 63–82). Washington, DC: American Psychological Association.

Lave, J., & Wenger, E. (1991). *Situated learning: Legitimate peripheral participation.* Cambridge: Cambridge University Press.

Leopold, A. L., Ratcheva, V., & Zahidi, S. (2016). *The future of jobs: Employment, skills and workforce strategy for the fourth industrial revolution.* World Economic Forum, Davos. Retrieved from http://www3.weforum.org/docs/WEF_Future_of_Jobs.pdf.

OECD (The Organisation for Economic Co-operation and Development). (2017). *Social and emotional skills: Well-being, connectedness and success.* Retrieved from http://bit.do/e2qpH.

Riedinger, K. (2015). Identity development of youth during participation at an informal science education camp. *International Journal of Environmental & Science Education, 10*(3), 453–475.

Robbing, P., & Aydede, M. (2009). A short primer on situated cognition. In P. Robbins & M. Aydede (Eds.), *The Cambridge handbook of situated cognition* (pp. 3–10). New York: Cambridge University Press.

Rogoff, B. (1990). *Apprenticeship in thinking: Cognitive development and social context.* London: Oxford University Press.

Roth, W.-M., & Jornet, A. G. (2013). Situated cognition. *WIREs Cognitive Science, 4,* 463–478.

Smith, E. R., & Semin, G. R. (2004). Socially situated cognition: Cognition in its social context. *Advances in Experimental Social Psychology, 36,* 53–117.

Tan, E., Calabrese-Barton, A., Kang, H., & O'Neil, T. (2013). Desiring a career in STEM-related fields: How middle school girls articulate and negotiate identities-in-practice. *Journal of Research in Science Teaching, 50*(10), 1143–1179.

Vosniadou, S., Loannides, C., Dimitrakopoulou, A., & Papademetriou, E. (2001). Designing learning environments to promote conceptual change in science. *Learning and Instruction, 11*, 381–419.

Vygotsky, L. (1978). *Mind in society: The development of higher psychological processes.* Cambridge, MA: Harvard University Press.

Wilson, B. G., & Myers, K. M. (2000). Situated cognition in theoretical and practical context. In D. H. Jonassen & S. M. Land (Eds.), *Theoretical foundations of learning environments* (pp. 57–88). Mahwah: Lawrence Erlbaum Associates.

Gultekin Cakmakci is a Professor of Science Education at Hacettepe University and has been teaching courses on STEM education and public engagement with STEM. His research interests focus on developing scientific literacy among students and the general public and on the design, implementation, and evaluation of STEM teaching. He is currently a board member of the *Public Communication of Science and Technology (PCST), EU STEM Coalition, Turkish STEM Alliance* and *Journal of Research in STEM Education.*

Mehmet Aydeniz is a Professor of Science Education at the University of Tennessee, Knoxville. Dr. Aydeniz's research focuses on students' appropriation of epistemic and social norms of science and science teachers' pedagogical knowledge to support student learning along these goals. Dr. Aydeniz is also interested in studying science teachers' pedagogical knowledge of engineering design and computational thinking. He is the editor of *Journal of Research in STEM Education.*

Amelia Brown has B.S. in Plant Sciences, and spent 10 years as a STEM professional in Food and Agricultural Sciences before deciding to pursue a M.S. in Science Education. After teaching middle school science for several years, Amelia decided to further her education. She is currently a Ph.D. candidate in Science Education at the University of Tennessee, Knoxville. Her research focuses on culturally responsive pedagogies.

Joseph M. Makokha is a Ph.D. student in engineering design at Stanford University. He is interested in the interaction between novice users and autonomous systems; and how design enables effective understanding between them. He has taught math and robotics to high school students, co-founded a technology startup, and still works as an academic technology specialist among others.

Chapter 21
Activity Theory—Lev Vygotsky, Aleksei Leont'ev, Yrjö Engeström

Tony Burner and Bodil Svendsen

Introduction

Commonly, various systems and institutions undergo a change in order to improve practices or make them more effective. However, there are few systematic theories that can be used in research to both study the changes and contribute to change and transformation of practice. Activity Theory can be used to study developmental change in systems and institutions such as hospitals and schools. It applies both a historical and a situational perspective; both an individual (micro) and a systemic (macro) perspective. In this chapter, we explain the history of Activity Theory and how it can be used in practical terms to understand change and development in general, and inquiry-based science teaching in particular.

Activity Theory has developed within the sociocultural approach to learning and development (Vygotsky, 1978; Wertsch, 1991, 1998), and pays attention to historicity, the present situation, to the individual, and the collective system. Research with human participants will to some degree involve intervention, and "the introduction of research instruments into practice, including dialogue between researcher and participants, is itself change-inducing" (Wardekker, 2000, p. 270). Activity Theory is about learning and change and is a suitable research and development approach in order to address the gap between theory and practice.

Activity Theory is based on theories developed by a group of revolutionary Russian psychologists in the 1920 and 1930s. The fundamental concept of the approach

T. Burner
Professor, Department of Languages and Literature Studies, University of South-Eastern Norway, Drammen, Norway
e-mail: tony.burner@usn.no

B. Svendsen (✉)
Associate Professor, Department of Teacher Education, Norwegian University of Science and Technology (NTNU), Trondheim, Norway
e-mail: bodil.svendsen@ntnu.no

© Springer Nature Switzerland AG 2020
B. Akpan and T. Kennedy (eds.), *Science Education in Theory and Practice*,
Springer Texts in Education, https://doi.org/10.1007/978-3-030-43620-9_21

was proposed by Lev Vygotsky (1896–1934), the founder of the school. The relationship between the individual and the social community appears to be a classic challenge in psychology. After the Russian Revolution in 1917 Russian psychologists tried to solve this issue, and it was not an easy task since a solution had to fit the philosophy of the Marxist doctrine. Vygotsky was a central character in this context and he tried to reconcile the philosophical side of Marxism with a psychology of human development and link socialization to the social individual. In particular, he stressed three key elements that were central to his thinking: First, that human mindset is influenced by its living conditions. There are common features in the environment around humans, resulting in a united mindset and how they understand each other. Second, artifacts surrounding humans impact their living conditions. Third, humans can attain more in life by collaborating than striving alone (Vygotsky, 1978).

Activity Theory is an object-oriented theory (Engeström & Sannino, 2010). According to Engeström (2001, pp. 136–137), Activity Theory can be summed up with five characteristics.

1. Prime unit of analysis: "A collective, artifact-mediated and object-oriented activity system, seen in its network relations to other activity systems, is taken as the prime unit of analysis" (Engeström, 2001, p. 136).
2. Multi-voicedness: "An activity system is always a community of multiple points of view, traditions and interests" (Engeström, 2001, p. 136).
3. Historicity: "Activity systems take shape and get transformed over lengthy periods of time. Their problems and potentials can only be understood against their own history" (Engeström, 2001, p. 136).
4. Contradictions: Contradictions play a central role as "sources of change and development…[They] are historically accumulating structural tensions within and between activity systems" (Engeström, 2001, p. 137).
5. Possibility of expansive transformations: "An expansive transformation is accomplished when the object and motive of the activity are reconceptualized to embrace a radically wider horizon of possibilities than in the previous mode of activity" (Engeström, 2001, p. 137).

Yrjö Engeström has together with colleagues at CRADLE (Center for Research on Activity and Learning) at the University of Helsinki used the theory to analyze and intervene in many settings and situations. Activity Theory is an approach that can be used to analyze human interactions and relationships within specific social contexts. It focuses on collective social practices and considers the complexity of real-life activity. It is being increasingly used to examine issues in teacher education, as well as in other fields.

Activity Theory has developed through the following three generations or schools (Engeström, 2001): The first school was developed by Vygotsky and later his students, contributing with the cultural historical aspects of Activity Theory. The second school was mainly Leont'ev's work, a student of Vygotsky, contributing to the differences between individuals and collective activity. The third and last school was developed by Engeström, with its networks of interacting activity systems.

Vygotsky and First School of Activity Theory

Vygotsky's ideas, developed during the 1920 and 1930s, were a response to what he called "a crisis in psychology", which was most evident in the study of "consciousness"—a synonym for "mind" (Bakhurst, 2007). It was a reaction towards a reductionist understanding of psychology, where human processes were reduced to physiology or neurology by proponents like Ivan Pavlov and Vlamidir Bekhterev. Pavlov, the winner of Nobel Prize in Medicine in 1904, developed the theory of conditional reflexes through his famous experiments with dogs (Van der Veer, 2007). He found that dogs would salivate not only when they got food, but also when various conditions preceding food reminded them of food. From this, Pavlov inferred that mental activity is reflexive. Bekhterev, the founder of reflexology, claimed that all human behavior consists of complex forms of reflexes (Van der Veer, 2007). Pavlov's findings inspired the American John B. Watson, who is considered to be the founder of the school of behaviorism, and his later colleague B. F. Skinner. Signalization, or stimuli, was at the core of Pavlov's theory. It meant that organisms learn that certain stimuli signal others (Van der Veer, 2007). However, Vygotsky considered this an inadequate description of human being's higher mental functions. He introduced the concept of signification, meaning that humans are not passively reacting to their environment but actively determine their behavior through signs (Van der Veer, 2007). Bakhurst (2007) explains it in this way:

> The cornerstone of Vygotsky's "dialectical method" is the idea that everything in time must be understood in its development. Accordingly, he argues that to understand the mature human mind, we must comprehend the processes from which it emerges. The higher mental functions, he argues, are irreducible to their primitive antecedents; they do not simply grow from the elementary functions as if the latter contained them in embryo. To appreciate the qualitative transformations that engender the mature mind, we must look outside the head, for the higher mental functions are distinguished by their mediation by external means (p. 53).

Vygotsky's identification of mediated action as a unit of analysis was revolutionary. It overcame the Cartesian individual and the untouchable societal structure split. Vygotsky based his findings on reading Marx's theories on changing social and material conditions. The foundational idea of dialectical materialism is that human beings, besides acquiring knowledge and being the result of the evolution of species, also produce and transform culture. Vygotsky extended Marx's theory to psychology, emphasizing that a unit of analysis has to pay attention to the history and developmental processes (Vygotsky, 1986). He claimed that "[…] humans personally influence their relations with the environment and through that environment personally change their behavior, subjugating it to their control" (Vygotsky, 1978, p. 51). Moreover, Vygotsky was influenced by Engel's writings on the centrality of tool and sign mediation in human functioning (Wertsch, 1985). He formed what is called "the basic triangle", illustrating that the subject cannot act on the object directly but through tool mediation. "This type of organization is basic to all higher psychological processes", according to him (Vygotsky, 1978, p. 40).

Mediation also provides "a link between social and historical processes, on the one hand, and individuals' mental processes, on the other… the focus is on how the inclusion of tools and signs leads to qualitative transformation" (Wertsch, 2007, p. 178). Thus, change is fundamental to understanding higher mental functions. From this point of view, the goal of instruction in schools is "to assist students in becoming fluent users of a sign system" (Wertsch, 2007, p. 186). Teachers try constantly to do this with their students—whether the sign system is reading literacy, ICT, classroom management or inquiry-based teaching.

Vygotsky (1986) emphasized cultural mediation and its importance for thinking: "The rational, intentional conveyance of experience and thought to others requires a mediating system" (p. 7). He argued that tools and signs mediate higher mental functioning and human action. Mediational means, particularly language, are products of cultural, historical, and institutional forces (Wertsch, 1991). In fact, Wertsch (1998) argues for mediated action as a unit of analysis in order to overcome the pitfalls of individualistic reductionism. As in the definition of "activity" in Activity Theory, he claims that the action is characterized "[…] by dynamic tension among various elements" (Wertsch, 1998, p. 27). As pointed out by Wertsch (1998), there are often resistance and tensions involved in mediated action through cultural tools. Now we turn to the second school of Activity Theory.

Leont'ev and Second School Activity Theory

Aleksei N. Leont'ev, Alexander Luria and other Soviet researchers developed Vygotsky's ideas into what is called the second generation of Activity Theory (Engeström, 2001). The focus then moved from the individual to the collective.

Leont'ev, one of Vygotsky's students, contributed with the concept of activity (Leont'ev, 1978, 1981). He criticized American psychology, which was mostly occupied with explaining what makes children what they are. Leont'ev distinguished between activity, action, and operation, and operated with collective activity as a key unit of analysis. The focus should, according to Leont'ev, be on the object and motive (Leont'ev, 1981). The *activity* of driving a car can be illustrative of these concepts. When one shifts gear while driving, the *action* is the shifting of gear from first to second gear. After one has learned to shift gears, the action becomes an *operation*. Thus, an activity is realized through actions. Activities have their own language, for example, teachers working in schools. For somebody who does not know what a school is, the activity will seem foreign. That is why one has to study an activity from the inside. Within a school, there are several activities, for example assessing student performance, which also has its own jargon. For someone unacquainted with assessment, the activity will not make so much sense. Teachers' work within an activity, for example, student assessment, becomes automatized. Their actions within the activity thus become operations. It is important to study the actions and verbalize the operations to understand the activity (cf. Vygotsky's idea of the social preceding the individual).

Leont'ev claims "[…] the main feature that distinguishes one activity from another is its object. After all, it is precisely an activity's object that gives it a specific direction", which also shows that there is always a need, a motive: "There can be no activity without a motive" (Leont'ev, 1981, p. 59). Moreover, Leont'ev formally operationalized the roles of communities, the rules that structure them, and the negotiation of tasks. He was much more concerned with practical *life* and *activity* than his predecessor Vygotsky, who was more concerned with genesis and the mediation of mind by cultural tools.

The second generation of Activity Theory is inspired mostly from Leont'ev's work. In his well-known example of "primeval collective hunt", Leont'ev (1981, pp. 210–213) explained the essential difference between an individual action and a collective activity. The distinction between activity, action, and operation became the basis for Leont'ev's three-level model of activity. The highest level of collective activity is driven by an object-related motive; the middle level of individual (or group) action is driven by a conscious goal; and the bottom level of sub-conscious operations is driven by the conditions and tools of the action at hand. The idea of internal contradictions as the energetic forces of change and development in activity systems was conceptualized by Il'enkov (1982) and started to grow as a guiding principle of empirical research. Cole (1988) was one of the first to outline the deep-rooted insensitivity of the second generation Activity Theory towards cultural diversity. Nevertheless, Leont'ev never graphically extended Vygotsky's original model into a model of a collective activity system, the graphical extension was done by Engeström (1987, p. 78). With this, we turn to the third school of Activity Theory.

Engeström and Third School Activity Theory

When Activity Theory went global, questions of diversity and dialogue between different traditions or perspectives gradually became serious challenges. It is these challenges that the third generation of Activity Theory deals with. It develops conceptual tools to understand dialogue, multiple perspectives and voices, and networks of interacting activity systems. In this mode of research, the basic model is expanded to include a minimum of two interacting activity systems.

The minimum elements in the activity system are: *Subject, mediating artifact, object, rules, community,* and *division of labor*. The upper triangle with subject, object, and mediating artifact as its nodes is Vygotsky's original triangle (Vygotsky, 1978, p. 40) turned upside down. The acting *subject* could be a person or a group; it is through the subject's eyes and interpretations the activity is constructed. *Mediating artifact* is what links the subject to the object in Vygotsky's original triangle. The *object* is the goal of the activity, whereas the *outcome* is the ultimate goal or vision of the activity. *Rules* include norms and conventions in the activity system, *community* refers to all the people involved in the activity system, and *division of labor* refers to the object-oriented actions that are conducted by the people involved in the activity system. All the nodes in the triangle interact.

The principles of Activity Theory for inquiry and development are useful in conducting and studying development. Conducting developmental research which uses the activity system as a starting point can add knowledge about the situation before and after an intervention. A significant goal of using an inquiry approach is to learn from the often unexpected ways in which the intervention reveals new understandings of both theory and practice.

According to Rantavuori, Engeström, and Lipponen (2016), when whole collective activity systems, such as work processes and organizations, need to refine themselves, traditional modes of learning are not enough. Nobody knows exactly what needs to be learned. The design of the new activity and the acquisition of the knowledge and skills it requires are increasingly intertwined. In expansive learning activity, they merge (Engeström, 2015). Earlier studies of expansive learning (e.g., Engeström, 2008, pp. 118–168) have demonstrated that features of expansive learning may be found when participants face an open-ended problem-solving task, such as a need to plan something that is new for them. In an expansive learning cycle, the initial simple idea is transformed into a complex object, a new form of practice.

Relying on Activity Theory, the theory of expansive learning is fundamentally an object-oriented theory where the object is both the resistant raw material and the future-oriented purpose of an activity (Rantavuori et al., 2016). The object is the true carrier of the motive of the activity. In an expansive learning activity, motives and motivation are not tailed predominantly inside individual subjects—they are in the object to be transformed and expanded. As pointed out by Rantavuori and colleagues (2016), a powerful object of learning has an expansive potential to go beyond the exchange value, being typically an open-ended problem or challenge that has relevance for the learners and is not limited to reproducing predefined correct answers. Expansive learning is understood as a circular process in which strategic actions based on contradictions drive new strategic actions and contradictions in a cyclic process (Engeström & Sannino, 2010, p. 2). Engeström's (1999) expansive cycle of learning is related to his activity system and shows the levels of action during formative interventions. This model assumes that development does not necessarily follow a linear pattern.

In expansive learning, learners learn something that is not yet there (Rantavuori et al., 2016). The learners construct a new object and concept for their collective activity and implement this new object and concept in practice. The theory of expansive learning is based on the dialectics of ascending from the abstract to the concrete (Engeström & Sannino, 2010). This is a method of grasping the essence of an object by tracing and theoretically reproducing the logic of its development, that is, its historical formation through the emergence and resolution of its inner contradictions. Contradictions are the driving force of transformation (Engeström & Sannino, 2010). Contradictions may create disorder and conflicts that can be perceived as a problem, but contradictions may also lead to change and new knowledge (Leont'ev, 1978). Through the process of the expansive cycle, the object and motive of the activity are reconceptualized to allow greater possibility and flexibility than the previous pattern of activity.

The idea of contradictions as a source of innovation was introduced by Il'enkov (1982) and is a guiding principle of Activity Theory, which is illustrated with Engeström's (1999) expansive circle. The contradictions between the various elements in the activity system are the starting point for development. When contradictions are identified, the development forms the formative cycle, which can be illustrated in the expansive learning cycle (Engeström & Sannino, 2010, p. 8). The expansive transformation is accomplished when the object and motive of the activity are reconceptualized to embrace a broader perspective of potentials than in the earlier means of the activity.

Vygotsky's concept of zone of proximal development (ZPD) is another central root to the theory of expansive learning. Vygotsky (1978, A full cycle of expansive transformation can be understood as a collective journey through the zone of proximal development of the activity (Engeström, 2000, p. 526; Engeström, 2001, p. 137). Meira and Lerman (2001) argue that the ZPD is not something that pre-exists; it is a symbolic space for interaction and communication where learning leads the development. They refer to Wertsch's (1985) statements about how the ZPD is not a measurable object. Nor is it only related to the interactional events which lead to cognitive change. According to Wertsch (1985), the ZPD is not just a property of the child, nor is it merely the result of inter-psychological functioning alone. As pointed out by Engeström, who has developed the individual understanding of Vygotsky's ZPD (1978, p. 174), "it is the distance between the present everyday actions of the individuals and the historically new form of the societal activity that can be collectively generated as a solution to the double bind potentially embedded in the everyday actions." The ZPD is redefined as space for expansive transition from actions to activity (Engeström & Sannino, 2010, p. 4). In the following, we turn to inquiry-based science teaching within a framework of Activity Theory.

Inquiry-Based Science Teaching and Mediating Artifacts

Inquiry-Based Science Teaching (IBST) is according to Linn, Davis, and Bell (2004) basically about teachers teaching students to obtain a better understanding of the world in which they work, communicate, learn, and live. Inquiry is the intentional process of diagnosing problems, critiquing experiments, and distinguishing alternatives, planning investigations, researching conjectures, searching for information, constructing models, debating with peers, and forming coherent arguments (Linn, Davis, & Bell, 2004).

Questioning and finding answers are extremely important in IBST as aids in effectively generating knowledge. Teaching strategies that actively engage students in the learning process through inquiries are more likely to increase conceptual understandings, and there can be variable amounts of direction from the teacher, in both open and guided inquiry. IBST is not only about asking questions but is a way of transforming data and information into valuable knowledge. As a tool for teaching inquiry, teachers can use the 5E model (Fig. 21.1). The 5E model (cf.

Fig. 21.1 The 5E model
(Svendsen, 2015)

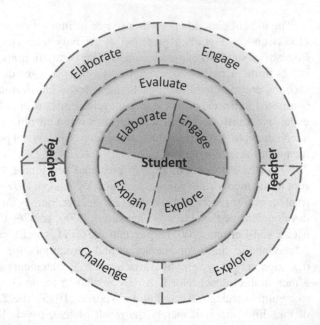

Chap. 4: Malone's intrinsic motivational theory) can be used to support teachers in planning, implementing, and evaluating teaching. The model has its origins in the Biological Sciences Curriculum Study (BSCS), in which American scholars developed educational programs and research on teaching and learning in science. The five Es are the initial letters in the words engage, explore, explain, elaborate, and evaluate. The intention of the model is to be used for planning, implementation, and evaluation of learning and teaching.

Teachers teach by engaging students with a starter. A startup should be both motivating and related to phenomena that students can relate to (like everyday phenomena). The students' prior knowledge is accessed by the teacher or the syllabus and helps students to become engaged in a concept through the use of short activities, or introduction to phenomena in order to endorse interest and provoke prior knowledge. The activities of this phase make connections to past experiences and expose students' misconceptions; they should serve to ease cognitive imbalance. Activity refers to both mental and physical activity (Bybee et al., 2006). Once the activities have engaged the students, they need time to *explore* the ideas. Inquiry-based activities are designed so that the students have common, concrete experiences upon which they continue formulating concepts, processes, and skills. Students work actively with the material (read, write, investigate, observe, etc.) and add knowledge and skills to reach new learning goals. This level is concrete and hands-on, and the use of touchable materials and concrete experiences is essential, but not necessary. The aim of inquiry-based activities is to establish experiences that teachers and students can use later to introduce and discuss concepts, processes, or skills. Explanation provides openings for teachers to directly introduce a concept, process, or skill. Students explain their understanding of the concept. An explanation from the teacher may guide them toward a deeper understanding, which is a critical part of their new understanding. By facilitating activities that build on the knowledge and skills students already possess, and allow students to reflect, discuss, read, and write to achieve the learning objectives, the teacher can introduce new concepts that challenge student's conceptual understanding (Bybee et al., 2006).

Teachers have a variety of techniques and strategies at their disposal to stimulate and develop student *explanations*. Once students have explanations and terms for their learning tasks, it is important to involve them in further experiences that extend, or *elaborate*, the concepts, processes, or skills. This level facilitates the transfer of concepts to closely related but new situations. Students' theoretical understandings and skills are challenged by their new experiences and by guidance of their teachers. They develop deeper and extensive understanding, more information, and adequate skills. Students apply their understanding of the concept by conducting supplementary activities. Elaborative activities provide further time and experiences that contribute to learning.

Evaluation should be continuous, varied, and be a part of all levels. Assessment is self-assessment, continuous assessment, and final assessment. It can be oral and written. Teachers need to evaluate their own learning in a reflective way. Students consider their own learning and understanding, and the teacher will assess student learning in relation to learning objectives in each subject or in an activity, and in relation to the objectives of the curriculum. Students might also benefit from collaborative learning when working inquiry-based. Students engaged in collaborative learning capitalize on one another's resources and skills, asking one another for information, evaluating one another's ideas, monitoring one another's work, etc. The importance of the difference between individual actions and collective activities is to be found within the second school of Activity Theory. When students interact across activity systems, conceptual tools to understand dialogue, multiple perspectives and voices, and networks of interacting activity systems need to be developed. In this mode, the basic model from the first school of Activity Theory is expanded to include a minimum of two interacting activity systems, known as the third school of Activity Theory.

In conclusion, the 5E model can be supportive in making inquiry-based teaching explicit and targeted. By shaping clear learning aims for teaching, teachers can use the model as a reflection tool for designing, planning, implementing, and evaluating their teaching sequences and in this way expand their professional learning. Rendering the activity system, the 5E model represents a mediating artifact on which teachers and students can act and create their own understanding of the model to enhance learning and understanding of science. Mediating artifact is what links the subject to the object in Vygotsky's triangle, and it is acted upon by the subject to the object. According to Leont'ev (1981), mediation is the subject's activity. The object refers to the "problem space" at which the activity is focused and which is formed and transformed into outcomes with the help of physical and symbolic, external and internal mediating instruments, including both tools and signs. The goal of the activity is the object, and the outcome is the goal of the activity, in this case, IBST.

Summary

- Activity Theory can be used to study the developmental change in systems and institutions.
- Activity Theory has developed within the sociocultural approach to learning and development.
- Activity Theory has developed through the following three generations or schools:
  - The first school was developed by Vygotsky. The important part here is the concept of mediation.
  - The second school was developed by Leont'ev. The important part here is the difference between individual actions and collective activities.
  - The third school was developed by Engeström. The important part here is the network of activity systems.
- Expansive learning is central in Activity Theory. In expansive learning, learners learn something that is not yet there.
- Science teaching can benefit from using mediating artifacts to understand the principles behind inquiry-based teaching and trigger a learning process.
- Inquiry-Based Science Teaching is about asking questions and a way of transforming data and information into valuable knowledge.

Recommended Resources

Books

Miettinen, R. (2009). *Dialogue and creativity. Activity theory in the study of science, technology and innovations*. Berlin: Lehmanns Media.

Journal

Mind, Culture, and Activity: http://www.tandfonline.com/loi/hmca20.

Internet Source

CRADLE (Center for Research on Activity, Development and Learning): http://www.helsinki.fi/cradle/index.htm.

References

Bakhurst, D. (2007). Vygotsky's demons. In H. Daniels, M. Cole, & J. Wertsch (Eds.), *The Cambridge companion to Vygotsky* (pp. 50–77). Cambridge: Cambridge University Press.

Bybee, R., Taylor, J. A., Gardner, A., Van Scotter, P., Carlson, J., Westbrook, A., et al. (2006). *The BSCS 5E instructional model: Origins and effectiveness*. Colorado Springs, CO: BSCS.

Cole, M. (1988). Cross-cultural research in the sociohistorical tradition. *Human Development, 31,* 137–151.

Engeström, Y. (1987). *Learning by expanding: An activity-theoretical approach to developmental research.* Helsinki: Orienta-Konsultit.

Engeström, Y. (1999). Innovative learning in work teams: Analyzing cycles of knowledge creation in practice. In Y. Engeström, R. Miettinen, & R.-L. Punamäki (Eds.), *Perspectives on activity theory* (pp. 377–406). Cambridge: Cambridge University Press.

Engeström, Y. (2000). Activity theory as a framework for analyzing and redesigning work. *Ergonomics, 43*(7), 960–974.

Engeström, Y. (2001). Expansive learning at work: Toward an activity theoretical reconceptualization. *Journal of Education and Work, 14*(1), 133–156.

Engeström, Y. (2008). *From teams to knots: Activity-theoretical studies of collaboration and learning at work.* Cambridge: Cambridge University Press.

Engeström, Y. (2015). *Learning by expanding: An activity-theoretical approach to developmental research* (2nd ed.). Cambridge: Cambridge University Press.

Engeström, Y., & Sannino, A. (2010). Studies of expansive learning: Foundations, findings and future challenges. *Educational Research Review, 5*(1), 1–24.

Il'enkov, E. V. (1982). *The dialectics of the abstract and the concrete in Marx's Capital.* Moscow: Progress.

Leont'ev, A. N. (1978). *Activity, consciousness and personality.* Englewood Cliffs: Prentice Hall.

Leont'ev, A. N. (1981). The problem of activity in psychology. In J. V. Wertsch (Ed.), *The concept of activity in Soviet psychology* (pp. 37–71). Armonk: M. E. Sharpe, Inc.

Linn, M. C., Davis, E. A., & Bell, P. (2004). Inquiry and technology. In M. C. Linn, E. A. Davis, & P. Bell (Eds.), *Internet environments for science education* (pp. 3–27). Mahwah, New Jersey: Lawrence Erlbaum Associates.

Meira, L., & Lerman, S. (2001). The zone of proximal development as a symbolic space. *Social Science Research Papers. 13.* London, South Bank University, Faculty of Humanities and Social Science.

Rantavuori, J., Engeström, Y., & Lipponen, L. (2016). Learning actions, objects and types of interaction: A methodological analysis of expansive learning among pre-service teachers. *Frontline Learning Research, 4*(3), 1–27.

Svendsen, B. (2015). Mediating artifact in teacher professional development. *International Journal of Science Education, 37*(11), 1834–1854.

Van der Veer, R. (2007). Vygotsky in context: 1900–1935. In H. Daniels, M. Cole, & J. Wertsch (Eds.), *The Cambridge companion to Vygotsky* (pp. 21–49). Cambridge: Cambridge University Press.

Vygotsky, L. S. (1978). *Mind in society: The development of higher psychological processes.* Cambridge: Harvard University Press.

Vygotsky, L. S. (1986). *Thought and language.* Cambridge: The MIT Press.

Wardekker, W. (2000). Criteria for the quality of inquiry. *Mind, Culture, and Activity, 7*(4), 259–272.

Wertsch, J. V. (1985). *Vygotsky and the social formation of mind.* Cambridge: Harvard University Press.

Wertsch, J. V. (1991). *Voices of the mind: A sociocultural approach to mediated action.* Cambridge: Harvard University Press.

Wertsch, J. V. (1998). *Mind as action.* Oxford: Oxford University Press.

Wertsch, J. V. (2007). Mediation. In H. Daniels, M. Cole, & J. Wertsch (Eds.), *The Cambridge companion to Vygotsky* (pp. 178–192). Cambridge: Cambridge University Press.

Tony Burner (Ph.D.) works as a Professor of English at the Department of Languages and Literature Studies at the University College of Southeast Norway, where he has taught in-service and pre-service courses for teachers and student teachers the last 13 years. He has broad international experience with research, among others from Finland, Australia, Iraqi Kurdistan, and Vietnam. His main research interests are English education, classroom assessment, research and development work, teacher mentoring, multilingualism, and professional development.

Bodil Svendsen (Ph.D.) works as an Associate Professor of Natural Science at the Department of Teacher Education at the Norwegian University of Science and Technology, where she has taught in-service and pre-service courses for teachers and student teachers since 2007. Bodil has teaching experience from elementary school, middle school, senior high school and adult education teaching Natural Science, Biology and Geography. She has international experience with research, among others from Finland, Scotland, Denmark, Sweden and England. Her main research interests are Natural Science education, school development, R&D work, teacher mentoring, professional development and gifted children research. She established and led the Center for Gifted and Talented in STEM in Trondheim, Norway, from 2016 until 2019.

Chapter 22
Multiliteracies—New London Group

Shameem Oozeerally, Yashwantrao Ramma, and Ajeevsing Bholoa

Introduction

The concept of 'multiliteracies' was, for the first time, articulated by the New London Group (NLG) in 1996 to raise awareness of the use of multiliteracy-based approaches to literacy pedagogy. Traditionally, literacy pedagogy has been related to 'formalised, monolingual, monocultural and rule-governed forms of language' (NLG, 1996, p. 61). The fulcrum of the reflection around multiliteracies is the changing social environment, rooted in change and dynamism, in a landscape of increasing cultural and linguistic diversity. In other words, one of the driving forces of this type of reflection is an increasingly heterogeneous linguo-cultural landscape, exacerbated by the expansion of technology and access to the Internet. Falling frontiers, as well as increasingly osmotic cultural-linguistic barriers can be considered as part of the factors stimulating the genesis of a multiliteracies-based philosophy. There are two defining aspects of multiliteracies:

i.   Multiplicity of communication channels and media and
ii. Increasing saliency of cultural and linguistic diversity.

Since the work of the New London Group, the term 'multiliteracies' now encompasses a multitude of disciplines, in particular, visual literacy (Bell, 2014; Drapper, 2015), oral vernacular genres (Newman, 2005), emotional literacy (Oksuz, 2016), information literacy (Rowsell & Walsh, 2011) and cultural literacy (Claassen, 2007). With the advent of digital technology, online critical literacy (Freebody, 2007) has

S. Oozeerally (✉) · Y. Ramma · A. Bholoa
Mauritius Institute of Education, Reduit, Mauritius
e-mail: s.oozeerally@mie.ac.mu; s.oozeerally@mieonline.org

Y. Ramma
e-mail: y.ramma@mie.ac.mu

A. Bholoa
e-mail: a.bholoa@mie.ac.mu

© Springer Nature Switzerland AG 2020   323
B. Akpan and T. Kennedy (eds.), *Science Education in Theory and Practice*,
Springer Texts in Education, https://doi.org/10.1007/978-3-030-43620-9_22

emerged to represent the basic skills that a learner displays when online. The concept of multiliteracies has since then been extended beyond the boundary of linguistics to other disciplines and includes science (Alvermann & Wilson, 2011) and mathematics (Chinnappan, 2008) multiliteracies.

One of the fundamental characteristics of multiliteracies being multimodality, we argue that knowledge is transformed through the filters of the 'subject-experiencer', who is not only a passive subject but an active agent. Robillard (2008b, p. 145) proposes the term '*acteur-L*', a 'socialised, historicised and reflexively-constructed being' (own translation from French). In this sense, the (science) teachers as well as the learners are seen as active meaning-makers who are able to navigate and negotiate information in order to 'achieve their various cultural purposes' (NLG, 1996, p. 64).

The multiliteracies approach also comes as a critique to the monocultural paradigms prevalent in the philosophy of education which considers 'literacy pedagogy [...] [as a] carefully restricted project—restricted to formalised, monolingual, monocultural and rule-governed forms of language' (NLG, 1996, p. 2). Dominant perspectives on literacy were based on linear, text-centric and 'language-centric' postures. As the linguistic dimension is central to the definition of multiliteracies, it is possible to draw parallels with recent literature in the epistemology of language studies. Literacy pedagogy, in the traditional sense, has been embedded in a form of reasoning that is analogous to that of the techno-linguistic perspective, which advocates a mono-dimensional and unimodal approach (Robillard, 2008a). Multiliteracies offer interesting perspectives where the sociolinguistic landscape is diverse (Carpooran, 2007).

The NLG demonstrates how the conception of a singular notional form of language (stable system based on rules), based on the assumption that we can discern and describe the correct usage, which corresponds to an authoritarian kind of pedagogy, reduces the very notion of literacy to a mechanistic process.

Learning Experience and Multiliteracies

Learning, in the multiliteracy paradigm, seeks to seamlessly integrate different communicative systems, akin to modes, in a complex process involving different sub-processes (Jewitt, Kress, Ogborn, & Tsatsarelis, 2001):

i.   Selection
ii. Adaptation
iii. Transformation

Learning is, therefore, a dynamic process of knowledge network formation mediated by multimodal realisations of language practices. Jewitt et al. (2001, p. 5) define the term 'mode' as 'organised, regular, socially-specific means of representation'. This is a fundamentally heterogeneous definition as it allows the flexibility of integrating communication systems in the actional, linguistic, paralinguistic, graphical and symbolic (among others) dimensions.

The research of Jewitt et al. (2001) shows that learning, through the multi-modal/multiliterate lens, is an active process of meaning-making and remaking. The reflexive dimension, whereby the subject-experiencers (*acteurs-L*) negotiate with their identities and cognitive schema, via transduction, is fundamental in knowledge transformation as well as reconfiguration. Hence, meaning emerges as a consequence of choice and experience. The notion of experience is a generic term that has two distinct meanings (Engel, 2007), the first being traditional empiricism, which considers experience as a set of data to be used for analytical purposes, from a controlled and controllable setting (experimentation). The second meaning states that not all events are controllable; humans live events prior to any form of logical or analytical reasoning, and not all experiences can or should be rationalised. Such a stance integrates elements like sensitivity, affectivity and imagination.

Science and Multiliteracies

In science, multiliteracies have largely been restricted to the multimodal nature of scientific texts (Alvermann & Wilson, 2011) and visual representations (Jewitt, 2005)—the latter being in conjunction with ICT.

Science deals with the study of nature and is itself ontologically and epistemologically multimodal. Alvermann and Wilson's (2011) depiction of experiencing science via an outing, where the students are learning physics concepts through conversation and experienciation, points to the fact that science (including physics) is intrinsically multimodal and deals with a wide variety of semiotic devices.

Multimodality, Science and Semiotics

Multimodality, in essence, concerns the possible varieties of communicational modes beyond 'verbal' language and therefore deals with semiotics, which is a field studying signs, symbols and their meanings. Signs can be in the form of phonemes (sounds), gestures, diagrams, graphics, etc. In his discussion about the multimodal nature of scientific language, Lemke (2004) goes in the same line of thought, stating how the science learner is de facto exposed to a wide variety of communicational modes. He also puts forth the fact that 'natural language' cannot express the wide range of information embedded within scientific thinking, explaining the need for heterogeneous modes of expression like diagrams and formulae. Lemke (2004), however, adopts a structural perspective on linguistics, evoking the combinatorial properties of 'discrete' semantic units. What he identifies as limitations in linguistics has already been extensively discussed in subsequent 'branches' of linguistics, like discourse linguistics. Benveniste (1966) as a linguist and semiotician, for example, already identified issues pertaining to structural perspectives that appeared to relegate the human actor to the background. Likewise, modern perspectives on linguistics (Robillard, 2008a) reinstate the human communicator as an important actor in meaning-making, and acknowledge that 'linguistics' move much beyond 'verbal language' and encompass the whole spectrum of meaning-making devices. In this sense, while we agree with Lemke's posture regarding the importance of semiotics, especially in science teaching, we also argue that our alignment on language experienciation moves beyond the structural perspective and aligns with the more modern take on language studies.

We argue also that reflections around multiliteracies, especially applied to science teaching and learning, are incompatible with traditional structural views of linguistics. This is fundamental as language is a central part of science. Lemke (1987) goes in the same direction, saying that talking about science goes beyond 'talking' and is more about 'doing' science through the medium of language. Additionally, two of the premises of Wellington and Osborne's (2001) book on language and literacy in science education are the fact that language is a major part of science education, and that language is a major barrier to most pupils in learning science.

The value of Lemke's (2004) arguments lies in the fact that it tallies with the first definition criterion of multiliteracies, i.e. multimodality, as he highlights the centrality of semiotics, as well as the possibilities offered by the use of a wide variety of communicational modes. However, multiliteracies are also about linguistic and cultural diversity. Notwithstanding the seemingly accepted multimodality encompassed within the scientific expression, our research demonstrated that the respondents (see below) did not make use of their multilingual resources in dealing with scientific concepts. This led us to interrogate the presence (or absence) of multiliteracies-based thinking in the scientific classroom.

Interest and Motivation in Light of a Multiliteracies-Based Approach

Researching the notion of 'interest' emanated at least partially from a criticism of the limits of a strictly cognitive/structural view of how individuals decode, understand, store and remember information. This is relevant to us as it allows us to link multimodal experiential learning which is central in a multiliteracies-based approach and the notion of interest, which we posit as being a constitutive dimension of multiliteracies.

Beyond being an important motivational variable, 'the term interest also refers to a relatively enduring predisposition to re-engage with particular content such as objects, events and ideas' (Hidi, 2006, 70). The centrality of engagement and re-engagement in this definition is useful for a multiliteracies-based approach, in the sense that the multimodality and experiential meaning-making imply a certain level of engagement with knowledge. The example of the outing (see Alvermann and Wilson, 2011 above) encompasses experiential meaning-making which can be linked to predisposition of the learners present during the event of the outing to engage with the knowledge constructed while being in the situation. The affective dimension is also important in the conceptualisation of interest. Hidi (2001, 2006) highlights the importance of the 'affective factor of interestingness of ideas' (Hidi, 2001, 192), notably on how learners processed discourse. She also states that despite the possibility of researching the emotional and cognitive dimensions separately, the two are actually mutually inclusive and feedback into each other.

The concept of 'situational interest' is also relevant to our reflection as it concerns a broader category of 'environmentally-triggered interest' which encompasses text-based interest. Hidi (2001, 2006) establishes a distinction between situational interest and individual interest. While the latter tends to develop slowly, the former is triggered by elements in the immediate environment. By stating that 'a person's interest can also be triggered by a visual stimulus such as play object, or viewing a picture, an auditory stimulus [...] or a combination of visual and auditory stimuli like a TV show' (Hidi, 2001: 192–193) Hidi categorises multimodal triggers of interest as being part of situational interests.

While providing scope for a wide variety of interest-triggers, whether individual or situational, Hidi's approach appears to be focused on text and reading. It is true that the text is one of the mediators of knowledge when it comes to science teaching and learning. It is, however, only one of the trajectories through which knowledge can be constructed. As mentioned earlier, science is intrinsically multimodal and multiliteracies allow the opportunity for the different actors in science teaching and learning to explore interests that go beyond the text. In other terms, knowledge could be potentially accessible to learners, even though their learning styles, preferences and interests might not be focused exclusively on texts. Multimodality, for example, allows the opportunity for the flourishing of different types of 'situational interests', like actional or visual and provides pathways for their legitimization as part of the multimodal learning processes that are central in science, notably in the acquisition of scientific concepts.

Integration of Multimodality in Physics

On the epistemological front, physics integrates mathematics; therefore, knowledge of physics and mathematics are interdependent. The same multimodal characteristic applies to the practical dimension: physics, like other sciences, integrates not only graphical and symbolic representations, but also hands-on activities where students carry out various experiments. The multiliteracy-based approach argues for a coherent integration of the hands-on and minds-on aspect, which relates to the cognitive dimension. As Holsterman, Grube, and Bögeholz (2010) state, learners display motivation and interest when hands-on and minds-on activities are carefully structured in the lesson.

We argue that two additional processes are fundamental in the conception of a multiliteracy-based approach in the teaching and learning of science: transduction (Jewitt et al., 2001) and negotiation. Transduction is a useful metaphor emanating from biology. In the biological sense, transduction is the process whereby a virus injects its DNA in a cell for reproduction. While the process remains parasitic, we argue that multiliteracies lay emphasis on how the learner reflexively transduces his own 'socio-cognitive DNA' into the process of meaning-making to create meaningful, assimilable and transferrable knowledge. As for negotiation, it not only concerns others (on how knowledge is to be conveyed and communicated to learners, for example) but also within the self. In other words, knowledge is filtered and infused with different identities of the self in order to be meaningful. The process of transduction and negotiation, combined with the three above-mentioned processes, are based on the conceptualisation of communication as being inherently multimodal. Knowledge construction is about the *acteur-L* operating a set of essentially multiple choices from a general conception of multiplicity.

A multiliteracy-based approach also integrates analogies and metaphors in reasoning, meaning-making, and communication as being central (Jewitt et al., 2001). The authors also argue for the need to move away from unimodal linear and exclusively linguistic-textual processes. In this sense, science needs to consciously attend to and integrate various modes of communication, inclusively and variably. Moreover, awareness needs to be developed on how these modes are coherently integrated into teaching, learning and general reflection. Correlatively, as there are different cognitive and representational demands associated to each mode, the teacher-experiencer, as well as the student-experiencer (and anyone involved in the process of meaning-making) needs to negotiate the suitability of these modes with respect to what is best adapted to the learning needs. Such negotiation can best be achieved while taking into consideration students' interest development (Hidi, 2006).

Methods

We adopted a qualitative approach designed to explore how multiliteracy skills were constructed by prospective physics educators during a physics task. The task required the physics trainee educators to describe the motion of a plain piece of paper and a similar crumbled one in a situation of free fall. The study was conducted in a teacher training institute during the course of a 15-week module in physics which was taught by the second author. The module which was delivered through the inquiry approach consisted of multimodal tasks, involving the trainees, to engage in meaning-making of physics concepts. Initially, five trainees registered for the course, but after three weeks, two dropped out. From the remaining three, data for the study were collected from two trainees who formally consented to being interviewed. Consent was obtained to videotape the session.

A semi-structured task-based interview was carried out by two interviewers. A think-aloud protocol (Charters, 2003) was considered to gather verbal data. The interview proceeded concurrently while the participants interacted with the task and with the interviewers. The task-based interview provided opportunities not only to explore and explain the multiliteracy skills of the participants, but it also provided opportunities to assess their knowledge and understanding of physics concepts. In this study, the interview involved both participants interacting with the researchers and with each other. The participants were given the free choice of individual or pair interviews, but both of them agreed to participate in the interviews in pairs, as they felt they would be more secure to express their opinions and their combined effort and the snowballing effect on the responses (Crawford, 1997) may generate a wider range of information, insights and ideas.

In addition, the first and second authors were the facilitators of the interview session. The session was held in the physics lab where the second author taught the physics module to the trainees. It was at the end of the module (that is, in the 15th week) that the task-based interview was conducted. Also, as the second author was known to the participants, his presence in the interview session created both a fruitful and candid atmosphere that maximised confidence among the trainees because of his position as an insider researcher (Costley, Elliot, & Gibbs, 2010).

The video think-aloud session was coded by the three authors during regular meetings. The meetings also served to review and agree on the final codes (see Fig. 22.1). The codes were classified into three themes, namely, linguistic communication [LC], actional [AC] and visualisation [VN], which relate to deductive analysis of the multiliteracy skills. A fourth theme emerged inductively from the data and was labelled 'Physics Conceptual Understanding' [PCU].

Conceptual Analysis of the Physics and Multiliteracy Tasks

The concepts associated with the task are organised in the concept map (Fig. 22.2)

In Table 22.1, we illustrate the generic information with respect to the four themes. In particular, we report on the (mis)use of linguistic features with respect to their relevance and articulation with scientific concepts.

Discussion

The discourse of the trainees was dominated by monolingual communication (Robillard, 2008a, b). All the explanatory elements were communicated in English, despite the students having the choice of languages among the three transcommunal vehicular languages of the local linguistic landscape (Carpooran, 2007). This translates into a singular notional form of language and communication (NLG, 1996). In this sense, only one of the defining criteria of multiliteracies was observed, i.e. multimodality. Consequently, the formation of knowledge networks was not dynamic or mediated by multimodal realisations of language practices, but was instead focused on a linear process with high reliance on the linguistic aspect.

> Whereas if I'm considering air resistance, they will experience different resistances due to their differences in shape, so F will not be the same; therefore I have different 'g', I have different 'a' for both, so if I place it in this equation again, 't' will be different.

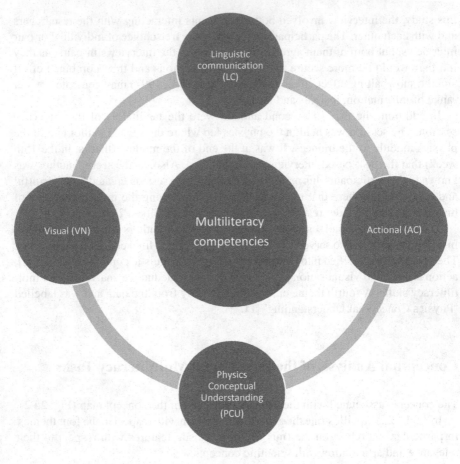

Fig. 22.1 Multiliteracy themes

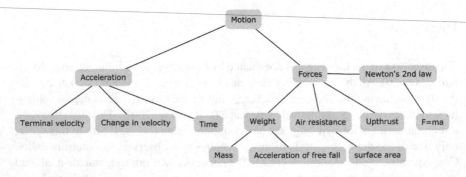

Fig. 22.2 Concept map—physics concepts

Table 22.1 Multiliteracy—linguistic features and physics conceptions

Linguistic features	Sub-category	Code	Examples	Comments
Syntax and conjunctions				Conjunctions are used to connect elements of a sentence. They can be used to express different logical relationships like addition, cause and effect, comparison, etc.
	Conjunction (causality) [1]	LC-S1	If-then	Used in conditional statements to express cause and effect
	Conjunction (Causality, comparison, conclusion) [2]	LC-S2	So	Depending on the context, it can be used to express cause and effect, comparison and conclusion
	Preposition [3]	LC-S3	On/upon	The phrasal verb act on v/s act upon is a point of discussion regarding directionality and vector with respect to a force applied (see discussion)
	Transition word (causality, conclusion) [4]	LC-S4	Therefore	Depending on the context, 'therefore' may indicate causality or conclusion
Reference	Anaphora and cataphora [5]	LC-R1	It	Pronoun used to refer to a thing/an object/an event previously mentioned (anaphoric)
	Semantic [6]	LC-R2	[polysemy, semantic ambiguity, etc.]	The semantic dimension deals with meaning

(continued)

Table 22.1 (continued)

Linguistic features	Sub-category	Code	Examples	Comments
Expressions and Pragmatics				The pragmatic dimension concerns language in use and with respect to contexts
	Probability [7]	LC-EP1	Probably, maybe	Adverbs denoting a possibility and indicating a posture where the speaker is not certain with his/her statements (see below)
	Language register [8]	LC-EP2	A little bit	(colloquial) loose quantifier
	Implicature [9]	LC-EP3		Meaning a speaker conveys or implies without directly expressing
Physics conceptual understanding	[10]	PCU		During the course of the interview, correct (CPCU) as well as incorrect physics understanding of concepts (IPCU) were detected

In the above example, the transition word 'therefore' is wrongly used because it does not express the conclusion drawn from previous statements. This is also linked and meshed into a misconception when the trainee teacher made reference to different values of acceleration of free fall. One of the consequences, as observed, was the misuse of linking words (connectors and conjunctions) to articulate their statements.

Similarly, as can be deduced from the statement, 'if I am going to say that the paper is going to fall ...' the term 'if' is not appropriate as it conjures up unwarrantable assumptions, such as there might be a possibility that when released the paper might not fall. It also implies a possibility that the respondent may not say that the paper is going to fall. Additionally, the formulation of 'if I am going to say' also indicates a colloquial transposition of linguistic structures from the first language of the trainee. There are several instances of the intricate meshing of the various aspects we observed (see Table 22.1). Issues can also be found at the level of implicature, which directly links to loose assumptions made by the students. In the example 'But only if it stays in that position, if it falls in that position, there won't be much air resistance' the ambiguous syntactic construction may lead the listener to believe that air resistance adapts to the object, which is conceptually inaccurate. Moreover, there was no active

Table 22.2 Think-aloud session

Concepts	Think-aloud task	Multiliteracy	Comments
Air resistance	S1: … they both will take the same time to fall a certain distance in vacuum. Nevertheless, the air resistance will be greater for the plain paper if it is not in vacuum … and we've seen an experiment on that	CPCU [minds-on]	Trainee S1 rightly inferred that both objects will take the same time to fall in vacuum. However, no mention was made as to why the plain paper will experience greater air resistance as compared to the crumbled paper. Insight into area of contact could have been discussed and supported by means of a diagram
Air resistance	S2: But only if it stays in that position, if it falls in that position, there won't be much air resistance	Expression and pragmatics [LC-EP3] implicature [Hands-on—partly]	Here, from the trainee's explanation, it is implied that it is the air resistance that adapts to the object. The sentence construction is ambiguous. Moreover, there was no attempt to let go of the paper from a certain height to validate their assumptions. Such an approach adopted by the trainees relates to a deductive way of reasoning, which is usually a recourse for less experienced problem-solvers (Llyod & Scott, 1994)

(continued)

Table 22.2 (continued)

Concepts	Think-aloud task	Multiliteracy	Comments
Air resistance	S1: Even if it stays flat, there will be a difference in air resistance because the size is different for both	LC-Reference[LC-R2]	It is unclear whether 'size' relates to surface area/shape/dimension. Here also, there was no attempt to release the plain paper
Time of fall	S2: But if [they are] in vacuum [they] will both fall at the same time, i.e. they will take the same amount of time to fall a certain distance	LC-Reference [LC-R2]	'at the same time' implies the release of the objects at the same moment, whereas here 'duration' of motion is of interest
Newton's Second Law	S1: If we were to explain the concept, so Newton's Second Law comes in handy, we have to apply F = ma, if there is no resistance, so the resultant force on both will be the weight, so the acceleration will be g, therefore it will take the same time. Then we apply the equation of motion	LC-Syntax and conjunctions [LC-S2] IPC LC-Syntax and conjunctions [LC-S2] V- Visual-symbolic	'so' is wrongly used as a conjunction to explain cause–effect relationships, especially given that it follows 'if' in the first two instances ('then' should be used instead) The third use of 'so' as a conclusive connector is correct. However, the conceptual links with the last statement whereby the student is proposing the application of the equation of motion, are loose, as the equation of motion is irrelevant. It is not clear what will the equation of motion bring to the motion under consideration. The expression is ambiguous First use of equation (F = ma) after prompt from interviewer

(continued)

Table 22.2 (continued)

Concepts	Think-aloud task	Multiliteracy	Comments
Newton's Second Law Area Viscosity	S2: One thing that S1 explained about Newton's Law and all that, but ultimately, everything should be tested experimentally as far as possible We will have to start by making a few assumptions, for example, if I am going to say that the paper is going to fall, I start it in this position, well it's not going to stay in that same position when it's falling, but if I were to consider all that, it's going to move and it's going to be too complicated, so I have to make a few assumptions that for example, there is no rotation when its moving or falling down. Make a few assumptions and then consider the area, the viscosity of air, and things like that I think we have to do the same thing about this [holding crushed paper], we cannot consider this complicated sheet, we have to consider it as a sphere, else it becomes too complicated to calculate, at least on paper, maybe simulations can be better	LC-Expression and pragmatics [LC-EP2] Expression and pragmatics [LC-EP1] Reference [LC-R1] IPC Actional	'All that' is a colloquial expression that lacks accuracy. The reference is also unclear whether the term is referring to what has been mentioned before (anaphoric) or what has been mentioned after (cataphoric) in the text The use of 'if I am going to say…' implies a posture whereby the speaker is using a 'safety net' strategy through the use of a hypothetical statement which indicates that he is unsure of his statement, in which case there is no legitimacy to say that this statement is wrong 'ultimately' also appears to be used in contradiction with 'as far as possible' The reference of 'it' is unclear following 'all that' Considering 'rotation' and 'viscosity' in the argument adds more variables to the simple problem and thus complicating the task further Attempt to have recourse to the actional dimension. However, the action was not completed (that is, the objects were not released). The concepts are normative (they are only being evoked) and the actional dimension appears to be limited, being present only as a spatial indicator (height, position [vertical/horizontal]) It is unfortunate that the trainee did not release the paper as this would have helped him to situate to what extent his hypotheses, in light of the assumptions he mentioned earlier, were valid

(continued)

Table 22.2 (continued)

Concepts	Think-aloud task	Multiliteracy	Comments
Mass Weight Newton's Second law Acceleration due to gravity Laws of motion Air resistance	S1: No, but I think that we can simply introduce the concept, at least for the 'A' level, I think we do that at HSC [Higher School Certificate]; just introduce the concept of both objects having the same mass and hence the same weight; so in vacuum we have only the weight acting on it; so the resultant force will be the weight, according to Newton's Second Law, it will be $F = ma$, F = weight (W), so we have $W = mg$, I have the acceleration g, I apply the Laws of Motion, I will have the same t, I have $S = ut + \frac{1}{2} at^2$, the distance is the same if they are at rest and dropping them, acceleration is the same so it will be the same. Whereas if I am considering air resistance, they will experience different resistances due to their differences in shape, so F will not be the same, therefore I have different g; I have different 'a' for both, so if I place it in this equation again, t will be different	LC-Syntax and conjunction [LC-S3] LC-Syntax and conjunctions [LC-S1] LC-Syntax and conjunctions [LC-S4] IPC	The weight is the force due to gravity that acts on a mass. Learners may be confused in this particular case: 'weight acting on it'. Kibble (2006) lays ample emphasis to revisit the use of language to allow learners to find their own voice, which should be in conjunction with the correct conception that physicists use. In this particular case, the weight is the force that acts downwards from the centre of gravity of the object The 'if' appears inconsistent with the students' arguments as they had the opportunity to experiment with the objects, which they did not do The transition word 'therefore' is wrongly used as it does not adequately express the conclusion drawn from previous statements. Furthermore, a misconception surfaced the moment she made reference to different values of acceleration of free fall

(continued)

Table 22.2 (continued)

Concepts	Think-aloud task	Multiliteracy	Comments
	S2: The resultant acceleration will be different		There is a tendency to mix up acceleration of free fall g with the net acceleration a
Resultant force Air resistance Weight Acceleration	S1: In the presence of air resistance, the weight remains the same; it's the resultant force acting on the body that changes; so 'a' changes. So, if I were to explain, I would simply, maybe I would say that in the absence of air resistance, the resultant force = weight. So the acceleration of the body = g; the acceleration due to gravity. In the presence of air resistance, weight still acts on both objects but the resultant force is not equal to the weight because I have other forces acting on the body. Then if I say that F resultant f_r = ma and I have weight which is the downward force, I'm considering linear motions that is along this direction. If I say that the resistive upward force is f_{res} so I would have f_r, the motion is downward f_r = downward force—upward force so I have f_r = mg − f_{res} so g is still remaining the same it is the 'a' which is changing. In the absence of air resistance f_{res} = 0 so I have ma = mg as I said here. So a = g	CPCU Expression and pragmatic [LC-EP3] LC-Syntax and conjunction [LC-S3] LC-Reference [LC-R2] Expression and pragmatics [LC-EP1]	Trainee S1 has now reviewed her initial version. Earlier, she claimed that g would be different In this case, simply and maybe appear to be contradictory. 'Simply' means 'without ambiguity' and cannot be followed by a tentative statement The notion 'weight still acts on ...' remains problematic. Furthermore, there is ambiguity in the interpretation of resultant force. It stands to reason that the resultant force is the vector sum of all the forces experienced by the object Here, the term 'resultant' is misused as it is semantically inaccurate: the student states other forces subsequently acting upon the body after having evoked resultant force 'If' again denotes a posture where the speaker is unsure of his statements
Air resistance Rotational motion	S1: We are considering linear motion and I'm not considering air resistance in other directions Maybe we should make it clear that the shape, the rotational motion will not be taken into consideration	Expression and pragmatics [LC-EP1]	'Maybe' denotes a hypothetical posture. The premise should be clearly established in terms of assumptions (e.g. assuming that)
Rotation	S2: What about the objects? Because we are moving from something that we are going to do practically and then we are going to explain it; so there must be some assumption, like I said because of it's the shape, it's not going to rotate—if I assumed all of that		Again, the trainee refers to 'less-than-fully established propositions' (Fortus, 2009, p. 87) as these propositions or assumptions are advantageous to them in the construction of their arguments

remaking of information; often, the students had recourse to a (loose form of) a priori and explicit/propositional knowledge, such as viscosity of air, and the application of the equation of motion. The statement made by one of the respondents, namely '… therefore I have different g …' was not challenged by her peer and is another example of fragmented knowledge or weak conceptual link (Kibble, 2006).

Moreover, it was also noted that there was an absence of reflexivity (Robillard, 2008a, b), notably through the absence of experiential knowledge. There was only one instance of reference to an external resource (Jewitt et al., 2001) where the trainee gave an answer based on a documentary film he had seen as he explained: '… because we have seen an experiment on that'. These situations may hinder learners in developing situational interest (Hidi, 2001, 2006) which may eventually affect performance (Hidi, 2001).

The subject-thinker-experiencer (Robillard, 2008b) was backgrounded (or completely erased); instead, there were attempts of linearly reproducing knowledge without transduction or transformation. The absence of links to lived experiences, coupled with the scarcity of using analogies and metaphors, further indicates the absence of transduction and meaning-making (Jewitt et al., 2001). Exclusive reference was made to experiment while the experience was absent. In other words, the trainees were unable to integrate conceptual and experiential knowledge in order to proceed with their own meaning-making process and the transmission of meaning. Holstermann, Grube, and Bögeholz (2010) also stress on the importance of experience, stating that hands-on activities, per se, are not the direct trigger of a scientific attitude. What was striking is that the minds-on engagement of the trainees was loosely associated with the corresponding hands-on. The trainees were operating within the traditional set-up, driven by instructions, which produce recipe approaches in science. Within this approach, there are limited opportunities that foster harmony between elements of minds-on and hands-on. We, therefore, posit the following model where adequate acquisition of multiliteracy competencies is essential for development of minds-on engagement, which in turn encourages hands-on connections (Fig. 22.3).

Fig. 22.3 Multiliteracy and hands-on, minds-on

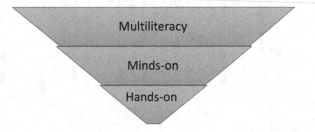

Conclusion

Based on the study, we have identified a number of shortcomings in trainees' argumentation in their understanding of a particular physics concept. There is evidence that problems developed in multiliteracy skills impinge significantly on the minds-on engagement of the trainees, and thus, limit their efforts for hands-on connection.

The interview has revealed that the trainees' thinking and reasoning are inconsistent with their attempt to make rational choices in physics within the premises of multiliteracy competencies (linguistic communication, actional, visual and physics conceptual understanding). The trainees demonstrated limited multiliterate competencies, as their interaction was founded on linear stimulus-response dynamics since they systematically operated on prompts. This created issues in coherence, particularly when they were unable to form meaningful links between their statements. This led to ambiguous conceptual links (see Table 22.1). Unimodality (Robillard, 2008a, b) was also one of the defining traits of their communication. The predominant mode was verbal; other modes (drawing a diagram, raising the items but not displaying the process of free fall) came only after the cues of the interviewers. Despite having been provided with the items (a sheet of paper and a crumpled one), there was no recourse to the actional dimension (Jewitt et al., 2001). In other words, the actional dimension was largely bypassed, which led to fragile connections among the physics concepts. The students experienced difficulties to conceptually relate resultant force to the mathematical equivalence of a vector sum of all the downward forces acting on the object.

While multiliteracies provide fertile perspectives in the teaching and learning of science, it is also true that English-centrism, and general language-centrism when it comes to assessment and evaluation, especially in multilingual contexts like Mauritius, is a potentially hindering factor. The former reduces the perspectives of exploring the integration of the diverse repertoire of multilingual speakers in scientific communication. Research in this direction is relatively scarce but also opens avenues for further research which could potentially feed-forward to reflections around multiliteracies. The latter is also somewhat paradoxical. While our above discussion acknowledged the multimodal nature of science, language-centrism de facto reduces possibilities for allowing more leeway to other forms of expression, including actional and gestural. However, multiliteracies remain a valuable theoretical pathway that could potentially have significant benefits in how learners approach the learning of science. Concept acquisition and development can potentially be reinforced through the simultaneous integration of multilingual resources and experiential strategies, where the learner can be encouraged to make use of his/her active-meaning capacities, in meaningful situations, via intelligible language. Concomitantly, the experiential dimension also corresponds to the 'practical' aspect, bridging the gap between laboratory-based, in vitro experimental practice and real-life, in vivo experiential practice, thereby providing opportunities to empower the learner to autonomously engage in scientific inquiry. Furthermore, a multiliteracies-based perspective would also have ramifications in the learning of other subject areas, like mathematics and even language, as factual, scientific knowledge feeds forward to certain topics which are in the syllabi of language studies. As such, language studies are also about situating real-life meaningful experiences and the expression thereof through linguistic devices. Notwithstanding the potential benefits of a multiliteracies-based approach, there remains significant work to be done on the integration of the multilingual resources of the learner in the teaching–learning of science. As we highlighted above, this provides opportunities for further research, notably in the context of heterogeneous multilingual contexts, which represent potential laboratories in exploring multiliteracies in the teaching and learning of science.

Chapter Summary

- The notion of multiliteracies encompasses a wide range of disciplines, like emotional literacy, oral vernacular and cultural literacy among others.
- In this study, four themes were identified with respect to multiliteracy skills: linguistic communication, actional, visualisation and Physics conceptual understanding.
- It was observed that trainees manifested low multiliteracy skills. Their discourse was dominated by monolingual communication, with attempts mainly geared towards linear reproduction of knowledge, which had direct implications on how physics concepts were understood and used.
- Physics trainees need to improve their multiliteracy competencies to be able to navigate between minds-on engagement and hands-on connections.

Recommended Resources

Ajayi, L. (2010). Preservice teachers' knowledge, attitudes, and perception of their preparation to teach multiliteracies/multimodality. *The Teacher Educator, 46*(1), 6–31. https://doi.org/10. 1080/08878730.2010.488279.

Charters, E. (2003). The use of think-aloud methods in qualitative research: An introduction to think-aloud methods. *Brock Education, 12*(2), 68–82.

Leander, K., & Boldt, G. (2012). Rereading "A pedagogy of multiliteracies": Bodies, texts, and emmergence. *Journal of Literacy Research, 45*(1), 22–46.

References

Alvermann, D. E., & Wilson, A. A. (2011). Comprehension strategy instruction for multimodal texts in science. *Theory and Practice, 50*(2), 116–124. https://doi.org/10.1080/00405841.2011. 558436.

Benveniste, E. (1966). *Problèmes de linguistique générale*. Paris: Gallimard.

Bell, J. C. (2014). Visual literacy skills of students in college-level Biology: Learning outcomes following digital or hand-drawing activities. *The Canadian Journal for Scholarship of Teaching and Learning, 5*(1), 1–13. https://doi.org/10.5206/cjsotl-rcacea.2014.1.6.

Carpooran, D. (2007). *Appropriation du francais et pédagogie convergente dans l'Océan Indien: Interrogations, applications, propositions.* Paris: Editions des Archives Contemporaines.

Charters, E. (2003). The use of think-aloud methods in qualitative research. An introduction to think-aloud methods. *Brock Education, 12*(2), 68–82.

Chinnappan, M. (2008). Productive pedagogies and deep mathematical learning in a globalised world. In P. Kell, W. Vialle, D. Konza, & G. Vogi (Eds.), *Learning and the learner: Exploring learning for new times* (pp. 181–193). University of Wollongong.

Claassen, G. (2007). Journalism and cultural literacy: An exploration towards a model for training journalism students. *South African Journal for Communication Theory and Research, 21*(1), 12–20.

Costley, C., Elliot, G., & Gibbs, P. (2010). *Doing work based research: Approaches to enquiry for insider-researcher.* Los Angeles: Sage Publications Ltd.

Crawford, I. (1997). *Marketing research and information systems.* Rome: Food and Agriculture Organisation of the United Nations.

Drapper, D. C. (2015). *Digital knowledge mapping as an instructional strategy to promote visual literacy: A case study,* D. M. Baylen & A. D'Alba (Eds.). New York: Springer. https://doi.org/10.1007/978-3-319-05837-5_11.

Engel, P. (2007). «*Experience*» *Dictionnaire des concepts philosophiques,* M. Blay (Ed.). Paris: Larousse & CNRS Editions.

Fortus, D. (2009). The importance of learning to make assumptions. *Science Education, 93,* 86–108.

Freebody, P. (2007). *Literacy education in schools: Research perspectives from the past, for the future.* Victoria: Australian Council for Educational Research.

Hidi, S. (2001). Interest, reading, and learning: Theoretical and practical considerations. *Educational Psychology Review, 13*(3), 191–209.

Hidi, S. (2006). Interest: A unique motivational variable. *Educational Research Review, 1,* 69–82.

Holstermann, N., Grube, D., & Bogeholz, S. (2010). Hand-on activities and their influence on students' interest. *Research in Science Education, 40,* 743–757. https://doi.org/10.1007/s11165-009-9142-0.

Jewitt, C. (2005). Multimodality, "reading" and "writing" for the 21st century. *Discourse: Studies in the Cultural Politics of Education, 26*(3), 315–331.

Jewitt, C., Kress, G., Ogborn, J., & Tsatsarelis, C. (2001). Exploring learning through visual, actional and linguistic communication: The multimodal environment of a science classroom. *Educational Review, 53*(1), 5–18.

Kibble, B. (2006). Understanding forces: What's the problem? *Physics Education, 41*(3), 228–231.

Lemke, J. (2004). The literacies of science. In E. Wendy Saul (Ed.), *Crossing borders in literacy and science instruction.* (pp. 33–47). Newark, DE: International Reading Association and Arlington, VA: NSTA Press.

Lemke, J. (1987). *Talking science: Content, conflict, and semantics.* Paper presented at American Educational Research Association meeting, Washington DC, 1987. Arlington VA: ERIC Documents Service (ED 282 402).

Llyod, P., & Scott, P. (1994). Discovering the design problem. *Design Studies, 15,* 125–140.

New London Group. (1996). A pedagogy of multiliteracies: Designing social futures. *Harvard Educational Review, 66*(1), 60–92.

Newman, M. (2005). Rap as literacy: A genre analysis of Hip-Hop ciphers. *Text, 3,* 399–436.

Oksuz, Y. (2016). Evaluation of emotional literacy activities: A phenomenological study. *Journal of Education and Practice, 7*(36), 34–39.

Robillard, D. (2008a). *Perspectives alterlinguistiques* (Vol. 1). Paris: Démons: l''Harmattan.

Robillard, D. (2008b). *Perspectives alterlinguistiques* (Vol. 2). Paris: Ornithorynque: L'Harmattan.
Rowsell, J., & Walsh, M. (2011). Rethinking literacy education in new times: Multimodality, multiliteracies, and new literacies. *Brock Education, 21*(1), 53–62.
Wellington, J., & Osborne, J. (2001). *Language and literacy in science education*. Buckingham: Open University Press.

Dr. Shameem Oozeerally is a Lecturer in the French Department at the Mauritius Institute of Education. His research interests gravitate around complexity theory and the epistemology of language sciences. He also conducts research in Creole studies and has worked on interdisciplinary research projects in the area of early childhood language experienciation, as well as ecolinguistic discourse analysis in the primary education and curriculum context of Mauritius.

Yashwantrao Ramma is Professor of Science Education and Chair of Research at the Mauritius Institute of Education. As a physicist, he has worked on several research projects related to technology integration and misconceptions of both physics teachers and students. Currently, he is leading research projects on exploring teachers' content knowledge and pedagogical content knowledge across various subject areas, on indiscipline and school violence in primary schools and also on students' transitions from secondary to university and teacher training.

Dr. Ajeevsing Bholoa is a Senior Lecturer in Mathematics Education at the Mauritius Institute of Education. He is currently the Programme Coordinator for pre-service B.Ed. honours and is also involved in curriculum development at the primary and secondary levels. His research interests are related to the integration of technology as a pedagogical tool in teaching and learning of mathematics and the identification of content knowledge and pedagogical content knowledge of teachers.

Chapter 23
Project and Problem-Based Teaching and Learning

Michael R. L. Odell and Jaclyn L. Pedersen

Introduction

Today's educators are tasked with preparing students for an uncertain and complex future. Traditional education approaches are not up to the task of preparing students to live and work in a global, information-based economy that is rapidly changing. In the report, *Rising Above the Gathering Storm* (National of Academy of Sciences, National Academy of Engineering, & Institute of Medicine, 2007, p. 6), the report emphasizes the need for "world-class science and engineering" as the principal means of creating "new jobs for U.S. citizenry as a whole as it seeks to prosper in the global marketplace of the 21st century" (p. 40) in light of increasing competition from emerging economies. Today's modern economies have steadily increased their capacity and ability to create and commercialize knowledge as a means for economic growth. Creating and commercializing knowledge reinforces the importance of quality STEM Education that prepares students for a rapidly changing future where innovations can be developed and introduced in short order, rendering current knowledge and skills in the workforce obsolete (National Research Council, 2000, 2012).

It should be noted, that preparing students for an uncertain future is not a new idea. In *My Pedagogic Creed*, John Dewey (1897) discussed providing students an education that prepares students for the "modern" world as the twentieth century dawned.

M. R. L. Odell (✉)
College of Education and Psychology, College of Engineering,
University of Texas at Tyler, Tyler, TX, USA
e-mail: modell@uttyler.edu

J. L. Pedersen
College of Education and Psychology, University of Texas at
Tyler University Academy, Tyler, TX, USA
e-mail: jpedersen@uttyler.edu

© Springer Nature Switzerland AG 2020
B. Akpan and T. Kennedy (eds.), *Science Education in Theory and Practice*,
Springer Texts in Education, https://doi.org/10.1007/978-3-030-43620-9_23

With the advent of democracy and modern industrial conditions, it is impossible to foretell definitely just what civilization will be twenty years from now. Hence it is impossible to prepare the child for any precise set of conditions. To prepare him for the future life means to give him command of himself; it means so to train him that he will have the full and ready use of all his capacities; that his eye and ear and hand may be tools ready to command, that his judgment may be capable of grasping the conditions under which it has to work, and the executive forces be trained to act economically and efficiently (Dewey, 1897, p. 77).

Teaching a finite set of knowledge and skills will not suffice to address the challenges of this century and beyond. Today's educators face the challenge of preparing students for jobs that have not yet been created and problems that are yet to arise (Bybee & Fuchs, 2006; National Science Teachers Association, 2011). Simply providing students with a predefined set of knowledge and skills is no longer viable, as these will be obsolete before the student enters the workforce. Educators in the fields of science, technology, engineering and mathematics (STEM) need to reexamine teaching and learning through a 21st century lens and utilize pedagogies that facilitate students' abilities to access, understand, and use knowledge. This includes an education approach that supports the 21st century skills of communication, creativity, critical thinking, and collaboration (Partnership for 21st Century Skills, 2015). Although not specific to PBL, research indicates that students are more successful at applying what they learn when instruction utilizes real-world contexts (Bransford et al., 1999).

Unfortunately, education systems are slow to respond and pedagogies utilized in most schools remains largely traditional. Although there have been efforts to provide schools with more technology, use of technology still mirrors past practices. If we are to prepare students for the 21st century and beyond, STEM education should be designed to prepare learners with the skills to confront new challenges (Boud & Feletti, 1997). In addition, STEM teacher education programs must prepare teachers to utilize 21st century pedagogies to facilitate student learning.

The foundational concept behind project-based learning (PBL) and problem-based learning (PrBL) is to develop students who can manage their own learning. Directing and managing one's own learning is a central tenet of 21st century pedagogy. See Chap. 32, *21st Century Skills,* for additional information. Students engaged in PBL/PrBL learn by designing and constructing solutions to real-world problems. PBL has five characteristics, including: (1) project outcomes tied to curriculum standards and learning goals; (2) driving questions and/or problems that are ill-defined and can lead students to conceptual understanding; (3) student knowledge building and inquiry through investigations; (4) students managing their own learning; and (5) projects based on real-world problems and questions (Trilling & Fadel, 2009) (Table 23.1).

In the STEM context, this allows students to recognize the interdisciplinary nature of complex problems and fosters critical thinking beyond disciplinary boundaries when approaching a problem. Working in teams also mirrors how STEM is carried out in the workplace. Similar to the workplace, students must take responsibility for different aspects of their project, collaborate for a common outcome, critique each other's work, and create professional quality products that in many cases will

Table 23.1 PBL and PrBL Similarities

Similarities of project- (PBL) and problem-based learning (PrBL)
Both PBL and PrBL
• Are inquiry approaches
• Focus on open-ended or ill-defined tasks
• Provide real-world applications of content and skills
• Build 21st century skills (Four C's)
• Foster self-management and independence
• Foster deeper learning
• Foster self-reflection
• Develop presentation skills

be judged by experts outside the school. In traditional school settings, the teacher controls time spent on an assignment. Please refer to Chap. 17, *Experiential Learning,* for additional insights.

Contrarily, problem-based learning (PrBL) fosters and motivates students to manage their own time while still being held accountable for deadlines; checkpoints for deliverables throughout the learning process. Students are encouraged, and often times required, to seek expert advice and critiques from professionals in the field, which pertains to the current topic in their project. Ultimately, students present their work in a formal setting designed to mirror the 21st century workplace equipping them with valuable communication skills necessary for success. Beyond employing the tenets of the Four C's (critical thinking, communication, collaboration and creativity), both PBL and PrBL foster deeper learning and relevance by allowing students to learn in a real-world context (Hmelo-Silver, 2004). While both PBL and PrBL utilize collaborative teams and require students to design and carry out projects or investigate problems that cross discipline boundaries, the amount of time spent on PrBL is shorter. The following chart summarizes the differences in the two approaches (Table 23.2).

Figure 23.1 illustrates the PBL model approach as practiced by STEM Academies

Table 23.2 Differences Between PBL and PrBL

Differences between project- and problem-based learning	
Project-based Learning (PBL)	Problem-based Learning (PrBL)
Often interdisciplinary using multiple subjects	Typically single subject. More common as a strategy in mathematics.
Long duration (weeks/months)	Shorter duration (Days/Weeks)
Focus on developing a product	Focus on solving a problem
Product addresses a real-world task	Product includes proposed solution to a problem

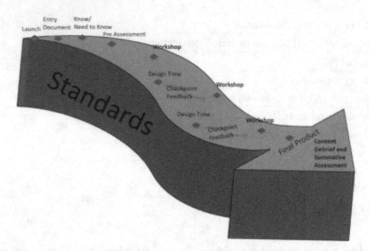

Fig. 23.1 The project-based learning approach

in the State of Texas in the U.S. The arrow indicates the path of a project. Each activity launches with an entry document.

The entry document can be a scenario or a task that is ill-defined, embedded with the standards, and that students must address to create a product or solve a problem to address a real-world issue. Students then begin an iterative process to determine what they know about the problem and what they need to know.

"Know" and "Need to Know", statements are documented and posted in the classroom where they are visible. One of the instructional goals is to address the "Need to Know" statements and move them to the "Know Statements" so that a final solution or product can be created. "Workshops", which are otherwise known as lessons, are designed based on the standards and address the "Need to Know" statements (Next Generation Science Storylines & STEM Teaching Tools, 2016).

As students work on their project, PBL pedagogy builds in checkpoints so that the teacher can monitor progress and provide coaching and feedback where necessary. During each checkpoint, some sort of "deliverable" is typically due. A "deliverable" is simply one piece of the overall final product. It is important to note that the standing checkpoints and deliverables throughout the process are the key pieces to not only holding students accountable and providing students with immediate feedback, but also provide the rationale that the process is considered PBL, versus a traditional project. Students are actually building the final product as they go by acquiring new knowledge in workshops, research or outreach to professionals, and then immediately applying that knowledge to their product in the form of a deliverable.

The important difference here is that in a traditional project, students would be taught all the information up front and then given time at the end of the lesson to build a project, typically of the teachers' choosing. If students are struggling, the teacher will develop extra "workshops" to address the content or skills needs of the students. Checkpoints also give students a chance to self-reflect, as they are then able to take the feedback they received from their checkpoint and revise their deliverables as needed. In many instances, this "just-in-time" learning provides students with immediate needed skills or information to address a "Need to Know" to work towards completion of their project or solution. This process is repeated until final products are ready for review. Presentations to an authentic audience are done when the product is polished and finalized. Students are then given critiques and provided with the opportunity to self-reflect. The products are all aimed at meeting the same parameters set by the teacher, however, the students are able to exhibit creativity and choice in their products. Students are given the autonomy to display their own voice and choice as long as they are still meeting the standards and constraints set by the project rubric.

PrBL derives from a theory originally described in 1977 titled the *information processing* approach to learning (Morris, Bransford, & Franks, 1977). This type of approach suggests that for students to effectively learn material, they need to be placed in situations in which they are required to restructure information they already know within a real-world context, all while gaining new knowledge. Students are then able to deepen their depth of knowledge by discovering new ways to manipulate the content, teaching the information to their peers, discussing the information in a broader context, and even being able to debate the content amongst their classmates. This type of teaching differs from more traditional types of instruction due to students being engaged in more self-directed pedagogical methods. Instructional methods like PBL and PrBL can be traced back to John Dewey's belief that students learn best by thinking and doing in settings that appeal to our natural instincts to investigate and create (Dewey, 1938). PrBL, more specifically, has its roots in medical school environments (Hutchings & O'Rourke, 2004). Problem-based learning was developed in order to teach doctors how to explore and solve medical cases. Today, this method is used in schools in order to accomplish the same type of learning, just in a non-medical context.

Figure 23.2 illustrates the PrBL approach. PrBL, like PBL, is a student-centered inquiry approach to learning. The process is very similar to PBL, however, PrBL takes much less time to execute in the classroom than PBL.

The PrBL process begins with the required curriculum standards, which are rewritten and posed in a problem-like context and presented to the students. The model can be described in phases when facilitated in the classroom. In the introduction phase, students initially approach the problem by going through a "know/need to know" process just as they did with PBL. They sort through their current knowledge and decide what it is that they are yet to know. Students are then encouraged to approach the problem individually and come up with possible ways to solve the problem and/or possible solutions to the problem. During this phase of the process, students may experience some struggle, which is a positive point and key component of the process. It is during this phase that students learn through cognitive dissonance, when two or more ideas may come to mind and seem to collide, but only one can be correct. This phase challenges students to think deeper because the answer will not be obvious. If students have always been "spoon-fed" answers, shown exactly how to solve every problem they face, and have never had to truly rely on their own problem-solving

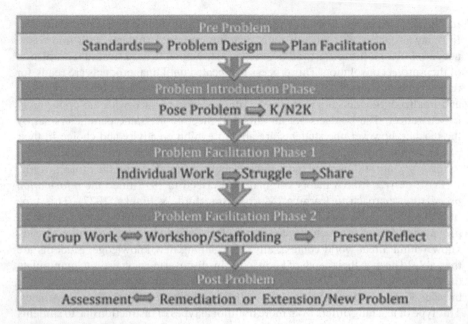

Fig. 23.2 The problem-based learning model

skills, then this phase will seem very uncomfortable at first. However, it is a necessary step in the overall process in order to achieve the most effective learning outcomes.

At the end of the individual work time, students share and discuss their possible methods for solving the problem as well as possible solutions if they have any. Students are then placed in groups for the next facilitation phase and use their ideas to come up with a common method and solution. During this phase, teachers also conduct "workshops" just as occurred during the PBL approach, and teach new concepts if needed or fill in any gaps the students may have that have prohibited them from being able to solve the problem thus far. Once a solution has been found, students share their method and solution with the class through a presentation. Post-problem, students are encouraged to reflect on their method, solution, the process they went through, the social-skills they have developed, and self-evaluate. This time can also be used to re-teach if needed, or extend thinking for students who need more of a challenge.

When compared to PBL, the overall process is very similar; however, in PrBL, the process generally consists of a period of one to three days, while implementing PBL in the classroom can typically take several weeks. PrBL has been shown to be an effective instructional method, in particular in math classrooms (Strobel & van Barneveld, 2009). It is important to remember that in PrBL, the process and the methods students use to solve the problem, are just as important (if not more) than the actual solution.

Reflection in PBL

PBL/PrBL-based pedagogies build the process of reflection. Students, through a process termed "critical friends", conduct an initial review of the solution or final product. The process focuses on developing collegial relationships, encouraging reflection, and collegial dialogue. In general, the critical friends' process serves as a peer evaluation structured to provide safe and meaningful feedback in a professional manner. This allows students to focus on the quality of their work and develops a mindset centered around continuous improvement. Based on the feedback received, students utilize the reviews from the critical friends' process to refine their final projects and develop solutions. Students present final projects and solutions in a formal setting and are assessed for a grade.

Benefits of PBL

There have been a number of studies comparing both PBL and PrBL-based learning to traditional direct instruction. Research studies have found that student outcomes in the learning of facts and basic skills are equal to or better than outcomes achieved using more traditional classroom instruction. Studies comparing student learning outcomes of PBL/PrBL when compared to traditional instruction indicate that when implemented well, students who experienced these approaches showed increases in long-term retention of content, showed equal or better performance on high-stakes tests, improved problem-solving and collaboration skills, and improved students' attitudes towards learning (Strobel & van Barneveld, 2009; Walker & Leary, 2009). In 2016, MDRC and the Lucas Education Foundation reviewed the research and literature found that the design principles utilized in PBL promoted deeper learning. In addition, there was evidence that PBL promoted higher level thinking skills, and intra/interpersonal skills in students (Boss et al., 2011; Condliffe et al., 2017). This chapter addresses PBL and PrBL in relation to STEM Education. Chapter 31 addresses the STEAM pedagogical approach and can provide additional insights.

Preparing Future Teachers in PBL/PrBL Pedagogies

Like many other countries, the United States of America has a perpetual shortage of STEM teachers. Every year, school districts struggle to fill positions in physical sciences, mathematics, engineering, and computer science classrooms. One of the contributing factors to the STEM teacher shortage was that universities in the U.S. over the past two decades had made it difficult for future STEM educators to seek teacher licensure while completing a Bachelor's Degree. This was caused by state governments limiting the number of credits for a bachelor's degree in order to lower the cost of higher education. Discipline departments responded by removing non-essential courses from STEM degrees. This included pedagogy courses for future STEM teachers and was an unintended consequence of higher education policy designed to help more students earn a college degree by reducing time to degree completion.

In response, the University of Texas at Austin developed the UTeach Program. The UTeach program is an innovative university-based teacher preparation program working to increase the number of qualified STEM teachers in U.S. secondary schools (Pérez & Romero, 2014). The program combines students earning STEM degrees to include secondary teaching certification without adding time or cost to earning a Bachelor's degree. The UTeach model has proven an effective solution and been replicated at over 40 universities across the U.S. Fig. 23.3 provides a map of the universities replicating the UTeach Program today.

One of the hallmarks of the UTeach Program is a focus on rigorous research-based instruction that embeds 21st Century skills and pedagogies. UTeach courses are designed to develop preservice teachers' deep understanding of STEM and build strong connections between mathematics and science. The courses are also designed to bring together educational theory and practice. This approach is unique to university-based STEM teacher preparation programs in the U.S.

All preservice teachers enrolled in the UTeach program complete one project-based instruction (PBI) course that includes components of PrBL as well. Preservice

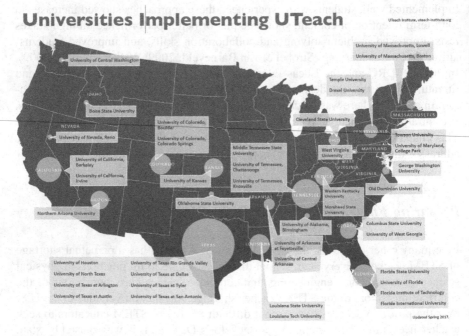

Fig. 23.3 UTeach Universities across the United States

teachers develop PBL or PrBL instructional units, and plan, implement, and analyze their teaching experiences in secondary school classrooms. In addition, preservice teachers enrolled in PBI courses are required to complete clinical hours in schools to practice PBL/PrBL methodology in a classroom setting and observe effective implementation of both from cooperating teachers trained in the pedagogies.

During their capstone course in the UTeach sequence, known as Apprentice Teaching, these future STEM teachers are provided a teaching assignment in a STEM classroom in a local school district for an entire semester. During this semester, they are encouraged to implement inquiry-based practices, which include PBL/PrBL. This practice has proven to not only prepare future teachers but also help share these innovative teaching models with current teachers in hopes of making changes in their own teaching styles. As more teachers are prepared in PBL/PrBL, there is optimism that schools will adopt the 21st century pedagogy as the primary instructional approach.

PBL as a School Reform Model

PBL can also provide an effective model for school reform. In the report *Rising Above the Gathering Storm,* it was recommended that if the U.S. is to remain competitive in the 21st century economy, there must be a serious effort to "enlarge the pipeline of students who are prepared to enter college and graduate with a degree in STEM" (National of Academy of Sciences, National Academy of Engineering, & Institute of Medicine, 2007, p. 6). This would be accomplished by increasing the number of students who complete and pass advanced STEM courses. A recommendation was made suggesting that states develop statewide specialty STEM high schools

> Specialty secondary education can foster leaders in science, technology, and mathematics. Specialty schools immerse students in high-quality science, technology, and mathematics education; serve as a mechanism to test teaching materials; provide a training ground for K–12 teachers; and provide the resources and staff for summer programs that introduce students to science and mathematics. (National of Academy of Sciences, National Academy of Engineering, & Institute of Medicine, 2007, p. 6).

Since distribution of this report, the implementation of dual-credit courses, where students receive both high school and college credit, has become commonplace in U.S. schools. The State of Texas created an initiative to develop specialty STEM schools similar to those described in *Rising above the Gathering Storm.* The Texas Science, Technology, Engineering and Mathematics (T STEM) Academies initiatives creates "rigorous secondary schools focusing on improving instruction and academic performance in STEM-related subjects and increasing the number of students who study and enter STEM careers." T-STEM Academies are demonstration schools and learning labs that develop innovative methods to improve STEM instruction. The primary instructional strategy of these academies is PBL/PrBL. T-STEM academies are open enrollment schools designed to provide a high-quality STEM Education for all students (Texas Education Agency, 2018).

The T-STEM Academy initiative is designed to prepare students to thrive in the 21st Century economy by providing students a course of study that allows them to enter into STEM majors during their university experiences, and ultimately into STEM fields critical to the economy of Texas after graduation. "The cornerstone of T-STEM Academy learning is student engagement and exposure to innovation and design in STEM-focused instruction and learning that models real-world contexts" (Texas High School Project, 2010, p. 2). Academies also serve as demonstration schools to inform math and science teaching and learning statewide. The initiative aims to closely align high school curriculum with admission requirements of competitive colleges and the STEM qualifications for 21st century jobs (Texas Education Agency, 2018).

Schools seeking T-STEM designation are required to apply and agree to implement the *T-STEM Academy Design Blueprint* (Texas Education Agency (2018). The Academies implement the T-STEM Design Blueprint, and use the T-STEM Rubric and Glossary, as a guidepost to build and sustain the academy. T-STEM schools are required to address seven benchmarks: (1) Mission-driven leadership; (2) school culture and design; (3) student outreach, recruitment, and retention; (4) teacher selection, development and retention; (5) curriculum, instruction, and assessment; (6) strategic alliances, and; (7) academy advancement and sustainability. It should be noted that benchmark 5 outlines a series of curriculum, instruction, and assessment indicators that are considered essential for 21st century success. These include ensuring that the STEM Academy

- Graduates students with a Distinguished Level of Achievement with a primary focus on a STEM Endorsement; and a Performance Acknowledgement;
- Develops a STEM-focused, integrated curriculum, assessment and instruction for the Academy;
- Ensures that students complete three years of STEM electives during their middle school (ages 11–14) experience, and four years of STEM electives during their high school (ages 15 = 18) experience;
- Provides extracurricular STEM activities, field experiences, clubs, and competitions
- Provides an Internship and/or capstone project; and
- Provides PBL/PrBL curriculum, instruction, and assessment.

Achievement data in 2011 indicated T-STEM Academies outperform their peer schools in meeting college-readiness benchmarks. T-STEM academies scored at a 12% higher rate and achieved a 21% higher completion rate in dual-credit- and advanced placement courses (Texas Education Agency, 2018). As of August, 2018, there are 132 T-STEM academies operating in the State of Texas.

Impact of T-STEM Designation and PBL/PrBL Pedagogies as a Reform Strategy

In 2014, a school district in East Texas, The Innovation Academy, in response to poor academic performance, adopted the PBL/PrBL instructional model, as well as the T-STEM Design Blueprint as a reform strategy. Schools in the district performed poorly on state assessments and were in danger of losing their accreditation. As a result, the district faced closure if the schools performed poorly for a third straight year. After significant research and deliberation, a school improvement plan was developed using the T-STEM Blueprint as the school improvement model and PBL was adopted as the primary instructional strategy for all subjects, and PrBL was adopted for all mathematics classes since the strategy was better aligned with the outcomes of the mathematics standards and assessments.

The improvement plan also included the creation of a professional development (PD) plan to re-train all teachers in PBL/PrBL. Teachers had already received training in both after they were initially hired, but the duration of the training and follow-up support post-training had not been adequate. In the revised PD plan, all teachers were required to complete two weeks of common PBL/PrBL training. Newly-hired teachers received additional training. Where possible, UTeach graduates were hired for open positions as their teacher preparation program had already provided them experiences in PBL/PrBL.

To provide post-training support, the district invested in PBL/PrBL coaches that would work with teachers throughout the school year as mentors. The coaches were selected by the district administration from the pool of existing veteran teachers who had demonstrated they implemented PBL/PrBL instruction in their classrooms with fidelity, and that their students performed well on state assessments. The district also created a common planning time for all teachers in addition to the teachers' normal planning period. The common planning time was facilitated by the coaches where teachers completed projects designed to improve school culture, student achievement, PBL instruction, and the development of 21st century skills in students. Progress was slow at first, but over the next three years following the implementation plan described above, positive results became evident.

Table 23.3 shows the impact of the approach over the past 5 years. The district test scores have shown continuous improvement while the state average on assessments has remained constant or declined. It is important to note, all subject area assessments improved and the district continues to outperform the state average on all assessments. The district also noted other benefits including better student engagement as measured by a decrease in the number of discipline referrals issued by teachers and an increase in the attendance rate from 95 to 97%. In the U.S., school funding is determined by a school's average daily attendance (ADA). The 2% increase in attendance resulted in significant additional funding that was used for increased STEM programming compared to previous years.

Table 23.3 State assessment scores before, during, and after PBL/PrBL STEM implementation

	2013	2014	2015	2016	2017
Math	Pre-intervention		Post-intervention		
State	75	75	81	75	79
District	44	56	75	80	85
Science					
State	79	78	78	80	79
District	54	67	78	88	85
Reading					
State	76	76	76	75	72
District	72	71	83	85	86
Writing					
State	70	71	70	70	67
District	52	54	78	76	78

Lessons Learned Implementing PBL/PrBL

In the case above, there were a number of challenges to implementing PBL in the district's three schools. Implementing PBL/PrBL-based approaches requires teachers to take on a new role. Teachers must transition as the "source of knowledge" to coaches that facilitate knowledge. This shift from teacher-centered learning to student-centered learning can be difficult for many teachers. The difficulty most often lies in the fact that they themselves were not taught in ways other than those supported by teacher-centered environments when they were in school, nor were they trained to teach in other ways during their teacher preparation programs.

In 2010, several East Texas middle schools were given grants to start PBL/PrBL instruction in their schools. Each school contracted with an outside team to train all sixth grade teachers and to have mentors on-site each day throughout the school year to coach and advise teachers through the transition. The thought was to start with the sixth grade class of students and follow them through their middle school experience, each year training the next set of teachers. Each school had its own set of challenges but the primary challenge was changing the instructional methods and thinking of the teachers. Some of the teachers had been teaching in their more "traditional" ways for twenty plus years, which made it harder to learn and implement an entirely new way of teaching. Most teachers made comments throughout that year that indicated they felt like first-year teachers all over again, but that they enjoyed the challenge of implementing PBL/PrBL.

In order to entirely change the way teachers are accustomed to teaching, two things should happen based on this particular experience. First, the teachers must "buy-in", in other words, the teachers must want to change their instructional practices. Second, the administration has to "buy-in" and truly understand PBL/PrBL approaches in order to be able to provide instructional feedback to their teachers. In other works, the administration has to fully support the PBL/PrBL model or there is little chance that the entire faculty can and will consistently implement PBL/PrBL strategies.

Conclusions

In this chapter, we have highlighted PBL/PrBL learning as a STEM pedagogy and as a means of school reform to improve student learning and prepare students in content and the context of 21st century skills. At a time when needed knowledge and skills are constantly changing due to the rapid technological developments in the global economy, PBL/PrBL provides students the opportunity to engage in critical thinking and problem-solving versus rote memorization of a defined set of knowledge and skills. PBL/PrBL pedagogies foster the development of 21st century skills. In addition, there is evidence to support that effective implementation of PBL/PrBL also develops content and skills as measured by high-stakes assessments. As more schools implement PBL/PrBL, there will be increased pressure on teacher preparation programs to produce new teachers prepared in these approaches. Evidence exists citing these approaches as effective reform tools to improve schools, increase relevance for students, and prepare students for the workforce. There is also a need for more research on the outcomes of PBL/PrBL after school programs to study the impacts in the workplace.

Chapter Summary

- Project- and Problem-based pedagogies are inquiry-based that promote deeper learning and develop 21st century skills.
- Integrating PBL/PrBL can be used as a reform strategy to improve student achievement in STEM.
- The development of specialty schools may be one solution to integrating PBL/PrBL as the primary instructional approach for STEM learning.
- There are STEM teacher preparation programs that are recognizing the importance of PBL/PrBL by providing preservice teachers with coursework and clinical experiences in these approaches.

References

Boss, S., Johanson, C., Arnold, S. D., Parker, W. C., Nguyen, D., Mosborg, S., Nolen, S., Valencia, S., Vye, N., & Bransford, J. (2011). The quest for deeper learning and engagement in advanced high school courses. *The Foundation Review, 3*(3), Article 3.

Boud, D., & Feletti, G. (1997). *The challenge of problem-based learning* (2nd ed.). London: Kogan Page.

Bransford, J., Brown, A. L., Cocking, R. R., & National Research Council (US). (1999). *How people learn: Brain, mind, experience, and school*. Washington, D.C: National Academy Press.

Bybee, R. W., & Fuchs, B. (2006). Preparing the 21st century workforce: A new reform in science and technology education. *Journal of Research in Science Teaching, 43*(4), 349–352. Retrieved from http://onlinelibrary.wiley.com/doi/10.1002/tea.20147/epdf.

Condliffe, B., Quint, J., Visher, M. G., Bangser, M. R., Drohojowska, S., Saco, L., & Nelson, E. (2017). Project-based learning a literature review working paper. MDRC. Retrieved from https://s3-us-west-1.amazonaws.com/ler/MDRC+PBL+Literature+Review.pdf.

Dewey, J. (1897). *Education today: My pedagogical creed*. New York: Putnam.

Dewey, J. (1938): Experience and education. In J. A. Boydston & S. M. Cahn, *John Dewey: The later works of John Dewey, 1925–1953: 1938–1939/Experience and Education, Freedom and Culture, Theory of Valuation, and Essays,* vol. 13 (pp. 1–62). Carbondale: Southern Illinois University Press.

Hmelo-Silver, C. E. (2004). Problem-based learning: What and how do students learn? *Educational Psychology Review, 16*(3), 235–266. Retrieved from https://link.springer.com/article/10.1023/B:EDPR.0000034022.16470.f3.

Hutchings, B., & O'Rourke, K. (2004). Medical studies to literary studies: Adapting paradigms of problem-based learning process for new disciplines. In M. Savin-Baden & K. Wilkie (Eds.), *Challenging research in problem based learning* (pp. 174–189). SRHE & OUP: Berkshire, UK.

Morris, C. D., Bransford, J. D., & Franks, J. J. (1977). Levels of processing versus transfer appropriate processing. *Journal of Verbal Learning and Verbal Behavior, 16*(5), 519–533.

National of Academy of Sciences, National Academy of Engineering, & Institute of Medicine. (2007). *Rising above the gathering storm: Energizing and employing America for a brighter economic future*. Washington, DC: The National Academic Press. Retrieved from https://www.nap.edu/catalog/11463/rising-above-the-gathering-storm-energizing-and-employing-america-for.

National Research Council. (2000). *Inquiry and the National science education standards*. Washington, DC: National Academy Press.

National Research Council. (1996). *National science education standards*. Washington, DC: The National Academies Press. https://doi.org/10.17226/4962.

National Research Council. (2012). *A framework for K–12 science education: Practices, crosscutting concepts, and core ideas*. Washington, DC: National Academies Press.

National Science Teachers Association. (2011). *Quality science education and 21st century skills*. Retrieved from http://www.nsta.org/about/positions/21stcentury.aspx.

Next Generation Science Storylines & STEM Teaching Tools. (2016). *Using phenomena in NGSS-designed lessons and units, (Achieve)*. Seattle, WA: STEM Teaching Tools, Institute for Science and Math Education, University of Washington.

Partnership for 21st Century Skills. (2015). *P21 framework definitions*. Retrieved from http://www.p21.org/storage/documents/docs/P21_Framework_Definitions_New_Logo_2015.pdf.

Pérez, M., & Romero, P. (2014). Secondary STEM teacher preparation as a top priority for the university of the future. *The Journal of the World Universities Forum, 6*(4), 21–36.

Strobel, J., & van Barneveld, A. (2009). When is PBL more effective? A meta-synthesis of meta-analyses comparing PBL to conventional classrooms. *Interdisciplinary Journal of Problem-Based Learning, 3*(1). Retrieved from https://doi.org/10.7771/1541-5015.1046.

Texas Education Agency. (2018). *Texas science, technology, engineering, and mathematics initiative (T-STEM)*. Retrieved from https://tea.texas.gov/T-STEM/.

Texas High School Project. (2010). *Texas science technology engineering and mathematics academies design blueprint, rubric, and glossary*. Austin, TX: Texas Education Agency.

Trilling, B., & Fadel, C. (2009). *21st century skills learning for life in our times*. San Francisco, CA: Wiley.

United States. National Commission on Excellence in Education. Department of Education. (1983). *A nation at risk: The imperative for educational reform: A report to the Nation and the Secretary of Education, United States Department of Education.* Washington, D.C.:The Commission: [Supt. of Docs., U.S. G.P.O. distributor].

Walker, A., & Leary, H. (2009). A problem based learning meta analysis: Differences across problem types, implementation types, disciplines, and assessment levels. *Interdisciplinary Journal of Problem-Based Learning, 3*(1). Retrieved from https://doi.org/10.7771/1541-5015.1061.

Michael Odell Ph.D., holds a joint appointment as Professor of STEM Education in the College of Education and Psychology and in the College of Engineering at the University of Texas at Tyler, United States of America. His research interests focus on STEM Education, STEM school design, school reform, and education policy.

Jaclyn Pedersen M.Ed., is the Curriculum Director for the Innovation Academies at the University of Texas at Tyler. Her research interests mathematics education, instructional coaching, and STEM teacher preparation.

Chapter 24
Radical Constructivism—von Glasersfeld

Radical Constructivism

Constructivism has been hugely influential in education in all disciplines for many years (Slezak, 2014; Young & Muller, 2010). The variant under discussion in this chapter, radical constructivism, has had considerable impact in science and mathematics education, since it was first developed by Ernst von Glasersfeld in the seventies (Lerman, 1996; Olssen, 1996; Riegler, 2001; Slezak, 2010). While constructivism may have abated in influence to an extent since its highpoint in the late nineties, it continues to underpin much thought, theory and pedagogy in science education (see for example Chap. 18: Social Constructivism; Chap. 19: Lev Vygotsky). Concepts of student developmental learning and hypothetical learning pathways that originated in radical constructivism (von Glasersfeld, 2007; Steffe, 2007) have heavily influenced the underlying philosophy of the recent Next Generation Science Standards (NGSS). In many respects, radical constructivist theories about student learning have now become accepted wisdom in science education research, teaching and learning.

Constructivism emerged as a reaction to the empiricism and behaviourist psychology that dominated educational theory in the twenties and thirties (see for example Chap. 6: Classical and Operant Conditioning), and in education has its roots in developmental psychology (Matthews, 2012; Olssen, 1996), particularly the work of Jean Piaget (see Chap. 10: Jean Piaget). Von Glasersfeld defined radical constructivism as a 'theory of knowing that provides a pragmatic approach to questions about reality, truth, language and human understanding' (von Glasersfeld, 1995, Abstract). The main application of radical constructivism in science education is in the realm of learning science, and how teachers can best support their students to acquire and develop scientific concepts. The use of the word 'radical' to describe this version of constructivism reflects his notion that it is a particularly controversial theory for

G. Walshe (✉)
University of Limerick, Limerick, Ireland
e-mail: Grainne.Walshe@ul.ie

© Springer Nature Switzerland AG 2020   359
B. Akpan and T. Kennedy (eds.), *Science Education in Theory and Practice*,
Springer Texts in Education, https://doi.org/10.1007/978-3-030-43620-9_24

educators to take on board. As he points out 'to introduce epistemological consid-erations into a discussion of education has always been dynamite' (von Glasersfeld, 1995, p . xi). Indeed the theory of radical constructivism has been quite controversial within the community of philosophers of science and science educators ever since it first emerged (Quale, 2008; Riegler & Quale, 2010).

This chapter begins with a brief biography of von Glasersfeld. It goes on to outline the epistemological issues he raised in his theory of radical constructivism, the implications of radical constructivism for learning about scientific practice, that is, how scientists come to develop new scientific knowledge, and its implications for teaching and learning science. The final section discusses some of the criticisms and limitations of radical constructivism.

Biography of Ernst von Glasersfeld

Ernst von Glasersfeld was born in Germany in 1917, and spent his early childhood in Austria. He initially studied mathematics in Zurich, and then moved to Vienna, where he was introduced to the work of Wittgenstein. He and his wife lived in Ireland during the Second World War, where he learned of the work of the Irish idealist philosopher Berkeley and of the philosopher Giambattista Vico. These thinkers, along with other philosophers, had a profound influence on his ideas. He moved back to Italy after the war, where he became part of a circle of intellectuals who were developing a theory of semantics. Von Glasersfeld went on to become one of the pioneers in the field of cybernetics, working on a project to develop machine translation. In 1967 he started working in the University of Georgia where he became interested in Jean Piaget's work on cognitive development, and became gradually more involved in the world of education, particularly in mathematics education. In 1987, he moved to work with a physics education group in the Scientific Reasoning Research Institute in the University at Amherst. He passed away in 2010.

The Traditional Epistemological View

In order to understand von Glasersfeld's radical constructivism, it is crucial to under-stand that he is arguing from the position that most people are tied to what he refers to as the traditional western epistemology. They believe that our knowledge faithfully reflects an ontological reality that exists independently of the observer (von Glasers-feld, 1995). Von Glasersfeld traces the ideas that underpin radical constructivism as far back as the Ancient Greek philosophers, right through to the present day, in the work of a variety of philosophers and theorists, including Vico, Kant, Berkeley, Darwin, and de Saussure. The main tradition of western philosophy is informed by metaphysics:

A metaphysical realist... is one who insists that we may call something 'true' only if it corresponds to an independent, 'objective' reality. ... most scientists today still consider themselves 'discoverers' who unveil nature's secrets and slowly but steadily expand the range of human knowledge; and countless philosophers have dedicated themselves to the task of ascribing to that laboriously acquired knowledge the unquestionable certainty which the rest of the world expects of genuine truth. Now as ever, there reigns the conviction that knowledge is knowledge only if it reflects the world as it is. (von Glasersfeld, 1984, p. 20)

This is the way in which most of us live our lives. We perceive objects or events with our senses, and we believe that what we perceive corresponds or matches to a physical reality that actually exists.

The Constructivist Epistemological View

Von Glasersfeld describes radical constructivism as being a departure from this traditional epistemology and from traditional cognitive psychology, in that it moots a different conception of the relation between knowledge and reality. Within the traditional notion, there is an iconic correspondence or match between knowledge and reality, whereas within radical constructivism, the relation is that of an adaptation or a functional fit of knowledge to reality, which can never be directly experienced. This is the constructivist aspect of his theory: we actively construct our world, our knowledge, from what we perceive, rather than passively receive sensory images of a pre-existing reality. However, this is not to say that radical constructivists deny the existence of an objective world, of reality. On the other hand neither do they say it exists. 'Radical Constructivism is agnostic' (Riegler, 2001, p. 1). It is not concerned with ontology, whether what we know actually exists, but rather how we come to know.

While we play an active part in constructing our reality, that does not mean that we can therefore construct any old conception of reality. It has to be viable. Similar to the theory of evolution put forward by Darwin, the notion of viability is not a free-for-all. Just as the environment places constraints on the living organism (biological structures) and eliminates all "variants that in some way transgress the limits within which they are possible or 'viable', so the experiential world, be it that of everyday life or of the laboratory, constitutes the testing ground for our ideas [cognitive structures]" (von Glasersfeld, 1984, p. 30). The analogy with knowledge von Glasersfeld makes is that knowledge is useful or viable if it stands up to experience and enables us to make predictions and to bring about or avoid particular events or experiences. If knowledge does not serve that purpose, it becomes questionable, unreliable or useless, and is eventually devalued as superstition. In other words, our ideas, theories, our laws of nature are structures which either hold up or not when exposed to the experiential world, from which they derive. These cognitive structures do not tell us how the objective world might actually be, rather a structure gives us one means to achieve a specific goal.

Von Glasersfeld summaries the fundamental principles of radical constructivism as:

1. Knowledge is not passively received but built up by the cognizing subject.
2. The function of cognition is adaptive and serves the organization of the experiential world, not the discovery of ontological reality (von Glasersfeld, 1995, p. 35).

Radical Constructivism and Jean Piaget

The work of the educational psychologist Jean Piaget, particularly Piaget's notion of 'genetic epistemology' was very influential in von Glasersfeld's theory of radical constructivism. Piaget suggested that we construct our concepts and our picture of the world we live in, developmentally (von Glasersfeld, 1995). Von Glasersfeld therefore utilizes the word genetic in the sense 'developmental'. In this perspective, knowledge does not exist there to be uncovered by the cognizing subject, but is constructed by them from their experiences (von Glasersfeld, 2001b).

The essence he takes from Piaget is that a cognizing organism has developed certain 'keys' or structures that allows it to achieve certain goals. The cognitive organism evaluates its experiences, and tends to repeat certain ones and to avoid others. We perceive certain regularities within the flow of our experiences, for example, that an apple is smooth and sweet and round, or that to touch a hot object is painful, and we adapt our behaviour to these experiences. It does not matter what an object might be in reality or from an objective point of view (if that were possible to have), rather what matters is whether or not it behaves as is expected of it; in other words does it 'fit' with our cognitive structures built up from our experiences (von Glasersfeld, 2001b).

Symbols and units in science and mathematics are an example of such mental constructions, or ways of organizing experience. The active experiencer creates the units, but also creates the discrete entities to be counted. The mind segments and coordinates the continuous flow of raw experiential material into such structures. We then assimilate further experiences to them, building endlessly on previous structures (von Glasersfeld, 2001b).

Radical Constructivism and Learning About Scientific Practice

Von Glasersfeld sees a number of implications for the discipline of science of radical constructivism (2001b). He argues that most philosophers would describe Piaget's theory as incorrect because it is based on what they call the 'genetic fallacy', that is, knowledge is developed over time, rather than simply there, waiting and available to be discovered by scientists. On the contrary, von Glasersfeld draws on the work of the philosophers of science, Karl Popper and Thomas Kuhn, who he says indicated in various ways in their writings that scientific knowledge does not simply emerge over time, as scientists happen to make more discoveries. Rather, scientific models are scientists' theoretical models of various mechanisms. They check the viability of their model to explain phenomena by doing experiments. Scientists use great creativity in their construction of scientific models. Non-scientists do the same in a less coherent and explicit way; in both cases the point is not to obtain a true picture of reality but rather to construct structures that allow us to manage our experiences and to explain natural phenomena (von Glasersfeld, 2001b). From a science education perspective, therefore, radical constructivism provides an explanation of how scientists develop new knowledge. It can help students to learn and understand about the nature of scientific practice, as well as providing an approach for learning science concepts.

Radical Constructivism and Science Teaching and Learning

The faculty of cognition is central in radical constructivist views of knowledge and knowing. The basic assumption of radical constructivism is that all knowledge is constructed by the individual learner for the purpose of making sense of their experiential world (Quale, 2008). Like other forms of constructivism, the emphasis is a move away from teacher-centred learning to a more student-centred focus (see Chaps. 16–26). The implications of radical constructivism for science education are that, therefore, 'the art of teaching has little to do with the traffic of knowledge, its fundamental purpose must be to foster the art of learning'(von Glasersfeld, 1995, p. 192).

This is the logical outcome of the radical constructivist notion that we, as cognizing subjects, develop or construct our own knowledge (von Glasersfeld, 2010). Creating concepts requires a form of construction; by which von Glasersfeld means reflection on mental operations: recognition of the connections made when the cognizing subject co-ordinates sensory elements or mental operations. We produce certain conceptions because of our tendency to look for something familiar in what we perceive. The significance of this for teaching is that students have to construct their concepts on the basis of their own thoughts, that is, their own mental operations and reflections, and that concepts cannot be directly conveyed by language, which is very open to misinterpretation (von Glasersfeld, 2001a). Therefore, forms of pedagogy that are centred on rote-learning or passive forms of learning are not good approaches to supporting students in developing their understanding of a given topic.

While the focus within radical constructivism is on how students learn, von Glasersfeld (2001a) provided some practical suggestions for how radical constructivism might be translated into teaching methods. The essence of a radical constructivist approach to pedagogy might be broadly encompassed by the now-familiar notion of active learning. Von Glasersfeld (2001a) suggests using conversations and asking students to verbalise their conceptual understanding as a way of both teachers understanding where the students are at, and as a learning strategy. His suggestions for creating a radical constructivist-informed pedagogy to promote student conceptual learning and development include:

- Creating opportunities for making students think.
- Teachers must have a range of didactic situations at their disposal to stimulate student creation of concepts.
- Do not tell students their work is wrong; recognize and support their efforts to learn, thereby motivating them.
- With regard to the relativity of words, teachers should pay particular attention to students' naïve conceptions, in order to influence a new train of ideas and to prevent students forming incorrect conceptions.
- Encourage students to verbalize their constructions and their thought processes in order to stimulate their thinking and creating of concepts. (von Glasersfeld, 2001a)

Initially von Glasersfeld's radical constructivism was very influential in mathematics education. He worked with a number of mathematics educators in the 1970s and 1980s on research that took a constructivist approach, in particular with Les Steffe on developing new approaches to the learning and teaching of arithmetic (von Glasersfeld, 1995). He also worked with and influenced science educators (Tobin, 2007), and wrote about teaching methods for more learner-centred or active approaches to teaching physics in the classroom. Radical constructivist methods of teaching provide students with opportunities to engage in scientific inquiry, through a process of reflecting and discussion on the outcomes of scientific activities. A practical example that he gives is that teachers could show students two routes by which a ball can travel through a chute, and ask the students which will arrive first. The counterintuitive correct outcome is that the ball arrives first by the longer route that has a steeper downhill slope for part of the route. Through discussion and exploration and reflection, the students can come to understand why this is so, and the physical concepts behind it, in a way that is not possible through simply providing them with the correct answer (von Glasersfeld, 2001a).

Supporters of radical constructivism in science education have tended to connect didactic modes of teaching directly to a belief in traditional western epistemology. Knowledge is viewed as:

> out there, residing in books, independent of a thinking being. … As a result, teachers implement a curriculum to ensure that students cover relevant science content and have opportunities to learn truths which usually are documented in bulging textbooks. (Lorsbach & Tobin, 1992, p. 1)

Therefore adopting a radical constructivist epistemology is seen by some to lead inevitably to more effective active and inquiry-based learning in the science classroom (Lorsbach & Tobin, 1992; Matthews, 1998), as in the example described above by von Glasersfeld for teaching physics concepts. Radical constructivist epistemology has hence been the inspiration for approaches in science education which focus on the learner and the role of language in negotiating meaning, both for students and for the professional development of teachers (Tobin, 2007). But for others, adopting a radical constructivist approach will have an even more dramatic effect. Andreas Quale argues that current problems in science education, such as decline of student enrolment in science subjects, can be addressed by taking the relativist epistemological and ontological perspective offered by radical constructivism. The traditional image of science projected to students is rooted in realism (there is an objective reality independent of human observation and reflection, and that it is the task of science to search for this true knowledge of this objective reality). In contrast, radical constructivism posits that all knowledge is constructed by the individual learner for the purpose of gaining understanding and control of their experiential world. Note that unlike von Glasersfeld himself, Quale does not reject relativism. Quale sees this as a more empowering position for learners that will therefore engage their interest and attention in science. If reality is not the ultimate arbitrator of truth, then humans themselves are solely responsible for their own decisions and actions. This means that students do not have to blindly accept the knowledge that is handed down to them by higher authorities, but can instead become active socio-political agents (Riegler & Quale, 2010). From this perspective, students would be empowered by the radical constructivist stance on scientific knowledge to take actions counter to traditional wisdom and authority, such as refusing to accept the unwillingness of those in power to tackle the causes of environmental degradation.

Indeed as Matthews (2012) and others point out, constructivism has had a very positive influence in science education in alerting teachers to the importance of students' prior learning and the need to be aware of their existing concepts in relation to learning new material. Radical constructivism stresses the importance of student understanding, which has fed into very progressive pedagogies that focus on engaging students in their learning. It also has highlighted the fallibility of science, the culturally determined and conventional aspects of scientific knowledge-production, the historicity of scientific concepts, and so on. While constructivism does not have a monopoly on these insights, it has certainly promoted them to the betterment of science education.

Radical constructivism has also had an impact beyond the development of active and engaging classroom pedagogies. Von Glasersfeld's collaborator Les Steffe developed the 'teaching experiment' approach to developing understanding of student learning (von Glasersfeld, 1995). This methodology, and the constructivist approach to student learning underpinning it, in turn lead to the development of mathematical learning trajectories (Clements & Sarama, 2004), a major innovation in mathematics curriculum development. Learning trajectories describe students' thinking and learning in a specific mathematical domain. They lay out a conjectured route through a set of instructional tasks 'designed to engender those mental processes or actions hypothesized to move children through a developmental progression of levels of thinking, created with the intent of supporting children's achievement of specific goals in that mathematical domain' (Clements & Sarama, 2004, p. 83). Other radical constructivists, such as Paul Cobb, one of Steffe's graduate students, went on to work with a number of eminent U.S. science educators in the further development of the teaching/design experiment methodology for developing hypothetical learning pathways of student thinking (Cobb, Confrey, Disessa, Lehrer, & Schauble, 2003). In science, these pathways are called learning progressions, the scientific equivalent of learning trajectories in mathematics (Duschl, Maeng, & Sezen, 2011). The NGSS are based on learning progressions, as outlined in the *Framework for K-12 Science Education* (National Research Council, 2012), showing the extent of the influence that radical constructivist ideas continue to have in science education.

Radical constructivism therefore may once have been a departure from the dominant theories of education that existed before the 1970s, but it now permeates most aspects of science education.

Criticisms and Limitations of Radical Constructivism

Slezak (2014) notes that there have been many critics of radical constructivism, who have argued that it has 'serious, if not fatal, philosophical problems, and further, it can have no benefit for practical pedagogy or teacher education' (p. 1024). Slezak (2010) highlights von Glasersfeld's allegiance to what he calls Berkeley's 'notorious' idealism in his advocacy of the recommendation that we give up the requirement that knowledge represents an independent world. Slezak insists that von Glasersfeld encourages the attribution of idealism through his misleading claims that the great physicists of the twentieth century did not consider their theories to be descriptions of an ontological reality. Slezak points out that Piaget himself, a major referent for von Glasersfeld's theories, does not deny the existence of an objective reality beyond our sense-data, arguing that von Glasersfeld misinterprets Piaget in this respect. Rather Piaget clearly states that the subject's thought processes depend both on an organism's internal mental constructions, but also on the fact that the organism is not independent of its environment but can only live, act or think in interaction with it. Slezak (2010) therefore states

> Thus, while von Glasersfeld is at pains on every occasion to emphasize the unknowability of reality and the need to abandon notions of objectivity and truth, Piaget by contrast, writes in an altogether different mood. ...it is evident that his version of constructivism is quite different from Piaget's. (p. 104)

Several critics note that this idealist turn in radical constructivism could lead to scientific knowledge being undervalued and discredited. There is a concern that if we construct our own knowledge, then 'anything goes'. This is relativism, that is, the notion that there are no grounds on which to decide that one version of reality or knowledge is any better than or more true than another. Scientists themselves are aware that theories can change, but they do not necessarily hold relativistic views about the nature of scientific knowledge (Harding & Hare, 2000). They believe scientific knowledge is true, and they use it as the basis of further investigations. They are open-minded about scientific knowledge, but not relativist. They could not operate otherwise (Harding & Hare, 2000). And indeed not all science educators who are constructivist agree with von Glasersfeld's position on the unknowability of reality. Taber (2006) highlights that the radical constructivist view of science knowledge is inappropriate as it 'sets learner's ideas to be of equal validity to currently accepted knowledge' (p. 199). Taber presents the debate as being a question of whether constructivism is seen as being about (a) how science learning occurs (von Glasersfeld called this trivial constructionism), or (b) the nature of human knowledge (the radical constructivist perspective). Radical constructivism, Taber suggests, goes too far in the direction of giving equal weight to learners' misconceptions as to accepted scientific theories and laws.

However, von Glasersfeld refutes the charge of relativism, or that radical constructivism rejects the idea that there is such a thing as reality; rather he says that it sidesteps this issue. His argument is that we trust in the permanence and stability of objects and conditions, such as, for example, that our front door will always be where it was the night before when we wake up afresh each morning, and that we could not live otherwise (von Glasersfeld, 2001a). In addition, he insists that radical constructivism gives agency to the knower/learner in that it puts emphasis on the active role we all have in constructing knowledge, thereby giving us responsibility for our actions (von Glasersfeld, 2010).

Nonetheless for some critics his strong emphasis on the individual construction of knowledge always risks a slide into a skeptical idealism, which must inevitably present problems for teachers (Matthews, 2012; Olssen, 1996). If, as von Glasersfeld suggests, there is no basis on which to be sure that any given mental construction reflects the world as it actually is, this in turn means that the advice given by radical constructivists to teachers to orient learners in particular ways is impossible to follow. This is because there are no grounds or criteria by which teachers can decide what orientations students' constructions should take (Olssen, 1996). While it is of course important that science teachers are interested in students' individual constructions of knowledge, teachers still want students to understand the basic theories of science (Harding & Hare, 2000).

Matthews (2012) recognizes the great positives that result for students because of the value that constructivism gives to active methods of learning. However, he suggests that its over-emphasis on the isolated nature of cognition, that is, its insistence that we all construct our own knowledge is misguided, and may simply be getting in the way of good teaching

> Why must learners construct for themselves the ideas of potential energy, mutation, linear inertia, photosynthesis, valency, and so on? Why not explain these ideas to students, and do it in such a way that they understand them? This process may or may not be didactic: it all depends on the classroom circumstance. There are many ways to explain science: didacticism is just one of them. (Matthews 2012), p. 38

Most students would find it impossible to re-construct for themselves the scientific knowledge that has been developed by many scientists over many centuries, and hence taken to its logical conclusion, radical constructivist pedagogy could do students a great disservice.

Finally, Slezak (2014) insists that there is a question mark over the relevance of much of the theoretical underpinnings of radical constructivism—the focus on epistemological issues—to education, saying that 'there is a sharp contrast between such esoteric philosophical matters and the practical recommendation taken to follow from them' (p. 1024). The kind of practical advice von Glasersfeld offers teachers, includes for example, 'Asking students how they arrived at their given answer is a good way of discovering something about their thinking' (Slezak 2014, p. 1028). As Slezak notes, such insights will be familiar to all teachers, and while these are sound recommendations, they are hardly revolutionary, a view reiterated by d'Agnese (2015).

There are limitations therefore to the usefulness of radical constructivism, at least in the extreme version that some of its adherents have advocated. If von Glasersfeld's ideas were taken to their logical conclusion in the classroom, it would be very difficult for teachers to know what to teach, or for students to learn the scientific knowledge that we would like them to know. If students were to encouraged to take a relativist stance on all knowledge, they might reject accepted and proven scientific knowledge, such as that underlying climate change and evolutionary theory. Nonetheless, radical constructivism raises issues that science educators and students should be concerned about, in relation to the nature of science, such as how we can evaluate claims of scientific truth and how knowledge development comes about. Even if we do not accept the relativism and skepticism that some say is inherent in radical constructivism, its insistence on the importance of the learner's role in making sense of their world can have a very positive impact on teaching and learning processes in the science classroom. Moreover, the impact of radical constructivist ideas in curriculum development, for example, in the now widespread acceptance of learning trajectories and learning progressions, is considerable.

Conclusion

Radical constructivism has been a major force for change in science education since the 1970s. The major difference with other forms of constructivism is von Glasersfeld's emphasis on the epistemological aspects of the learning process. The basic tenets of radical constructivism are that knowledge is not passively received through the senses, but is actively constructed by the cognizing subject, the learner, and that the function of cognition is organization of the experiential world rather than discovery of an independent reality. This highlighted the need for more active methods of teaching and learning science, as opposed to the notion that students should rote-learn a body of scientific facts. Radical constructivism was instrumental in bringing about the great revolution that ushered in progressive pedagogies in the late twentieth century. However, critics of radical constructivism have argued that it places too much emphasis on the unknowability of reality, leaving it open to the charge of relativism and potentially undermining the basis on which teachers could know which scientific ideas and theories to teach students. Nonetheless, radical constructivism opened the door for teachers and students to free themselves from very rigid approaches to teaching and learning, particularly in the area of science education, where absorption of facts taught didactically was once the order of the day. Radical constructivism continues to have lasting impact through the focus on learners actively making sense of the natural world that underpins the vast majority of scientific educational research, curriculum development and teaching practice today.

Chapter Summary

- The two main principles of radical constructivism are that knowledge is actively constructed by the learner, and that the function of cognition is organization of the experiential world rather than discovery of an independent reality.
- Von Glasersfeld called for more active and engaging teaching methods to be used to assist students to constructing their scientific knowledge.
- Radical constructivism has also been very influential in the development of learning progressions in science curricula.
- Criticisms of radical constructivism include that it undermines the basis on which teachers can decide what scientific knowledge is most important for students to learn, and that it over-emphasizes the isolated nature of cognition.

Resources

von Glasersfeld, E. (1995). *Radical constructivism: A way of knowing and learning*. London and Washington DC: The Falmer Press.

von Glasersfeld, E. (2001). Radical constructivism and teaching. *Prospects, 31*(2), 161–173. doi:10.1007/bf03220058

von Glasersfeld, E. (2007). *Key works in radical constructivism*, M. Larochelle (Ed.). Rotterdam: Sense Publishers.

Matthews, M. R. (2012). Philosophical and pedagogical problems with constructivism in science education. *Tréma, 38*, 40–55.

References

Clements, D. H., & Sarama, J. (2004). Learning trajectories in mathematics education. *Mathematical Thinking and Learning, 6*(2), 81–89.

Cobb, P., Confrey, J., Disessa, A., Lehrer, R., & Schauble, L. (2003). Design experiments in educational research. *Educational Researcher, 32*(1), 9–13.

d'Agnese, V. (2015). 'And they lived happily ever after': The fairy tale of radical constructivism and von Glasersfeld's ethical disengagement. *Ethics and Education, 10*(2), 131–151. https://doi.org/10.1080/17449642.2014.999425.

Duschl, R., Maeng, S., & Sezen, A. (2011). Learning progressions and teaching sequences: A review and analysis. *Studies in Science Education, 47*(2), 123–182. https://doi.org/10.1080/03057267.2011.604476.

Harding, P., & Hare, W. (2000). Portraying science accurately in classrooms: Emphasizing open-mindedness rather than relativism. *Journal of Research in Science Teaching, 37*(3), 225–236. https://doi.org/10.1002/(SICI)1098-2736(200003)37:3%3c225:AID-TEA1%3e3.0.CO;2-G.

Lerman, S. (1996). Intersubjectivity in matematics learning: A challenge to the radical constructivist paradigm? *Journal for Research in Mathematics Education, 27*(2), 133–150. https://doi.org/10.2307/749597.

Lorsbach, A. W., & Tobin, K. (1992). Constructivism as a referent for science teaching. In F. Lorenz, K. Cochran, J. Krajcik, & P. Simpson (Eds.), *Research matters ...to the science teacher. NARST Monograph, Number Five.* Manhattan, KS: National Association for Research in Science Teaching.

Matthews, M. R. (1998). The nature of science and science teaching. In B. J. Fraser & K. Tobin (Eds.), *International handbook of science education* (Vol. 2, pp. 981–999). Dordrecht and Boston: Kluwer Academic.

Matthews, M. R. (2012a). Philosophical and pedagogical problems with constructivism in science education. *Tréma, 38,* 40–55.

National Research Council. (2012). *A framework for K-12 science education: Practices, crosscutting concepts, and core ideas.* Washington, DC: The National Academies Press.

Olssen, M. (1996). Radical constructivism and its failings: Anti-realism and individualism. *British Journal of Educational Studies, 44*(3), 275–295. https://doi.org/10.1080/00071005.1996.9974075.

Quale, A. (2008). *Radical constructivism: A relativist epistemic approach to science education.* Rotterdam: Sense Publishers.

Riegler, A. (2001). Towards a radical constructivist understanding of science. *Foundations of Science, 6*(1–3), 1–30.

Riegler, A., & Quale, A. (2010). Can radical constructivism become a mainstream endeavor? *Constructivist Foundations, 6*(1), 1–5.

Slezak, P. (2010). Radical constructivism: Epistemology, education and dynamite. *Constructivist Foundations, 6*(1), 102–111.

Slezak, P. (2014). Appraising constructivism in science education. In *International handbook of research in history, philosophy and science teaching* (pp. 1023–1055). Berlin: Springer.

Steffe, L. P. (2007). Radical constructivism and 'school mathematics'. In M. Larochelle (Ed.), *Key works in radical constructivism* (pp. 279–289). Rotterdam: Sense publishers.

Taber, K. S. (2006). Constructivism's new clothes: The trivial, the contingent, and a progressive research programme into the learning of science. *Foundations of Chemistry, 8,* 189–219.

Tobin, K. (2007). The revolution that was constructivism. In M. Larochelle (Ed.), *Key works in radical constructivism* (pp. 291–297). Rotterdam: Sense Publishers.

von Glasersfeld, E. (1984). An introduction to radical constructivism. In P. Watzlawick (Ed.), *The invented reality* (pp. 17–40). New York: Norton.

von Glasersfeld, E. (2007). *Key works in radical constructivism,* M. Larochelle (Ed.). Rotterdam: Sense Publishers.

von Glasersfeld, E. (1995a). *Radical constructivism: A way of knowing and learning.* London and Washington, DC: The Falmer Press.

von Glasersfeld, E. (2001a). Radical constructivism and teaching. *Prospects, 31*(2), 161–173. doi:10.1007/bf03220058.

von Glasersfeld, E. (2001b). The radical constructivist view of science. *Foundations of Science, 6*(1), 31–43. doi:10.1023/a:1011345023932.

von Glasersfeld, E. (2010). Why people dislike radical constructivism. *Constructivist Foundations, 6*(1), 19–21.

Young, M., & Muller, J. (2010). Three educational scenarios for the future: Lessons from the sociology of knowledge. *European Journal of Education, 45*(1), 11–27. https://doi.org/10.1111/j.1465-3435.2009.01413.x.

Gráinne Walshe is Director of the Science Learning Centre at the University of Limerick. She teaches undergraduate introductory physics. Her research interests include science and mathematics integration, curriculum development for STEM education at second- and third-level, with a focus on physics education, and supporting gender balance in science.

Chapter 25
Constructive Alternativism: George Kelly's Personal Construct Theory

Keith S. Taber

Introduction

George Kelly proposed a perspective that he called constructive alternativism, and from within this developed Personnel Construct Theory (PCT). This chapter offers an introduction to PCT and its relevance for practice and research in education. Kelly's background and motivation for developing his ideas are briefly considered, and then the grounds for considering PCT as a theoretical framework in education are discussed. The nature of PCT as a constructivist theory is discussed, highlighting its similarities and points of difference with other constructivist theories that are commonly adopted in education. Kelly developed practical tools to apply his theory—the method of triads and repertory grid. The potential of these tools to those working in education is considered.

George Kelly (1905–1967)

George Kelly was something of a polymath, a renaissance man in a time of specialists. For his undergraduate degree he studied physics and mathematics. For his master's degree he chose sociology. He became interested in education, and so went back to college to read for a first degree in that subject. He then took both a master's and then a doctoral degree, in psychology, before taking up work as a therapist. In a critical review of the dominance of constructivist thinking in science education, Solomon (1994, p. 7) described Kelly as "a psychologist who studied patients locked away in the solitary world of the schizophrenic".

K. S. Taber (✉)
Faculty of Education, University of Cambridge, Cambridge, UK
e-mail: kst24@cam.ac.uk

© Springer Nature Switzerland AG 2020
B. Akpan and T. Kennedy (eds.), *Science Education in Theory and Practice*,
Springer Texts in Education, https://doi.org/10.1007/978-3-030-43620-9_25

Kelly himself had been trained in the therapeutic methods of Freudian psycho-analysis. The Freudian perspective posits a structure to the mind and mechanisms by which early life experience could lead to various neuroses. Kelly found the system unsatisfactory as a basis for offering practical support for his clients. Kelly was deal-ing with people who often were deeply distressed in terms of how they understood their lives and he judged that Freud's theory did not offer him tools that were useful in helping his clients. He therefore came to a new way of thinking about patients' prob-lems that he considered had more potential to be productive. He codified his system as PCT, which he included in a technical book to support other therapists who might want to adopt his methods. The account of the theory was then later republished as 'A Theory of Personality: The psychology of personal constructs' (Kelly, 1963).

Constructive Alternativism

Kelly came to "a philosophical position" that he labelled "constructive alternativism" (Kelly, 1958/1969a, p. 64): "the notion that one does not have to disprove one propo-sition before entertaining one of its alternatives" (p. 55). Kelly was arguing that given that there is generally some uncertainty about our existing understandings, we should be open to considering other options, alternative conceptualisations, even when they seem inconsistent with aspects of our current thinking.

This reflected Kelly's work with his clients, many of whom had developed ways of making sense of their worlds—their relationships, their lives, their role in the workplace—which were unproductive and impacting them in negative ways. Kelly thought that "no one needs to paint themselves into a corner; no one needs to be completely hemmed in by circumstances; no one needs to be the victim of his or her biography" (Kelly, 1963, p. 15).[1]

Personal Construct Theory

Kelly set out his theory as a set of principles or tenets, described as a basic postulate and a series of corollaries (Kelly, 1963), reproduced in Table 25.1. Kelly's theory is constructivist in the way that it suggests that an individual person understands the world through developing a system of constructs that are personal to that individual, and which are the basis for interpreting experience. A construct had broad application as it was "an abstraction and, as such, can be picked up and laid down over many, many different events in order to bring them into focus and clothe them with personal meaning" (Kelly, 1958/1969b, p. 87). For Kelly such constructs encompassed the cognitive, affective and conative (Kelly, 1963, p. 130) and were bipolar continua. Examples might be 'large–small' or 'up–down'—that is dimensions which are each defined in terms of two poles that can be considered 'opposites' but which allow of intermediates. However, whereas we can all appreciate 'large–small' and 'up–down'

Table 25.1 The key tenets of Kelly's PCT (based on Kelly, 1963)

Principle label	Principle posits
The basic postulate	A person's processes are psychologically channelised by the ways in which he or she anticipates events
The construction corollary	We conservatively construct anticipation based on past experiences
The experience corollary	When things do not happen as expected, we change our constructs. This changes our future expectations
The dichotomy corollary	We store experience as [bipolar] constructs, and then look at the world through them
The organisational corollary	Constructs are connected to one another in hierarchies and networks of relationships. These relationships may be loose or tight
The range corollary	Constructs are useful only in limited ranges of situations. Some ranges are broad, others narrow
The modulation corollary	Some construct ranges can be 'modulated' to accommodate new ideas. Others are 'impermeable'
The choice corollary	We can choose to gain new experiences to expand our constructs or stay in the safe but limiting zone of current constructs
The individuality corollary	As everyone's experience is different, their constructs are different
The commonality corollary	Many of our experiences are similar and/or shared, leading to similarity of constructs with others. Discussing constructs also helps to build shared constructs
The fragmentation corollary	Many of our constructs conflict with one another. These may be dictated by different contexts and roles
The sociality corollary	We interact with others through understanding of their constructs

as we all share similar meanings for the labels and all use such discriminations (Lakoff & Johnson, 1980), many personal constructs would be more idiosyncratic, and would not always have communicable labels that would be readily understood by others. Indeed, a key feature of many personal constructs is that as well as not having explicit labels, the very construct itself may be tacit. That is, we may be applying discriminations without even being aware of doing so—personal constructs may be part of our implicit cognition. This links with work in science education on the role of implicit knowledge elements in cognition (Brock, 2015; diSessa, 1993; Taber, 2014a—see also Chap. 26), and more widely with the idea of two complementary systems of thought (Evans, 2008) acting within human cognition: faster and intuitive (preconscious), and slower and deliberative (conscious).

It is worth recalling that Kelly's theory was first presented in the 1950s, as much of it now seems mainstream given the widespread influence of constructivist thinking in education. As one example, Kelly's notion of looking at the world through one's constructs is reflected in constructivist work using the metaphor of people putting on different glasses to see the world (Pope & Watts, 1988). Kelly shares with such constructivist thinkers as Piaget (see Chap. 10) and Vygotsky (see Chap. 19) an assumption that frameworks for thought are developed iteratively over time such that each individual builds up a personal apparatus for modelling the world. Kelly's conception of constructs suggests somewhat discrete highly focused elements, where Piaget's theory (1970/1972) was based around the construction of domain-general structures of cognition that are largely under developmental control (albeit dependent upon opportunities to engage in, and derive feedback from, action in the environment). Like Vygotsky (1934/1986), however, Kelly's system did not posit completely independent elements (peas in a pod, in Vygotsky's simile), but a *system of* constructs (the organisational corollary—see Table 25.1).

In terms of social conditions, Kelly's theory makes an interesting complement to Piaget and Vygotsky. Piaget acknowledges social influences (see, for example, Piaget, 1959/2002, Chap. VI), but has been widely criticised for seeming to underplay them in much of his writing, whereas for Vygotsky (1978) the social context is critical as development of higher psychological functions relies on the modelling available from others. Kelly seems to stand in a somewhat intermediate position. Most of his principles (see Table 25.1) can be read as concerning how the individual interprets experience to produce a system for making sense of the world, and so anticipating the future. However, discussion and intersubjectivity also put in appearances (the commonality and sociality corollaries): suggesting that for Kelly social interaction was one aspect of a more general process by which constructs are derived. Moreover, in Kelly's theory there is no substantive distinction between constructs based on interaction with the physical environment and constructs deriving from enculturation, nor between those which are open to explicit reflection and those that channel tacit cognition (distinctions which are important in Vygotsky's theory).

Kelly's theory offers a good fit with many of the results of the research into what was called children's science or the alternative conception movement (Taber, 2009). This work highlighted the wide range of—sometimes idiosyncratic—alternative conceptions students presented that were alternative to the target concepts presented in the school curriculum. Piaget's theory explained in general terms why building the canonical (often abstract) concepts of formal science was challenging for students, and Vygotsky's theory explained why cultural mechanisms for reproducing knowledge were compromised by spontaneous thinking, but both of these approaches could be seen as deficit models: failures of logic or failures of cultural transmission—or indeed in some (judged to be) less developed social contexts, a society collectively lacking the resources for higher cognitive development (Luria, 1976).

Some researchers in science education wanted a theoretical base more in keeping with an ethnographic frame for exploring learners' ideas: that is, for seeking to characterise and understand the nature and internal logic (i.e. derivation) of alternative conceptions, rather than simply their failure to match up to formal scientific concepts.[2] Kelly's theory, which did not posit personal constructs as essentially limited or flawed, fitted this stance. In this regard, Kelly's theory has much in common with Glasersfeld's (1993) 'radical constructivism' where a person's understanding of the world is seen as a construction of reality based on that person's current interpretation of experience—with its necessarily limited access to the external world (see

Chap. 24). Glasersfeld's constructivism suggests that we can never have unmediated access to an objective reality, but—to the extent that new experiences can offer opportunities to better understand the world—we can refine our constructions. From the perspective of PCT, personal constructs are not second-class versions of canonical ways of thinking, but rather all human conceptualisation occurs in terms of individuals' systems of personal constructs. So, the ideas of Darwin, Einstein, Freud, de Beauvoir, Keynes, Marx—and so forth—are as much products of personal construing as those of any science undergraduate, school pupil or toddler.

Kelly's system also linked well with the motivations of those exploring the nature of students' ideas. Much of the importance attached to alternative conceptions by science educators was in their potential to be impediments to learning of canonical ideas. There were active debates about whether learners' conceptions were theory-like (coherent principles applied consistently) or not, stable or not, readily discarded when challenged or not, commonly held or idiosyncratic (Taber, 2009). The evidence available, or certainly the published interpretations of it, supported different views. It seems more obvious now that such debates were over-simplistic as people's ideas vary along such dimensions (Taber, 2014b), and so more useful research questions asked about the particular conditions *when* student ideas seemed to be theory-like or not, and so forth.

Kelly's theory can encompass the range of empirical findings from research into learners' ideas. Constructs could be more or less tightly arranged into hierarchies (organisational corollary); could have limited or more extensive ranges of application (range corollary), and could be more or less coherent (fragmentation corollary); could be more or less readily modified (modulation corollary); could be more or less like those of their peers (individuality and commonality corollaries). Of course, such an inclusive theory has limited predictive power unless it explores *when* (under what conditions) constructs have particular qualities—but this framework provides a suitable language for discussing the phenomena of learners' ideas in science. In terms of Lakatos' (1970) model of scientific research programmes, PCT (a) offers a hard core of commitments (i.e. Table 25.1) for a research programme and (b) suggests a positive heuristic for developing a belt of auxiliary theory to provide tools for diagnostic assessment and to develop teaching approaches (Taber, 2009). One of the most influential science education research groups in the 1970–1980s, the Personal Construction of Knowledge Group based at Surrey University (UK), adopted this perspective (Pope, 1982).

Kelly's own professional concern was in the extent to which people could change the way they construed their own realities, by positing, testing and adopting alternative constructions. This clearly has parallels with the key focus in science education on conceptual change. It might be argued that a difference is that in science education the teacher wants to shift thinking towards a canonical target, where in therapy the aim was to help the individual see the world in a way that they themselves could be more comfortable with; however, in both cases the outside agent is supporting a client in making changes that the client themselves might in principle somewhat desire (assuming they have entered therapy or class voluntarily) yet might resist because such changes may seem threatening or nonviable. Interestingly, one of the most influential general books produced by those researching student ideas, Driver's (1983) '*Pupil as Scientist?*' reflects Kelly's key metaphor for the person construing their world.

People as (Informal) Scientists

Driver (1983, Preface) wrote that "pupils, like scientists, view the world through the spectacles of their own preconceptions, and may have difficulty in making the journey from their own intuitions to the ideas presented in science lessons". Driver's title was posed as a question: a question Kelly had also posed. Kelly asked if it was possible to apply more universally (to people generally) his notion of being a scientist, one who:

> observes, becomes intimate with the problem, forms hypotheses inductively and deductively, makes test runs, relates data to predictions, controls experiments so that he or she knows what leads to what, generalises cautiously, and revises thinking in the light of experimental outcomes…our model of a person is that of person-the-scientist and our questions will revolve about the issue of whether a person can be understood in this manner, both in the floodlight of history and in the dark of his or her closet (Kelly 1958/1969a, pp. 62–63).

According to the widely discussed falsificationist model of science championed by Popper (1989), experimental outcomes can falsify the hypothesis being tested and then a good scientist should happily acknowledge that the hypothesis has been refuted and seek an alternative. Huxley (1870) had famously described 'the slaying of a beautiful hypothesis by an ugly fact' as 'the great tragedy of science'. However, taking a view that fits with more recent descriptions of 'science in action' (Latour, 1987) or 'science-in-the-making' (Shapin, 1992), Kelly recognised that the scientist had various options available. These included changing the predictions (perhaps revisiting what should be anticipated on the basis of a current construction); or the grounds for making predictions (perhaps switching to other constructs in the personal system); or the operational pattern of the constructs being used (which might be considered parallel to rejecting the instrumental theory rather than the substantive one: as when Galileo's contemporaries refused to admit what he saw through the telescope and justified this by rejecting that such a device could offer a valid image of the heavens); adopting a new construct to make sense of the findings—or even rejecting the results: the scientist "may refuse to accept the verdicts given by the data and ignore them, distort the perception of them, or manipulate them in such a way that they will appear to confirm the hypothesis" (Kelly, 1961/1969, pp. 110–111). Some of these options may seem illogical, unprofessional or counterproductive, but (notwithstanding Huxley and Popper), in practice, scientists can adopt a wide range of strategies to avoid letting some inconvenient datum spoil an elegant theory. Indeed, the philosopher Lakatos argued that a naive adherence to falsificationism was not even logically justifiable. Often, the most rational thing for a scientist to do with an apparently uncooperative result produced in an otherwise productive research programme is to 'quarantine' (Lakatos, 1970) it as a puzzle to return to later, and then to, for the time being at least, carry on regardless.

Science as a Process of Knowledge Construction

Kelly posited two models of how science might be imagined to proceed. One he described as 'accumulative fragmentalism', which saw science as analogous to a collective endeavour to complete a vast jigsaw puzzle, where each piece in turn needed to be found and carefully verified and fitted into its right place, before moving on to the next piece. This matched a commonly held image (perhaps even caricature) of the work of science, but Kelly preferred a different description, indeed a 'philosophical position', that he called 'constructive alternativism'. This perspective

> is a constructive one. We understand our world by placing constructions on it. And that is the way we alter it too. There is no finite end to the alternative constructions we may employ; only our imagination sets the limits. Still, some constructions serve better than others, and the task of science is to come up with better and better ones. Moreover, we have some handy criteria for selecting better ones; at least we think we have, and they, too, are subject to reconstruction. (Kelly, 1964/1969, p. 125).

In this model, there is no sense that we might soon finish the jigsaw picture of nature, as an "ultimate correspondence" between our constructions and reality was "an infinitely long way off" (Kelly, 1961/1969, p. 96). Kelly thought "that reality is subject to many alternative constructions, some of which may prove more fruitful than others" and that progress comprised of inventing new constructions that would seem useful for a while, but would ultimately be found unsatisfactory, and so come to be replaced.

For Kelly, science was not an inevitable march of progress to a realistically achievable end, but rather a process we could have reasonable confidence was, on the whole, shifting in "in the right direction" (Kelly, 1961/1969, p. 96). This view again seems to reflect a contemporary perspective of the nature of science as offering 'reliable knowledge' (Ziman, 1978/1991), if not absolute truth corresponding to an objective external reality. This then was a constructivist notion of how science proceeds, and indeed for how people should proceed more generally. A person "develops his ways of anticipating events by construing—by scratching out his channels of thought" (Kelly, 1958/1969b). This can therefore offer a perspective for thinking about how individual learners may slowly modify their constructs within science classes in response to the experiences provided for them to construe. If nothing else, Kelly's insights can be valuable in both warning science teachers to be prepared for students to sometimes be slow in shifting from their alternative conceptions, but also reassuring them that in time such shifts can be achieved. A student trying to make sense of the implications of Newton's first law of motion, or seeking to come to terms with the immense timescale over which life on earth has evolved, needs time to 'scratch out' new channels of thought.

The Construct System as a Framework

The importance of considering a person's constructs as forming a system is that in a well-integrated system a single component cannot be changed as if in isolation from the rest of the system. Kelly described the hierarchical system as "a network of constructs" reflecting the relational model of concepts (Gilbert & Watts, 1983) that considers each concept to sit in a multi-dimensional net of linked concepts—the 'content' of a concept is "the full range of meanings due to its associations within the wider web or net of concepts" (Taber, 2019, p. 31).[3] If our understanding of one concept changes, this has potential repercussions for all those other concepts that are linked to it—and so through a 'conceptual inductive effect' (Taber, 2015) to those linked *indirectly* through those other concepts. The term alternative conceptual 'framework' is sometimes used to label those closely related student conceptions which have been organised into extensive structures based on some key alternative conception. For example, the common misconception that chemical change is motivated by atoms seeking full shells can be at the core of an extensive network of ideas (some more canonical, many contrary to scientific principles) about chemical stability, chemical reactions, chemical bonding, ionisation energies and so forth (Taber, 2013a).

Kelly used the metaphor of a person building their own 'maze', or 'labyrinth'—an ongoing building project where the structure was subject to perpetual revision, but where "the complex interdependent relationships between constructs in the system often makes it precarious for the person to revise one construct without taking into account the disruptive effect upon major segments of the system" (Kelly, 1958/1969b). This seems reminiscent of Khun's (1996) notion of scientists generally working within the framework of a particular familiar, indeed encultured, disciplinary matrix with its associated paradigm channelling thought. However, in this sense, Kelly's views better aligned with Popper (1994) who rejected what he called the 'myth' of the framework: that a person's current commitment need act as a kind of thought-prison. With Popper, Kelly thought that people build their own mazes (prisons) and so could ultimately deconstruct them. Kelly did, however, realise that this might be difficult work—which was where the therapist could hopefully help. Similarly, in a science classroom, the teacher can support a student in breaking out of the 'prison' of a tenacious conceptual framework that has come to be habitually used to make sense of a whole class of events—such as identifying a force with the direction of motion; or considering chemical change to occur so that atoms can fill their shells with electrons; or assuming that biological species are fixed types. Kelly did apply his work in education. He described how in working with teachers,

> ...the teacher's complaint [about a pupil] was not necessarily something to be verified or disproved by the facts of the case, but was, rather a construction of events in a way that, within the limits and constructions of her or his personal construction system, made the most sense to her or him at the moment. (Kelly, 1958/1969b, p. 76).

So, for example, Kelly found that teachers' complaints of 'lazy' students usually referred to pupils who needed more support to cope with classroom demands, and

Kelly worked at 'reorientation' of the teacher's perception of the situation as "it usually happened that there was more to be done with [the teacher] than the child...- Complaints about motivation told us much more about the complainants than it did about their pupils" (Kelly, 1958/1969b, p. 77).

A Methodology for PCT

The therapist would, however, need tools, and in particular would need to help the client make explicit their current ways of construing the world as a first step to appreciating that other alternative construals could be viable. The parallel with science education is clear here, with the recommendations in constructivist literature that teachers must elicit students' alternative conceptions to understand students' current ways of thinking, as part of the process of developing teaching to shift learners' thinking towards the scientific models represented in the curriculum (Driver & Oldham, 1986; Russell & Osborne, 1993).[4]

From the perspective of educational research, this means that Kelly's work offers a system that includes both a theory offering metaphysical commitments and associated methodology. Kelly tells us constructs are like this (ontology)—in particular, that they often act through implicit thought without being open to immediate conscious inspection—and this means there are certain challenges in identifying them (epistemology); and he then proposes an approach to proceed accordingly (methodology). Kelly offered two related tools that have since found widespread use: the construct repertory test (CRT) and the repertory grid (Fransella & Bannister, 1977).

The Construct Repertory Test: The Method of Triads

The basis of Kelly's CRT, also known as the method of triads, is to provide an activity where the person is asked to make discriminations (i.e. to construe the world) without necessarily having to explicitly apply criteria. This means that implicit constructs may be used, whereas a task that relies on a reflective activity (such as giving a verbal description) cannot directly tap intuitive thought. Kelly considered that, as his system concerns bipolar constructs, the simplest approach was to present three elements to be discriminated and to ask the client "to think of some important way in which you regard two of them as similar to each other but in contrast to the third" (Kelly, 1961/1969, p. 106)—in effect, which two fit together best; and which is the odd one out?

Kelly would prepare a deck of cards for this activity, from which various triads could be selected for presentation. He would first ask his clients to tell him about significant people in his life so the cards would have the names or roles of parents, siblings, spouse, boss, colleagues, neighbours or whoever. This version of the approach is known as the Role CRT. The method of triads therefore elicited some of the ways a person made discriminations, and therefore drew upon the constructs the person used (whether aware of them of not) to interpret the world. A verbal label could be put at one pole of the elicited construct ('kind', 'hurtful', 'loving', 'bossy', 'cold', etc.). The implicit pole might be given a label if the person readily offered one, or might just be an unnamed contrast.

Kelly's method can be applied widely. The 'elements' (as Kelly called what was presented, such as names of people or roles) need not be about people: indeed, in published research, objects or images of various kinds have been presented—for example, the names of museums and art galleries, or planets, or pictures of different designs of writing pens. An example from science education asked students to make discriminations among triads of cards showing representations of submicroscopic structures such as atoms, ions and molecules (Taber, 1994). Of course, in Kelly's original work, the elements were selected to be of significance for the client so the act of making discriminations had ecological validity—it linked to a client's own concerns.

In such therapeutic work the practice is ideographic, concerned with the nature of the individual (Taber, 2013b). Kelly's method can be used in more nomothetic research looking to test a population in their response to a common set of elements. If those being tested have no strong interest in the elements presented then the method loses some of its essential nature. One precaution to avoid asking for meaningless discriminations is to precede the presentation of triads by a screening stage. So, for example, if people were to be presented with triads of the names or images of famous scientists, the researcher could first go through the pack and ask the study participant to sort the elements into those they did or did not recognise as scientists. The test would then proceed with only the scientists recognised by that participant included in the bespoke deck.

For many purposes, the CRT can suffice as a method for exploring student conceptions. The researcher can gain insights into student thinking by exploring the choices a person makes and how they describe their constructions. The results of applying the method can be a grid of the form shown in Fig. 25.1. Such a grid may have diagnostic value in science teaching if discriminations are made which seem contrary to conventional science, and it can offer insights into the imaginative and idiosyncratic thinking of an individual.

The Repertory Grid

The repertory grid moves beyond the CRT. Having elicited labels for personal constructs, the participant is asked to then rate each element on each construct on a numerical (e.g. 5 or 7 point) scale. The outcome of this would be a grid with an entry in each cell (such as in the hypothetical case shown in Fig. 25.2). The strength of this

Element presented: Elicited constructs:	Vygotsky	Piaget	Kelly	Ausubel	Glasersfeld	Kuhn	Popper	…
focus on social	✓	X	✓					
natural scientist	X	✓	✓					
psychologist		✓		✓		X		
focus on stability					X	✓	X	
admits relativism					✓	✓	X	
…								

Fig. 25.1 The form of the outcome of the construct repertory test

Element presented: Elicited constructs:	Vygotsky	Piaget	Kelly	Ausubel	Glasersfeld	Kuhn	Popper	…	
focus on social	7	2	5	4	2	7	2	…	focus on individual
natural scientist	3	6	6	4	4	7	7	…	[no label given]
psychologist	7	7	7	7	5	1	1	…	not psychologist
focus on stability	2	4	2	4	5	6	3	…	focus on change
admits relativism	2	4	6	4	6	5	1	…	denies relativism
…		…		…		…		…	…

Fig. 25.2 The form of the repertory grid

type of data is that it allows a systematic analysis, to reflect aspects of the structure of a person's constructs (cf. the organisational corollary, Table 25.1).

The quantitative data generated allows tree diagrams to be constructed similar to those used in cladistics to show the relationships among different species: these can both reflect the degrees of perceived similarity among the elements and also the degrees of similarity among the elicited constructs applied. It is important (given the apparent precision of numbers) to recognise that any representation of the construct system produced is a model subject to the limitations of the methodology (Taber, 2013c). Any particular administration of the CRT is sampling from a vast repertoire of potential triads that could be presented. Moreover, the discriminations made in relation to a particular triad need not exhaust possible discriminations based upon available constructs. Just as the same interview questions could potentially access different responses from the same person, construct elicitations and ratings of elements should not be considered definitive. However, the analysis can offer a basis for identifying significant shifts between administrations potentially due to conceptual change.

Classroom Application

Whilst the repertory grid is mainly a technique for research or detailed work with individual clients, the CRT has much potential to be used both in science education research and teaching. The elicitation can take the form of a research interview mediated by the use of triads as a focus for discussion—avoiding the formality of a psychometric test (Taber & Student, 2003). The process of selecting triads can be used for real-time hypothesis-testing as the researcher seeks to interpret the participant's thinking, and PCT offers a complement to approaches such as interview-about-instances (White & Gunstone, 1992).

There is also potential for the method of triads to be used as a teaching activity to initiate group discussion among students. Even quite young students can engage in choosing the 'odd one out'. Despite the strong links between Kelly's ideas and thinking about both the nature of science and students' science learning, there has been limited application of CRT to science teaching. Teachers could have multiple packs of 'elements' (which might be names/images/symbols for different organisms/habitats/organs/cell types/compounds/circuit components, etc.) which could be used in classroom starter or review activities. The approach can also be used in conjunction with other techniques. For example, students producing a revision concept map of a topic could be given a set of relevant cards and told that at any point where they feel they have exhausted their ideas they should pause and spend a few minutes playing the odd-one-out game (i.e. the method of triads) in pairs. This 'oblique strategy' is likely to help bring other features to mind.

The technique can also be used to encourage creative thinking in science. Despite imagination being an essential complement to logic in scientific work (Taber, 2011), this is often not sufficiently emphasised in school science. All scientific discoveries begin as imagined possibilities that are then empirically tested, and some of the most significant discoveries have involved imagining possibilities not entertained by scientific peers at that time. It has been suggested that later science learning is supported by early rich conceptualisation—that is, the ability to think up many possibilities is more valuable than coming up with canonical ideas (Adbo & Taber, 2014). Kelly's triads are intended to explore the manifold nature of a person's conceptual system, and so multiple responses are encouraged.

For example, students could be given a pack of elements showing the names/images of a range of types of animals. How many ways, for example, can the triad elephant/ant/dolphin, or the triad bat/snail/seahorse, be construed? There are clearly a great many possibilities. The activity would likely not only engage diverse biological knowledge (habitat, diet, geographical range, reproduction…), making it a suitable occasional activity for reviewing prior learning, but would, as well as revealing alternative conceptions, likely elicit conjectures that were uncertain that could motivate new learning. Similarly, if students were asked to construe a triad of elements [sic, elements as the triad elements!]—say, sulphur/magnesium/uranium—there are a great many potential responses. Too often science is taught as a very close-ended activity—where there is a right answer, a right way of thinking—which does not fully reflect the practice of science itself, yet when students are asked to be imaginative in a context where idiosyncratic responses are not subject to censor, they can respond with creativity and enthusiasm (Taber, 2016). The method of triads offers an accessible, flexible and engaging classroom activity, yet underpinned by a substantive psychological theory.

Notes

1. Because of the norms of the time he was writing, Kelly tended to use the male pronoun, referring to man, his, him, etc. This seems anachronistic, if not sexist, to a contemporary reader, and quotations are here updated to be gender neutral.
2. It is sometimes useful to distinguish the conceptualisations of individuals (as conceptions) with the canonical conceptual structures of academic science (concepts)—then personal constructs relate to conceptions rather than concepts. However, a concept is empty unless it is applied by someone (and so is their conception), suggesting that this distinction uses 'concept' as a referent for an ideal with which real conceptions could (in principle) be contrasted (see Taber, 2013c).
3. Although this is certainly not the only way in which concepts may be understood, this conceptualisation seems to underpin (deliberately or inadvertently) the common use of tools such as concept maps to elicit and represent conceptual structures.
4. 'Represented' in the curriculum, because it is assumed that the target knowledge in school science is a curriculum model which often simplifies the actual scientific model to provide a realistic target for teaching/learning that offers the essence of the scientific model—that is, an 'intellectually honest' (Bruner, 1960) simplification.

Further Reading

Bannister, D., & Fransella, F. (1986). *Inquiring man: The psychology of personal constructs* (3rd ed.). London: Routledge.
Kelly, G. (1963). *A theory of personality: The psychology of personal constructs*. New York: W. W. Norton & Company.

References

Adbo, K., & Taber, K. S. (2014). Developing an understanding of chemistry: A case study of one Swedish student's rich conceptualisation for making sense of upper secondary school chemistry. *International Journal of Science Education, 36*(7), 1107–1136. https://doi.org/10.1080/09500693.2013.844869.

Brock, R. (2015). Intuition and insight: Two concepts that illuminate the tacit in science education. *Studies in Science Education, 51*(2), 127–167. https://doi.org/10.1080/03057267.2015.1049843.

Bruner, J. S. (1960). *The process of education.* New York: Vintage Books.

diSessa, A. A. (1993). Towards an epistemology of physics. *Cognition and Instruction, 10*(2&3), 105–225.

Driver, R. (1983). *The pupil as scientist?.* Milton Keynes: Open University Press.

Driver, R., & Oldham, V. (1986). A constructivist approach to curriculum development in science. *Studies in Science Education, 13,* 105–122.

Evans, J. S. B. T. (2008). Dual-processing accounts of reasoning, judgment, and social cognition. *Annual Review of Psychology, 59*(1), 255–278. https://doi.org/10.1146/annurev.psych.59.103006.093629.

Fransella, F., & Bannister, D. (1977). *A manual for repertory grid technique.* London: Academic Press.

Gilbert, J. K., & Watts, D. M. (1983). Concepts, misconceptions and alternative conceptions: Changing perspectives in science education. *Studies in Science Education, 10*(1), 61–98.

Glasersfeld, E. V. (1993). Questions and answers about radical constructivism. In K. Tobin (Ed.), *The practice of constructivism in science education* (pp. 23–38). Hilsdale, New Jersey: Lawrence Erlbaum Associates.

Huxley, T. H. (1870). Biogenesis and abiogenesis. In *Collected essays* (Vol. VIII, Critiques and Addresses, pp. 229–271).

Kelly, G. (1958/1969a). Clinical psychology and personality: The selected papers of George Kelly. In B. Maher (Ed.), *Clinical psychology and personality: The selected papers of George Kelly* (pp. 46–65). New York: Wiley.

Kelly, G. (1958/1969b). Man's construction of his alternatives. In B. Maher (Ed.), *Clinical psychology and personality: The selected papers of George Kelly* (pp. 66–93). New York: Wiley.

Kelly, G. (1961/1969). A mathematical approach to psychology. In B. Maher (Ed.), *Clinical psychology and personality: The selected papers of George Kelly* (pp. 94–113). New York: Wiley.

Kelly, G. (1963). *A theory of personality: The psychology of personal constructs.* New York: W. W. Norton & Company.

Kelly, G. (1964/1969). The strategy of psychological research. In B. Maher (Ed.), *Clinical psychology and personality: The selected papers of George Kelly* (pp. 114–132). New York: Wiley.

Kuhn, T. S. (1996). *The structure of scientific revolutions* (3rd ed.). Chicago: University of Chicago.

Lakatos, I. (1970). Falsification and the methodology of scientific research programmes. In I. Lakatos & A. Musgrove (Eds.), *Criticism and the growth of knowledge* (pp. 91–196). Cambridge: Cambridge University Press.

Lakoff, G., & Johnson, M. (1980). The metaphorical structure of the human conceptual system. *Cognitive Science, 4*(2), 195–208.

Latour, B. (1987). *Science in action.* Cambridge, Massachusetts: Harvard University Press.

Luria, A. R. (1976). *Cognitive development: Its cultural and social foundations.* Cambridge, Massachusetts: Harvard University Press.

Piaget, J. (1959/2002). *The language and thought of the child* (3rd ed.). London: Routledge.

Piaget, J. (1970/1972). *The principles of genetic epistemology* (W. Mays, Trans.). London: Routledge & Kegan Paul.

Pope, M. L. (1982). Personal construction of formal knowledge. *Interchange, 13*(4), 3–14.

Pope, M., & Watts, M. (1988). Constructivist goggles: Implications for process in teaching and learning physics. *European Journal of Physics, 9,* 101–109.

Popper, K. R. (1989). *Conjectures and refutations: The growth of scientific knowledge* (5th ed.). London: Routledge.

Popper, K. R. (1994). The myth of the framework. In M. A. Notturno (Ed.), *The Myth of the framework: In defence of science and rationality* (pp. 33–64). Abingdon, Oxon: Routledge.

Russell, T., & Osborne, J. (1993). Constructivist research, curriculum development and practice in primary classrooms: Reflections on five years of activity in the science processes and concept exploration (SPACE) project. In *Paper Presented at the Third International Seminar on Misconceptions in the Learning of Science and Mathematics.* Ithaca: Cornell University.

Shapin, S. (1992). Why the public ought to understand science-in-the-making. *Public Understanding of Science, 1*(1), 27–30.

Solomon, J. (1994). The rise and fall of constructivism. *Studies in Science Education, 23*, 1–19.

Taber, K. S. (1994). Can Kelly's triads be used to elicit aspects of chemistry students' conceptual frameworks? In *Paper Presented at the British Educational Research Association Annual Conference*, Oxford. http://www.leeds.ac.uk/educol/documents/00001482.htm.

Taber, K. S. (2009). *Progressing science education: Constructing the scientific research programme into the contingent nature of learning science.* Dordrecht: Springer.

Taber, K. S. (2011). The natures of scientific thinking: Creativity as the handmaiden to logic in the development of public and personal knowledge. In M. S. Khine (Ed.), *Advances in the nature of science research—Concepts and methodologies* (pp. 51–74). Dordrecht: Springer.

Taber, K. S. (2013a). A common core to chemical conceptions: learners' conceptions of chemical stability, change and bonding. In G. Tsaparlis & H. Sevian (Eds.), *Concepts of matter in science education* (pp. 391–418). Dordrecht: Springer.

Taber, K. S. (2013b). *Classroom-based research and evidence-based practice: An introduction* (2nd ed.). London: Sage.

Taber, K. S. (2013c). *Modelling learners and learning in science education: Developing representations of concepts, conceptual structure and conceptual change to inform teaching and research.* Dordrecht: Springer.

Taber, K. S. (2014a). The significance of implicit knowledge in teaching and learning chemistry. *Chemistry Education Research and Practice, 15*(4), 447–461. https://doi.org/10.1039/C4RP00124A.

Taber, K. S. (2014b). *Student thinking and learning in science: Perspectives on the nature and development of learners' ideas.* New York: Routledge.

Taber, K. S. (2015). The role of conceptual integration in understanding and learning chemistry. In *Chemistry education: Best practices, opportunities and trends* (pp. 375–394): Wiley-VCH Verlag GmbH & Co. KGaA.

Taber, K. S. (2016). 'Chemical reactions are like hell because…': Asking gifted science learners to be creative in a curriculum context that encourages convergent thinking. In M. K. Demetrikopoulos & J. L. Pecore (Eds.), *Interplay of creativity and giftedness in science* (pp. 321–349). Rotterdam: Sense.

Taber, K. S. (2019). *The nature of the chemical concept: Constructing chemical knowledge in teaching and learning.* Cambridge: Royal Society of Chemistry.

Taber, K. S., & Student, T. A. (2003). How was it for you?: The dialogue between researcher and colearner. *Westminster Studies in Education, 26*(1), 33–44.

Vygotsky, L. S. (1934/1986). *Thought and language.* London: MIT Press.

Vygotsky, L. S. (1978). *Mind in society: The development of higher psychological processes.* Cambridge, Massachusetts: Harvard University Press.

White, R. T., & Gunstone, R. F. (1992). *Probing understanding.* London: Falmer Press.

Ziman, J. (1978/1991). *Reliable knowledge: An exploration of the grounds for belief in science.* Cambridge: Cambridge University Press.

Keith S. Taber is the Professor of Science Education at the University of Cambridge. Keith trained as a graduate teacher of chemistry and physics, and taught sciences in comprehensive secondary schools and a further education college in England. He joined the Faculty of Education at Cambridge in 1999 to work in initial teacher education. Since 2010 he has mostly worked with research students, teaching educational research methods and supervising student projects. Keith was until recently the lead Editor of the Royal Society of Chemistry journal 'Chemistry Education Research and Practice', and is Editor-in-Chief of the book series 'RSC Advances in Chemistry Education'. Keith's main research interests relate to conceptual learning in the sciences, including conceptual development and integration. He is interested in how students understand both scientific concepts and scientific values and processes.

Chapter 26
Knowledge-in-Pieces—Andrea A. diSessa, David Hammer

Danielle B. Harlow and Julie A. Bianchini

Introduction

In this chapter, we present a perspective on learning that views a conception or understanding (novice, intermediate, or expert) not as a fully formed whole, but as a set of small, integrated ideas that are activated in coordination. This conception of learning was introduced by Andrea diSessa in 1993 as an alternative to one of the prevailing learning theories in science education at that time, one largely derived from the philosophy of science, conceptual change (see Posner, Strike, Hewson, & Gertzog, 1982).

What Is Knowledge-in-Pieces?

To understand what is meant by multiple, small ideas activated in coordination, imagine a high school marching band changing formations on a football field, a school of fish swimming together to look like a large fish, a flock of geese flying in a v-shape, or a marquee or road sign that uses small lights to spell out words. Each of these items can be thought of as a single object (e.g., the school of fish) or as individual pieces of the collective whole (e.g., one fish). To make the whole object, each individual piece must act appropriately. As one example, in the marching band, each musician must walk to a specific place for the formation to look correct from the audience's perspective. As a second example, consider a sign along the road made up of small lights that read "Slow Down: Construction Ahead." The word SLOW is

D. B. Harlow (✉) · J. A. Bianchini
Department of Education, UC-Santa Barbara, Santa Barbara, CA 93106-9490, USA
e-mail: danielle.harlow@ucsb.edu; dharlow@education.ucsb.edu

J. A. Bianchini
e-mail: jbianchi@education.ucsb.edu

© Springer Nature Switzerland AG 2020   389
B. Akpan and T. Kennedy (eds.), *Science Education in Theory and Practice*,
Springer Texts in Education, https://doi.org/10.1007/978-3-030-43620-9_26

made up of independent lights. For this word to appear, all the correct lights must be activated at the same time and all others must remain off or unactivated (see Fig. 26.1).

However, if some of the lights are not activated or if some of the lights are on that should not be, the sign might look like it says something else. For example, just two extra lights (shown in gray) turn the S into an 8 (see Fig. 26.2).

As we describe in this chapter, the difference between thinking about an idea as a fully formed concept (analogous to the word SLOW in Fig. 26.1) and thinking about the smaller pieces that make up the word (the individual lights) has important implications for how we understand students' thinking and, thus, how we teach. Davis, Horn, and Sherin (2013) explained the advantage of adopting a knowledge-in-pieces theory of learning to inform education:

> The crux of the problem is this: A basic tenet of all constructivist theories of learning maintains that new knowledge is built from existing knowledge (Piaget 1978; Vygotsky 1978), and so learning only takes place at the edges of what is already known. Thus, any account of learning which has the form *delete the old knowledge, replace it with the right knowledge* is no account of learning at all. A useful account of learning must chart a path from novice to expert that builds on useful aspects of a novice's knowledge, and gradually reshapes that knowledge into the expert form. (Smith, diSessa, & Roschelle, 1994, p. 35)

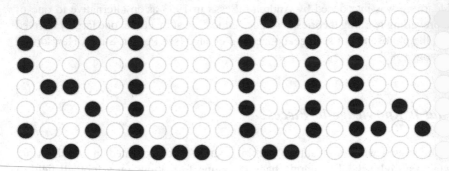

Fig. 26.1 Sign made up of many small lights reading "SLOW"

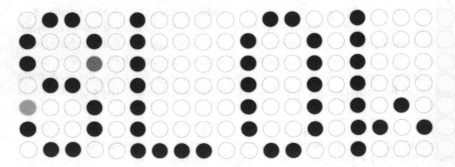

Fig. 26.2 Sign made up of many small lights reading "8LOW"

In sum, the knowledge-in-pieces perspective (KiP, for short) provides a framework for building accounts of scientific learning that is consistent with the larger, overarching theory of constructivism.

Chapter Map

This chapter begins by describing the development of a knowledge-in-pieces perspective on learning and how it differs from other theories of learning. We follow with implications of this theory for teaching. We then conclude with a summary of the main points of the chapter and recommended readings.

The Development of the Knowledge-in-Pieces Perspective

Andrea diSessa: Phenomenological Primitives

In 1993, Andrea diSessa wrote a paper titled "Toward an Epistemology of Physics" in which he described physics students' intuitive ideas about how things work, termed *mechanisms*, and about which sort of events cause other sorts of events. Prior to this paper, other researchers of students' physics ideas (e.g., McCloskey, 1983) had noticed that students develop ideas about the world that are consistent with their everyday interactions, but inconsistent with the ideas that physicists have. For example, students who observe a box pushed across a floor likely notice that once pushed, it almost immediately begins to slow down and eventually comes to a stop. Students may explain this as a result of the force *running out* of the box. This idea is in contrast to that held by physicists, who would explain that the surface (floor) is pushing against the box in the opposite direction of motion and that this opposing force is slowing the box down. McCloskey and others of his time viewed students' understandings as robust and coherent, even if inconsistent with science ideas. These types of ideas are commonly referred to as *misconceptions*.

In his paper, diSessa (1993) presented a different idea. He described a cognitive structure made up of small units he called *phenomenological primitives*, or *p-prims* for short. The two words in the name indicate that the p-prims both are tied to interpretations of experiences in the real world (phenomena) and are minimal elements of memory. These p-prims are cued by contexts and, often, cued along with other p-prims. In this paper, diSessa identified several initial p-prims. One is *Ohm's p-prim*: the idea that "more effort implies more result; more resistance implies less result; and so on" (p. 126). This p-prim can be generalized from the experience of pushing a box across a rug and many other experiences of moving items across surfaces. Another p-prim identified in this early paper is *force as a mover*, or the idea that "pushing an object from rest causes it to move in the direction of the push" (p. 129). What is important is that these ideas are neither correct nor incorrect in and of themselves. Rather, in some contexts, they are appropriate, while in others they are not. This paper was met with challenges from other scholars (e.g., Chi & Slotta, 1993) who agreed with many of the ideas presented, but disagreed with the assertion that there was little underlying structure in intuitive physics ideas.

A second paper by diSessa (diSessa & Sherin, 1998) was published five years later. In this paper, "What Changes in Conceptual Change?", diSessa and Bruce Sherin built on the idea of p-prims to provide a mechanism for learning that challenged a prevailing theory of that time, conceptual change. To understand diSessa and Sherin's contribution, it is first important to understand the conceptual change model of learning (e.g., Posner, Strike, Hewson, & Gertzog, 1982). The key premise of the conceptual change model is that misconceptions held by novices can be *replaced* by more expert views. This replacement occurs either by understanding new relationships between *concepts*, such as that between mass and force (weak conceptual change); or by changing the understanding of the concepts themselves, such as what constitutes a force (strong conceptual change). In conceptual change theory, concepts are described using single words, such as mass, force, and friction.

In a more recent paper, diSessa and Sherin (1998) introduced the idea of *coordination classes*, which they described as a knowledge system—as an alternative to the idea of concepts. In other words, their interpretation of a concept was much fuzzier than in conceptual change and could not be described in terms of a single word. Instead, coordination classes consist of different types of cognitive elements and create a system of strategies for gaining information from the world. Some of these strategies include narrowing attention and selecting and combining information to determine what is observed. These were called *readout strategies*. Other cognitive elements relate to reasoning, connecting new observations with other information. This second type of cognitive element makes up the *causal net*. According to diSessa and Sherin, readout strategies and the causal net evolve together as a student learns.

David Hammer: Resources

David Hammer (2004) continued this thread of thinking about ideas as knowledge-in-pieces, which he referred to as a *manifold ontology*. Manifold means "many parts" and ontology, in this case, means "understandings of what sorts of entities that 'exist' in minds" (p. 1). Like diSessa (1993), Hammer described this view as an alternative to the prevailing ideas of the time—that students have *conceptions* that must be replaced and that cognitive dissonance and accommodation, derived from Piaget, must take place for learning to occur. In conceptual change, the assumption is that students must become dissatisfied with existing (incorrect) conceptions, experience conflict (cognitive dissonance), and then find a new conception that resolves that conflict. Hammer argued that this last step could not occur if concepts were assumed to be cognitive structures because the students, using existing conceptual structures to process the new information, would not necessarily see their way to a new conception. Like diSessa, Hammer also proposed the existence of small cognitive units. However, unlike diSessa, he did not hypothesize about their specific nature and used the term *resources* rather than p-prims for these small constituent cognitive units.

As did diSessa (1993), Hammer (2004) first emphasized that resources are neither correct nor incorrect in and of themselves. Rather, the usefulness of the resources depends on the situation. Secondly, he noted that they are context-dependent, meaning that resources that are activated in one context may not be activated in another. Hammer also expanded the idea of resources to include a class of resources he called *epistemological resources*. These resources help explain how students approach learning—and likely how teachers approach teaching. These resources include ideas such as knowledge as propagated stuff, knowledge as fabricated stuff, and so on. Like the physics resources that students draw on to make causal inferences about how the world works, students may activate different epistemological resources in different contexts. This accounts for the observation that the same student may think that knowledge is invented in some contexts but that knowledge comes from authority in other instances.

Hammer and colleagues (Hammer, Elby, Scherr, & Redish, 2005) emphasized the context-dependent nature of resources in a paper on knowledge transfer; the researchers examined a student they called Sherry. During a class discussion, when asked how big a mirror needed to be to see her whole body, Sherry stated that the mirror must be as tall as she was, defending this idea against classmates who claimed that a mirror needed to only be half as tall as the individual. The following week, Sherry revealed to the class that she had a mirror at home that was half her height and in which she could see her whole body. Clearly, Sherry had the (daily) experience with mirrors that would have allowed her to correctly state that a mirror half her height would allow her to see her entire body. And, at home, she did know this. But in the class discussion, she drew on some other resources to reason a full-length mirror was needed. The example of Sherry highlights how knowing what students know is insufficient for predicting whether they will be able to activate and apply the appropriate ideas in any given context, explaining the contextual nature of knowledge. Sherry's experience clearly exemplifies that students may have ideas that they activate and use in one context and not in others. The knowledge-in-pieces perspective on learning suggests that this is likely to happen in science classrooms all the time, even when it is not as obvious as Sherry's example.

David Hammer and Tiffany-Rose Sikorski: Using Knowledge-in-Pieces to Inform Learning Progressions

Whether one calls these small cognitive elements p-prims, resources, or something else, these knowledge-in-pieces perspectives highlight the complexity and context-dependent nature of students' science ideas. In 2015, David Hammer and Tiffany-Rose Sikorski encouraged researchers to embrace this complexity and context-dependency in their efforts to construct learning progression frameworks. A learning progression is a description of student thinking about an important, disciplinary-specific idea that increases in coherence and sophistication over time (National Research Council, 2007); it is considered an integral learning construct in the *Next Generation Science Standards* (*NGSS*; NGSS Lead States, 2013). Collectively, the different levels of a learning progression describe the conceptual pathways students are likely to follow in their progression toward mastery of a topic. It is anchored at the lower end by what we know about the concepts and reasoning of young children entering school, and at the upper end, by what disciplinary experts identify as appropriate scientific knowledge and practices. A learning progression framework is grounded in both the disciplinary knowledge of the field and research on student learning.

Hammer and Sikorski (2015) argued that existing learning progression frameworks are too simplistic: Rather than embrace the multiplicity of students' ideas and identify numerous possible pathways, most describe a single, levels-based sequence or present a small number of alternative sequences. More specifically, they noted that students' efforts to achieve coherence in their ideas can differ dramatically from each other and from scientifically accepted understandings; the myriad ideas that students can decide to pursue, assess, and refine are not adequately captured in most learning progression frameworks. Hammer and Sikorski also noted that, when researchers aggregate student data to construct a learning progression, they tend to dismiss variations in students' ideas as conceptually insignificant noise rather than to incorporate the idiosyncratic particularities of each classroom into their frameworks. In short, they recommended researchers construct learning progressions that more closely attend to students' prior ideas and experiences, how each student chooses to engage (or not) with the learning experience and, thus, the idiosyncrasies of classroom contexts. In the following section, we further explore how the knowledge-in-pieces perspective on student learning intersects with the *NGSS* (NGSS Lead States, 2013), in particular, how it helps to inform reform-based science teaching.

Connections Between the Knowledge-in-Pieces Perspective and Reform-Based Science Teaching

As introduced above, understanding learning through a knowledge-in-pieces lens leads to particular ideas about students and how best to teach them. A key difference between viewing students' ideas as full conceptions and viewing their ideas as small cognitive pieces that are activated in particular contexts is how teachers work with students' ideas that differ from accepted scientific knowledge. Viewing ideas as fully formed conceptions leads to the assumption that students have fully formed incorrect ideas, often called misconceptions, that must be ferreted out and replaced. Viewing ideas as p-prims or resources, in contrast, encourages teachers to value rather than dismiss students' ideas, working with students to build from their existing ideas toward accepted scientific understanding. It also encourages teachers to help students learn to apply their ideas in appropriate contexts—to learn the difference between appropriate and inappropriate contexts for a given p-prim or resource.

Current policy documents, including the *NGSS* (NGSS Lead States, 2013), expect that all students will develop deep understandings of a few key ideas, called Disciplinary Core Ideas, in four domains: Physical Science, Life Science, Earth and Space Sciences, and Engineering. Further, students are expected to do so by engaging in questions and phenomena that relate to concepts that cross disciplinary boundaries using the practices of science and engineering.

Below, we begin with a vignette of what using a knowledge-in-pieces perspective to inform reform-based teaching looks like. We then provide four specific recommendations for teachers.

Ms. Carter and the Teaching of Magnetism

Ms. Carter, a fourth-grade teacher, serves as an example of how a knowledge-in-pieces perspective on student learning can be used to inform instruction. Ms. Carter began her unit on magnetism by asking her students to check to see if a nail was attracted to a paperclip (it was not). Students then rubbed the nail with a magnet and tested whether it was able to pick up other paperclips (it was). Next, she had students draw their ideas about what was going on inside the nail in order for it to behave like a magnet. Her students shared their ideas with a partner. Finally, Ms. Carter shared some specific student ideas with the whole class.

Ms. Carter: ... I want to tell you about what Julia said. She said, "When you're rubbing it, then magnetic dust is on the outside of the nail and that's giving it that temporary magnetism and then that rubs off and that's why it goes away." Sean, however, has a different idea—this is pretty interesting. I wanted to share it with the class.

[Pause]

Ms. Carter: All right, [Sean], tell us about your drawing.
Sean: I think there's like little pieces of steel inside the nail and when you rub
              the magnet against it, the steel pieces get really active and they start
              to bounce around and then they get like a magnet and when you stick
              them together they stick and then sooner or later they like run around,
              so they, so they slow down and they'll stop and drop.

Ms. Carter was surprised by Julia and Sean's responses. She had expected her students to express a common, scientifically incorrect idea that she herself and her fellow
teachers had expressed when conducting this same set of activities during a professional development program. More specifically, she had expected her students to
think that two different types of charges separated into the two poles of a magnetized
nail and that these two different types of charges mixed together in a non-magnetized
nail. She knew, from her professional development, how to respond to the idea that
charges mixed and separated: She had planned to implement an activity that would
result in evidence that challenged this particular idea. However, as stated above, her
students' ideas were different from what she had expected.

Nevertheless, as suggested by looking at learning through a knowledge-in-pieces
lens, Ms. Carter recognized that there were pieces of the children's ideas that were
valuable to act on. She recognized that one of her students, Julia, had proposed that
something on the *surface* of the nail was changing and that another student, Sean,
had proposed that something *inside* the nail was changing. Rather than dismiss these
ideas as incorrect and attempt to replace them with canonical science knowledge,
she developed an activity that tested just these small parts of a larger idea. She had
students rub the "dust" off the surface of the nail and test whether that changed the
magnetic properties. The students discovered that even after rubbing the dust off
the nail, it still behaved as a magnet and concluded that something *inside* the nail
changed when it was rubbed with a magnet. Her efforts to build on her students'
ideas resulted in a powerful learning opportunity (for a more complete description
see Harlow, 2010).

Instructional Idea 1

**Teachers must recognize that students may use ideas to construct responses
that are scientifically inaccurate, but useful in their everyday life (e.g., closer is
more, earth is flat).** As illustrated in the above vignette, teachers should recognize
that pieces of knowledge (we refer to them as resources for the remainder of this
section) are neither correct nor incorrect in and of themselves, but are either accessed
appropriately or inappropriately depending on the context. A classic example is the
idea that 'closer is more': the idea that the closer one is to a source, the stronger its
influence is. For example, the closer one is to a heat source, such as a candle or fire,
the stronger the heat. This is a reasonable idea to draw on when a child decides to stay

away from a fireplace to avoid being too hot. However, many students incorrectly draw on this idea to explain why it is warmer in the summer than in the winter: They think that the distance between the sun, a source of heat, and the earth varies between seasons and, thus, explains the difference in temperature between summer and winter. In this case, rather than attempt to replace students' concept of seasons caused by distance, science teachers can help students understand that they are applying a useful resource in an inappropriate context.

Instructional Idea 2

Teachers must recognize students' initial ideas as useful and productive for building understanding that is consistent with canonical knowledge. Again, as illustrated by the vignette, perhaps the most important implication of a knowledge-in-pieces perspective for science teaching is that students' ideas are useful and productive—not only for interacting in everyday life, but for developing ideas that are consistent with canonical science knowledge. That is, eliciting and embracing students' ideas do not mean sacrificing the development of scientific knowledge, even when students' initial ideas do not match with scientists' ideas. Starting with students' existing ideas, helping them to value their ideas, and then building on them through engaging in the practices of science and engineering can lead to sophisticated ideas that are consistent with canonical science ideas. Below we describe two related strategies teachers must use to help their students build on their existing knowledge: learning what their students' ideas actually are and (re)designing their instruction to build on these ideas.

Instructional Idea 3

Teachers must elicit students' existing ideas through a variety of means, including formative assessments, predictions, and modeling. To effectively build on students' ideas requires teachers to know what resources they bring to bear in a given context. The stage of eliciting students' ideas is critical to the learning process, not only so that teachers know which ideas or resources their students activate, but so that the students themselves articulate and own these ideas. One method of eliciting students' ideas is to have them make a prediction about a phenomenon and to explain their reasoning. For example, a question that asks if a container full of ice weighs more, less, or the same after melting (adapted from Keeley, Eberle, & Farrin, 2005) elicits students' initial ideas about mass, heat, and changes of state. Another effective method of eliciting students' ideas is to have them propose an initial model of the unseen mechanisms that drive a phenomenon. For example, students might be asked to draw all the forces on a kicked soccer ball after the ball has left the foot (from Goldberg, Robinson, & Otero, 2007). This modeling activity elicits students' existing ideas about forces, energy, gravity, and motion.

The recommendation that teachers elicit students' ideas each time they introduce a new unit or topic implies another challenge—that instruction must be individualized for each and every student. In classrooms of 30 or more students, such individualized instruction may seem beyond teachers' reach. However, individual student differences do not mean that teaching groups of students is impossible. Research on students' ideas in science demonstrates that there is often a handful of common ways that students interpret a given phenomenon. Further, teachers can use the various ideas articulated by students during instruction. Students can be prompted to examine and critique each other's ideas—to collectively determine which ideas are supported by evidence and which ones are not. Such an activity aligns with the practices of constructing explanations and engaging in argument from evidence articulated by the *NGSS* (NGSS Lead States, 2013).

Instructional Idea 4

Teachers must design their instruction, often in the moment, around their students' ideas. Once teachers are aware of students' ideas, they must design instruction that values these ideas as productive and moves students toward more sophisticated understandings. Teaching science in this way is difficult because teachers cannot assume passive transmission of knowledge. The knowledge-in-pieces perspective helps teachers to understand why treating students' ideas as large concepts that can be simply replaced through lectures or explanations is likely to be problematic. The complexity of knowledge means that, even if students are able to correctly repeat the expected canonical response, they may not actually understand the idea; they are instead repeating memorized responses. As the knowledge-in-pieces perspective on learning helps us understand, students do not passively learn ideas that teachers tell them. Rather, it is vital that students grapple with the ideas they hold and test these ideas against real phenomena to build toward accepted scientific understanding.

Campbell, Schwarz, and Windschitl (2016) offered suggestions for strategies to help teachers engage their students in learning science in ways that allow them to grapple with their prior ideas and to develop science knowledge consistent with the new standards. These suggestions included the following:

- Include some level of uncertainty in students' science activities rather than using them to confirm authoritative science ideas.
- Engage students in using their own ideas and experiences to construct and revise explanations of phenomena or to solve problems.
- Model out loud how a scientist reasons about ideas (comparing ideas, changing them in response to evidence). Invite students in small groups to rehearse conversations about evidence and explanations.

- Emphasize collective sense-making as an important goal (Carlone & Smithenry 2014). Ask students or student groups who have contrary explanations to share their thinking in whole-class settings (p. 70).

All of the above strategies ask teachers to use the varied and complex ideas that students bring to a classroom while engaging them in the practices of science and engineering described in the *NGSS* (NGSS Lead States, 2013) in order to develop understandings of disciplinary core ideas.

Summary

In brief, teachers should view the ideas discussed here as a model of student learning that can be used to understand students' developing ideas. The knowledge-in-pieces perspective was originally proposed to explain concept development in physics and has been extended for use in understanding concept development in other science disciplines, science practices, students' understanding of what counts as knowledge (epistemological resources), and teaching (pedagogical resources). This perspective also resonates with other constructivist theories of learning (see the other chapters in Sect. IV of this volume).

In this chapter, we described the development of a knowledge-in-pieces perspective on learning and its application to reform-based science instruction.

- A knowledge-in-pieces perspective views ideas as composed of small cognitive units, for example, p-prims or resources, that students activate in concert in particular contexts. These cognitive units are neither correct nor incorrect in and of themselves.
- A knowledge-in-pieces perspective can be understood as a response to conceptual change theory and other theories of learning that conceptualize ideas as large pieces or concepts that must be replaced.
- Although a knowledge-in-pieces perspective was first developed in the 1990 s, it resonates with ideas put forth in the recent *NGSS* (NGSS Lead States, 2013), in particular, the ideas of learning progressions and effective science instruction.
- A knowledge-in-pieces perspective has clear implications for reform-based science instruction. It suggests that teachers value rather than dismiss students' ideas—that they build from students' existing ideas toward accepted scientific understandings.

Recommended Resources

For those interested in learning more about a knowledge-in-pieces perspective of learning, we recommend reading volume 10, number 2/3 of *Cognition and Instruction*. This issue is where the original paper by diSessa was published as well as challenges to his ideas and diSessa's responses to these challenges. We also recommend a more recent article written by Campbell et al. (2016), which clearly describes the connection between a knowledge-in-pieces perspective on learning and its implications for reform-based science instruction.

References

Campbell, T., Schwarz, C., & Windschitl, M. (2016). What we call misconceptions may be necessary stepping-stones toward making sense of the world. *Science and Children, 53*(7), 28–33.

Carlone, H., & Smithenry, D. (2014). Creating a "We" culture: Strategies to ensure all students connect with science. *Science and Children, 52*(3), 66–71.

Chi, M. T., & Slotta, J. D. (1993). The ontological coherence of intuitive physics. *Cognition and Instruction, 10*(2–3), 249–260. https://doi.org/10.1080/07370008.1985.9649011.

Davis, P. R., Horn, M. S., & Sherin, B. L. (2013). The right kind of wrong: A "knowledge in pieces" approach to science learning in museums. *Curator: The Museum Journal, 56*(1), 31–46. http://dx.doi.org/10.1111/cura.12005.

diSessa, A. A. (1993). Toward an epistemology of physics. *Cognition and Instruction, 10*(2–3), 105–225. https://doi.org/10.1080/07370008.1985.9649008.

diSessa, A. A., & Sherin, B. L. (1998). What changes in conceptual change? *International Journal of Science Education, 20*(10), 1155–1191. https://doi.org/10.1080/0950069980201002.

Goldberg, F., Robinson, S., & Otero, V. (2007). *Physics and everyday thinking*. Armonk, NY: It's About Time.

Hammer, D. (2004). The variability of student reasoning, lecture 3: Manifold cognitive resources. In *The Proceedings of the Enrico Fermi Summer School in Physics, Course CLVI*. Italian Physical Society.

Hammer, D., Elby, A., Scherr, R. E., & Redish, E. F. (2005). Resources, framing, and transfer. In J. P. Mestre (Ed.), *Transfer of learning from a modern multidisciplinary perspective* (pp. 89–119). Greenwich, CT: Information Age Publishing.

Hammer, D., & Sikorski, T. R. (2015). Implications of complexity for research on learning progressions. *Science Education, 99*(3), 424–431. https://doi.org/10.1002/sce.21165.

Harlow, D. B. (2010). Structures and improvisation in inquiry-based instruction: A teacher's adaptation of a model of magnetism activity. *Science Education, 94*(1), 142–163. https://doi.org/10.1002/sce.20348.

Keeley, P., Eberle, F., & Farrin, L. (2005). *Probing students' ideas in science* (Vol. 1). Arlington, VA: NSTA Press.

McCloskey, M. (1983). Naive theories of motion. In D. Gentner & A. L. Stevens (Eds.), *Mental models* (pp. 299–324). Hillsdale, NJ: Lawrence Erlbaum Associates.

National Research Council. (2007). *Taking science to school: Learning and teaching science in grades K-8*. Washington, DC: The National Academies Press.

NGSS Lead States. (2013). *Next generation science standards: For states, by states*. Washington, DC: The National Academies Press.

Piaget, J. (1978). *Behavior and evolution*. New York, NY: Random House.

Posner, G. J., Strike, K. A., Hewson, P. W., & Gertzog, W. A. (1982). Accommodation of a scientific conception: Toward a theory of conceptual change. *Science Education, 66*(2), 211–227. https://doi.org/10.1002/sce.3730660207.

Smith, J. P., III, diSessa, A. A., & Roschelle, J. (1994). Misconceptions reconceived: A constructivist analysis of knowledge in transition. *The Journal of the Learning Sciences, 3*(2), 115–163. https://doi.org/10.1207/s15327809jls0302_1.

Vygotsky, L. S. (1978). *Mind in society: The development of higher psychological processes*. Cambridge, MA: Harvard University Press.

Danielle B. Harlow is a professor of science education at the University of California, Santa Barbara (UCSB). Her research investigates elementary school students' understanding of science and engineering and elementary school teachers' ideas about how to teach science and engineering in ways that engage students in authentic practices of these disciplines. Prior to moving to UCSB, she earned a doctorate in science education from the University of Colorado at Boulder and served in the Peace Corps as a physics teacher.

Julie A. Bianchini is a professor of science education at the University of California, Santa Barbara (UCSB). Her research investigates prospective, beginning, and experienced teachers' efforts to learn to teach science in equitable ways. Her recent Noyce Teacher Scholarship Programs focus on ways to support preservice mathematics and science teachers in learning to teach the *Common Core in Mathematics* and the *Next Generation Science Standards* to English learners. She serves as Faculty Director of UCSB's CalTeach/Science and Mathematics Initiative and as chair of her department.

Part V
Intellectually-Oriented and Skill-Based Theories

Chapter 27
Multiple Intelligences Theory—Howard Gardner

Bulent Cavas and Pinar Cavas

> *An intelligence is the ability to solve problems, or to create*
> *products, that are valued within one or more cultural settings.*
> *Howard Gardner—Frames of Mind (1983).*

Introduction

Approaches to education are closely related to how the learning is perceived by people who are in charge of educational programs. Yet, educational researchers have not yet fully explored the dynamics that result in learning in individuals. The concept of learning has dynamic and unique characteristics, which is in keeping with its complexity and can change according to the nature of each individual. Multiple intelligences (MI) theory emerges in this perspective as one of the best theories in solving problems that human beings encounter in the learning process.

In this chapter, MI theory is introduced along with its recent modifications. The educational implications of the MI theory are presented next and supported by the science education research on the theory. The uses of MI theory in science education are provided with examples, and suggestions for implementation of MI theory are given. Finally, the chapter ends with a discussion on the critique of the theory and the modifications to the theory in response to the critique.

B. Cavas (✉)
Buca Faculty of Education, Dokuz Eylül University, Izmir, Turkey
e-mail: bulentcavas@gmail.com

P. Cavas
Faculty of Education, Ege University, Izmir, Turkey
e-mail: pinarcavas@gmail.com

© Springer Nature Switzerland AG 2020
B. Akpan and T. Kennedy (eds.), *Science Education in Theory and Practice*,
Springer Texts in Education, https://doi.org/10.1007/978-3-030-43620-9_27

Multiple Intelligences Theory

Multiple intelligences theory was put forward by Professor Howard Earl Gardner in the late 1970s. Professor Gardner is a psychologist in the department of Cognition and Education at Harvard Graduate School of Education. He is recognized for his studies on the conception of learning from the perspective of Multiple Intelligences theory.

Howard Earl Gardner is an American developmental psychologist born on July 11, 1943 in Scranton, Pennsylvania. He defined himself as "a studious child" who loved to read and play the piano, and later he became a gifted pianist. Gardner took his bachelor's degree in social relations in 1965 and a doctoral degree in developmental psychology in 1971 from Harvard University. Howard Gardner is the John H. and Elisabeth A. Hobbs Professor of Cognition and Education at the Harvard Graduate School of Education and also an adjunct professor of psychology at Harvard University. Gardner currently serves as the Chairman of Steering Committee for Harvard Project Zero since 1995 and senior director of this project since 2000. He was inspired by the works of Piaget, Erikson, Riesman, and Bruner to investigate human nature and human cognition. In 1983, he developed the theory of multiple intelligences which has influenced many fields including education.

The concept of intelligence has continued to change throughout the years. Until 1980s, cognitive psychologists defined intelligence as the ability to solve problems or answer items on standard IQ tests. The works of well-known psychologists in the early 1900s such as Binet and Spearman in the area of intelligence served as a basis for developing more than 70 IQ tests. These IQ tests were generally designed to reveal the students' levels of knowledge in specific areas like mathematics or language and assumed that a score of 100 would indicate an average intelligence. Yet, in the late 1970s, Howard Gardner proposed the theory of multiple intelligences (hereafter referred to as MI), and this theory has impacted many important fields especially psychology and education.

In 1983, Gardner wrote a book entitled *Frames of Mind* where he put forth a new understanding of the construct of intelligence. The book has been translated into more than 20 languages, and the tenet of MI has been spreading around the world. In his book, Gardner rejected and changed the accepted idea that there is a single intelligence measured objectively and reported by a single score. The theory gained importance due to its emphasis on diverse intelligences, which had not been measured by standardized tests like IQ tests or tests applied in schools. This approach was revolutionary considering that the cognitive scientists heavily studied the mind from the perspective of traditional conceptions of intelligence formulated in the early twentieth century, and based on the studies of cognitively oriented psychologists like Jean Piaget (Davis, Christodoulou, Seider, & Gardner, 2011).

Gardner examined some research conducted in biology, psychology, anthropology, and neurology to identify the nature of intelligence. Then, he claimed that each person has a number of different intellectual capacities and tendencies in different areas and as such each individual has several types of intelligences. Based on his examination, he formulated key criteria including eight factors that had to be met to be classified as a full-fledged intelligence. These factors are given below:

1. Potential isolation by brain damage;
2. The existence of savants, prodigies, and other exceptional individuals;
3. A distinctive developmental history and a definable set of experts' "end-state" performances;
4. An evolutionary history and evolutionary plausibility;
5. Support from psychometric findings;
6. Support from experimental psychological tasks;
7. An identifiable core operation or set of operations;
8. Susceptibility to encoding in a symbol system (Hoerr, 2000).

Based on these factors, Gardner (1999a) defined intelligence as "a biopsychological potential to process information that can be activated in a cultural setting to solve problems or create products that add value in a culture" (p. 34). This definition is very important in understanding students' abilities and potentials. In accordance with this definition, intelligence cannot be seen or counted since environmental factors, cultural values, education, and personal efforts can affect intelligence. The intelligences individuals possess define the ways for people in creating products or solving problems in relation to all the factors they experience (Gardner, 1993).

Howard Gardner identified seven intelligences in his studies in psychology, human cognition, and human potential. These intelligences were named by Gardner as linguistic intelligence, logical–mathematical intelligence, spatial intelligence, bodily–kinesthetic intelligence, musical intelligence, interpersonal intelligence, and intrapersonal intelligence. *Linguistic intelligence* is the capacity to use words effectively, whether orally or in writing. It involves the mastery of spoken and written language to express oneself or remember things. *Logical–mathematical intelligence* is the capacity to use numbers effectively, detect patterns, think logically, reason deductively, and carry out mathematical operations. These two kinds of intelligences are typically the abilities that are expected by the traditional school environments to support and assess most IQ measures or tests of achievement. *Spatial intelligence* is the ability to perceive the visual–spatial world accurately and involves sensitivity to color, line, shape, form, space, and the potential for recognizing and manipulating the patterns of spaces. *Bodily–Kinesthetic* intelligence includes an expertise in using one's whole body or parts of the body to express ideas and feelings; and solve problems or create products. *Musical intelligence* is the capacity to perceive, discriminate, transform, and express musical forms, and use them for performance or composition. *Interpersonal intelligence* is the capacity to perceive and make distinctions in the moods, intentions, motivations, and feelings of other people. The last intelligence is the *intrapersonal intelligence,* and this intelligence is about self-knowledge and the ability to act adaptively on the basis of that knowledge.

According to Gardner (1991):

> we are all able to know the world through language, logical-mathematical analysis, spatial representation, musical thinking, the use of the body to solve problems or to make things, an understanding of other individuals, and an understanding of ourselves. Where individuals

differ is in the strength of these intelligences - the so-called profile of intelligences -and in the ways in which such intelligences are invoked and combined to carry out different tasks, solve diverse problems, and progress in various domains (p. 12).

In the mid-1990s, Gardner proposed that one more intelligence, *naturalistic intelligence*, met the criteria for identification as an intelligence as well. Naturalistic intelligence involves high expertise in recognition and classification of the numerous species—the flora and fauna—of the environment. More recently, Gardner has added an additional intelligence, the *existential intelligence*. He defines this intelligence as "the capacity to locate oneself with respect to the furthest reaches of the cosmos—the infinite and the infinitesimal—and the related capacity to locate oneself with respect to such existential features of the human condition as the significance of life, the meaning of death, the ultimate fate of the physical and the psychological worlds, and such profound experiences as love of another person or total immersion in a work of art" (Gardner, 1999a, p. 60).

The intelligences, according to Gardner (2006), are demonstrated at different aptitudes in different individuals. For example, a person might possess high spatial intelligence but may not necessarily be good at naturalistic intelligence. This does not necessarily mean that the person has only spatial intelligence, but we all demonstrate some intelligence better than others due to our experiences with the world around us or genetic factors. The acceptance of this theory also conflicts with the deeply rooted assumptions in education. Gardner (2011) explains that multiple intelligences theory "challenges an educational system that assumes that everyone can learn the same materials in the same way and that a uniform, universal measure suffices to test student learning" (p. 26).

Educational Implications

The book, *Frames of Mind*, has a deeper impact on educational communities than it has on psychology (Gardner, 2011). The change in the conception of intelligence challenged the way education scholars conceive learning. However, Gardner is a psychologist and does not provide an educational model. This led to many educators misinterpreting the intelligences as learning styles. The MI theory assumed that the intelligences operate independently of one another, and a teaching method might support much intelligence at different levels. The misunderstanding was realized when educators often thought that if students do not understand a concept using a particular teaching method, the teacher is expected to change it to address a different intelligence. In fact, this is mainly changing the learning style and not the intelligence. Therefore, Gardner clarified later that intelligence is not the same as learning style. While learning style is the different ways through which a student approaches a learning task, intelligence is defined as a capacity people have with different strengths.

The MI theory found its way through educational policies in a variety of ways. For example, The Association for Supervision and Curriculum Development (ASCD) and its "teaching the whole child" initiative shed a light on the multiple intelligences theory and interdisciplinary teaching support. ASCD clearly indicates that students having this support can be fully engaged and challenged academically (Martin, Bishop, Ciotto, & Gagnon, 2014). Based on this view, the multiple intelligences theory has been implemented in many different countries' curricula in teaching and learning of science as key initiatives. Curricula were renewed in many countries to adapt to the new ways of thinking (e.g., Science education curriculum for elementary grades in 2003; Schools such as EXPO Elementary School in Minnesota and Key Learning Community in Indiana; Schwert, 2004). The reason for including MI theory in curricula is the belief that such an approach would enhance the individual strengths of every child (Campbell & Campbell, 1999). Educational research also supported this position. It has been reported that academic achievement, motivation, and meaningful learning have increased significantly in the classrooms where most of the researches apply MI theory.

Science Education Researches on the Effectiveness of Multiple Intelligences Theory

In MI theory, it is stated that each student's undiscovered hidden powers and potentials can be revealed, and MI teaching strategies and these latent powers can be revealed in appropriate forms. It is understood from the quality and quantity of the work done in this field that MI theory can be used very effectively especially in science education.

Multiple intelligences theory has thus been used effectively in science teaching and learning environments in addition to the other disciplines such as physical education and music education. The literature review shows how multiple intelligences theory has been used to improve the students' cognitive skills, in terms of understanding, remembering, applying and expanding knowledge; to explore the relationships among multiple intelligences, between music intelligence and mathematical intelligence; to see effects on the teaching–learning process within the technology lessons; to develop students' academic achievement in science courses; and finally to assist the memory as an aid in remembering.

Before implementing MI theory in science education, it is better to understand its impact on other fields such as physical and music education. Both fields are directly connected with science education. For example, the studies conducted by Martin and McKenzie (2013) and Blumenfield-Jones (2009) have shown that the MI theory-based activities can be effective in teaching sports, dance, and tennis units which can be adapted and applied in teaching science. In addition to physical education mentioned above, music provides new opportunities in the science concepts and discourses (Gardner, 1999a). The use of effective music as an addition to the course implementation supports students' memory (Crowther, Williamson, Buckland, & Cunningham, 2013) and makes recall easy (Schulkind, 2009).

Cognitive skills connected with recognizing, remembering, applying, and building new knowledge from previous knowledges can be easily improved by the implementation of multiple intelligences theory. For example, Lai and Yap (2016) worked on the assessment of cognitive skills when multiple intelligences theory was implemented in the teaching and learning of chemistry at university level. The researchers implemented longitudinal research method in order to examine differences in students' cognitive skills. The findings of their study were really useful in understanding the effect of multiple intelligences theory on students' cognitive skills. They found out that the students' multiple intelligences profile showed a shift from intrapersonal dominance in the previous years, 2011–2013 to interpersonal, kinaesthetic, and naturalistic intelligences in subsequent years, 2014–2015.

Administrators, teachers, and families are crucial stakeholders at all educational levels. If they are properly integrated and coordinated in the educational system, the teaching and learning would be enhanced. A study (Richards, 2016) which explored the position of administrators, teachers, and parents on applying the multiple intelligences theory (MI) in the early childhood curriculum showed that administrators would not play important roles in this process even when they indicated that they were aware of the effectiveness of the multiple intelligences theory. The teachers, in addition, mentioned that the teaching was better, and it was exciting and fun. The teachers also suggested that more time is needed for effective planning and implementation of multiple intelligences theory. The families also held multiple intelligences theory-based teaching and learning as a "great concept". The study endorses MI integration, and differentiated instruction posits that MI integration allows the diverse learners to display how they are using their "smarts". It enhances and builds on children's strengths and love for learning.

Nowadays, science, technology, engineering, mathematics (STEM) education is very popular all over the world. STEM educators argue that rather than trying to teach these different but interrelated four disciplines as separate subjects, STEM educators try to integrate them into a cohesive learning paradigm based on real-world applications (Hom, 2014). In this framework, a study was conducted to explore how Gardner's MI theory would be in the teaching–learning process within the technology lessons. The study mainly focussed on the MI theory as an explanation variable of the emotional response within the different educational parts in the Spanish technology curriculum. The study postulated that a different intelligence style (IS) will orient the student to a vision of the engineering and technology. The participants of this study consisted of 135 students from Compulsory Secondary Education level and reported their predominant IS and on the emotions that aroused them. The findings of this study revealed that only those with a logic–arithmetic or environmental IS were not affected by the syllabus units. The researchers added that best teaching and learning practices are required for encouraging further engineering studies (Martín, Gragera, Dávila-Acedo, & Mellado, 2017).

Another STEM-based study, conducted to understand the relationship of people's music and math ability, explored relationships among multiple intelligences, between music intelligence and mathematical intelligence, between musical intelligence and temporal–spatial reasoning, and among the scores of music theory, mathematical intelligence, and temporal–spatial reasoning. The study consisted of 83 elementary school students from 3rd and 5th grades. The results of the study revealed that there is a significant moderate positive correlation among the eight intelligences; musical intelligence has a moderate correlation with mathematical intelligence, but that may be due to temporal–spatial reasoning ability; among music theory, mathematical intelligence, and temporal–spatial reasoning, there are significantly low correlations, and the relationship between the scores of music theory and mathematical intelligence may be due to temporal–spatial reasoning abilities. Belonging to the music class at school can best predict achievement in music theory (Junchun, 2015).

Abdi, Laei, and Ahmadyan (2013) investigated the effects of MI theory as teaching strategy on the students' academic achievement in science courses. The researchers worked with 40 students from 5th-grade elementary schools. In the study, data from an experimental group where MI teaching strategy used is compared with data from a control group that was traditionally instructed. At the end of eight weeks, the results showed that students who were instructed through a teaching strategy based on MI achieved higher scores than the ones which were instructed through the traditional approach. This situation provides important outcomes in support of the claim that MI-based teaching strategy is an effective way to increase students' achievements in science courses.

Goodnough (2001) presents a case study's results based on the data collected from a teacher using action research. In the study, the MI theory was used to make decisions on the learning outcomes that are focused on the space and astronomy unit. The results showed that by using the MI theory in teaching and learning environment, the teacher could provide better student-centered opportunities to students according to their learning needs. The author also highlighted that MI theory can provide unique science activities to make science accessible to students and to assist them in achieving high levels of scientific literacy.

In the study conducted by Ozdermir, Güneysu, and Tekkaya (2006), MI-based teaching and traditional teaching method were compared. In the study, 4th-grade students were selected as the target group and "diversity of living things" was used as a science topic. The experimental group was instructed through multiple intelligence strategies, while the control group employed traditional methods. As data collection tools, they used Diversity of Living Things Concepts Test (DLTCT) and the Teele Inventory of Multiple Intelligences (TIMI). The results showed that MI-based instruction produced significantly greater achievement in the understanding of diversity of living things concepts and on students' retention of knowledge.

Suprapto, Liu and Ku (2017) analyzed some empirical studies in multiple intelligence, the interpretive perspective, MI in critical view, and the own-personal view about MI theory. The authors depict the lesson from implementation of the theory in school (Taiwan) in terms of compliance with the criteria of intelligence by Howard Gardner and how MI theory in science classrooms could be implemented. The study stated that the implementation of multiple intelligence in the science classroom integrates the existence of science, technology and/or engineering, mathematics (STEM), and arts.

The studies mentioned above show that MI theory provides very important benefits in science education. However, in order to use MI theory effectively and efficiently in science learning and teaching environments, the recommendations given at the end of this chapter require careful consideration. The important aspects that a science teacher should apply in the classroom environment for a better implementation of MI theory are given below.

The Classroom Implementation of Multiple Intelligences Theory

Although MI theory is a psychology-based theory, it makes greatest contribution to education by proposing that educational environments need to rearrange taking into consideration students' needs and capabilities. The MI theory does not provide any strong or the best teaching methods, but it expects that teachers should expand their repertoire of teaching techniques, tools, and strategies, and develop their teaching skills. One of the most important goals in educational research is to identify factors that provide long-lasting and meaningful learning for students. Once these factors are identified, it is expected that higher levels of learning will be achieved by making necessary changes in classroom settings. In many researches, it has been reported that implementing MI theory in the classrooms have significantly increased students' motivation, academic achievement, attitudes, and support meaningful learning.

Although MI theory is very crucial to improve students' knowledge and skills, there are some challenges implementing MI theory effectively in the classroom. One of the biggest challenges for teachers having great number of students in the classroom is knowing and implementing the correct educational strategy to provide effective instruction to all students with diverse intelligences. Enhancing students learning in an equally effective manner to reach the desired learning outcome is great challenge for teachers all over the world. The teaching and learning environments and related activities for the students with diverse intelligence may be easily organized if the teachers have enough knowledge and experiences on the MI theory. The teachers should take a step to this kind of teaching and learning environments knowing their students' MI characteristics and their preferred learning styles. With this important information about their students, the teachers can decide how to use effective instruction to reach defined goals in the curriculum. Using the advantages of the MI theory, the teachers can get improved learning outcome by their students. Additional advantages of MI theory for the classroom implementation are as follows:

- MI theory provides different potential opportunities for teachers to use it in their classroom environments.
- The teachers can use students' preferred learning style to increase their motivation for science courses.
- More fun and exciting teaching and learning environments can be created using effective educational strategies based on MI theory.

- Students' negative attitudes toward science courses can be changed to positive using their preferred learning styles connected with their preferred intelligence characteristics.
- Students' hidden powers can be discovered when they are asked to perform tasks in the classroom based on multiple intelligences theory.
- The theory and its implementations can be used in all educational sectors from kindergarten to graduate school.

MI theory does allow teachers to use and implement the most suitable intelligence way to plan and organize the classroom settings. It does not mean that teachers need to teach or learn in all nine intelligence ways. It is just a decision-making process in which the most effective teaching strategy with connected learning tools according to the learning environment should be used.

Table 27.1 shows how MI theory can be used for the teaching of specific science concepts.

Suggestions for Science Teaching

The success of MI theory depends on some important factors. One of these is the readiness of teachers. It should not be expected that teachers will be able to do all above-mentioned teaching methods for intelligence within the scope of MI theory. They need to have knowledge and skills of MI theory and classroom applications. It is not anticipated that teachers who do not have enough experience can do effective MI teaching practice. For this reason, it should be emphasized that in their undergraduate and in-service training, teachers should develop knowledge and skills for the design of effective teaching and learning environments where MI theory is in use. Depending on the adjustment process of the students in the class, it is expected that the teacher will apply different teaching methods for the appropriate intelligence type in the class.

The main purpose of MI theory is to provide opportunities for students to learn in environments in which they prefer to learn. For this reason, provision of learning environments that support different intelligences is recommended depending on the predefined type of students' intelligence. Particular attention should be given to recognizing opportunities for students to discover their own potentials and weaknesses. The students, who feel comfortable, take care to reveal the hidden power that exists in them. This situation allows the students to feel valued and develop greater freedom in their learning choices. They fulfill their duties and responsibilities more willingly. Therefore, it is very important to carry out activities to uncover the intellectual capacities that the students have during the learning process and to observe them in the process.

In the implementation of MI theory, learning and teaching environments where activities are to be carried out should be carefully selected. This could be a regular classroom environment, a nature walk, augmented reality presentations to class, or

Table 27.1 Using multiple intelligences theory in science teaching

The type of intelligence	What	How (example)
Verbal–linguistic	Words, sentences	Students are asked to read related texts from important sources and present the topic using language effectively and persuasively
Logical–mathematical	Mathematical equations and formula	Students are asked to work with the numbers and formulae to come up with the solutions
Visual–spatial	Pictures, 3D structures	Students are asked to work with 3D structures, models using augmented reality environments
Musical–rhythmic	Music, poems	Students are asked to write a musical demonstration for a science concept and then perform it with a musical instrument in the classroom
Intrapersonal	Self-reflection	Students are asked to perform an activity of thinking about their own feelings and behavior connected with science phenomena
Bodily–kinesthetic	Physical activities or games	A physical activity/game is designed for students to learn a science concept
Interpersonal	Social interactions	Students are asked to work with others to identify, define, and solve problems, which includes making decisions together
Naturalistic	Activities in the natural environment (forest, zoo, etc.)	Students are asked to observe a science concept in the natural environment and give real examples
Existential	Inquiring	Students are asked to inquire about the beginning of the universe, purpose of life, etc.

collaborative work environments where they can learn through social interaction. In these learning environments, students feel comfortable and are able to show their individual intellectual strengths.

Criticisms of MI Theory and Responses

The multiple intelligences theory has been criticized in several respects:

- Some criticized Gardner arguing that the intelligences work independently from one another, but not giving a clear explanation when asked about an activity that uses several intelligences at once. In response to this criticism, Gardner and Walters (1993) suggested that there might be a communication mechanism that connects the intelligences. However, this was in conflict with Gardner's first assumption of independent intelligences.
- There is a view that MI theory may not be easily implemented in practice. For instance, if the intelligences are independent of each other, teachers would need extra effort to address the multiple intelligences in various lessons throughout the day. Similarly, a class full of different intelligences might not benefit a single method of teaching. Thus, some teachers may be discouraged from implementing the theory in their classrooms, especially with large class sizes and little time. Klein (1997) asserted that the pedagogy based on MI theory is problematic. The problem is mainly related to the assessment of individual intelligences. The MI theory recommends that the teachers focus on strengthening students' weak intelligences while providing support for the strong intelligences. However, since there is no reliable way to identify the existence of individual intelligences, teachers would not be able to determine the progress of students.
- Morgan (1996) criticized the concept of multiple intelligences as being the old cognitive styles. Many psychologists are opposed to the use of the term *intelligences* with similar arguments. According to these psychologists, the right term is *talent* since it is shown through the tasks that people use talents to solve problems. However, Gardner (1993) refuted this conception claiming that not all intellectual capacities are talents. He asserted that it is just another form of old belief about being smarted measured through the proficiency in language and logic. Gardner argued that using kinesthetic or naturalistic intelligences is not just talents. The brain scan images in neuroscience research support Gardner's argument. It was shown that subjects performing physical or art activities show that these activities impact on neural systems in different ways as language and mathematical skills do (Viadero, 2003).

The MI theory is still a popular approach in education despite the above criticisms. Researchers still find support in favor of educational settings that use MI.

It is clear from all pedagogical studies that we still have limited knowledge of how humans learn. We have not fully explored yet the process of learning in different individuals. The concept of learning has dynamic and unique characteristic that is in keeping with its complexity and can be changed according to each individual. MI theory emerges as one of the best theories in solving problems that human beings encounter in the learning process. It is clear that MI theory, thanks to the ability to renew and update itself, will be able to shed light on the understanding of the complex learning and mental processes of humans in the contemporary world.

Summary

- This chapter introduces Howard Gardner's multiple intelligences theory and its implementation in science teaching.
- Multiple intelligences theory was put forward by Howard Gardner based on the postulate that humans have various intelligences. Gardner identified nine different intelligence areas in his theory: "musical–rhythmic", "visual–spatial", "verbal–linguistic", "logical–mathematical", "bodily–kinesthetic", "interpersonal", "intrapersonal", "naturalistic", and "existential intelligence.
- The theory has shown how relevant activities can reveal hidden potentials of learners.
- The chapter also presents how MI theory may be implemented in science teaching and learning
- Advantages and criticisms of MI theory discussed have been highlighted.

Recommended Resources

Gardner, H. (1983). *Frames of mind: The theory of multiple intelligences*. New York: Basic Books.

Gardner, H. (1999a). *Intelligence reframed: Multiple intelligences for the 21st century*. New York: Basic Books.

Gardner, H. (1999b). *The disciplined mind: What all students should understand*. New York: Simon & Schuster.

Gardner, H. (2004). Frequently asked questions—Multiple intelligences and related educational topics. Retrieved March 9, 2018, from http://multipleintelligencesoasis.org/wp-content/uploads/2013/06/faq.pdf.

Gardner, H. (2006). *Multiple intelligences: New Horizon*. New York: Basic Books.

Gardner, H. (2011). *The unschooled mind: How children think and how schools should teach*. UK: Hachette.

References

Abdi, A., Laei, S., & Ahmadyan, H. (2013). The effect of teaching strategy based on multiple intelligences on students' academic achievement in science course. *Universal Journal of Educational Research, 1*(4), 281–284.

Blumenfield-Jones, D. (2009). Bodily-kinesthetic intelligence and dance education: Critique, revision, and potentials for the democratic idea. *Journal of Aesthetic Education, 43*(1), 59–76.

Campbell, L., & Campbell, B. (1999). *Multiple intelligences and student achievement: Success stories from six schools*. Alexandria, VA: Association for Supervision and Curriculum Development. Retrieved from https://goo.gl/Y7Cz3w.

Crowther, G. J., Williamson, J. L., Buckland, H. T., & Cunningham, S. L. (2013). Making material more memorable with music. *American Biology Teacher, 75,* 713–714. https://doi.org/10.1525/abt.2013.75.9.16.

Davis, K., Christodoulou, J. A., Seider, S., & Gardner, H. (2011). The theory of multiple intel-ligences. In R. J. Sternberg & S. B. Kaufman (Eds.), *Cambridge handbook of intelligence* (pp. 485–503). New York: Cambridge University Press.

Gardner, H. (1983). *Frames of mind: The theory of multiple intelligences.* New York: Basic Books.

Gardner, H. (1991). *The unschooled mind: How children think and how schools should teach.* New York, NY: Basic Books.

Gardner, H. (1993). *Multiple intelligences: The theory in practice.* New York, NY: Basic Books.

Gardner, H. (1999a). *Intelligence reframed: Multiple intelligences for the 21st century.* New York, NY: Basic Books.

Gardner, H. (1999b). *The disciplined mind: What all students should understand.* New York: Simon & Schuster.

Gardner, H. (2006). *Multiple intelligences: New horizon.* New York, NY: Basic Books.

Gardner, H. (2011). *The unschooled mind: How children think and how schools should teach.* United Kingdom: Hachette.

Gardner, H., & Walters, J. (1993). Questions and answers about multiple intelligences theory. *Multiple intelligences: The theory in practice* (pp. 35–48). New York: Basic Books.

Goodnough, K. (2001). Multiple intelligences theory: A framework for personalizing science curricula. *School Science and Mathematics., 101*(4), 180–193.

Hoerr, R. T. (2000). Becoming a multiple intelligences school. Retrieved from http://www.ascd.org/publications/books/100006.aspx.

Hom, E. (2014). What is STEM education? Retrieved from https://www.livescience.com/43296-what-is-stem-education.html.

Junchun, W. (2015). To explore the relationship of temporal spatial reasoning between music and mathematics by an inventory based on the multiple intelligence theory. *Education and Psychological Research, 38*(3). https://doi.org/10.3966/102498852015093803002.

Klein, P. (1997). Multiplying the problems of intelligence by eight: A critique of Gardner's theory. *Canadian Journal of Education, 22,* 377–394.

Lai, H. Y., & Yap, S. L. (2016). Application of multiple intelligence theory in the assessment for learning. In S. Tang & L. Logonnathan (Eds.), *Assessment for learning within and beyond the classroom* (pp. 427–436). Singapore: Springer.

Martin, M. R., Bishop, J., Ciotto, C., & Gagnon, A. (2014). Teaching the whole child: Using the multiple intelligence theory and interdisciplinary teaching in physical education. In *The chronicle of kinesiology in higher education* (*Special Edition*, pp. 25–29).

Martín, J. S., Gragera, G. J. A., Dávila-Acedo, M., & Mellado, V. (2017). What do K-12 students feel when dealing with technology and engineering issues? Gardner's multiple intelligence theory implications in technology lessons for motivating engineering vocations at Spanish secondary school. *European Journal of Engineering Education, 42*(6), 1330–1343.

Martin, M., & McKenzie, M. (2013). Sport education and multiple intelligences: A path to student success. *Strategies, 26*(4), 31–34.

Morgan, H. (1996). An analysis of Gardner's theory of multiple intelligence. *Roeper Review, 18,* 263–269.

Ozdermir, P., Güneysu, S., & Tekkaya, C. (2006). Enhancing learning through multiple intelligences. *Journal of Biological Education, 40*(2), 74–78.

Richards, D. (2016). The integration of the multiple intelligence theory into the early childhood curriculum. *American Journal of Educational Research, 4*(15), 1096–1099. Retrieved from http://pubs.sciepub.com/education/4/15/7.

Schulkind, M. D. (2009). Is memory for music special? *Annals of the New York Academy of Sciences, 1169,* 216–224. https://doi.org/10.1111/j.1749-6632.2009.04546.x.

Schwert, A. (2004). *Using the theory of multiple intelligences to enhance science education.* Unpublished Master's Thesis. The University of Toledo, USA. Retrieved from http://utdr.utoledo.edu/cgi/viewcontent.cgi?article=1487&context=graduate-projects.

Suprapto, N., Liu, W. Y., & Ku, C. K. (2017). The implementation of multiple intelligence in (science) classroom: From empirical into critical. *Pedagogika, 126*(2), 214–227.

Viadero, D. (2003). Staying power. *Education Week, 22*(39), 24.

Dr. Bulent Cavas completed graduate studies in the field of science education at Dokuz Eylul University, Faculty of Education, and Science Teacher Training Programme in 1998. He made his Post-Doc in Middle East Technical University in Ankara, Turkey. He has attended many international, European, and National Projects as a researcher or principal investigator. He has over 150 national and international publications and written 10 books on science and science education. He has attended and organized many international symposia, congresses, and workshops in different countries. Currently, his research interests are Responsible Research and Innovation, Open Schooling, Inquiry-Based Science Education, and Virtual Reality in Science Education. He is the current President of International Council of Associations for Science Education (ICASE—www.icaseonline.net). Currently, he is working as Director of Distance Education Application and Research Center and Professor of Science Education at Dokuz Eylul University (www.deu.edu.tr) in Izmir, Turkey.

Dr. Pinar Cavas was born in Izmir, Turkey, in 1976. She received the B.S. degree in Physics in 1998, and B.A. and Ph.D. degrees in primary education in 2005 and 2009, respectively. She is working as an Associate Professor at Faculty of Education, Ege University (www.ege.edu.tr), in Izmir, Turkey. She joined many national and international projects related to science, math, and technology. She is also qualified in elementary teacher training. She has many national and international publications related to science education and elementary teacher training. Her research fields are Scientific Literacy, Competences of Elementary Teachers, Inquiry-Based Science Education, and Motivation to Learn Science. She is married and has two children.

Chapter 28
Systems Thinking—Ludwig Von Bertalanffy, Peter Senge, and Donella Meadows

Bao Hui Zhang and Salah A. M. Ahmed

Introduction

This chapter was created for current and future science teachers who intend to implement systems thinking in science education. We spent tremendous efforts to present the theoretical and historical background of systems thinking, a new paradigm in science education. Hope this helps readers to understand well the practical aspects of using systems thinking in science teaching and learning contexts.

Background

It is merely an axiom that the world's systems have various sorts because of their degree of complexity. One purpose of science is to provide clear descriptions, explanations, and/or predictions of behaviors of such complex phenomena in both natural sciences and social sciences. Unfortunately, for science, only some world's systems are static and simple ones that have foreseeable, reproducible, and reversible behaviors. The rest are with dynamic and ordered complexity. The classical scientific approach known as the analytic approach is based on reductionism for studying any science phenomenon. Reductionism sees systems as static, closed, mechanical, linear, and deterministic. However, that reflects only a small picture of the world because most of the systems include ordered complexity. Real world's systems have a fluid and flow equilibrium, and they are open systems that have unforeseeable, irreproducible, and irreversible behaviors. Reductionism cannot describe how such

B. H. Zhang · S. A. M. Ahmed (✉)
School of Education, Shaanxi Normal University, Xi'an 710062, China
e-mail: salah784@gmail.com

B. H. Zhang
e-mail: baohui.zhang@snnu.edu.cn

© Springer Nature Switzerland AG 2020
B. Akpan and T. Kennedy (eds.), *Science Education in Theory and Practice*,
Springer Texts in Education, https://doi.org/10.1007/978-3-030-43620-9_28

419

complex systems work. Thus, an alternative view of the world uses a holistic approach that views a system as a whole and is more than the sum of its parts. This approach focuses more on the interactions and relations between the system's components. We refer to it as systems thinking that is a universal mode of thinking; this form of thinking is based on a holistic view. Systems thinking is not limited to any domain of knowledge; it integrates both analytic and synthetic approaches. To understand systems thinking, we first define "system". This term "system" has been used for multi-purposes in different areas. For instance, people frequently mention communication system, education system, solar system, social system, economic system, transport system, or ecological system, and the like. The term "system" came from a Greek word σύστημα meaning "(a) whole compounded of several parts or members" (Rose, 2012, p. 9). The first use of this term was in the eighteenth century by German philosopher Immanuel Kant in the book *Critique of Pure Reason* (Reynolds & Holwell, 2010). According to Merriam-Webster's online dictionary, a system is "a regularly interacting or interdependent group of items [elements] forming a unified whole (n.d.)". Bertalanffy defined a system "as a complex of interacting elements" (Von Bertalanffy, 1969, pp. 55–56). That means the elements are standing in interrelations. Jackson (2003) defined a system as "a complex whole where the functioning of which depends on its parts and the interactions between those parts" (Jackson, 2003, p. 3). A system can exist in any format. For example, hard systems include physical systems like river systems; soft systems include more malleable systems like biological, sociological, and economic systems.

A system usually includes three essential components: elements, interconnections, and functions or system goals (Meadows & Wright, 2009). Nonetheless, systems are perceivable objects. In some cases, we can only recognize some particular components; other components are hard to define. A system's boundary is such an example. Different system boundary conditions may significantly change the system behaviors. A system's boundaries can be defined according to our view of the system itself. For example, devices (e.g., iPad, Laptop, and the like) that you are using to read an e-book (assuming you detached it from other systems, the Internet, and power) can be considered as a system. We can outline its elements, interconnections, and system functions that enable this machine to process and present the data. Where are the boundaries of your device? Using the mechanistic system, we can sufficiently define its boundary as the device itself; the body represents the physical boundaries. In contrast, if you connect it to an electricity source, now ask yourself where is the boundary of this machine? What about these data? Where do these data come from? Consider these data as virtual elements that come from an external source; there are also other systems like the Internet through the input terminals. If someone used this device, we should also consider the human–computer interaction (HCI). The system then includes more than one system. The boundaries are essential components in order to draw important interactions. However, deciding a system's boundary is a significant challenge. It is difficult to imagine these innumerable series boundaries, because most of these boundaries would be worthless for many reasons. Therefore, we only try to define the best boundaries by making a decision such as extending our investigation to the individual parts. A system is usually presented in the form of models which represents a real thing or science phenomenon. A model might have different parts/variables that are interrelated; one variable can affect other variables and might also be affected. A model as a whole highlights certain aspects of a system. Modeling is the process of designing, testing, and revising/abandoning models (Zhang, Liu, & Krajcik, 2006). Meadows and Wright (2009) emphasized that one of the most troubling functions of modeling is defining the systems boundaries, especially in behavioral and social systems, as they stated:

> ...Systems rarely have real boundaries. Everything, as they say, is connected to everything else, and not neatly. There is no clearly determinable boundary between the sea and the land, between sociology and anthropology, between an automobile's exhaust and your nose. There are only boundaries of words, thought, perception, and social agreement-artificial, mental-model boundaries (p. 95).

If systems are not perceptible or sensible objects, how can we know the most important parts of a system? Meadows explained that by illustrating the effects of changing system components, the largest impacts come from changing systems' functions. For instance, as Meadows illuminated, if we consider a football team as a system with parts such as players, ball, field, coach, and the like, one of its interconnections is the rules of the football game. The system's goal is to win football games. If change occurred in the low level of system elements, such as changing some or even all the players, we obtain less effect on the system; we still call it a football team. Similar things happen if we look at an automobile as a system. Replacing some parts does not change the whole; it is still a car. When we move up, a change occurs at the interconnections level, and we can recognize some effects. For example, if we used the same elements like team players in this case, but used rules of basketball instead of those of football, we have a new game. However, the big impacts occur in changing a system's functions or goals. For example, if we changed the purpose of the football team from winning games to losing the games, other components, such as the elements and the interconnections, remain the same, and the results might be reversed. Therefore, it is obvious that system functions are the important component of a system; a small change in system function can cause a significant change in the whole system (Meadows & Wright, 2009).

Systems Thinking Theories

Historical Background

The beginning of the last century witnessed scientific revolutions that were not limited to modern theories in physics, such as the theory of quantum mechanics and relativity. Revolutions also extended to the science of biology. As a result of these scientific revolutions, scientists have changed their views of the world. The new contribution to biology came from Ludwig von Bertalanffy. In his General System Theory (GST), he was seeking the unity of science. He looked at the world as whole. However, the wholeness views were not new. These views have historical roots from spiritual traditions of Hinduism, Buddhism, Taoism, sufi-Islam, ancient Greek philosophy (Reynolds & Holwell, 2010) to the modern systems thinkers, such as Nicholas of Cusa, Gottfried Wilhelm Leibniz, and Johann Wolfgang von Goethe (Drack, Apfalter, & Pouvreau, 2007). These individuals, along with others who came later, influenced GST.

We start this section by repeating one famous quotation that describes systems thinking. Churchman said "A systems approach begins when first you see the world through the eyes of another" (Churchman 1968, p. 231, as cited in Reynolds & Holwell, 2010). The term of systems thinking is still new. It was coined by Barry Richmond in 1986. After much thought, Richmond came up with the term "systems thinking" that is nested in the old term "structural thinking", when he was preparing his first user's guide for his software STELLA (Structural Thinking, Experiential Learning Laboratory with Animation) (Richmond, 1994). In this instance, "systems" in plural seems to indicate the nested nature of thinking. Systems thinking is a holistic paradigm that assists in understanding complex phenomena. Complex problems tend to be linked to different problems and seldom exist individually out of the same context. Peter Senge defined systems thinking as "a discipline for seeing wholes, as a framework for seeing interrelationships rather than things, for seeing patterns of change rather than static" (Senge, 2006, p. 68). Systems thinking can help link pieces together in order to see the big picture that might lead to understanding the situation, despite its complexity. Barry Richmond considered systems thinking "as the art and science of making reliable inferences about behaviors by developing an increasingly deep understanding of underlying structure" (Richmond, 1994).

Ludwig von Bertalanffy and the System Theory

Karl Ludwig von Bertalanffy was born on September 19, 1901 to a Catholic family. The roots of his family date back to the nobility of Hungary during the sixteenth century. General Systems Theory (GST) was formulated in the 1920s when Bertalanffy attempted to explain the functioning of biological living systems. Bertalanffy grounded GST based on the wholeness or Gestalt. The wholeness in GST referred not only to the sum of its parts, but also extended to the parts' relations (Drack et al., 2007).

Bertalanffy (1969) first recognized living organisms as open systems. He called a system "closed" if no materials entered or left. The system is "open" if there were "import and export of material" (p. 121). Having empirical knowledge in related disciplines like biology, or physics, Bertalanffy built his theoretical model of open systems with "steady states", "dynamic equilibrium", "equifinality", and the like. He outlined the set of mathematical equations that articulated the relationships. A system can maintain itself and constantly exchanging matter and [energy] with a surrounded environment (Von Bertalanffy, 1969). This new thought was a revolution because it sought the unification of science. This interdisciplinary perspective produced a new kind of scientific knowledge. It shifted the classical view from steady systems to dynamic systems, from isolation to openness, from traditional linear thinking that focused on the parts, to see the whole. In his book (Von Bertalanffy, 1969), Bertalanffy described the aims for his GST:

(1) There is a general tendency toward integration in the various sciences, natural and social.
(2) Such integration seems to be centered in a general theory of systems.
(3) Such theory may be an important means for aiming at exact theory in the nonphysical fields of science.
(4) Developing unifying principles running "vertically" through the universe of the individual sciences, this theory brings us nearer the goal of the unity of science.
(5) This can lead to a much-needed integration in scientific education (von Bertalanffy, 1969, p. 38).

We can summarize the core ideas of system theory as follows:

- System theory seeks the laws of unity among diverse phenomena; it aims to find the common aspects instead of focusing on a single system. A system's entities represent the whole of natural, behavioral, or social phenomena, but the whole is more than the sum of the entities; it included the interrelations among them.
- According to Bertalanffy, the biological, behavioral, and social systems are essentially open systems that can be divided into small closed/open systems with respect to the connection with the surrounding environment.
- Any open system with its environment constantly exchange substance, energy, or even information as the input and output through a living communication channel; the channel can decrease noise to a higher degree than another lifeless communication channel (Von Bertalanffy, 1969, p. 98).
- In the open systems model, the system is dynamic over time. Along with system's life cycle, it is constantly involved in building up and breaking down as self-renewing processes (Von Bertalanffy, 1969, p. 39); such self-maintenance process drives the system toward higher heterogeneity and organization (Von Bertalanffy, 1969, p. 143).
- The boundary's function is to outline the system from its surrounding environment and any other subsystems of the entire system as a whole.
- The feedback plays an essential role in leading the system actions, and behaviors toward its goals.

Peter Senge and the Theory of Systems Thinking

Senge was born in 1947. At Stanford University, he studied both engineering and philosophy. In 1970, he received his first degree from Stanford University in Aerospace engineering. Two years later at Massachusetts Institute of Technology (MIT), Senge finished his master's degree in social systems modeling. Then he continued working with Jay Forrester as a researcher at MIT until he earned his Ph.D. degree in management in 1978. His dissertation focused on "a comparison between aspects of economic modeling through the System Dynamics National Model". After graduation, he started his career as a lecturer at MIT Sloan School of Management (Ramage & Shipp, 2009).

In the 1950s, a massive movement in systems theory occurred when Peter Michael Senge, one of the systems thinking leaders, illustrated systems thinking language using system dynamics that was founded by Jay W. Forrester.

Senge named systems thinking as "The Fifth Discipline" in his book. He clearly described how organizations could learn, and how systems thinking could accelerate this learning. Of course, systems thinking in this learning process was not alone; there were four other aspects: "personal mastery, mental models, shared vision, and team learning" (Ramage & Shipp, 2009, p. 121). Systems thinking was integrated into each of them. They synergistically worked together. For example, systems thinking and mental models both were necessary for each other; one helps us to discover and test covert assumptions and the other one guides us to reorganize those assumptions to unearth causes that shaped complex problems (Senge, 2006). Systems thinking is necessary not only to recognize the salient variables but also to discover time delays and critical feedback relations. Without systems thinking, "most of our mental models are systematically flawed" (Senge, 2006, p. 203).

In this subsection, we have tried to enumerate rather than illuminate some Senge contributions. One contribution was elucidating the language of systems thinking as particular rules that control systems diagrams, such as systems archetype and feedback structures. Senge's systems archetypes are used to observe, explain, and predict the complex events. All systems archetypes in the Fifth Discipline or in other literature "Systems Archetypes as Structural Pattern Templates" seek to shift one's mental model (mindset) to systematic thinking. Another important contribution by Senge is explaining systems thinking laws (refer to his book *The Fifth Discipline*, Chap. 4, or Chap. 12-Part 2 in Ramage & Shipp, 2009).

Donella Meadows and the Theory of Systems Thinking

Donella H. Meadows was born on March 13, 1941 in Illinois, USA. Dr. Donella H. Meadows is well known as a systems analyst, an organic gardener, an eco-village developer, and a syndicated journalist. She was a professor of Environmental Studies at Dartmouth College until her death in 2001. She started her career as a scientist. She received her B.A. in chemistry from Carleton College in 1963, and she received her Ph.D. in Biophysics in 1968 from Harvard University (Ramage & Shipp, 2009). She then joined an international system dynamic team with Jay Forrester at MIT. Donella employed the tools of system dynamics, like computer modeling, to deliberate global problems such as the relationship between population, economic growth, and the earth resources. She used the concepts of stocks/flows and feedback loops to construct a detailed analysis of leverage points.

Donella recognized herself as a systems thinker, working with dynamic systems tools. Both Donella and Senge agreed with Jay Forrester that systems thinking did not necessarily give you the best viewpoint. It could just give you a unique view of the phenomena like other thinking paradigms. It shows some events and patterns that reflect the behaviors and complex relationships behind this order. Donella said "like any viewpoint, like the top of any hill you climb, it lets you see some things you would never have noticed from any other place, and it blocks the view of other things" (Meadows, Randers & Behrens, 1972, p. 2).

Systems Thinking and Its Relation to Science Teaching

In this section, we are going to diagnose the current situation of science education, and then provide a simple guide for science teachers and practitioners for implementing systems thinking in science education. We will provide some real examples supported by scientific research. Finally, we discuss the advantages and challenges of using this approach in science education.

Why Is Systems Thinking Important in Science Education?

One of the goals of science education is preparing our students for future challenges by enhancing their capacity for solving problems. In the late twentieth century, science education experts realized that one of the most important issues facing educational systems was using reductionism and mechanistic thinking. The world is made of systems with nonlinearity; decentered control is chronic in world complex systems. Traditional science curricula deliberately simplify and reduce complexity of nature that is strongly interconnected (Forrester, 1993).

Current science curricula deal with many topics superficially, using linear and analytic methods to simplify complex systems. Students study fragments of knowledge about any natural problem in different science subjects at different grades. They may not help students form broad pictures of any phenomena. Most science curricula are not able to develop a systems foundation for problem-solving. Traditional science curricula tend to simplify such a complex phenomenon. Peter Senge clearly supported this dilemma in science education. He states:

> From a very early age, we are taught to break apart problems, to fragment the world. This apparently makes complex tasks and subjects more manageable, but we pay a hidden, enormous price. We can no longer see the consequences of our actions; we lose our intrinsic sense of connection to a larger whole (Senge, 2006, p. 3).

Natural phenomena and global issues may relate to different subjects (physics, chemistry, biology, and the like) at the same time. The conventional approach usually separates science into separate domains (e.g., physics, biology, chemistry). Bertalanffy had also criticized traditional education. He commented that the demand for science education was training science learners to become "scientific generalists" in the field (Von Bertalanffy, 1969). Jay Forrester addressed this issue in many situations. Although the behaviors and events of various natural phenomena are controlled by the same natural laws, current science curricula ignore this fact and offer science in a fragmented form like physics, biology, and chemistry, which appear to be innately separated from one another (Forrester, 1993). According to Senge (2006), research has shown that many young children acquired thinking skills very quickly. This indicates that students have innate systems thinking skills. However, instead of developing these skills, traditional education suppresses them by using mechanical or linear thinking.

Therefore, science educators already made some efforts to shift from mechanistic to holistic or systems thinking. Systems thinking embedded in the new science standards such as the Next Generation Science Standards (NGSS). In this new learning approach, science instructors should shift their roles to be knowledge facilitators rather than being knowledge transmitters. Students should be involved in cooperative and competitive group work, and use non-routine problem-solving and non-linear thinking. More fortunately, the new reform of science education that made under an umbrella of STEM (science, technology, engineering, and mathematics) education has made move to bring systems thinking into K-16 curricula in the U.S. and elsewhere (e.g., Duschl & Bismack, 2016). Integrating systems thinking in STEM education can help students to develop a meaningful scientific literacy. To reach that goal, learners should collaboratively inquire and try to solve complex problems. They might be able to apply systems thinking skills to recognize the interdependence of natural and social phenomena, uncover patterns, and build concepts of systems that help them to obtain better understandings of complex world problems (National Research Council, 2012). Providentially, with the rapid growth in the capabilities of computers and mobile devices, integrating simulation and modeling to study complex systems has become more available. Over the decade, there have been many efforts to integrate systems thinking tools in science learning (Jacobson, Kim, Pathak, & Zhang, 2013; Zhang, Liu, & Krajcik, 2006). Still, there are challenges to teaching science in this new way and requires paradigm change and likely overhaul of the current school and university curricula.

Systems Thinking in Science Curricular Standards

Systems thinking is essential to any learning organization. Pioneer science educators realized the importance of integrating systems thinking concepts, skills, and tools in science standards. The first attempt began in the 1960s. The Science Curriculum Improvement Study (SCIS) developed some science curriculum units in elementary school level that included some of the systems thinking concepts such as system, sub-systems, interactions, and variables (Chen & Stroup, 1993). There are also current attempts to incorporate systems thinking into the science and STEM curricula which developed based on the new standards. We will limit ourselves to examples of some initiatives that include systems thinking concepts from the NGSS for K-12 science education in U.S. We make such a decision because such a move in the U.S. has been influential. The three-dimensional framework of the NGSS made a major revision by integrating systems thinking practice to include "Science and Engineering Practices (SEPs)", "Crosscutting Concepts" (Ccs), and "Disciplinary Core Ideas" (DCI). These recommendations aim to achieve systems thinking practice of science. The domain of Science and Engineering Practices (SEPs) emphasizes the practice that includes creating and testing models. SEP also highlighted the key sets of engineering practice that are designing systems (National Research Council, 2012, p. 30). More explicitly, the framework of crosscutting concepts (CCCs) includes "patterns; cause and effect; scale, proportion, and quantity; systems and system models; energy and matter; structure and function; stability and change" (NGSS Lead States, 2013, p. 79). Furthermore, systems thinking practices distributed across the third dimension that described the "Disciplinary Core Ideas" (DCI) that include four major domains: engineering, technology, and applications of science; the life sciences; the physical sciences; the earth; and space sciences (National Research Council, 2012, p. 31).

The framework begins in kindergarten. NGSS emphasized that students must know how to better identify issues, recognize patterns, and develop understanding of the natural phenomena around them. For example, in (K-ESS2-1) about the Earth's systems, students should think systematically and conduct some quantitative and qualitative observations of the local weather to "describe patterns over time". In crosscutting concepts, the world's systems are combined from parts that work together, patterns are observable, and they can be used as evidences to explain natural phenomena (NGSS Lead States, 2013). Similarly, in (K-ESS3-1) about Earth and human activity, students should use their prior knowledge to develop a model "to represent the relationship between the needs of different plants or animals (including humans) and the places they live" (NGSS Lead States, 2013, p. 8). They should also "use a model to represent relationships in the natural world" (p. 8).

Another example of life science is from middle school disciplinary core ideas (MS-LS1-3: From Molecules to Organisms: Structures and Processes); students should develop "basic understanding of the interaction of subsystems within a system and the normal functioning of those systems" (NGSS Lead States, 2013, p. 67). For example, they may recognize the basic roles of cells in body systems and understand how those systems work together to support the life functions of the organism (NGSS Lead States, 2013). In the same way, system thinking concepts clearly appeared in (MS-LS2-3.) as science learners are expected to understand the interdependences in ecosystems, matter's cycles, and energy exchange among living and non-living parts of an ecosystem. Students are also expected to be able to define the system's boundaries (NGSS Lead States, 2013, pp. 65, 70).

Similarly, high school students in Earth and human activity (HS-ESS3-3) should be able to use computational simulation to demonstrate the relations among factors that influence the controlling of natural factors that impact on the sustainability of human populations, and biodiversity. In (HS-ESS3-6), students should use a computer model to illuminate the relationships among earth systems like (hydrosphere, atmosphere, cryosphere, geosphere, and/or biosphere) and how the human activities impact those relationships (NGSS Lead States, 2013, p. 125). We applaud this effort as it will influence science education internationally. Similar efforts have also been demonstrated in some international comparative studies such as PISA (Program for International Student Assessment). The PISA framework "uses the term 'systems' instead of 'sciences' in the descriptors of content knowledge. The intention is to convey the idea that citizens have to understand concepts from the physical and life sciences, and earth and space sciences, and how they apply in contexts where the elements of knowledge are interdependent or interdisciplinary. Things viewed as subsystems at one scale may be viewed as whole systems at a smaller scale" (OECD, 2016, p. 27, https://www.academia.edu/12015821/PISA).

Developing and Assessing Systems Thinking Skills

In order to implement science standards and facilitate the systems thinking in science education, science instructors should use well-suited methods to assist science learners to obtain and develop the essential systems thinking skills. There are different concepts in systems thinking, for example, the ability to identify patterns, actions, and recognize circular cause–effect relations (Sweeney & Sterman, 2000). Based on their literature review, Assaraf, Dodick, & Tripto (2013, p. 36) classified systems thinking skills into eight hierarchical characteristics or abilities at three sequential levels as follows:

1. Identifying the components and processes of a system (level A).
2. Identifying simple relationships among a system's components (level B).
3. Identifying dynamic relationships within a system (level B).
4. Organizing systems' components, their processes, and their interactions, within a framework of relationships (level B).
5. Identifying matter and energy cycles within a system (level B).
6. Recognizing hidden dimensions of a system (i.e., understanding phenomena through patterns and interrelationships not readily seen) (level C).
7. Making generalizations about a system (level C).
8. Thinking temporally (i.e., employing retrospection and prediction) (level C).

The three levels are

Level A (analyzing, ability: 1),
Level B (synthesizing, abilities: 2, 3, 4, and 5), and
Level C (implementation, abilities: 6, 7, and 8).

The new trend in science and STEM education supports systems thinking skills. For instance, in the common investigation, learners start by identifying the system's parts and describing the parts' interactions. Then learners use their available data to develop a model of undertaken complex systems. Later, after testing and optimizing the model, they may apply systems thinking approach to evaluate and provide possible solutions for the related global challenges. This hierarchical order of the previous systems thinking skills facilitates the teaching and assessment of systems thinking skills. The degree of difficulty of systems thinking skills is ranked on a continuum from easiest to the most difficult. Lower level skills must be acquired first in order to master the highest level skills (Assaraf, Dodick, & Tripto, 2013; Rose, 2012, p. 19). Another effort that should not be ignored is a comprehensive set of systems thinking skills by Arnold & Wade (2017) that can be used either to guide the design of systems thinking materials/rigorous assessment rubric or to assess system thinking competencies. This contribution should increase the reliability and the accuracy of assessing systems thinking skills for different disciplines.

Tools for Systems Thinking

Systems thinking is considered an advanced complex cognitive skill in which science learners have to be involved in higher order cognitive processes (Hung, 2008). Systems thinking advocates suggest some tools that might help learners to develop systems thinking skills. Among the suggested tools are systems modeling and simulations that help science learners "to simulate the behavior of systems that are too complex for conventional mathematics, verbal descriptions, or graphical methods" (Forrester, 1993, p. 185). Monat, & Gannon (2015) listed some of these tools as follows:

1. Systems Archetype.
2. Behavior Over Time Graphs (BOT),
3. Causal Loops Diagrams with Feedback and Delays,
4. Systemigrams,
5. Stock and Flow Diagrams (including Main Chain Infrastructures),
6. System Dynamics/Computer Modeling,
7. Systemic Root Cause Analysis (RCA), and
8. Interpretive Structural Modeling (ISM) (pp. 21–24).

These tools support different tasks. For example, "behavior over time graphs (BOT) were used at the beginning to understand system behaviors" (Monat & Gannon, 2015, p. 21). Causal loops diagrams were used after BOT to outline the interrelationships among system's parts. Systemic diagrams were used "to translate a system problem (expressed as structured text) into a storyboard-type diagram describing the system's principal concepts, actors, events, patterns, and processes". Systemic Root Cause Analysis (RCA) "is a set of problem-solving methods that help to find a fault's first or root cause" (Monat & Gannon, 2015, pp. 21–24). Regarding modeling and simulation, as stated above, sophisticated computers are able to run and solve many complex simulations. In the past, science educators used different systems tools like

STELLA, Model-It, ThinkerTools, and some agent-based models (ABMs) environ-ments programmed by NetLogo, StarLogo. Presently, more user-friendly interfaces including web-based simulations such as the PhET Interactive simulation (website: https://phet.colorado.edu/), or modeling visualized interfaces, e.g., ViMAP (http://www.vimapk12.net), and Scratch (https://scratch.mit.edu/) provides a wide range of tools for science learners, without any prerequisites. Using these types support the development of systems thinking. That said, students still need additional basic scientific skills, such as basic mathematics to design and test their models. Students should be able to collect and interpret data, as well as draw and explain data graphs (Sweeney & Sterman, 2000). More advance technique is to conduct real-time model-ing using some new data collectors (sensors) linked to Arduino or any other hardware, then analysis and visualize the data with the modeling software, e.g., NetLogo. Some examples are presented in the following section.

Interestingly, some findings assert that after practicing systems thinking skills learners could develop some systems thinking habits that encompass multiple think-ing strategies. Sweeney and Waters Foundation labeled systems thinking habits. For more examples, refer to the suggested Resources/websites section.

Examples of Linking Theory and Practice

Many studies have emphasized the significance of importing systems thinking approach into education. Using systems thinking in science learning could help science learners connect their acquired knowledge. It could also increase student understanding of interdependence between systems. Systems thinking could enhance learners' awareness toward unifying science disciplines. Good implementation is to use some sensors to collect real-time data, produce a model, and deliver good expla-nations and predictions. The following photos (Fig. 28.1) present an example of integrating modeling in real-lab experiments. Through "bifocal modeling", students are able to study both physical real-lab and simulated virtual experiments (Blikstein, Fuhrmann, & Salehi, 2016, p. 513). In such advanced modeling, learners conduct an experiment and collect the data via terminal devices or sensors. Then they apply their scientific and mathematical skills and knowledge to design a virtual model to

Fig. 28.1 Examples of bifocal modeling: gas laws (*left*) & Newton's cradle (*right*) (replicated from Blikstein, Fuhrmann, & Salehi, 2016, p. 515)

compare and reconcile between the experiment data and the simulation outcomes of the same phenomenon (Blikstein, Fuhrmann, & Salehi, 2016).

Before ending this section, it is necessary to present some of the other efforts inspired by the work of Forrester and Meadows on system dynamics (Rose, 2012). For example, to implement NGSS in the real setting, NASA established The GLOBE Program, an international science and education program. GLOBE aims to provide students and the public worldwide with the opportunities to participate in data collection and the scientific process. The significance of this project is in linking worldwide science educators, researchers, students, scientists, workers, and citizens to participate in data collection and the scientific process, and develop student's scientific understanding of the Earth system and global environment. See the GLOBE website.

Similarly, the Waters Foundation project "Systems Thinking in Schools" facilitated schools in Arizona to integrate systems thinking into their education programs. For example, Borton Primary Magnet School implemented systems thinking as a teaching method. Pima County Schools achieved content standards and skills by utilizing systems thinking (Graefe, 2010). Orange Grove Middle School integrated system thinking in their science curriculum "since the fall of 1988 in a program called Directed Learning" (Rose 2012, p. 25).

Some European schools implemented systems thinking in elementary schools. For example, in Switzerland, the Pedagogical University of St. Gallen (PHSG) was involved in developing student systems thinking. *Systemdenken* is a German word meaning "systems thinking". A handbook was developed for elementary and middle school teachers to develop systems thinking competencies using action-oriented activities (see http://www.iue.ch/publikationen/systemdenken-foerdern). We have certainly seen some of the advantages of integrating systems thinking in science education. However, serious challenges remain.

Challenges of Integrating Systems Thinking in Science Education

The implementation of systems thinking in science education presents challenges to systems practitioners. These challenges are faced by both teachers and students. One of the biggest barriers is to find well-designed and tested instructional materials that foster students' systems thinking (Booth Sweeney & Sterman, 2000). Science teachers' lack of understanding of systems thinking is another barrier. Implementing systems thinking requires a high intellectual level and requires some technical skills. Unlikely, most science instructors need more training programs to be familiar with the systems thinking tools, especially the advance ones, e.g., Netlogo. Moreover, they should know "how to effectively facilitate students' systems thinking" (Hogan & Weathers, 2003). Science teachers need to shift their role from being the source of knowledge to learning facilitators. Yet, there is a need for empirical research concerning teacher professional development programs for systems thinking. Due to its inherent complexity and nonlinearity, systems thinking requires some higher order cognition abilities (Assaraf, Dodick, & Tripto, 2013). Students have difficulties in imperceptible causal relationships. Learners at the elementary levels seem to be lacking of understanding a system's hidden variables and their hidden relations. Even some new modeling environments like ViMAP, CTSiM, and Scratch are based on drag and drop add more coding scaffolds. Other systems structure scaffolding is also strongly needed in order to facilitate learning complex systems.

Summary

Systems thinking is a universal mode of thinking. From a holistic viewpoint, systems thinking has its significance in different disciplines. Systems thinking has also been incorporated into scientists' and engineers' work. Systems thinking has been the underpinning of various scientific breakthroughs and offers many powerful, interactive tools to aid students to systematize, analyze, synthesize, and visualize the data to understand complex systems. Thus, people can use it differently in their daily life to solve the problems and to understand how the world works.

To think systematically requires knowing about (a) system (e.g., the conceptual knowledge means understanding of the essential elements, cause-and-effect relationships among systems parts, discovering patterns, and interdependences between systems); (b) modeling systems behaviors in different ways (e.g., using systems thinking tools) (Meadows & Wright, 2009; Senge, 2006); (c) systems laws; and (d) learning disabilities (Senge, 2006). In addition, implementing systems thinking has some benefits in the K–12 education context, such as (1) helping science instructor to drive the learning process toward learner-centered approach; (2) increasing students' engagement; (3) providing learners more relevant experiences that lead to more effective learning; (4) improving problem-solving skills (Graefe, 2010); (5) changing teachers' and students' perspectives about the world; (6) facilitating learning by following basic system's rules; (7) increasing learners' collaboration and team working; sharing thoughts and solutions of the complex problem; and (8) facilitating and designing solutions, creating strategies, solving problems, while keeping the outcome/vision in mind at all times (Haines, 1998). We call for more attention, research, and exploration in implementing systems thinking in K-12 and higher education in order to develop students for the twenty-first century.

Recommended Resources

Books

1. Haines, S. G. (1998). *The manager's pocket guide to systems thinking & learning.* Amherst, MA: HRD Press.
2. Flood, R. L. (2002). *Rethinking the fifth discipline: Learning within the unknowable.* London: Routledge.
3. Jackson, M. C. (2003). *Systems thinking: Creative holism for managers.* Chichester: Wiley.
4. Meadows, D. L., Randers, J., & Behrens, W. W. (1972). *The limits to growth: A report for the club of Rome's project on the predicament of mankind.* New York: Universe Books.
5. Kim, D. H., & Lannon, C. (1997). *Applying systems archetypes.* Pegasus Communications Waltham.
6. Meadows, D. H., & Wright, D. (2009). *Thinking in systems: A primer.* London: Earthscan.
7. Ramage, M., & Shipp, K. (2009). *Systems thinkers.* London: Springer.
8. Senge, P. M. (2006). *The fifth discipline: The art and practice of the learning organization.* New York: Doubleday.

Journals

1. System Dynamics Review: Edited By: Yaman Barlas; Impact Factor: 1.448 (2018); Online ISSN: 1099-1727 (https://onlinelibrary.wiley.com/journal/10991727).
2. Systems Research and Behavioral Science: Edited By: M. C. Jackson OBE; Impact Factor: 1.052 (2018); Online ISSN: 1099-1743 (https://onlinelibrary.wiley.com/journal/10991743a).
3. Systemic Practice and Action Research, Impact Factor: 0.754 (2018); Available 1988–2018; ISSN: 1094-429X (Print) 1573–9295 (Online) (https://link.springer.com/journal/11213).

Websites

1. A repository of 800+ articles on various dimensions of Systems Thinking over the past 20+ years https://thesystemsthinker.com/.
2. The Bertalanffy Center for the Study of Systems Science (BCSSS) is an internationally Austrian independent research institute http://www.bcsss.org/.

3. The International Society for the Systems Sciences http://isss.org.
4. Donella Meadows Project http://donellameadows.org.
5. The Global Learning and Observations to Benefit the Environment (GLOBE) https://www.globe.gov/about/overview; GLOBE on Twitter https://twitter.com/globeprogram.
6. Linda Sweeney websites, which mixing complex systems theory, systems mapping, storytelling. http://www.lindaboothsweeney.net; Systems thinker habits: http://www.lindaboothsweeney.net/thinking/habits.
7. The Creative Learning Exchange (CLE): http://www.clexchange.org.
8. The Waters Foundation's Systems Thinking in education aims to enhance systems thinking practices for educators, students, and administrators as well. http://watersfoundation.org; Habits of a Systems Thinker https://waterscenterst.org/systems-thinking-tools-and-strategies/habits-of-a-systems-thinker/.
9. The Way of Systems is a good website for introducing systems thinking concepts, tools…etc. http://www.systems-thinking.org.
10. Collected videos for various dimensions of Systems Thinking https://www.youtube.com/user/systemswiki/videos.
11. Society for Organizational Learning (SoL) http://www.solonline.org/.
12. NetLogo is one of the best agent-based modeling environment (http://ccl.northwestern.edu/netlogo/). For more research articles see (http://ccl.northwestern.edu/netlogo/resources.shtml). For web-NetLogo: http://netlogoweb.org.
13. ViMAP is a new visual-programming language and modeling platform based on NetLogo (http://www.vimapk12.net/). For More research articles using ViMAP see the Publications section.

References

Arnold, R. D., & Wade, J. P. (2017). A complete set of systems thinking skills. *Insight, 20*(3), 9–17. https://doi.org/10.1002/inst.12159.
Assaraf, O. B.-Z., Dodick, J., & Tripto, J. (2013). High school students' understanding of the human body system. *Research in Science Education, 43*(1), 33–56. https://doi.org/10.1007/s11165-011-9245-2.
Blikstein, P., Fuhrmann, T., & Salehi, S. (2016). Using the bifocal modeling framework to resolve "Discrepant Events" between physical experiments and virtual models in biology. *Journal of Science Education and Technology, 25*(4), 513–526. https://doi.org/10.1007/s10956-016-9623-7.
Chen, D., & Stroup, W. (1993). General system theory: Toward a conceptual framework for science and technology education for all. *Journal of Science Education and Technology, 2*(3), 447–459.
Drack, M., Apfalter, W., & Pouvreau, D. (2007). On the making of a system theory of life: Paul A Weiss and Ludwig von Bertalanffy's conceptual connection. *Quarterly Review of Biology, 82*(4), 349–373. https://doi.org/10.1086/522810.
Duschl, R. A., & Bismack, A. S. (2016). *Reconceptualizing STEM education: The central role of practices.* New York: Routledge.
Forrester, J. W. (1993). System dynamics and the lessons of 35 years. *A systems-based approach to policymaking* (pp. 199–240). Springer.

Graefe, A. N. (2010). Assessing the potential benefits of learning about environmental issues through a systems thinking pedagogy. Master's thesis. Retrieved from http://commons.lib.jmu.edu/master201019/421.

Haines, S. G. (1998). *The manager's pocket guide to systems thinking & learning*. Amherst, MA: HRD Press.

Hogan, K., & Weathers, K. C. (2003). Psychological and ecological perspectives on the development of systems thinking. In A. R. Berkowitz, C. H. Nilon, & K. S. Hollweg (Eds.), *Understanding urban ecosystems: A new frontier for science and education* (pp. 233–260). New York: Springer.

Hung, W. (2008). Enhancing systems-thinking skills with modelling. *British Journal of Educational Technology, 39*(6), 1099–1120. https://doi.org/10.1111/j.1467-8535.2007.00791.x.

Jackson, M. C. (2003). *Systems thinking: Creative holism for managers*. Chichester: Wiley.

Jacobson, M. J., Kim, B., Pathak, S., & Zhang, B. H. (2013). To guide or not to guide: Issues in the sequencing of pedagogical structure in computational model-based learning. *Interactive Learning Environments, 23*(6), 715–730.

National Research Council. (2012). *A framework for K-12 science education: Practices, crosscutting concepts, and core ideas*. Washington, DC: The National Academies Press.

NGSS Lead States. (2013). *Next generation science standards: For states, by states*. Washington, DC: The National Academies Press.

Meadows, D. L., Randers, J., & Behrens, W. W. (1972). *The limits to growth: A report for the club of Rome's project on the predicament of mankind*. New York: Universe Books.

Meadows, D. H., & Wright, D. (2009). *Thinking in systems: A primer*. London: Earthscan.

Monat, J. P., & Gannon, T. F. (2015). What is systems thinking? A review of selected literature plus recommendations. *American Journal of Systems Science, 4*(1), 11–26.

PhET Interactive Simulations. (n.d.). Retrieved from https://phet.colorado.edu/.

Organisation for Economic Cooperation and Development. (2016). *PISA 2015 assessment and analytical framework: Science, reading, mathematic and financial literacy*. Paris: OECD Publishing Retrieved from https://doi.org/10.1787/9789264255425-en.

Ramage, M., & Shipp, K. (2009). *Systems thinkers*. London: Springer.

Reynolds, M., & Holwell, S. (2010). *Systems approaches to managing change: A practical guide*. London: Springer.

Richmond, B. (1994). System dynamics/systems thinking: Let's just get on with it. In *Paper Delivered at the 1994 International Systems Dynamics Conference in Sterling*, Scotland.

Rose, J. (2012). Application of system thinking skills by 11th grade students in relation to age, gender, type of gymnasium, fluent spoken languages and international peer contact. Master's thesis, University of Vienna, Vienna, Austria. Retrieved from http://ubdata.univie.ac.at/AC10497456.

Senge, P. M. (2006). *The fifth discipline: The art and practice of the learning organization*. New York: Doubleday.

Sweeney, L. B., & Sterman, J. D. (2000). Bathtub dynamics: Initial results of a systems thinking inventory. *System Dynamics Review, 16*(4), 249–286. https://doi.org/10.1002/sdr.198.

Von Bertalanffy, L. (1969). *General system theory: Foundations, development, application (revised ed.)*. New York: George Braziller.

Zhang, B. H., Liu, X., & Krajcik, J. S. (2006). Expert models and modeling processes associated with a computer modeling tool. *Science Education, 90*(4), 579–604.

Dr. BaoHui Zhang is Qujiang Scholar Professor, former dean of School of Education, Shaanxi Normal University, Xi'an, China. He received his Ph.D. degree from the University of Michigan, USA. He has work experiences in China, USA, and Singapore. He is president-elect of the Associations for Science Education (ICASE). His teaching and research has been at the intersection of science education, educational technology, and the learning sciences. He and his collaborators care the most about how to develop, implement, and sustain efforts in using ICT to facilitate student learning in science. More information can be found here: http://zhangbaohui.snnu.edu.cn/index.asp.

Salah A. M. Ahmed is a Yemeni student. He obtained his master's degree in Science from Shaanxi Normal University (SNNU), China, and received his bachelor's degree in Physics from education school at Ibb University, Yemen in 2006. Upon his graduation, he worked as physics and science teacher in different schools for 9 years. His research interests are learning sciences and educational technology that include the application of ICT in science, and other STEM disciplines.

Chapter 29
Gender/Sexuality Theory—Chris Beasley

Steven S. Sexton

Introduction

In 2005, Chris Beasley published *Gender & Sexuality: Critical Theories, Critical Thinkers*. In this one text, Beasley brought together both the history and theorists concerning gender and/or sexuality. Beasley begins by stating gender theory is focussed on sex and power. More specifically, gender theory is about the ways in which current social arrangements position the dynamics of power in regards to sex. Beasley highlighted the changing and evolving definition of sex and noted it has become more commonly associated with physical activity. As a result, her critical analysis of the sociopolitical theories and thinkers of gender and/or sexuality notes gender theory is composed of two categories: gender (the sexed categories, e.g. men and women) and sexuality (the sexual categories, e.g. hetero and homosexual). Like Beasley rather than the term gender theory, I believe it is more appropriate to use the term gender/sexuality theory and this will be used throughout this chapter.

Within gender/sexuality theory's two categories there are three subfields: Feminists studies and Masculinity studies (under gender) and Sexuality studies (under sexuality). This chapter is in support of Beasley's (2005) claim that, 'this interdisciplinary field of gender/sexuality theory assumes that sex is ineluctably a matter of human organisation—that is, it is political, associated with social dominance and subordination, as well as capable of change' (p. 9). As a New Zealand teacher educator with a strong social justice stance, I am concerned with how both students and teachers are positioned within the educational system.

This chapter will begin with a brief summary of Beasley's (2005), *Gender & Sexuality: Critical Theories, Critical Thinkers*. Then how the three subfields of gender/sexuality theory have been evidenced in my New Zealand science education programme in an undergraduate primary education programme. Most importantly,

S. S. Sexton (✉)
College of Education, University of Otago, Dunedin, New Zealand
e-mail: steven.sexton@otago.ac.nz

© Springer Nature Switzerland AG 2020   437
B. Akpan and T. Kennedy (eds.), *Science Education in Theory and Practice*,
Springer Texts in Education, https://doi.org/10.1007/978-3-030-43620-9_29

this chapter will report on research with student teachers challenging normative attitudes, values, and beliefs of schools.

Gender & Sexuality: Critical Theories, Critical Thinkers

Beasley's (2005) *Gender & Sexuality: Critical Theories, Critical Thinkers* is an extremely well-written account of the field of gender/sexuality theory. She provides an informative, systematic, and balanced perspective of Feminist, Masculinity, and Sexuality studies. Beasley highlighted how gender is a social process. As such, people and social practices are typically divided along with a binary division based upon sexed identities. This has resulted in many western societies of one gender placed in opposition or in a subordinate power position to the other. Feminist studies is a critique of these mainstream taken for granted normative attitudes, values, and beliefs of gender identities and power. It is a critical stance against the assumption of male superiority and centrality with a focus on the marginal. Masculinity studies is also a critical stance on the mainstream. While Masculinity studies has become more diverse in its focus it is still primarily focussed on the mainstream white middle-class heterosexual male. Sexuality studies is a critical stance concerned with theorising and/or political action in order to promote social change.

Gender & Sexuality: Critical Theories, Critical Thinkers is not an attempt to reach a final answer. It is an analytical overview of critical frameworks in this field. Beasley notes her overview is an examination of the debates between the three sub-fields rather than a description of perspectives. For example, in Beasley's overview chapter of gender and masculinity studies, she notes Masculinity studies differs from Feminist and Sexuality studies in the attitude expressed towards the subject matter. She highlights how Masculinity studies theorists often keep a critical distance from masculinity which is the subject of their field while Feminist and Sexuality studies theorists champion the marginalised subjects of their fields. Beasley works throughout her book to draw attention to the usefulness of developing and maintaining a critical unease rather than an acceptance of all views.

New Zealand Primary Educational Context

In New Zealand, it is illegal to discriminate based on a person's gender, religion, ethnicity, disability, age, politics, ethnicity, sexual orientation as well as a person's employment, family, and marital status. Issues arise because like many western societies, New Zealand is a multicultural society (Manning, 2013). While the numeric majority are Pākehā/European (Pākehā is a term referring to non-Indigenous New Zealanders) there is a significant Māori (Indigenous New Zealanders) population with legally recognised customs and traditions. However, nearly 80% of all teachers are Pākehā/European of which in primary education 82% are female with an average age of 50 (Ministry of Education, 2005).

I am a vocal advocate of social justice in education. I position social justice as the 'respect for differences between groups and between individuals and the dialectical overcoming of conditions of oppression and inequality' (Pereira, 2013, p. 163). Specifically, this chapter uses gender/sexuality theory as a social justice challenge to the forms of oppression that derive from harmful social, political, and/or cultural beliefs about a student teacher's gender, age, and physical appearance (Connell, 2011). It should be noted, this chapter positions gender as a social phenomenon derived from sociological relationships (Bradley, 2013).

Most importantly for this chapter, Bradley (2013) noted the school setting was just one of an individual's social institutions in which the gendering of individuals and relationships takes place. Gendered practices in school are a complex mixture of formal and informal educational, cultural, social, and political discourses about male and female identifications. This chapter is not focussed on the masculinity and femininity dichotomy of sex-role socialisation theories (Driessen, 2007). This chapter is focussed on gender/sexuality theory in primary science education as a means to promote, support, and facilitate the social justice of initial teacher education (ITE) student teachers.

Promoting, supporting, and facilitating social justice of student teachers requires initial teacher education to challenge a school's, teacher's, or educational community's normative attitudes, values, and beliefs regarding gender, age, and physical appearance (Cochran-Smith et al., 2009). Primary students want good teachers and are able to tell you what makes a good teacher for them. Both international and domestic research studies have noted that students assess their teachers according to the quality of their teaching, not by their biological classification (Connell, 2011; Haig & Sexton, 2014; Watson, Kehler, & Martino, 2010). Therefore, ITE should promote, facilitate, and scaffold student teachers as they learn and practice how to be 'effective' teachers; not 'male' or 'female' teachers.

Background

The student teachers reported on in this chapter were from two consecutive years (2015 and 2016) of an undergraduate primary education programme at the University of Otago. These six student teachers were all over the age of 18 at the time of their participation and all voluntarily agreed to be included in this study. As part of their final year of ITE, these student teachers were assigned to a primary classroom for the year. These student teachers began with the start of the New Zealand school year in late January with a two-week block in their assigned classroom to see how their mentor teacher established routines, their programme, and behaviour management. Then these student teachers continued in this classroom one-day per week before

undergoing a three-week sustained teaching experience. They returned to this classroom in their second semester of university for three weeks of one-day per week
before completing a five-week sustained teaching experience.

This chapter highlights the multidimensionality of identity. Specifically, it presents
six student teachers as examples of how they challenged the normative attitudes,
values, and beliefs of the classroom, school culture, and science. Amy (all names are
pseudonyms) and John self-identified as a lesbian woman and gay male, respectively.
While their sexuality was a part of their identity, it was not the only aspect and both
wanted to be seen as a teacher, not the lesbian or gay teacher. Jenny came to this
ITE programme as a mature 51-year-old adult after having raised her own family
and was working towards a new career. Emma was a 20-year-old from a small rural
community that was not supportive of her applying for either university study or an
ITE programme. Brad was a white, middle-class, heterosexual young man of average
build; while Luke had a commanding physical presence. He stood just under 190 cm
and weighed approximately 90 kg. While rugby was a personal passion, he wanted
to be a primary teacher who brought his skills in rugby to the school as an asset.

Gender/Sexuality Theory Challenging Normative Attitudes, Values, and Beliefs

As their science education facilitator, I physically present to the student teachers as
a white, middle-aged male, my identity politics do not arise. I introduced myself as
a primary teacher who works for a university's College of Education. Similarly, my
student teachers' identity politics are not an issue until a classroom, school, or educational community attempts to limit, marginalise, or place any of them on the periph-
ery. Amy, John, Jenny, Emma, Brad, and Luke presented a gender/sexuality theory challenge to their classroom's, school's, or educational community's normative
perceptions of how student teachers are identified.

Amy and John

Amy walks into a room expecting most people to form opinions about who she is
based solely on her physical appearance. She has short, cropped hair, almost never
wears make-up, and has piercings in her eyebrow, nose, and upper-lip. John walks
into a room and expects almost no one to notice him. He is of average height and
weight. He dresses and grooms himself conservatively. Nothing about him would
stand out in a crowd, that is until he starts talking. He acknowledges the tone, pitch,
and volume of his voice combined with the overactive use of hand gestures while
talking does support some of the commonly held stereotypes about gay men. Both

of these student teachers, however, felt compelled to challenge the formal and informal educational, cultural, social, and political discourses about normative male and female identifications. In discussions with how and why they saw the need to challenge their schools' taken for granted normative perceptions, both would align with Beasley's (2005) Race/Ethnicity/Imperialism. Amy from the Feminist studies and John the Sexuality studies perspective.

Amy knows gender has been a barrier to learning as some subjects like science have unconscious acceptance of gender-role stereotypes (Bailey, Scantlebury, & Letts, 1997). Amy grew up experiencing a school system that located masculinity and femininity separately in boys and girls. Her schooling experiences had paired boys with boys and girls with girls. Amy firmly believed it was one of her roles as a teacher to challenge not only her own attitudes, values, and beliefs but also those of her mentor teacher, school, and students. She introduced herself to her 29 Year 6 (students aged 11) class through activities around how everyone was unique highlighting what each student had about themselves that made them special. When planning her science unit, she wanted to combine these activities with change of state. For Amy a unit on the changes of state of water offered her the opportunity to build on her introduction activities, 'What makes you special?' while not only challenging what students think they know about water as a solid, liquid, and gas but also with whom they can work. For example, her students used closed systems (water in sealed containers) to demonstrate evaporation, condensation, and precipitation to their Year 2 (students aged 7) reading buddies. The Year 6 students then lead discussions with their Year 2 buddies about how solid water was different from liquid water and gaseous water as well as what made each state of water special.

As a teacher, John was a firm believer in using Learning without Limits (Hart, Drummond, & McIntyre, 2006) as a means to actively include and involve all students. John ensured that every one of his 26 Year 2 students in his class was aware of what they were doing and the reasons behind school rules. John used agar jelly activities as the focus of his unit. In the student teachers' ITE course, this activity was used as a way to show why there are school rules about health and hygiene. John wanted his students to come to the realisation that they themselves needed to rewrite the school rules from their perspective so that these rules were not power structures imposed on them. Therefore, John's students investigated tabletops, food dropped on the ground, shoes in the classroom, and covering their mouths while coughing or nose while sneezing. To ensure his students understood it was the ground, desktop, or shoe that was being investigated, the week before this unit began John's students washed their hands with soap and water and wiped fingers across agar dishes that were then labelled, sealed, and set aside over the weekend. Then the following week when this unit was introduced, these dishes were used to show how clean their hands were. John ensured his students understood no one was being singled out as 'dirty' or 'unclean' it was what they were investigating. John's students washed their hands after these activities to re-enforce this point. Over the next several days, the students checked the agar dishes to see what was happening from each surface investigated. To enhance their capacity for learning, John referred back to a previous unit on pets where students investigated what pets needed to be healthy and happy: food, water, shelter, warmth, and care. This allowed his students to make connections between previous activities and they were able to explain how the agar dishes provided them shelter, food, and water necessary for the mould and bacteria to grow. John's explicitly designed activities based on the Learning without Limits pedagogical principles of everybody, co-agency, and trust (Hart et al., 2006). These activities allowed his students to understand for themselves the health and hygiene explanations behind why the school had certain rules. As a result, the class co-created a new class charter to include we: sit on chairs or the mat, put dropped food in the compost bin, take shoes off at the door, and cover our mouth or nose when coughing or sneezing.

Amy deliberately planned for the disruption of normative attitudes, values, and beliefs in her school's buddy system. She reorganised the buddy system for gender equity: Year 6 boys with Year 2 girls and boys, Year 6 girls with Year 2 boys and girls rather than the established boy–boy and girl–girl arrangement. Amy agrees with Connell (2011) that it is through men and women who have grown up as boys and girls working together for gender equity change to happen. John led his students through a series of learning opportunities that facilitated them realising why they sit on chairs not the tabletops, put dropped food in the compost bin not eating it, and cover their mouths/nose. John knows men and boys are not isolated individuals and that they live in social relationships with women and girls. He supports Connell's (2011) idea that relational interests in which both boys and girls should grow up with opportunities to fulfil their talents. Both Amy and John challenged how their schools perceived them as identified lesbian (Amy) and gay (John) rather than the capable and confident student teachers putting into practice effective teaching pedagogies.

Jenny and Emma

Jenny knows she looks like a grandmother because she is one. She brought to her ITE programme a much more diverse and extensive set of life experiences than the typical student teacher did. It was not a surprise for her to discover that her own children were older than most of her student–teacher colleagues. After raising her family and helping her own children with their children, her family was now supporting her in gaining teacher qualification. In almost every demographic, Emma was the opposite of Jenny. Emma had just turned 20 at the start of her third and final year of study. She grew up in a small rural town with a population at the last census of 542. I was fortunate enough to be on her interviewing panel for admission to the College of Education. This interview almost did not happen as her confidential school report noted that they did not believe she had the capabilities to cope with either university-level study or being a teacher. Thankfully, we interviewed her anyway. She presented herself well and quite capably explained why she wanted the opportunity to be a teacher. She has held down a full-time job since being accepted for this University as she is supporting herself. While Jenny was a firm advocate of Beasley's (2005) Feminist Social Constructionism, Emma saw herself aligning more comfortably with the relational power position of postmodern Sexuality studies.

For Jenny, the essential needs that shape motivation, development, and learning are the fundamental needs of emotional and physical safety, being in close and supportive relationships, and being connected and belonging to a community. Jenny sees her role as the teacher to build students' cognitive development in an intentional and systematic manner by engaging them in challenging and meaningful activities. Jenny used Explosions: How and why you can set students on fire as the perfect opportunity to blend how to meet the fundamental needs of her 18 Year 5/6 (students aged 10–11) children in meaningful activities. Through a science unit titled, 'How amazing is water!' Jenny challenged her students to develop their self-esteem and willingness to step outside their comfort zone. To facilitate this, Jenny's students investigated the heat absorption capabilities of water. Throughout this unit, Jenny's students not only built up their scientific knowledge about water and its properties through collaborative discussions but also their willingness to take risks. The unit began with low-risk activities to build trust and culminated in all students willingly soaking one arm (from the tips of their fingers up to their elbow) in water before taking a handful of LPG-filled bubbles and having the bubbles set alight so they could toss fire (the flaming bubbles) into the air. This may have been a hands-on activity but also educational as her students were also able to explain how and why water's absorption of heat prevented them from being burnt.

Emma's unit title 'Forces and Motion: Our choices have consequences' was her opportunity to show her 31 Year 4 (students aged 9-10) class that they do have the ability to influence what happens. She is a firm believer that schools and teachers need to promote students' self-efficacy beliefs (Bandura, 1997). Self-efficacy is the belief in one's own ability to accomplish something successfully. From personal experience, she knows that many students will not attempt things they believe they will fail. Teaching for her is not putting obstacles in her students' path preventing them from being successful. Likewise, her classroom will not be a place of the stress, anxiety, worry, and fear that she experienced.

Emma wanted her students to experience that through science students learn to ask questions that will often lead to further investigations. While these investigations may challenge what they think they know, science allows students the opportunity to change their answers as they learn more. Emma began with students using the school's playground equipment to discuss in small groups: what makes it move, how we can stop it from moving, what do we have to do to keep in moving. Through this introduction activity that uses familiar equipment that is outside the classroom, Emma was able to gauge students' understandings of forces and motions. Then a return activity back outside with the slide set the stage for using groups to use ramps and marbles in the classroom to explore the effect of changing angle of ramp, weight of object, or friction (rolling on carpet, linoleum, towels, etc.). The students' self-selected groups and chose which activity to explore. In their groups they used markers to record their data. Each group then reported back to the rest of the class what they had done and what they had observed. It was through these whole class discussions that students realised while other groups were exploring similar activities, each group reported back different observations. The students questioned each activity as a whole class to explore how and why similar activities resulted in different observations.

Jenny refused to accept any barrier from teaching in a way that resonates socially (classrooms that are safe and supportive of each other) and emotionally (classrooms that build self-esteem through scaffold risk-taking) with her students. She was confident enough in her own abilities that her school's and educational community's initial concerns regarding her combined age and gender were not going to prevent her from teaching. Jenny saw herself as a modernist feminist (Bradley, 2013) who refused to be marginalised from what she knew she was capable of doing. Emma had real trouble seeing herself as a feminist and while she agreed with Jenny in many ways, she felt more comfortable within Sexuality studies rather than Feminist studies. Emma who rejects the essentialist position on identity categories still could not see herself as a postmodern feminist (Bradley, 2013) as central to her teaching philosophy was her student's agency and self-efficacy. When asked how her position differed from the postmodern feminist position that seeks a, 'deeper and more detailed understanding of the different shades of relationships, not only between men and women, but also women and themselves' (Bradley, 2013, p. 76) responded that as she not a minority in ethnicity or sexuality and sees social change as important; she therefore felt more comfortable in Sexuality studies rather than Feminist studies.

Brad and Luke

Brad was a single, white, heterosexual, middle-class young man of average size aged 22. In many ways, it could be argued that the education system of New Zealand was designed to support him. Unlike Brad, Luke took up space when he was in a room. He was 24 years old and is physically large. He looks like a rugby player because he is one. Like Brad, he was also a single, white, middle-class, heterosexual young man. When Brad interviewed for admission into this ITE programme he was told he would have no trouble getting a job as a primary teacher as schools are looking for strong male teachers. For his placement in his first year of the programme, he was assigned a Year 6 class. He admitted this was not what he wanted and this almost resulted in him pulling out of the programme. He reported that his mentor teacher treated him like a big child who could not teach the real subjects like Literacy and Numeracy. He was given the Physical Education classes as then he could take the boys for rugby. He did not want to be a rugby coach; he wanted to be a teacher. In his second year, he was assigned a Year 1 (students aged 6) primary class and knew that teaching the lower primary classes was what he wanted. He requested another Year 1 placement for his final year. Year 1 was where he felt he was best at building supportive interpersonal relationships with students and creating the classroom environment that promoted positive academic attitude, values, and beliefs.

When I went through teacher education, 3% of my cohort were male and all of us had to address issues raised by schools and colleagues with being male primary teachers. We were always labelled as the 'male' primary teachers, never just primary teachers. In Brad and Luke's year group, 21% of the cohort were male of which almost none have reported any issue with them being a primary teacher. Brad and Luke were two of those who did raise concerns as both have self-selected to be lower primary teachers (teachers of New Entrant—Year 2, students aged 5–7). Bard wants to be a New Entrant teacher (students aged 5). New Entrant classes are bridging classes in primary schools between early childhood centres and the start of compulsory education (Year 1, students aged 6). As stated, Luke sees himself in Year 1.

Brad is not on any political campaign to champion men's rights nor does he have any social agenda to promote male teachers of young children. His biggest issue was having to explain repeatedly to schools and parents he has no politics, no agenda, no campaign. He just wants to be a teacher, so what is the big deal. Thankfully, he has no personal experience with the politics, agendas, and campaigns both for and against male teachers, especially male teachers of young learners (see Martino, 2008; Mills, Martino, & Lingard, 2004; Skelton, 2003 for an international perspective and Haig & Sexton, 2014; Hood, 2001 for a domestic perspective). What Brad has experienced is having to demonstrate his ability to be a teacher and work in an educational system that until he entered ITE, had only supported him.

Brad believes he has been observed more times and more closely than most student teachers. He feels he has had to justify his intentions and reasons in more detail than most of his colleagues. As a result, he believes he is far better prepared than he might have been without this closer and more constant observation of his practice. His planning is grounded firmly in *The New Zealand Curriculum's* (Ministry of Education, 2007) effective pedagogies. He plans all his teaching opportunities around students being in a safe learning environment that facilitates shared learning as students make connections to their prior learning and experiences. Brad did this in his science unit of Day and Night for his New Entrant class. Brad began by taking his class outside and in pairs tracing a shadow of a tree, bush, or building on the edge of the Netball courts, outside the play area so as not to interfere with other classes who might come out to play on them. Once the students began, he waited until most were about halfway through their tracing and called them together to talk about how it was going and what parts were easier to trace. After his discussion, the students returned to their traces and tried to continue. Most students noticed the shadows had moved and were no longer the same. This lead into discussion on how and why the shadows moved. The next day the students went back outside to trace an object but this time they were given time to complete the shadow in one colour at the start of the day (8:50) and then just before playtime (10:30) they traced the shadow again in a different colour and then again after playtime (11:00). These activities lead into discussions about how the shadows were formed and why they changed as the day went from start of school to playtime to after playtime. Further sessions explored how the students understood Day and Night.

Luke used the topic of gardens to show his students that what they do has consequences. Specifically, he used an activity of growing 'wheat grass heads' (students put a photo of an animal on a cup, and as the wheat grass grew, it became the hair of the animal) to show his students how care, attention, and thinking about one's actions lead to better results. Over the week, Luke and his students tended to their wheat grass heads talking about and discussing sunlight, when/how much to water, temperature, and handling. Central to how Luke sees his role as the teacher is through reciprocal imitation (Zhou, 2012). In this Year 1 class, Luke knew it was his actions and behaviour as a caring, responsible, and effective teacher that his students imitated. This imitation offered Luke the opportunity to express his concept of 'self' as a teacher through his actions, experiences, and emotions (Zhou, 2012) instead of what 'others' may expect from a rugby player.

Both Brad and Luke like the concept of Liberal Human Rights from Sexuality studies (Beasley, 2005) as they would see themselves holding an anti-discrimination position. They stated they would also not only support gays and lesbians as members of the universal 'Human', but also women, ethnic minorities, and anyone else for that matter. Both had some difficulty with Masculinity studies and outright refused the idea of them being feminised men. They agreed that masculinity is a socially, historically, and culturally derived concept but were very uncomfortable with the concept of hegemonic masculinity. In further discussions, I drew their attention to Hearn's (2004) 'hegemony of men' and how I had used it in exploring friendship circles (Sexton, 2017). They agreed with Hearn's idea that hegemonic masculinity is too restrictive as the focus on masculinity is too limited. Hearn's hegemony of men, 'seeks to address the double complexity that men are both *a social category formed by the gender system and dominate collective and individual agents of social practice*' (Hearn 2004, 59, italics in original). Hegemony of men necessitates a critical look at the social category of men. These investigations require addressing the formation of the social categories and how these become the 'taken-for-grantedness' (Hearn 2004). Brad and Luke out of unexpected necessity have been required to examine their social position as gatekeepers for gender equality (Connell, 2011). As white, middle-class, heterosexual men they understand how their views on masculinity are 'socially constructed patterns of gender practice' (Connell, 2011, p. 10). While they have not had to struggle publicly or privately with Sexuality studies debate concerning their heterosexual desires, they have gained a better understanding of Feminism's need to critique the mainstream's presumption of what counts as normative.

Final Thoughts

Gender/sexuality theory concerns a complex issue that is often extremely personal to those involved. As a result, Beasley (2005) notes how it, 'may be passionately interpreted in sharply different ways' (p. 8). The six student teachers reported here support this claim. Amy and John are both members of the queer community but would argue being queer is only one part of who they are. They do not see a need to be politically radical but will challenge issues of identity politics that they do not agree with. Jenny and Emma came from two different generations. Jenny came of age in the late 70s while Emma has just now reached this milestone. Jenny was expected to marry and become a mother to her children, which she did. Emma was expected to stay in her local community and accept whatever future she was offered, which she did not. Both are trying to be the person they want to be; however, Jenny sees herself as a Feminist while Emma does not. It would not be hard to argue Brad and Luke are members of the mainstream as both are white, middle-class, heterosexual men. Like many in gender/sexuality theory, they do not see their gender or sexuality as an issue to them being the teacher they want to be. Also, like many in gender/sexuality theory, they are tired and frustrated with having to justify who they are and what they want to be just because of their gender and sexuality. Both know the importance of other men but more importantly their students seeing a range of possibilities for their own lives (Connell, 2011).

Our world is full of diverse individuals and therefore it is crucial that our classrooms provide an environment where everyone is safe, supported, and welcomed. This is for both students and their teachers. New Zealand teachers and schools are directed by the Ministry of Education policy to embrace diversity and inclusion (Ministry of Education, 2007). Those educational communities that do not see the variations within gender and sexuality as a hindrance are able to build responsive, reciprocal, and corroborative relationships needed to enrich each individual's education. Teachers should be encouraging differences as a means to learn from one another.

The purpose of initial teacher education is to prepare student teachers with the skills, knowledge, and behaviours needed to be effective teachers for all their students. The six student teachers in this chapter show how they challenged their school's, classroom's, or educational community's normative attitudes, values, and beliefs regarding gender, age, and physical appearance. These six student teachers explicitly sought to trouble the taken-for-grantedness of the primary education system in New Zealand. Amy disrupted her school's gendered buddy system. John eliminated the power structure behind school rules allowing his class to co-create a new class charter based on equality. Jenny and Emma disrupted the perceptions of age and gender while Brad and Luke disrupted the perception of gender, sexuality, and physical appearance. These student teachers have highlighted how we are all unique individuals and should focus on the normalisation of difference. As such, schools need a more sophisticated notion of normality, knowledge, and learning. We, as teachers, should question the taken-for-grantedness assumptions about what teachers and students should do. We have the ability to go against these messages.

Summary

- Initial teacher education should prepare all student teachers to be effective teachers no matter what their gender, age, or sexuality.
- The classroom must be an environment where both students and teachers feel safe, supported, and welcome.
- Diversity and Inclusion should be embraced as normative.

References

Bailey, B., Scantlebury, K., & Letts, W. (1997). It's not my style: Using disclaimers to ignore gender issues in science. *Journal of Teacher Education, 48*(1), 29–36.

Bandura, A. (1997). *Self-efficacy: The exercise of control.* New York, NY: W. H. Freeman.

Beasley, C. (2005). *Gender and sexuality: Critical theories, critical thinkers.* London: SAGE Publications.

Bradley, H. (2013). *Gender* (2nd ed.). Cambridge, UK: Polity Press.

Cochran-Smith, M., Shakman, K., Jong, C., Terrell, D. G., Barnatt, J., & McQuillan, P. (2009). Good and just teaching: The case for social justice in teacher education. *American Journal of Education, 115,* 347–377.

Connell, R. (2011). *Confronting equality: Gender, knowledge and global change.* Cambridge, UK: Polity Press.

Driessen, G. (2007). The feminization of primary education: Effects of teachers' sex on pupil achievement, attitudes and behaviour. *Review of Education, 53*(2), 183–203.

Haig, B., & Sexton, S. S. (2014). Primary students' perceptions of good teachers. *Set, 3,* 22–28.

Hart, S., Drummond, M. J., & McIntyre, D. (2006). Learning without limits: Constructing a pedagogy from determinist beliefs about ability. In L. Florian (Ed.), *The SAGE handbook of special education* (pp. 499–514). London: SAGE.

Hearn, J. (2004). From hegemonic masculinity to the hegemony of men. *Feminist Theory, 5*(1), 49–72.

Hood, L. (2001). *A city possessed: The Christchurch civic centre creche.* Dunedin, NZ: Longacre Press.

Manning, B. (2013, 11 December). Census 2013: More ethnicities than the world's countries. *New Zealand Herald.* Retrieved from http://www.nzherald.co.nz/news/article.cfm?c_id=1&objectid=11170288.

Martino, W. J. (2008). Male teachers as role models: Addressing issues of masculinity, pedagogy and the re-masculinization of schooling. *Curriculum Inquiry, 38*(2), 189–223.

Mills, M., Martino, W., & Lingard, B. (2004). Attracting, recruiting and retaining male teachers: Policy issues in the male teacher debate. *British Journal of Sociology of Education, 25*(3), 355–369.

Ministry of Education. (2005). *Education counts—teacher census.* Retrieved from http://www.educationcounts.govt.nz/publications/schooling/teacher_census.

Ministry of Education. (2007). *The New Zealand curriculum.* Wellington, New Zealand: Learning Media Ltd.

Pereira, F. (2013). Initial teacher education for social justice and teaching work in urban schools: An (im)pertinent reflection. *Alberta Journal of Educational Research, 59*(2), 162–180. Retrieved from http://search.ebscohost.com/login.aspx?direct=true&db=ehh&AN=95635307&site=ehost-live&scope=site.

Sexton, S. S. (2017). The intersection of self and school: How friendship circles influence heterosexual and self-identified queer teenage New Zealand boys' views on acceptable language and behaviour. *Gender and Education, 29*(3), 299–312.

Skelton, C. (2003). Male primary teachers and perceptions of masculinity. *Educational Review, 55*(2), 195–209.

Watson, A., Kehler, M., & Martino, W. (2010). The problem of boys' literacy underachievement: Raising some questions. *Journal of Adolescent & Adult Literacy, 53*(5), 356–361. Retrieved from http://www.jstor.org/stable/25614569.

Zhou, J. (2012). The effects of reciprocal imitation on teacher-student realtionships and student learning outcomes. *Minds, Brain, and Education, 6*(2), 66–73.

Steven S. Sexton is a senior lecturer at the University of Otago, College of Education. He obtained his Ph.D. from the University of Sydney in 2007. He has been a classroom teacher in Japan, Thailand, Saudi Arabia, Australia, and New Zealand. Currently, he delivers science education papers in both the undergraduate initial teacher education primary programme and the Master of Teaching and Learning programme. His research interest areas are in relevant, useful, and meaningful learning in science education, teacher cognition, and heteronormativity in schools.

Chapter 30
Indigenous Knowledge Systems

Constance Khupe

Introduction

Colonisation created an unplanned interface of knowledge systems, and an assumption that the knowledge of the conqueror was more beneficial for the conquered than the knowledge on which their lives were actually grounded. Subsequent to political conquest, colonial powers introduced to Indigenous populations a form of education that was for purposes of assimilation, and research that was unrepresentative of Indigenous perspectives. The recognition of Indigenous knowledge systems emanated from persistent calls for democracy, human rights, social justice, and inclusivity and for the decolonisation of institutions of knowledge production. Indigenous knowledge systems theory calls for the recognition of Indigenous knowledges and ways of living as valid resources for education and development, and as a valid framework for research among and with Indigenous peoples. In this chapter, I add my voice to the already established conversation for Indigenous people-centred education and research. Firstly, I use the genre of story to provide a historical context for the development of IKS theory. In presenting the theory, I outline factors that the theory addresses, and then present the nature of Indigenous knowledge systems. Lastly, I discuss with examples, application and related challenges for IKS theory in curriculum design, teaching and learning, and in research.

Historical Background

Once upon a time, different groups of humans found themselves inhabiting different parts of planet Earth, living off the land. Each locale had different environmental

C. Khupe (✉)
University of the Witwatersrand, Johannesburg, South Africa
e-mail: Constance.Khupe@wits.ac.za

© Springer Nature Switzerland AG 2020   451
B. Akpan and T. Kennedy (eds.), *Science Education in Theory and Practice*,
Springer Texts in Education, https://doi.org/10.1007/978-3-030-43620-9_30

conditions, which the humans began to adapt to for their survival. Although resultant ways of living varied according to setting, the human-nature relationship was largely guided by a view of nature as supplying human needs. Value systems emerged which governed this relationship, and for the greater part it was one of respect for nature, even to the point of reverence. Continual exploration deepened human understanding of intimacy with and reverence for the environment, and a striving to strike a balance between human survival and conservation.

Values and knowledge were passed on for the survival of posterity. The home language was the language of teaching and learning, and teachers used familiar points of reference within lived experience. Education was structured in ways that we would today describe as formal, informal and non-formal. The curriculum enabled continuing education, ensured relevance for prevailing needs and the roles that the learners played in the community, and was deeply embedded in culture. The teachers were native to the learning contexts: parents, siblings, grandparents, extended family and local experts in various fields. Knowledge was socially constructed, and was a reflection of the social context of origin. The same applied to inquiry. Inquirers did not (inappropriately) distance themselves or feign a neutral relationship with what they studied. This order persisted for centuries.

The different communities did not necessarily view their knowledge systems as closed, and they were not unreceptive of other knowledges. Knowledge was shared and exchanged on respectful terms across groups. This advanced the knowledge systems over time.

At some point, some of the people (mostly those who resided in Europe) began to build cities. Perhaps it was for convenience and for protection. Their connection to the land weakened as they began to invent ever quicker ways of getting things done. Soon enough they even thought of themselves as better than their relations who still lived off the land, and depended on labour-intensive processes to make a living. Many decades later, the machine people began to 'tour' other lands. They had apparently become so engrossed with themselves as to think they were the only legitimate inhabitants of Earth. Wherever they 'toured', they encountered the 'other' relations. They chose not to recognise them. Maybe it was not their fault. None of these 'relations' were like them. They were dark-skinned and spoke gibberish. Were they really human? The machine people could not resist the impulse to get these primitive people and their systems out of the way.

Migration had been common practice in human history, but the inter-continental movements and invasive tendencies of the machine people were way beyond normal. Their encounters with the dwellers of the distant lands resulted in varying degrees of physical and epistemic violence. They disrupted ways of living and enslaved their hosts. They expropriated both knowledge and natural resources. They obliterated languages and cultures, and separated children from parents. They introduced an education that elevated the settler perspective and demeaned the local, and conducted research in ways that objectified locals. In places, they even decimated local populations. The conquerors perceived themselves as superior to their hosts—politically, ideologically and epistemologically. They expropriated as theirs all the knowledge that they viewed as useful, and rejected and suppressed that which they thought was not. Knowledges, languages and cultures were debased and lost through the hegemony which prevailed through centuries of colonialism. Political freedom (where this has been achieved) failed to undo the deep-seated epistemological disenfranchisement among these formerly colonised peoples.

Science education and research—as brought into the conquered lands—were at best inconsistent with local systems of knowing and forms of inquiry. At worst, they were a representation and a vehicle of mental colonisation through which truth and reality were only defined in the terms of the powerful. Science education and research served colonial interests of power and domination, and thus silenced local knowledge systems. The cognitive distance of science learning from the realities of students created dissonance and cognitive conflict which has partly explained the poor academic outcomes among students from non-European backgrounds. The situation has not been different for research. Mainstream literature did not address issues faced by Indigenous researchers and Indigenous participants, neither were prescribed methods consistent with the ways of living among the researched. Colonial (and post-colonial) research and education did not achieve a happy ending for conquered peoples.

The construct of 'Indigenous' as a descriptor of peoples and knowledges emerged in the 1970s as a result of the struggles of Indians in North America (Smith, 2012). The term is now generally used with reference to many of the world's colonised peoples. The collective knowledge and values, ways of living in nature and cultural practices is referred to as Indigenous knowledge systems. In the following sections, I present IKS as a more culturally fitting perspective for science education and research. I will draw on literature to argue for applications of IKS theory with a view to promote discussion that contributes to sustained applications of IKS theory in science education and research.

Development of IKS Theory

Although resistance to colonisation and its related consequences has been ongoing for centuries, voices have been louder in the last few decades. Scholars have increasingly voiced against discordant education and research experiences that result from the hegemony of Eurocentric perspectives over Indigenous knowledge systems and cultural identities. Scholars, such as Bagele Chilisa, Catherine Odora Hoppers, Linda Tuhiwai Smith, Marie Battiste, Ngugi wa Thiong'o and Shawn Wilson, advocate for the transformation of curriculum, pedagogy and research for the decolonisation and emancipation of oppressed peoples. Linda Tuhiwai Smith's book *Decolonizing Methodologies: Research and Indigenous peoples* is among those that have become seminal in IKS education and research. I will draw from Smith's work as representative of the voices of Indigenous scholars who contributed to the development of IKS theory. A Maori by birth, Smith drew her reflections from her indigenous identity and experience, as well as her research experience. Her reflections and dissent developed within 'resistance' movements surrounding land dispossession as well as through activism by urban Maori. Among the demands of the latter was teaching of Maori language in schools. Smith is among the founders of Maori elementary school (Smith, 2012). Writing from critical and feminist perspectives, Smith counters the dominant Western assumption that research and research methods are culture-free. She calls on Indigenous peoples to re-imagine the role of knowledge, knowledge production and knowledge hierarchies. While for many societies, colonisation at a political level has ended, there is still a need for decolonisation at the intellectual and social structural levels. The world over, Indigenous people suffer dehumanising experiences especially in educational institutions. It is only through intellectual decolonisation that the colonising and dehumanising research practices and related education systems could be questioned, and more appropriate methods developed.

IKS theory foregrounds culture, language and worldview in education, research and development work among Indigenous peoples. The theory acknowledges the validity of identities and worldviews that inform the ways of being and ways of living from the vantage point of Indigenous peoples. IKS theory provides space for epistemic justice and the transformation of education and research to become aligned with Indigenous culture and ethics.

Why Indigenous Knowledge Systems Theory?

Addressing Socio-Political Injustice

Colonisation, dispossession and related knowledge subjugation were at the heart of the 'Othering' by colonial settlers in all the Indigenous communities where they landed. Post political freedom, Indigenous peoples are still battling for social justice, human rights, recognition and inclusivity of their cultures and knowledge systems. Education and research have been too intimately linked to the colonial project and its assumptions of superiority over Indigenous peoples and have directly contributed to their continued subjugation rather than emancipation—hence the legacy of 'scientific racism' (Langer-Osuna & Nasir, 2016). What was published as research findings on Indigenous peoples were sometimes broad generalisations drawn from superficial encounters, with little effort for in-depth understanding of the peoples (Smith, 2012). Colonialism thrived on expropriation of knowledge and knowledge products, paradoxically rejecting the people who developed them. The colonial era brought with it a system of intellectual property rights that is based on individualism and commercialisation—often not compatible with collectively held ownership of knowledge and resources that is characteristic of the way of being among Indigenous people. Indigenous languages, cultures and knowledge frameworks were denied survival (Smith, 2012). It is these and other social justice matters that prompted Indigenous scholars to call for the decolonisation of research methodologies, the acknowledgement of Indigenous languages as valid, and the decolonisation of curricula (as in the case of South African higher education). Democracy and political freedom achieved justice only at a superficial level. There is still need in many contexts to disrupt colonising education and research practices through recognition of local perspectives.

Countering Knowledge Monopoly

Through mainstream education, the Western knowledge system dominated, and continues to dominate other knowledge systems. Mainstream education dominates curriculum design, development and implementation through an erroneous assumption of culture-free curriculum and assessment. Resultant forms of education, and particularly science education, hardly have any representation of local perspectives. For ages, schools have presented as symbols for the exotic and the powerful—islands of knowledge that are all-knowing and whose sole purpose is to impart knowledge to those who are presumed not to have any. School science (as already mentioned) is founded on Western values. As a result, students from non-Western backgrounds often feel alienated from the subject because of teaching and learning processes that do not adequately engage their lived realities and ways of being. Instead of interrogating the appropriateness of the education offered to Indigenous children, the simplistic way out is often to label the students as incompetent and even incapable of learning. The questions that Smith (2012) asks about research are relevant for determining the perspective a particular education system serves. Such questions probe issues of ownership and interests being served, who the designers and beneficiaries are, and who will implement. The assimilatory nature of colonial and post-colonial education has been found to give rise to negative curriculum experiences for students, manifesting through lack of interest, lack of meaningful learning, discomfort with content and method, poor academic performance and outright rejection.

Although developments in teaching and learning increasingly place emphasis on student-centred teaching, Indigenous students' knowledges and experiences are often viewed as barriers to learning rather than as potentially valuable resources. Science education estranges Indigenous students from their own cultures and not much is done to assist students negotiate cultural borders between their Indigenous and Western science knowledge (Veintie, 2013). Curricula that are responsive to local knowledges and worldviews have been slow in coming. IKS theory comes with the need to disrupt the monopoly of Western knowledge for the purpose of cognitive justice, and to contribute to the development of inclusive education (Shava, 2016).

Research and Ethics

Colonially framed research has been characterised by dismissal of the cultural rights and languages of Indigenous populations, and an unwillingness to consider what is ethical from the perspectives of Indigenous participants. At present, even though research proposals go through ethics committees for approval, there is not always adequate representation of indigenous perspectives to provide guidance on the appropriateness of proposed methods of engaging communities.

The Nature of Indigenous Knowledge Systems

Assumptions

The following assertions (and assumptions) guide Indigenous knowledge systems theory:

- Indigenous people are legitimate, and so are their languages, cultures and knowledge.
- Knowledge is socially constructed and therefore cannot be understood apart from the knower.
- Knowledge is embedded in culture and language.
- Knowledge is characteristically diverse.

Research and education as knowledge production and transmission are not culture-free.

An outside perspective may easily view Indigenous knowledges systems as singular because of similarities in the worldviews from which the knowledge develops. However, Indigenous knowledges are place and culture-specific, and hence should not be construed as homogenous (Mbiti, 1969; Smith, 2012). The histories and cultures of different peoples are profoundly diverse; hence they produce different knowledges (Kincheloe & Steinberg, 2008). Firstly, the variation in knowledge systems corresponds with spatial variation in environmental conditions, and secondly it is a product of the culture. I need to quickly emphasise that understanding IK only as environmental response would be misleading. Reality in Indigenous worldview is made up of physical, social and spiritual dimensions. In different communities, there is sacred significance attached to mountains, rivers, caves and trees, for example. People live in respectful co-existence and interdependence with living, once-living and non-living elements (Khupe, 2014). The way of living and being—which informs Indigenous knowledge—is therefore a three-way relationship of the social, the physical and the spiritual.

Indigenous knowledge systems are open systems. They are evolving historical and social constructions that can safely be referred to in the present tense. They are knowledge-in-process. It is only because of the assimilationist nature of knowledge associated with Western worldviews that Indigenous knowledges were denigrated and replaced, and hence viewed as only belonging to an uncivilised past. IKS theory recognises the role of the following:

People
Knowledge is a social construction and cannot be separated from the knower. Elders (both the elderly and experts in different fields) are the centre of the development,

storage and transmission of Indigenous knowledge. Elders are the repositories of both information and experience in Indigenous communities.

Place

Indigenous knowledge is established in connection to the land. Indigenous cultures are subsistence. People live off the land and have an intimate relationship with their physical environment. *Place* is a main differentiating factor of the knowledge of different Indigenous groups, providing the necessary cultural experience that contributes to developing a people's sense of identity.

Relationships

Indigenous knowledges are relationship-oriented. These relationships manifest within the social dimension, between social and natural and between social and spiritual dimensions. The individual is almost always viewed in relationship. As a result, relationships are characterised by collective co-existence, ownership and responsibility. In some African contexts, this collective relationship has been described as ubuntu (Letseka, 2013). Hamminga (2005) uses the metaphor of a tree to describe this collective co-existence, asserting that no part of a tree can choose an individual existence. All individual effort is meant to point towards the communal project of the generation, preservation and transmission of knowledge and accompanying beliefs.

Language

Just like any other form of knowledge, IK is transmitted across time and space. Language is the medium of both IK representation and transmission across generations. Understanding Indigenous people's knowledge systems cannot be complete if the language in which the knowledge is coded is not understood (Khupe, 2017). Language represents culture and the collective memory of the people who speak it.

Holistic and Experiential

Indigenous knowledges are not limited to sensory perceptions. They include intuition, dreams and supernatural revelations and experiences. By their nature, Indigenous knowledges are not amenable to categorisation into the subject disciplines that characterise Western knowledge. They cut across disciplines covering (among others), agriculture, medicine, botany, art, craft, music, governance. Indigenous knowledge is experiential, not theoretical. It is a living process that has to be learned, understood and lived (Battiste, 2002).

Application of IKS Theory

As explained in previous sections, calls for the inclusion of Indigenous knowledges in science education come from a range of rationales. From the advent of colonialism, education has been a vehicle of the transmission of validated western information from teacher to student. There has been little willingness to push knowledge boundaries and to think beyond the prescriptions of Western epistemology. IKS theory calls on teachers to engage students in interpretive exploration of different knowledge systems and ways of coming to know. This assists students to not only appreciate the existence of multiple realities, but to also critically engage with their own assumptions.

Research and education as knowledge construction and transmission occur within broader historical, political and cultural contexts. We have already established that when research and education disregard the local context, whether intended or not—the process is alienating and dehumanising for Indigenous community stakeholders. It erodes trust. IKS theory calls for research and education that are centred around the priorities of Indigenous people, where space is opened for collective decision-making, power and knowledge sharing with local communities. A deficit perspective is not consistent with IKS theory. Rather, boundaries of education and research are opened up for the nurturing of Indigenous cultural identities. When appropriately applied, IKS theory results in community-centred education and research. Figure 30.1

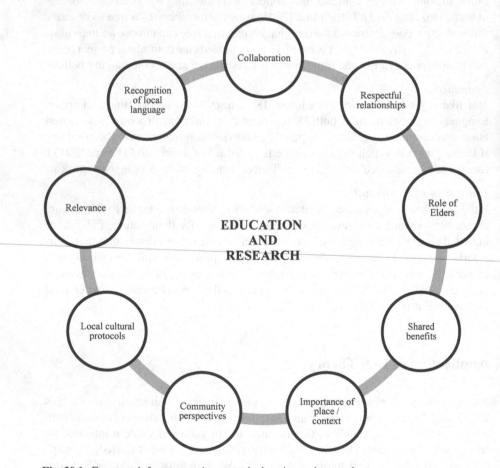

Fig. 30.1 Framework for community-centred education and research

suggests a framework for community-centred education and research. Details may vary according to contexts.

The framework underscores the appropriateness of education and research that are guided by the counsel of active community engagement (represented by the circular formation). I deliberately use the circle to symbolise collective engagement, consultation, deliberation and sharing which are common among Indigenous peoples.

Science Teaching and Learning

Application of IKS theory begins with a curriculum design that connects with the languages, cultures and holistic life context of the population that the curriculum is intended to serve. Teaching and learning resources and assessment methods draw from students' lived experiences and worldviews. The knowledges, cultures, languages values and experiences that students bring into the classroom and continue to draw on as they process knowledge from school science are made explicit in both the intended and implemented curricula. In cases where national curricula offer little support, local level curricula are informed by a deliberate focus on reconstructing indigeneity through ongoing collaboration and knowledge sharing with Indigenous peoples. Science education that is based on IKS theory recognises and engages with students' realities (knowledge, culture, language(s) and spirituality), not from a deficit perspective, but through deliberative dialogue. The aim will be to bridge the knowledge distance between local and school knowledge, and to give opportunity for cultural and cognitive safety when students cross borders between Indigenous and school knowledge.

IKS theory recognises the role of community elders, who are the custodians of local knowledge, language and culture. Elders become co-teachers and resource centres. IKS-based science education acknowledges the cultural and power-related dimensions of education, and actively seeks power-balance and inclusivity. An example of such school-community collaboration is the *Rekindling Traditions*—a cross-cultural science education project among the Cree people in Saskatchewan, Canada (Aikenhead, 2001; 2002). In *Rekindling Traditions*, teachers and Elders work together as cultural guides assisting students to appreciate both the nature of scientific knowledge and the nature of Indigenous knowledge through cross-cultural science teaching. A broader application of IKS theory is represented in Kaupapa Maori—education that incorporates a Maori world view. Kaupapa Maori initiatives cover pre-school to tertiary education all based on Maori worldview and cultural principles. Kaupapa Maori is based on what the Maori decide is best for their children. Few Indigenous communities have reached that stage of autonomy—allowing them to make their own decisions regarding education.

The application of IKS theory in science education requires the adoption of critical and creative teaching and assessment strategies that continually draw on students' contexts and experiences. Student experiences may include knowing languages that are neither spoken nor acknowledged in school, and having a perspective of reality that is not necessarily consistent with that of scientific knowledge. A common example of the latter could be students understanding spiritual elements as playing an active role in daily realities. Literature abounds where these and other students' lived experiences are presented as barriers to learning, especially in science. Teachers often describe multilingual Indigenous students as having 'language problems' simply because their home language is not the language of instruction. Negatively highlighting students' circumstances distracts teachers from imagining possibilities of drawing on the students' multilingual abilities as a resource for learning science.

Community-centred, context-sensitive Indigenous science education research studies point to the richness of local contexts as resources for science teaching. Based on findings from an IKS-science education research study in South Africa, Khupe (2014) provides suggestions for drawing on Indigenous knowledge and world-view in science teaching (Table 30.1). Such applications are only possible where the curriculum has penetrable boundaries that allow for integration of different ways of perceiving reality, and where teachers are given space for creativity. Often, teachers are constrained by examination-driven work environments that promote uniformity and chorus responses but unfortunately stifle imagination.

An IKS-based curriculum employs a variety of teaching and learning approaches: telling, showing, learning by participation, story-telling, outdoor learning, decentralising the teaching and learning process away from the teacher, and spreading the net of expertise to include the community. Story-telling is especially central to Indigenous culture, and teachers can draw on story to engage students more in science learning. While the actual teaching and learning activities vary according to context, the overarching aim will be to strengthen Indigenous language, affirm cultural identity and encourage community participation (Smith, 2012). I acknowledge the difficulty of implementing unconventional teaching methods in contexts where teachers are pressured to complete set content, and where assessment has not transformed enough to recognise the value of culturally sensitive approaches. However, the likely benefits of increased interest and engagement in science, deeper and more meaningful science learning, and an appreciation of diversity of cultural knowledge and experience, far outweigh the costs.

Research

The application of IKS theory in research is based on the establishment and development of respectful relationships. Research relationships that are characterised by parity often last beyond the research project's timeframes because of trust. Decision-making is a result of appropriate consultation processes, aiming for the representation of multiple voices, shared responsibility and co-participation of academic and community researchers/participants. In addition, IKS theory requires appropriate application of the following according to context:

- prioritising use of indigenous languages preferred by the local community.
- making the recognition of culture a fundamental consideration in the research design, the data generation and interpretation and the dissemination of findings. Guidance from people familiar with both local language and culture (cultural guides) ensures appropriately negotiated entry into, and respectful engagement with the community.
- ensuring adequate understanding and accurate representation of people's lived experiences rather than merely seeking to generalise.

Table 30.1 Application of local knowledge in science teaching and learning

Indicator of Knowledge and worldview	Possible classroom application
• Relationships based on *ubuntu* (emphasis is on respectable relationships) • Use of collective forms of possessive pronouns when referring to the place (e.g. our mountain, our river) • Closeness to and dependency on nature • Appreciation of the beauty of nature and expressing the need to care for and respect nature • Understanding reality as consisting of both the physical and spiritual elements	• Practical environmental science education activities based on respect and love for nature • Encouraging respectful learner–teacher and learner–learner conversation in the classroom • Encouraging deliberative argumentation on topical issues • Using project work where students consult with Elders and/or study phenomena in the local environment • Drawing on wisdom from Elders to facilitate discussions on issues that require a cultural perspective • Promoting classroom dialogue through giving space for use of local languages • Using code-switching as a language-brokering strategy to bridge between home language and language of science • Using locally found and valued examples as basis for science teaching and learning • Using role-plays to acknowledge potential roles for scientific and spiritual perspectives • Incorporating field-based learning experiences especially in topics such as life processes in plants and animals, diversity, change and continuity • Structuring teaching and learning from an issues-based approach, giving room for local perspectives • Giving opportunity for students to express their understanding of topics/concepts in the form of story, songs/poems and other Indigenous ways of knowledge transmission. Poems can reflect the extent to which students link concepts, as well as students' feelings, attitudes and spiritual connection towards the topic learnt • Including issues of importance to community in assessment

Adapted from (Khupe, 2014)

• promoting openness to learning.

Researchers need to consider how to protect communities from exploitation of their knowledge and resources. Both individual and collective rights to consent to participate (or refuse to do so) need to be based on honest disclosure of the purposes and intended outcomes of the research, as well as the ownership and dissemination of the findings. Researchers have the responsibility to explore and duly respect all layers of consent, whether individual, collective or both. They (researchers) need to think about how benefits from the study will be shared, and of appropriate ways to appreciate participants. It is crucial to respectfully represent participants in reports and in the dissemination of findings. In addition to these imperatives, researchers who are not familiar with the context of their intended studies have to develop a good understanding of the culture and ethics, and put plans in place for language mediation and ongoing guidance from the community.

Challenges to Application of IKS Theory

In a context where mainstream Western-oriented worldviews continue to dominate curriculum design, development and implementation, a primary challenge is that of balancing the valuing of local knowledge and appropriately preparing students for participation at a global scale. The easier path is often to ignore IK systems completely or to give them token recognition by mentioning them when it is convenient without necessarily doing anything about them. The pervasiveness of Western knowledge and ways of thinking often results in rejection of Indigenous knowledges and cultures even by Indigenous teachers and communities, who tend to view IK-based curricula as counter-developmental and as taking students backwards. As Western languages become increasingly dominant in communities, preference for Indigenous languages wanes and so does the ability to think beyond Western frameworks. IK-based curriculum innovations suffer the danger of being ignored by mainstream curriculum. In South Africa, although the inclusion of IKS is an imperative of post-democratic education, teachers hardly get pre-service preparation in IKS-based education. IKS theory is often considered in postgraduate courses, but even then, the courses are more for research purposes rather than intended for teacher education for its own sake. The extent to which these postgraduate IKS courses translate into sustained classroom practice is debatable.

Similar challenges are faced in IKS research. Funding bodies are less inclined to appreciate costs related to long-term engagement with communities, where results generated are neither quantifiable nor generalisable. In cases where research projects are for qualification purposes, students are under enormous pressure to complete in timelines that are consistent with conventional studies. The nature of IKS research studies as interpretive lends itself to criticism of not being 'scientific'. As a result, IKS theory suffers distrust from those in conventional research who are not willing to engage it on equal terms.

Summary

- The application of IKS theory is about redress. A rational starting point for that redress in science education and in science education research is acknowledging the socio-political history that led to IK systems being misunderstood, overlooked and marginalised.
- Education and research based on IKS theory requires respectful understanding of the cultural context, and the space for Indigenous peoples to provide guidance on decisions regarding teaching and research methods.
- The application of IKS theory to science teaching and learning requires thinking beyond the level of looking for bits of Indigenous knowledge that fit with science content. IKS must be validated not against the standards of Western science but on terms that fit its nature.
- Overcoming challenges of applying IKS theory in both education and research requires overcoming rigidity of thought in order to accommodate ways of being that are outside of Western frameworks.

Recommended Resources

African Forum for Children's Literacy in Science and Technology (AFCLIST). (2004). *School science in Africa: Learning to teach, teaching to learn*. Lansdowne: Juta Gariep.

Battiste, M. (2011). *Reclaiming Indigenous voice and vision*. Vancouver: UBC Press.

Chilisa, B. (2012). *Indigenous research methodologies*. Los Angeles: Sage.

Chinn, P. (2007). Decolonizing methodologies and Indigenous knowledge: The role of culture, place and personal experience in professional development. *Journal of Research in Science Teaching, 44*(9), 1247–1268.

Denzin, N. K., Lincoln, Y. S., & Smith, L. T. (2008). *Handbook of critical and Indigenous methodologies* (pp. 135–156). Thousand Oaks: Sage.

Goduka, N. M. (2013). Creating spaces for eziko sipheka sisophula theoretical framework for teaching and researching in higher education: A philosophical exposition. *Indilinga—African Journal of Indigenous Knowledge Systems, 12*(1), 1–12.

Kaupapa Maori: Do it Right. Retrieved June 28, 2017 from http://whatworks.org.nz/kaupapa-maori/

Mertens, D. M., Cram, F., & Chilisa, B. (2013). *Indigenous pathways into social research: Voices of a new generation*. Walnut Creek: Left Coast Press.

Ministry of Education. (2017). About Māori-medium education Retrieved June 28, 2017 from: https://education.govt.nz/ministry-of-education/specific-initiatives/nga-whanaketanga-rumaki-maori/about-maori-medium-education/.

Odora Hoppers, C. A. (Ed.). (2002). *Indigenous knowledge and the integration of knowledge systems: Towards a philosophy of articulation*. Claremont: New Africa Education.

Wa Thiong'o, N. (1994). *Decolonising the mind: The politics of language in African literature*. East African Publishers.

Rekindling Traditions: Cross-Cultural Science & Technology Units (CCSTU) Project. https://www.usask.ca/education/ccstu/welcome.html

Sutherland, D., & Henning, D. (2009). Ininiwi-kiskanitamowin: A framework for long-term science education. *Canadian Journal of Science, Mathematics and Technology Education*, 173–190.

Vakalahi, H., & Taiapa, J. (2013). Getting grounded on Maori research. *Journal of Intercultural Studies*, 1–11.

References

Aikenhead, G. (2002). Cross-cultural science teaching: Rekindling traditions for aboriginal students. *Canadian Journal of Science, Mathematics and Technology Education, 2*(3), 287–304.

Aikenhead, G. (2001). Integrating western and aboriginal sciences: Cross-cultural science teaching. *Research in Science Education, 31*(3), 337–355.

Battiste, M. (2002). *Indigenous knowledge and pedagogy in First Nations education: A literature review with recommendations*. Ottawa: Indian and Northern Affairs.

Hamminga, B. (2005). Epistemology from the African point of view. In B. Hamminga (Ed.), *Knowledge cultures: Comparative Western and African epistemeology* (pp. 57–84). Amsterdam: Rodopi.

Khupe, C. (2014). *Indigenous knowledge and school science: Possiblities for integration*. Doctoral dissertation, University of the Witwatersrand, South Africa. Retrieved from http://mobile.wiredspace.wits.ac.za/bitstream/handle/10539/15109/C.%20Khupe%20Thesis.pdf?sequence=2.

Khupe, C. (2017). Language, participation, and indigenous knowledge systems research in Mqatsheni, South Africa. In P. Ngulube (Ed.), *Handbook of research on theoretical perspectives on indigenous knowledge systems in developing countries* (pp. 100–126). IGI Global.

Kincheloe, J., & Steinberg, S. (2008). Indigenous knowledges in education: Complexities, dangers and profound benefits. In N. Denzin, Y. Lincoln, & L. Smith (Eds.), *Critical and indigenous methodologies* (pp. 135–156). Los Angeles: Sage.

Langer-Osuna, J., & Nasir, N. (2016). Rehumanizing the "Other": Race, culture, and identity in education research. *Review of Research in Education, 40*(1), 723–743.

Letseka, M. (2013). Educating for ubuntu/botho: Lessons from indigenous education. *Open Journal of Philosophy*, 337–344.

Mbiti, J. (1969). *African religions and philosophy*. London: Heinemann.

Shava, V. (2016). The application/role of indigenous knowledges in transforming the formal education curriculum: Cases from southern Africa. In V. A. Msila (Ed.), *Africanising the curriculum: Indigneous perspectives and theories* (pp. 121–139). Cape Town: SUN Press.

Smith, L. T. (2012). *Decolonizing methodologies: Research and indigenous peoples* (2nd ed.). London: Zed Books.

Veintie, T. (2013). Practical learning and epistemological border crossings: Drawing on indigenous knowledge in terms of educational practices. *Diaspora, Indigenous, and Minority Education, 7,* 243–258. https://doi.org/10.1080/15595692.2013.827115.

Chapter 31
STEAM Education—A Transdisciplinary Teaching and Learning Approach

Jack Holbrook, Miia Rannikmäe, and Regina Soobard

Chapter Overview

The purpose of this chapter is to recognise the need for a wider view of the science education domain, above and beyond the traditional biology, chemistry and physics, thus seeking the need to encompass technology, engineering, mathematics and also other societal important areas such as art (in its multiple conceptions). This view is seen as STEAM-ED. The approach to STEAM-ED, however, rejects a disciplinary focus and seeks to promote transdisciplinarity, emphasising transdisciplinary skills within a sustainable, world view, based on inquiry learning and using an approach based on social constructivist theory within an 'education through science' frame.

Introduction

Science education has undergone many changes over the last century in response to differing perceptions of the role of education and its purpose in the school setting. While 'science for scientists' can be taken to represent an intellectual, factual and conceptual approach, perhaps heavily embedded in a historical development, other approaches such as science-technology-society (STS) can be seen as more functional and bringing science learning closer to the realities of everyday life (Aikenhead, 1994). According to Holbrook and Rannikmäe (2009), scientific literacy (SL), or the

J. Holbrook (✉) · M. Rannikmäe · R. Soobard
Centre for Science Education, University of Tartu, Tartu, Estonia
e-mail: jack@ut.ee; jack.holbrook@ut.ee

M. Rannikmäe
e-mail: miia@ut.ee

R. Soobard
e-mail: regina.soobard@ut.ee

© Springer Nature Switzerland AG 2020   465
B. Akpan and T. Kennedy (eds.), *Science Education in Theory and Practice*,
Springer Texts in Education, https://doi.org/10.1007/978-3-030-43620-9_31

more society-related, scientific and technological literacy (STL) seek to bridge this divide, seeing both an academic scientific conceptual challenge and a 'science for all' vision. These literacies are also geared to include the development of personal and social skills for responsible citizenship and the acquisition of employability skills (Holbrook & Rannikmäe, 2014). However, whether science education is a prerequisite for emulating a scientist, or for functioning in everyday life, there is always the concern that the science learning is confined to solely scientific ideas and lacks coherence to technology, both as useful science applications in society and as tools to aid science learning. In regard to engineering, science underpins the design, creation and improvement of constructions or products, within the artificially created world as we know it, and towards underpinning all this through mathematical applications. These concerns can be envisaged within or across local, national and global environments, and extend to socio-scientific issues involving creative learning associated with social studies, perhaps focussing on sustainability, artistic endeavours, ethical aspects and other social interactions.

The approach to science education has moved away from a behaviourist learning base. This view assumes a learner is essentially passive, responding to environmental stimuli and behaviour is shaped through positive or negative reinforcement for which the goal is a permanent change of behaviour often translated into memory recall, or comprehension of isolated scientific concepts. Much emphasis is being placed on promoting an inquiry-based or problem-solving learning frame, and in seeking ways to make the learning within science education more intrinsically motivational for students. Strongly encouraged in this area are student-centred practices, based on self-determination theory (Ryan & Deci, 2000) and differential learning within the so-called zone of proximal development (see Chap. 19). Yet there is still a preponderance of a subject specific, content focus, often in individual sub-disciplines within the science field, yet desirably building on a 'simple to complex' vision of learning within the subject itself, but only allowing societal links as a distant after-thought.

This chapter is seeking to re-examine the role of science education in today's changing world. It is based on the realisation that science education is far different in reality from science itself. Science, as a philosophical endeavour, has an ancient history, heavily based on observation and explanations, and formulating theories and laws. However, in the twentieth and twenty-firstcenturies, numerous technologies have emerged which are invading our lives and, through acting as aids within society, have promoted the frontiers of changing lifestyles of people today. While technology may be difficult to define, and artistic design may enable some technologies to become more attractive and popular than others, the interrelationship between science and technology is ever-present. This is even more so when reflecting on the role of engineering and designing within today's created world. There is little doubt that the industrial revolution, arguably the beginning of modern technology, started by the ability to make available cheap and abundant energy (in this case steam power), stimulated further scientific progress. As a result, the developments in technology played a strong role in the new discoveries and developments in science, which in turn, furthered the developments in technology, of which engineering, as the design creation and improvement of technology, played a major role.

But alas, technology, and hence the interrelated science, can be both good and bad. Whether this is related to facilitating today's lifestyle (e.g. modern means of transport, supporting the increasing longevity of human life, the development of the digital world and artificial intelligence), or raising concerns about global warming, environmental degradation, sustainable ways of life, or the engineering of new technologies, all having a foundation in scientific advancements and yet have led to major social dilemmas in today's world. Furthermore, while design can take on an aesthetic dimension, it can lead to ethical and moral dimensions, all indicating the interrelationship of scientific endeavours with the social world. In short, in today's world, the science education dimension, encompassing learning associated with technological/engineering developments, is intertwined with the human or social dimension, especially so when seeking a sustainable world. There is a growing recognition that the learning associated with decision making within socio-scientific issues cannot be ignored, both in societal debates and also in the school science curriculum.

In recognition of the changing world, science education, or the education processes related to an understanding of the natural and artificial world, are, by necessity, also changing. Science education is widening to take note of technological links, societal involvements, creative problem-solving abilities and, inevitably, related to such developments, to make informed decisions, related to both technological choice and socio-scientific reasoning. Science education, even at the school level, can no longer function as isolated training in the promotion of higher level cognitive skills, or even that accompanied by creative, practical endeavours associated with cognitively driven psychomotor skills. Science education cannot ignore its mathematical base, and its interrelationship with the natural and artificial world. Science education has a role to play in preparing today's youth for a future, changing society, preparing students to relate to changing career opportunities and the need to recognise conflicting societal values, whether these are linked to religious intolerance, ways of life, or individuals' freedoms and limitations.

Enhancing STL is a multi-faceted vision and needs to relate to education, itself a moving frontier. With a multi-faceted vision of the role of education, it is inevitable that the older need to keep up with coverage of the ever-expanding scientific knowledge is very much diminished. The needs of society, whether associated with issues related to the local, national or global environment, or the understanding of interactions between the natural and technological world, suggest a demand for strong socio-scientific interlinking. With this, there is a growing need for bringing together scientific ideas, technological endeavours and engineering practices linked to social priority choices. The latter encompasses social, even artistic values, leading to a suggested intermix of scientific endeavours, through the enhancement of transdisciplinary skills, with wider areas of learning, seeing the whole (the overall education gains) as greater than the component parts (the education within subject sub-divisions).

Thus, this chapter proposes that science education is in need of moving from science disciplinary education (SE), from a science and technology education (STE), and even from a science-technology-society education (STSE), to a new vision of

scientific and technological literacy (Holbrook & Rannikmäe, 2009). A transdisciplinary scientific and technological literacy (STL) is encapsulated in terms, such as
STEM (science, technology, engineering, mathematics), or even more, in recognition of the social or artistic direction, STEAM (science, technology, engineering, art,
mathematics), not forgetting that STEAM in relation to education, especially within
a transdisciplinary view, can lead to the realisation of STEAM-ED as the new STL
goal.

The Theoretical and Philosophical Underpinning of STEAM-ED

The approach to learning within science education is grounded on constructivist
theory (see Chaps. 16–26, especially Chap. 18). In interrelating science conceptual
learning with socio-scientific relevance, social constructivism is very much favoured.
Linked to this is the growing recognition of the importance of the context in which
the relevance of the learning for the learner is enhanced.

The recognition of the education emphasis within a vision for science education
can be captured through the expression 'Education through Science (EtS)' (Holbrook
& Rannikmäe, 2007). The term 'Education through Science' can thus be proposed
as a philosophy. This recognises that the learning approach to education relates
to the issues and concerns of society (both present and futuristic), although just
because something is related to everyday life does not automatically mean it is seen
as relevant to students. The relevancy is likely to be linked to the immediate concern
or issue of the society, expressed in the media and impinging on the students' daily
life (from a local, national or global perspective). The emphasis on relevancy and
issues and concerns is seen to be important. However 'education through science'
portrays science education learning as appreciating the nature of science, developing
within student intellectual development as well as positive attitudes and aptitudes,
and acquiring skills associated with society development especially those linked to
interpersonal relationships and in making informed socio-scientific decisions within
society (Holbrook & Rannikmäe, 2007).

Yet as a philosophy, 'education through science' goes further. It recognises the
need to undertake academic challenges, preparation for the world of work and the
need to promote responsible citizenship. It seeks to encompass key learning competences for education, and thus provides a focus for the needs of students in learning
'how to learn' through the gaining of science and technology/engineering competences (accompanied by mathematical competences), interrelated with the importance of promoting social, cultural, entrepreneurial and digital competences enhancing personal and social attribute development and the need to further promote communication abilities in verbal, written, symbolic, graphic as well as digital aspects.
In this, it contrasts with the more standard view of science education, with its focus
on lessons labelled science, or a sub-division and organised by subject content.

Formulating a STEAM-ED Approach

In a quest to conceptualise STEAM-ED the following questions can be put forward:

(1) Is STEAM a more meaningful term to represent science education in a wide transdisciplinary sense, or does it need to be seen as a movement to reflect on a multidisciplinary vision of science education, merely seeking to replace STL as the philosophical pinnacle?

(2) Can STEAM-ED be seen as a more meaningful vision at the operational, or curriculum level to promote a skills enhanced, science education focus reflecting a new way of thinking about inquiry that includes a wide range of societal perspectives, or for considering, designing and implementing tangible solutions to "real world" problems?

Refuting a Disciplinary Approach in Education

Mahan (1970) criticised both the compartmentalisation of the traditional disciplines and ideals of detachment and aloofness associated with any disciplinary inquiry. The disciplines, as both intellectual and social constructs, are nothing more than organisational pillars within a system. Disciplinarity, which may be defined as the compartmentalisation of learning into system-defined units, can be perceived as essential for an understanding of an organisation of particular directions of knowledge. However, the knowledge specified in a discipline, thus defining the discipline, only relates to a small part of a larger picture. This leads to the need for a matrix of disciplines with specific methodologies, paradigms and inherited problem areas. Disciplinary thinking is in danger of becoming all-pervasive when one starts viewing and speaking about everyday matters in terms of specific disciplinary concepts and priorities.

Examining Multidisciplinarity, Interdisciplinarity, Transdisciplinarity

To overcome the limits of a given discipline, multiple disciplines can be grouped together, each giving separate isolated inputs, but still giving rise to the danger of omitting a valued discipline to tackle problems associated with a changing world, whether this is in connection with responsible citizenship, employability preparation, or conceptualising advancements. A Multidisciplinary approach involves the collecting of inputs from different disciplines putting them together without synthesis. Interdisciplinarity finds favour in grouping disciplines together leading to collaboration between researchers from different disciplines aimed at a synthesis and

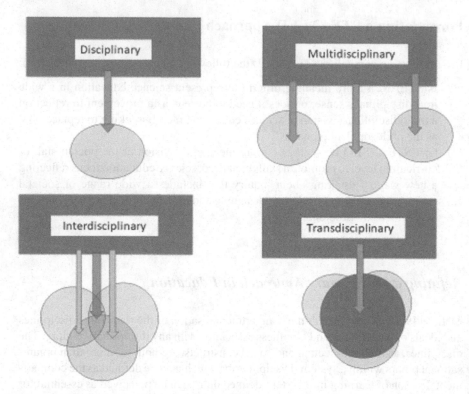

Fig. 31.1 Comparing the four different approaches

integration of knowledge, but the approach is still discipline-led and each discipline employs skills and directions suited to that discipline.

While the distinction between transdisciplinarity, multidisciplinarity and interdisciplinarity need not be sharp or absolute, transdisciplinarity generally rejects the separation and distribution of topics and scholarly approaches into disciplinary compartments (Choi & Pak, 2006).

Below is an illustration of the four different approaches (Fig. 31.1).

A Transdisciplinary Focus for STEAM Education

Transdisciplinarity is a new way of thinking about, and engaging in, inquiry (Montuori, 2008). It goes *beyond* disciplines and identifies with a new knowledge about *what is between, across, and beyond disciplines* (thus the term *trans*)" (McGregor, 2015) and includes approaches from ethical, metaphysical, and even mystical perspectives aiming at designing and implementing tangible solutions to "real world" problems. On the one hand, it can emphasise a concept of the human life world and lived meanings, while on the other, it can emphasise skills to tackle the interface between science, society, and technology in the contemporary world (ibid).

Transdisciplinarity seeks the framing of a topic/big idea as an overarching theme for the inquiry process. It moves instruction beyond just the blending of disciplines and links concepts and skills through a real-world context. Within science education, it aims at creating an engaged socially responsible science. For example, the notion of sustainability has evolved from a concept to a movement involving not only science, government, and industry, but citizen participation, including input from religious leaders, consumer awareness, boycotts and protests, and much more (Cardonna, 2014). Furthermore, with concerns voiced about a possibly dying planet, the need to prevent catastrophe lends a sense of transdisciplinary urgency to this work, with a necessity to not only to raise awareness, but provoke an informed change of behaviour. Evans (2015) has written of a sustainability crisis and thinks educators need to situate their discussions of sustainability in terms that are not only scientific, but ethical, involving "intergenerational fairness extending over long timeframes and on the health and integrity of human societies and the natural world." Sustainability can this be considered as an example of the expected direction of science education learning within transdisciplinary STEAM education?

Transdisciplinary education re-values the role of intuition, imagination and sensibility in the transmission of knowledge. Transdisciplinarity is sometimes described, at least in part, as a response to the increased complexity of contemporary problems in science and technology. Indeed, complexity itself could be a problem area for transdisciplinary studies. Complexity is not exactly synonymous with complicatedness, since a complicated system may be understandable in terms of its components, while in a complex system the individual components interact with each other and with their environment in such a way that the system as a whole cannot be explained in terms of its parts.

What sets transdisciplinarity apart from other approaches, and what assures its role in twenty-first-century education, is its acceptance of, and its focus on, the inherent complexity of reality. This is realised when one examines a problem, or phenomenon from *multiple angles* and *dimensions,* with a view towards discovering hidden connections between different disciplines (Madni, 2007). It is in using this multidimensional complexity to analyse problems and communicate and teach lessons about them that the novel contribution of transdisciplinarity lies (Bernstein, 2015).

Transdisciplinarity does not necessarily need to be seen as applied or practical. Macdonald (2000) insists that transdisciplinarity is as much about the liberal arts, and about cultural symbolisms, as it is about the so-called social and natural sciences, or professions like medicine, engineering, or law. Nevertheless, transdisciplinarity can be viewed as utilising skills for knowledge production, involving knowledge developed for a particular application and involving the work of experts drawn from academia, government, and industry.

Characteristics of Transdisciplinary Inquiry

According to Bernstein (2015), transdisciplinary inquiry-based focus is expected to involve:

(a) transcending disciplinary boundaries in an attempt to bring continuity to inquiry and knowledge;
(b) attention to comprehensiveness;
(c) context and frame of reference of inquiry and knowledge;
(d) interpenetration of boundaries between concepts and disciplines;
(e) exposing disciplinary boundaries to facilitate understanding of implicit assumptions;
(f) processes of inquiry and resulting knowledge;
(g) humanistic reverence for life and human dignity; and
(h) a desire to actively apply knowledge for the betterment of individuals and society.

A further property is *emergence*. Emergence, explained by Holland (2014) through the wetness of water, is seen as a characteristic of 'wetness' which cannot reasonably be assigned to individual molecules. Thus, the 'wetness' of water is not obtained by summing up the wetness of the constituent H_2O molecules—rather 'wetness' emerges from the interactions between the molecules.

A further key characteristic of transdisciplinarity is the tendency to think laterally, imaginatively and creatively, not only about solutions to problems, but on the combination of factors that need to be considered. Thus, inputs from the arts and humanities can transform research and education in sustainability, or other topics that are traditionally viewed as scientific, into an entirely new kind of product (Clark & Button, 2011). This leads to seeing desirable attributes to be developed, abilities to think in a complex, interlinked manner, and engaging in new modes of thinking and taking action.

According to Yakman (2008), STEAM education is an example showing how the boundaries between subjects can be removed and more integrated approaches to science teaching can take place in school settings. Integrating Arts into STEM, the fields of science, technology, engineering and mathematics education, supports students' cognitive, emotional and psychomotor growth, critical thinking, problem-solving skills, creativity and self-expression (Ge, Ifenthaler, & Spector, 2015). Art is an essential component, adding the construction of meaning, expressing observations and creativity (Ge et al., 2015), and working with other to develop transferable skills and abilities to deal with complex problems innovatively and creatively.

STEAM education seeks to relate to careers in truly "helping" professions that build communities and transform nations. These professionals are in charge of solving the complex problems of today's world and its future. They are working to find solutions for global warming, cancer, third world hunger, disappearing habitats, and an interdependent world economy. Yesterday's stereotype of the 'geek' in a lab coat is not representative of today's STL vision, where economists work with researchers on technical transfer and engineers build the state-of-the-art equipment for businesses working with cutting-edge technologies.

The STEAM-ED Emphasis on Skills

Much attention in science education has focused on the development of twenty-first century skills (see Chap. 32). Although these can be defined, grouped and determined differently in specific settings, an overview can be identified as

- Critical thinking, problem-solving, reasoning, analysis, interpretation, synthesising information;
- Research skills and practices, interrogative questioning;
- Creativity, artistry, curiosity, imagination, innovation, personal expression;
- Perseverance, self-direction, planning, self-discipline, adaptability, initiative;
- Oral and written communication, public speaking and presenting, listening;
- Leadership, teamwork, collaboration, cooperation, facility in using virtual workspaces;
- Information and communication technology (ICT) literacy, media and internet literacy, data interpretation and analysis, computer programming;
- Civic, ethical, and social-justice literacy;
- Economic and financial literacy, entrepreneurialism;
- Global awareness, multicultural literacy, humanitarianism;
- Scientific literacy and reasoning, scientific methods;
- Environmental and conservation literacy, ecosystems understanding; and
- Health and wellness literacy, including nutrition, diet, exercise, and public health and safety.

These skills, grouped into five main areas and more specifically categorised can be considered transdisciplinary skills, as illustrated in Fig. 31.2.

Examples of STEAM education in a Transdisciplinary frame

Example 31.1 *Water* (Bernstein, 2015)

A subject such as water falls between the various disciplines and is easily ignored or taken for granted by scholars since it seems on the surface to be neutral. A feature of the landscape, water is something used by animals and plants or that gets combined with other substances, something that makes everything else work, but that seems rather lacking in character in its own right, even though life itself could not exist without it. It has a chemical basis and can be studied from a chemical or physical perspective (hydraulics and hydrology); it is also important in technology, engineering, manufacturing, and equally important, the culinary arts, since there could be no food or drink without water. It is a component of nutrition, digestion, physiology and health; there are sanitation and purity considerations in using water and having it in our environment. There are cultural and religious aspects of water and it is a theme in all the arts. Geographers, geologists, economists and agricultural scientists could

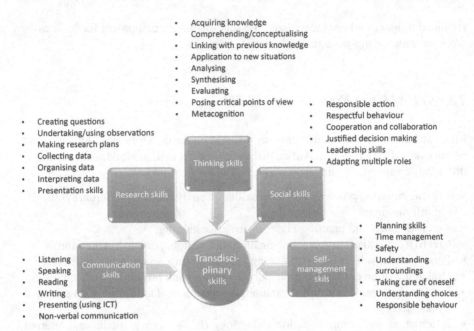

Fig. 31.2 Transdisciplinary skills

study water as a resource. Obviously, the sustainability of water as a resource is an issue, as in the problem of waste caused by packaging in disposable water bottles. There are even political aspects to an important resource such as water, shortages of which can lead to famine, war, revolution, or other vast socio-political changes. One could continue *ad infinitum* about the innumerable facets of water that need to be studied. Questions about water bring together the social sciences, humanities, physical sciences, biological sciences and practical arts and sciences in ways that can be enlightening for educational purposes on the interaction between disciplines.

Example 31.2 Water *related to a STEAM-ED focus*

In this example, A STEAM education approach is used in developing an optional course for 10–12 grade students in Estonia. The whole course includes eight modules, each starting with a motivational scenario introducing a relevant, science-related problem in a society setting to students. Following this comes a science learning component (using an inquiry approach to conceptualise new science content knowledge) and this is followed-up by students relating their new knowledge (science, mathematics) and skills (scientific, engineering, technology, etc.) to interact further with the initial issue faced in the scenario and seek to make a justified decision, based on evidence available for them. In this part, students also need to demonstrate creativity, design skills (for example, product design) and self-expression (oral or written communication, for example). Therefore, the three stages brings together a multitude of disciplines from SSI to STEM to creativity design and self-expression, which now can be referred as STEAM education. This skill-driven approach focuses heavily on social and self-management skills, while the frame is driven by research skills involving all relevant components of STEAM that lead to focused and justified decision making across a transdisciplinary spectrum.

Conclusion

STEAM-ED is seen as a way of interrelating science education to the relevance and issues of society at a local, national or global level, or even beyond into the universe, or even the world of fantasy. Such interrelatedness is seen as problematic, if viewed at a single discipline level and seen as the compartmentalisation of learning into system-defined units perceived as essential for an understanding of the organisation of knowledge. Transdisciplinarity, on the other hand, promotes across curriculum skills and leads to a new way of thinking about knowledge and inquiry. It identifies with a new knowledge about *what is between, across, and even beyond disciplines*.

By and large, the term STEAM or STEAM-ED merely replaces the term STL in its philosophical and societal-related considerations, both seeing science education as wider than the science disciplines and seeing education relating to societal values, employability needs and sustainability at an international level. It is proposed that STEAM-ED can seek more acceptability if it portrays science education in a wide transdisciplinary sense, going beyond, but interconnecting, the individual disciplines. Even more the term STEAM-ED, recognising the education thrust, can be favourably considered in reflecting on new ways of thinking about inquiry that include a wide range of societal perspectives, and for considering, designing and implementing tangible solutions to "real world" problems.

Summary

- Meaning of STEAM-ED
- Social Constructivist Theory
- Issue of Disciplinarity in Education
- Contrasting Multidisciplinary, Interdisciplinary and Transdisciplinary
- Introduction to Transdisciplinarity
- A STEAM-ED approach through transdisciplinarity
- Transdisciplinary inquiry
- Transdisciplinary skills
- Example of Transdisciplinary STEAM-ED.

Recommended Resources

(Nil).

References

Aikenhead, G. (1994). What is STS science teaching? In J. Solomon & G. Aikenhead (Eds.), *STS education: International perspectives on reform* (Chapter 5). New York: Teachers College Press.

Bernstein, J. H. (2015). Transdisciplinarity: A review of its origins, development, and current issues. *Journal of Research Practice, 11*(1), Article R1. Retrieved from http://jrp.icaap.org/index.php/jrp/article/view/510/412.

Cardonna, J. L. (2014). *Sustainability: A history*. Oxford, UK: Oxford University Press.

Choi, B. C. K., & Pak, A. W. P. (2006). Multidisciplinarity, interdisciplinarity, and transdisciplinarity in health research, services, education, and policy: 1. Definitions, objectives, and evidence of effectiveness. *Clinical Investigative Medicine, 29*(6), 351–364.

Clark, B., & Button, C. (2011). Sustainability transdisciplinary education model: Interface of arts, science, and community. *International Journal of Sustainability in Higher Education, 12*(1), 41–54.

Evans, T. L. (2015). Transdisciplinary collaborations for sustainability education: Institutional and intragroup challenges and opportunities. *Policy Futures in Education, 15*(1), 2015, 70–97.

Ge, X., Ifenthaler, D., & Spector, J. M. (2015). Moving forward with STEAM education research. In: X. Ge et al. (Eds.), *Emerging technologies for STEAM education, full STEAM ahead* (pp. 383–395). Switzerland: Springer International Publishing.

Holbrook, J., & Rannikmäe, M. (2007). Nature of science education for enhancing scientific literacy. *International Journal of Science Education, 29*(11), 1347–1362.

Holbrook, J., & Rannikmäe, M. (2009). The meaning of scientific literacy. *International Journal of Environmental & Science Education, 4*(3), 275–288.

Holbrook, J., & Rannikmäe, M. (2014). The philosophy and approach on which the PROFILES project is based. *Center for Educational Policy Studies Journal, 4*(1), 9–21.

Holland, J. H. (2014). *Complexity: A very short introduction*. Oxford, UK: Oxford University Press.

Macdonald, R. (2000). The education sector. In M. A. Somerville & D. J. Rapport (Eds.), *Transdisciplinarity: Recreating integrated knowledge* (pp. 241–244). Oxford, UK: EOLSS.

Madni, A. M. (2007). Transdisciplinarity: Reaching beyond disciplines to find connections. *Journal of Integrated Design and Process Science, 11*(1), 1–11.

Mahan, J. L., Jr. (1970). *Toward transdisciplinary inquiry in the humane sciences*. Doctoral dissertation, United States International University. UMI No. 702145. Retrieved from ProQuest Dissertations & Theses Global.

McGregor, S. L. T. (2015). The Nicolescuian and Zurich approaches to transdisciplinarity. *Integral Leadership Review, 15*(2). Retrieved from http://integralleadershipreview.com/13135-616-the-nicolescuian-and-zurich-approaches-to-transdisciplinarity/.

Montuori, A. (2008). Foreword: transdisciplinarity. In B. Nicolescu (Ed.), *Transdisciplinarity: Theory and practice* (pp. ix–xvii). Cresskill, NJ: Hampton.

Ryan, R. M., & Deci, E. L. (2000). Self-determination theory and the facilitation of intrinsic motivation, social development, and well-being. *American Psychologist, 55*(1), 68–78.

Yakman, G. (2008). STEAM education: An overview of creating a model of integrative education. Presented at the Pupils' Attitudes Towards Technology (PATT-19) Conference: Research on Technology, Innovation, Design & Engineering Teaching, Salt Lake City, Utah, USA.

Jack B. Holbrook is a visiting professor at the Centre for Science Education, University of Tartu, Estonia. Initially trained as a chemistry/maths teacher in the UK (University of London), Jack spent 5 years as a secondary school teacher before moving into teacher training, first in the UK followed by Tanzania, Hong Kong and Estonia. Currently, Jack is involved in guiding science education Ph.D. students, European science education projects and being an International Consultant in Curriculum, Teacher Education and Assessment. Jack's qualifications include a Ph.D. in Chemistry (University of London), FRSC from the Royal Society of Chemistry (UK) and Past President and Distinguished Award Holder for ICASE (International Council of Associations for Science Education). Jack has written a number of articles in international journals and as a co-editor a book entitled 'The Need for a Paradigm Shift in Science Education in Post-Soviet Countries.'

Miia Rannikmäe is Professor and Head of the Centre for Science Education, University of Tartu, Estonia. She has considerable experience in science education within Estonia, Europe and world-wide (Fulbright fellow—University of Iowa, USA). She is an honorary doctor in the Eastern University of Finland. She has a strong school teaching background, considerable experience in pre- and in-service teacher education and has strong links with science teacher associations worldwide. She has been a member of a EC high-level group publishing a report on 'Europe needs more Scientists'. She has been running a number of EC-funded projects and Estonian research grants. Her Ph.D. students are involved in areas such as scientific literacy, relevance, creativity/reasoning, inquiry teaching/learning and the nature of science.

Regina Soobard is a research fellow in the Centre for Science education, University of Tartu, Estonia. She earned her Ph.D. in science education at the University of Tartu (2015) on gymnasium students' scientific literacy development based on determinants of cognitive learning outcomes and self-perception. She is teaching at the MSc level and holding the position of director of the gymnasium science teacher programme, as well as co supervising Ph.D. students in science education and educational sciences. She has been awarded BAFF a scholarship for research in Michigan State University, USA.

Chapter 32
21st Century Skills

Teresa J. Kennedy and Cheryl W. Sundberg

Introduction

This chapter seeks to provide an overview of 21st century skills (21CS) as presented in a number of education and policy documents related to science, technology, engineering, and mathematics (STEM) education. Toward the end of the last century, and into the 21st century, the education community has been challenged with developing educational programming to keep pace with accelerating change in the economy and advances in technology. The rapid state of change in the workplace continues to increase demands on the education system to prepare students for an evolving workforce. Traditional approaches to education that focus on knowledge acquisition are not typically seen as meeting the needs of employers in the current global information economy context. Promising pedagogies such as Project-Based Learning (PBL) and Problem-Based Learning (PrBL), described in Chap. 23, as well as Phenomenon-based Learning (PhBL), help students learn information and address 21CS. PBL, PrBL, and PhBL pedagogies allow students to choose projects, problems, and phenomena to study as outlined in Chap. 3, Glasser's Choice Theory, and produce a product, as described in Chap. 7, Bandura's Social Learning Theory. Promising new frameworks for organizing education, such as through various STEM applications, along with the transdisciplinary approach to science education referred to as STEAM (science, technology, engineering, the arts, and mathematics) as discussed in Chap. 31, break down the artificial barriers of disciplines enabling students to understand the connected nature of knowledge utilizing critical skills leading to success in the 21st century economy. In addition, integrating new media approaches

T. J. Kennedy (✉)
College of Education and Psychology; College of Engineering, University of Texas at Tyler, Tyler, USA
e-mail: tkennedy@uttyler.edu

C. W. Sundberg
The Ronin Institute, Montclair, USA
e-mail: sundbergrc@bellsouth.net

© Springer Nature Switzerland AG 2020
B. Akpan and T. Kennedy (eds.), *Science Education in Theory and Practice*,
Springer Texts in Education, https://doi.org/10.1007/978-3-030-43620-9_32

to teaching and learning, as well as utilizing technological innovations in the science classroom, further connects students to real-life situations and creates opportunities to meaningfully construct concepts and relationships in context (see Chap. 9). Such frameworks allow students to focus on necessary content acquisition as well as 21CS and may provide schools the opportunity to keep pace with the increasing changes occurring in the workplace.

Education is often a reactive institution in responding to the needs of society. In the current century, the pace of advancements in society and the workforce are creating stress on the education system. Unlike the past, workforce needs change so fast that once education systems update curriculum and technology they become obsolete. Breaking down traditional disciplinary barriers in education and embracing pedagogies that include transferable skills to the workplace could help education keep pace with the rapid changes occurring. In other words, education systems, especially those promoting STEM education and supporting education for innovation, must become proactive and focus on meaningful learning (see Chap. 12). Since it is difficult to predict future change, and given that the costs associated with keeping up with technological developments tend to be prohibitive for schools, focusing on transferable 21CS makes sense. We already know the limitations of "re-active" education policies based on history of education reform efforts. Reactive education policy and practices cannot address the rapid changes that are occurring in the economy. Focusing on 21CS and school models that support 21st century STEM learning will allow educators to place emphasis on the future by aligning how students learn using the skills needed to survive and thrive in the modern workplace, rather than focusing on a defined set of knowledge that is quickly out of date.

Background

"21st century skills," "21st century learning," and "college and career readiness" are currently common phrases in the field of education. The push to embed workforce-related skills into STEM education can be traced back to the late 1970s. Ultimately, these efforts laid the foundation for the identification and promotion of 21CS into STEM education. Since the early 1980s, academia, government agencies, non-governmental organizations, and industry have invested in research to identify academic skills and competencies needed for the current and future workforce. Although the identification of specific 21CS in need of implementation in the workplace initiated in the United States, much attention to these goals spread across the globe as economies have expanded. In addition to efforts in the U.S., a number of reports released by international organizations, such as the Asia Pacific Economic Cooperation (APEC), the United Nations Educational, Scientific, and Cultural Organization (UNESCO), as well as the Organization for Economic Co-operation and Development (OECD), and their Business and Industry Advisory Committee, have made

substantial recommendations concerning 21CS. In addition, there have been significant publications from Australia, Canada, New Zealand, and the United Kingdom,
among other countries.

A Nation at Risk

Looking back, in 1981, then U.S. Secretary of Education, Terrel Bell, under President
Ronald Reagan, convened the National Commission on Excellence in Education,
tasked with examining the overall quality of education in the United States and
making recommendations for future educational improvements, including content
covered and mastery of skills. The commission issued the report entitled *A Nation at
Risk: The Imperative for Educational Reform* (National Commission on Excellence
in Education, 1983). This report had a profound impact on education in the United
States and, as a result, several reform efforts originated to improve education. A key
recommendation stated that educational reform should focus on creating a "Learning
Society." The report defined a learning society as follows:

> In a world of ever-accelerating competition and change in the conditions of the workplace,
> of ever-greater danger, and of ever-larger opportunities for those prepared to meet them,
> educational reform should focus on the goal of creating a Learning Society. At the heart of
> such a society is the commitment to a set of values and to a system of education that affords
> all members the opportunity to stretch their minds to full capacity, from early childhood
> through adulthood, learning more as the world itself changes. Such a society has as a basic
> foundation the idea that education is important not only because of what it contributes to
> one's career goals but also because of the value it adds to the general quality of one's life.
> Also at the heart of the Learning Society are educational opportunities extending far beyond
> the traditional institutions of learning, our schools and colleges. They extend into homes and
> workplaces; into libraries, art galleries, museums, and science centers; indeed, into every
> place where the individual can develop and mature in work and life. (The Learning Society
> section, para. 1).

Although a *Nation at Risk* did not address 21CS directly, the report did lay the
foundation for future reports and symposia from around the globe issuing similar
recommendations while addressing their own context. A few of the most influential
reports follow:

- *The Hobart Declaration on Common and Agreed National Goals for Schooling
  in Australia.* (Australian Education Council, 1989);
- *Learning: The Treasure Within.* (United Nations Educational, Scientific, and
  Cultural Organization, 1996);
- *Learning for the 21st Century.* (Partnership for 21st Century Skills, 2003);
- *21st Century Skills Realising Our Potential: Individuals, Employers, Nation.*
  (Crown Copyright, UK, 2003);
- *Rising Above the Gathering Storm: Energizing and Employing America for a
  Brighter Economic Future.* (National Academy of Sciences, National Academy
  of Engineering, & Institute of Medicine, 2007);

- *21st Century Learning: Research, Innovation and Policy, Directions from recent OECD analyses.* (OECD, 2003, 2005, 2008);
- *Education to Achieve 21st Century Competencies and Skills for All: Respecting the Past to Move Toward the Future.* (Asia Pacific Economic Cooperation, 2008a, 2008b, 2008c);
- *A Framework for K-12 Science Education: Practices, Crosscutting Concepts, and Core Ideas.* (National Academies Press, 2012);
- *AAAS Science Assessment.* (American Association for the Advancement of Science, 2015);
- *Shifting Minds 3.0: Redefining the Learning Landscape in Canada.* (C21 Canada, 2015); and
- *The Futures of Learning 1: Why Must Learning Content and Methods Change in the 21st Century?* (Scott, 2015).

Comparisons of International Student Assessments

International comparisons of student achievement scores revealing the need for improved academic success were included in many of the reports citing 21CS. Assessments such as the Programme for International Student Assessment (PISA), launched in 1997 by member countries of the OECD (2014a, 2014b), were used to monitor if students had acquired the knowledge and skills necessary for full participation in society. PISA scores became an increasingly important tactic to draw attention to the case that students in other countries may be outperforming students in their own country. Given the competitiveness of our global economy, PISA results have been used as an indicator of how well education systems are preparing students in mathematics and science, STEM subjects that are critical to the innovation economy.

In addition to assessing students' knowledge and skills, the OECD examined student competencies and identified three "Competency Categories" (OECD, 2003, 2005). These categories included: (1) using tools interactively, (2) interacting in heterogeneous groups, and (3) acting autonomously. "Using tools interactively" was defined as acquiring and using language, information, and knowledge, including using the tools of technology. Simply having access to technology does not guarantee students have the technical skills to use the technology. For example, many schools provide students with computers but do not equip teachers and students with the skills to use them in a meaningful context.

Interacting in "heterogeneous groups" refers to relating well to others. These skills are often referred to as "soft skills" and include the ability to be cooperative and work in teams. In addition, they also require students to manage and resolve conflicts, which involves using advanced communication skills. The final category is "acting autonomously," which includes knowing how the individual fits within the greater context of society. Acting autonomously is important, since unlike the past where one's position was well-defined, that is no longer the case today. The ability to plan

Table 32.1 Organization For Economic Cooperation and Development competency categories

Using tools interactively	Interacting in heterogeneous groups (soft skills)	Acting autonomously
Utilizing technology	*Relating to others*	*Fitting into society*
• Language • Symbols • Texts • Information • Knowledge • Technology	• Be cooperative • Work in teams/groups • Manage conflicts • Resolve conflicts • Communication skills	• Form and conduct life plans • Conduct personal projects • Define projects and set goals • Identify and evaluate resources • Prioritize and refine goals • Balance resources/meet goals • Learn from past actions • Monitor progress/make adjustments • Defend/assert one's rights, interests, limits, needs

and carry out life plans and coordinate personal projects is especially important, but it also has implications for science education. Similar to the project-based learning approach, this competency assumes individuals are able to: (a) define projects and set goals, (b) identify and evaluate both the resources they have access to as well as the resources they need, (c) prioritize and refine goals, (d) balance resources to meet multiple goals, (e) learn from past actions and plan future outcomes, and (f) monitor one's progress, adjusting goals as the project unfolds. Acting autonomously also addresses the need to defend and assert one's personal goals and needs (OECD, 2005). See Chap. 5, *Bildung* Theory and Chap. 13, discovery learning in a cultural context. Table 32.1 lists the three OECD competency categories and their associated skills.

Defining Skills for Success

The National Education Association (NEA) of the United States, working in cooperation with educators, education experts, and business leaders in the U.S., defined the skill sets necessary for success in work, life, and citizenship in 2002 known as the *Framework for 21st Century Learning* (National Education Association, 2012). The report highlighted 18 skills that schools could refer to when building standards, professional development, and assessments. Ten years later, many believed the framework was too long and complicated, and therefore these skills were further refined into four primary skills termed the "Four C's": critical thinking, communication, collaboration, and creativity (Partnership for 21st Century Skills, 2010, 2015).

In 2012, the U.S. National Research Council (NRC) categorized the 21CS to more clearly show the relationships between skills, as well as summarized evidence collected in support of 21CS development. One of the concerns that the NRC addressed was the lack of specific definitions for the terms used to describe each skill. They further organized the skills into three broad competency domains: (1) the cognitive domain, including thinking and reasoning; (2) the intrapersonal domain, involving self-management and the ability to regulate one's behavior and emotions; and (3) the interpersonal domain, focusing on self-expression, interpretation of messages, and appropriate response (NRC, 2012). A content analysis of the existing lists of 21CS was completed and the skills were further grouped within the three domains. Table 32.2 displays the three NRC domains and their associated competencies/skills. The Four C's (critical thinking, communication, collaboration, and creativity) are clearly visible within this framework.

The World Economic Forum's report a *New Vision for Education: Unlocking the Potential of Technology* highlighted the growing deficit in 21CS development in our youth and included strategies focused on addressing this gap through technology. Sixteen skills were identified for student success in the 21st century and emphasized the need for "lifelong learning" (World Economic Forum, 2015, p. 3). These 16

Table 32.2 National Research Council competency categories

Cognitive domain	Intrapersonal domain	Interpersonal domain
Cognitive processes/strategies	*Intellectual openness*	*Teamwork/collaboration*
• Critical thinking • Problem solving • Analysis • Reasoning and argumentation • Interpretation • Decision-making • Adaptive learning	• Flexibility • Adaptability • Artistic and cultural appreciation • Personal and social responsibility • Appreciation for diversity • Adaptability • Continuous learning • Intellectual interest and curiosity	• Communication • Collaboration • Cooperation • Teamwork • Coordination • Interpersonal skills
Knowledge	*Work ethic/conscientiousness*	*Leadership*
• Information literacy • ICT literacy • Oral and written communication • Active listening	• Initiative • Self-direction • Responsibility • Perseverance • Grit • Career orientation, ethics • Integrity • Citizenship	• Responsibility • Assertive communication • Self-presentation • Social influence with others
Creativity	*Positive core self-evaluation*	
• Creativity • Innovation	• Self-monitoring • Self-evaluation • Self-reinforcement • Physical and psychological health	

Table 32.3 World Economic Forum competency categories

Foundational literacies	Competencies	Character qualities
Applying core skills to everyday tasks	*Approaching complex challenges*	*Approaching their changing environment*
• Literacy • Numeracy • Scientific literacy • Information and communications technology literacy • Financial literacy • Cultural and civic literacy	• Critical thinking/problem-solving • Creativity • Communication • Collaboration	• Curiosity • Initiative • Persistence/grit • Adaptability • Leadership • Social and cultural awareness

skills were divided into three categories, (1) Foundational Literacies, defining how students apply core skills to the tasks they are faced with on a day-to-day basis; (2) Competencies, defining how students solve complex problems and challenges; and (3) Character Qualities, defining how students react to the environment around them. Table 32.3 shows the World Economic Forum 21CS. Once again, the Four C's emerge.

The Queensland Curriculum and Assessment Authority (2017) assembled an explanation of 21CS providing teachers with a clear picture of supportive student expectations. Creative pedagogy is crucial for 21st century learning and skill development. Instructional interventions and techniques such as team-teaching, social constructivist game design/game play, as well as uses of social media including wikis and online communications, help to equip students with 21CS (Chu, Reynolds, Tavares, Notari, & Lee, 2017). See Chap. 4 for information on gaming design and Chap. 9 regarding New Media Technologies.

The Four C's are present in a number of 21CS policy documents from around the globe in one fashion or another, often times including social and emotional intelligence, technological literacy, and problem-solving skills (See Chap. 27, Multiple Intelligences and Chap. 28, Systems Thinking). Application of knowledge and moving beyond rote memorization is required, thus PBL/PrBL/PhBL approaches push students to be more creative, use multiple technologies in their projects, and develop the higher-level thinking skills needed in higher education and the workplace. Creativity is a motivation for learning (see Chap. 2).

Crockett, Jukes, and Churches (2011) further sought to refine the classification of 21CS. Based on New Zealand's Ministry of Education (2007) definition for the five key competencies for living and lifelong learning (thinking; using language, symbols, and text: managing self; relating to others; participating and contributing), along with the International Baccalaureate Program (2010) desired student outcome skills (inquirers; knowledgeable; thinkers; communicators; principled; open-minded; caring; risk-takers; balanced; reflective), Crockett, Jukes and Churches contend that students need to develop transparency-level skills in the following six areas: problem-solving; creativity; analytical thinking; collaboration; communication; and ethics, action and accountability. Further to this, Crockett and Churches (2018) push for schools to play a significant role in preparing students for life beyond the classroom. They postulate that purposefully teaching skills that focus on effective and ethical participation in online and offline communities promote critical thinking and the development of global digital citizenship (GDC) practices, resulting in students who are contributors and valuable citizens.

For the purposes of this chapter, we are interested in and will focus on 21CS for STEM Education. STEM is a critical component supporting our technology-rich global economy. To maintain technological development, there is a need to prepare students in the STEM disciplines within a 21st century context. STEM education intersects with 21CS to include rigorous core content with critical thinking skills (Bybee & Fuchs, 2006).

21st Century STEM Education: A Second Renaissance?

Driven by economic forces and global competition, refocusing education on the development of skills required for success in our rapidly changing digital society has become an international quest. Many of these skills are associated with inquiry learning described in numerous STEM education reform documents. These documents also place priority on mastering skills in analytic reasoning and complex problem-solving (see Chap. 11, Mastery Learning), along with working collaboratively as a team player, rather than solely developing traditional academic skills based on content knowledge. 21CS and effective STEM pedagogy appear to have a number of related goals.

Historians generally date the first Renaissance in Western culture to have occurred around the 11th century (Montgomery & Kumar, 2015). Descriptors for the First Renaissance typically involve the convergence of the arts with science, technology, engineering, and mathematics (STEAM), as described in Chap. 31. However, traditional education has isolated STEM subjects into distinct disciplines. The artificial barriers between STEM disciplines are disappearing, as each of the components of science, technology, engineering, and mathematics are necessary for innovation. In this sense, we see a resurgence of the convergence of STEM disciplines, non-STEM disciplines, and 21CS such as the Four C's. For example, stereotypes of engineers as expert loners promulgated by the media are counterproductive. In stark contrast, high tech multimedia cartoonists and science fiction illustrators function as teams of artisans, musicians, and technicians (Disney, 2016). Science educators assert the importance of the role of science in today's society. From the WWII Generation, Baby Boomers, and Millennials, our global society is united by technology, and thus, we have life-long education mediated by the latest gizmo technology, from tweeted homework reminders, cell phones, and tablet-based educational games and apps, to self-driving vehicles.

In the STEM context, what is meant by the term 21CS? In general, essential skills in STEM education include rigorous core content (biology, chemistry, Earth sciences, engineering, mathematics, physics, and technology) combined with critical thinking skills (Bybee & Fuchs, 2006). In addition, business and education leaders increasingly tout the importance of *soft skills* (effective verbal and written commu- nication, career readiness, and emotional I.Q.) for the workforce in a global, high technology economy (Broadening Advanced Technological Education Connections, 2013). Members of today's workforce need teamwork and collaboration skills and the ability to apply critical thinking and problem-solving skills within day-to-day scenarios (The Conference Board, Inc., 2006).

A central barrier to reform is the dichotomy between the importance of *hard skills*, referring to rigorous core content, and the need for soft skills, such as col- laboration, creativity, and work ethic. Student success relies on "greater connection between traditional core courses, hard skills, and soft skills like social skills and workforce readiness" (Voogt, Erstad, Dede, & Mishrass, 2013, p. 410). In addi- tion, Silva (2009) valued not only cognition of core constructs but emphasized the essence of 21CS to increase students' capacity for applying knowledge to real-world scenarios. PBL/PrBL/PhBL provide an inquiry-based pedagogy for the inclusion of 21CS. These pedagogical approaches promote student-centered opportunities for deeper learning. Collaboration is essential for problem-solving because the com- plexity of real-life problems require a variety of skills. Özdemir (2019) postulated entrepreneurship often begins with a problem from an individual who seeks out col- laboration from others with diverse expertise, which Özdemir termed *soloborative learning, solo thinking, collaborative tinkering* (see Chap. 8 Connectionism, Edward Thorndike).

PBL/PrBL/PhBL also embed critical thinking, problem-solving, and many of the soft skills included in the Four C's into the STEM classroom. Beers (2016) listed creativity and innovation as crucial skills involved in the processes of problem- solving and the creation of new products and services. In addition to academic skills, Beers noted 21st century students need cultural awareness to succeed in the global economy.

In a position statement, the National Science Teaching Association (2011) indi- cated science has many inherent connections to 21CS including complex communica- tion skills and unique problem-solving scenarios/strategies. In an analogous manner, Beers (2016, p. 5) reported the emphasis on STEM intersects with 21CS, stating that STEM is "inherently cross-curricular." In STEM professions, a team approach is used to solve complex problems and, often, workable solutions are mediated with technology. STEM professionals use a wide variety of technology tools and software applications in problem-solving. For example, spreadsheets are used to mine and analyze data. Data collecting technologies, such as satellite imaging of atmospheric phenomena, are submitted to large databases, analyzed by computer software, and the information gleaned from the analyses is utilized in the building of complex mod- els and hypotheses, such as weather forecasting. See The GLOBE Program, https:// www.globe.gov, for more information about involving students in large-scale inter- national data collection activities, and refer to Chap 9, New Media Technologies and Information Processing Theory—George A. Miller and Others.

In 2015, a landmark research study in the U.S., conducted by the Research Consortium on STEM Career Pathways, surveyed high school students in STEM classes and concluded that "creativity is an essential skill for 21st century students" (Educational Research Center of America, 2016, p. 3). The study reported that increasing student efficacy for confidence in the ability to learn and select a career in STEM strongly correlates with a creative component in STEM classes in primary and secondary schools. While this study noted that males were consistently more confident of their STEM abilities than females, it also determined that when including a creative component in the curricula, students, both female and male, were approximately twice as likely to report confidence in their ability to learn STEM. The study concluded that "creative learning matters," further suggesting that "greater access to STEM learning environments which students themselves see as creative might boost the STEM confidence of a generation" (Educational Research Center of America, 2016, p. 6).

This same study also noted that in addition to the gender gap described above, historically marginalized groups in the U.S. (women, African Americans, Hispanics, and Native Americans) often did not plan to pursue a STEM career, despite students acknowledging STEM is important for future career aspirations. According to data collected by the Educational Research Center of America, female students were 38% less likely to select a STEM profession and historically marginalized groups were not provided with an adequate number of advanced courses in STEM. To address equity, the Educational Research Center of America recommended including creativity learning in all STEM courses and increasing access of advanced STEM courses for historically marginalized students, significantly underrepresented in STEM professions. To augment the traditional classroom, after school STEM clubs, maker spaces, and camps can provide opportunities for increased representation of historically marginalized students in STEM professions (Educational Research Center of America, 2016, p. 8). See Chap. 29 for additional information addressing the gender gap in STEM education.

What Makes an Effective 21st Century STEM Education Program?

According to the National Research Council Committee on Highly Successful Schools or Programs for K-12 STEM Education (NRC, 2011), one factor for achievement in highly successful schools for students ages 5–18 is the implementation of a STEM blueprint, guiding a college preparatory curriculum with focused emphasis on college readiness for all students. Along with committed educators and community leaders, the curricula should involve active student learning, linking prior knowledge to new knowledge in learning tasks where students are engaged in the practices of science.

Many effective programs promoting 21st century STEM skills have been function-ing for years in countries around the world. In addition to the International Baccalau-reate program previously mentioned, Montessori Schools and their K-12 Academies, Reggio Emilia Schools, and schools within the Waldorf Education System, have also earned global reputations. Short descriptions of each follow.

Montessori Schools adhere to a method of early childhood education building on the way children learn naturally. Developed in 1907 by Italian physician, educator and innovator Maria Montessori, the schools encompass a child-centered educa-tional philosophy involving multi-age grouping in the classroom, mirroring real-world interactions between people of all ages. The Montessori methods congruent with 21CS include encouraging students to think critically, work collaboratively, and act boldly (American Montessori Society, 2018).

The North America Reggio Emilia Alliance (NAREA), based on the early child-hood education philosophy termed *Reggio Emilia*, was developed by Loris Malaguzzi in 1945 in Italy. The NAREA early childhood education approach centers on the phi-losophy that intelligent children deserve intelligent teachers (NAREA, 2018). 21st century STEM skills supported by NAREA schools include relating to others in cooperative experiences and *Progettazione,* projects designed by teachers in cooper-ation with their students who in turn share the results of their projects with the larger group to promote learning from one another (Reggio Emilia Australia Information Exchange, 2018).

Waldorf Education, established by Emil Molt and Rudolf Steiner in 1919, has its foundations in Anthroposophy, the belief that humanity has the wisdom to transform itself and the world, through one's own spiritual development (Waldorf Education, 2018). Waldorf schools integrate arts in all curricular areas, and currently there are more than one thousand Waldorf/Steiner schools in over 60 countries. The education philosophy of Waldorf schools support 21CS, and reports show "94% of Waldorf graduates attended college or university with almost half selecting a STEM major" (Mitchell & Gerwin, 2007; Montgomery & Kumar, 2015, p. 16).

Many charter and private schools in the U.S. take on specific disciplinary themes such as STEM and other content areas. For example, High Tech High (HTH), an inte-grated network of thirteen charter schools in San Diego, California, serves students age 5–18. HTH has its own teacher certification program, and focuses on curricula involving PBL supported by state-of-the-art technology. While the numbers of total alumni attending and/or completing a university education are impressive, 30% of HTH graduates enroll in a STEM field compared to 17% of high school graduates across the U.S., with 35% of their graduates attending university as first-generation college students (HTH, 2016). Why is HTH so successful? HTH's philosophy centers on the concept of *Teacher as designer*, involving teacher teams designing the courses they teach and providing students with opportunities to work on real-life problems that are meaningful and important to them in order to increase student engagement (Cernavskis, 2015; HTH, 2016).

Another U.S. public charter school STEM network, the Denver School for Science and Technology (DSST), includes fourteen schools on eight campuses in Denver, Colorado, and utilizes similar philosophies related to project and problem-based pedagogies. DSST schools increase student achievement for all students at a significantly faster rate than comparable schools (Carroll, 2015).

Critics of traditional high schools often recommend competitive enrollment for effective STEM schools. One unique facet of DSST is the policy of open enrollment, which is often atypical for other STEM-focused schools (Cernansky, 2013). However, despite the open enrollment policy, Cernansky reported student achievement on state assessments is high, citing that students graduating from DSST enroll in STEM programs at three times the national average, with a diverse student population (high percentage of traditionally marginalized students in STEM, girls, and low-socioeconomic students). What makes the difference in achievement? School culture is cited as one factor for the success of DSST. In addition, students in their junior year at DSST are required to complete a STEM-focused internship offering students a needed conceptual bridge from school into higher education and the work force.

In the U.S., there are several emerging STEM school models that promote 21st century STEM skills. Manor New Technology High School (MNTH) in Texas was created as part of a statewide STEM initiative entitled *The Texas High School Project*; creating inclusive STEM high schools where students are accepted in the program based on interest in STEM as opposed to traditional acceptance criteria, high aptitude or prior achievement (Lynch et al., 2017). Opportunity structures are built into the design of the MNTH curricula, providing students with support in pursuing STEM opportunities and careers. Evaluation of program success cites almost 100% of MNTH students graduated high school and have been accepted into post-secondary programs. In addition, scores on the 8th grade assessment conducted during the 2007 school year indicated that the students enrolled at MNTH scored above average in science, 65% students meeting standard, compared to 53% their peers meeting standard in other schools in the district (Lynch et al., 2017).

The Texas High School Project was established by the 79th Texas Legislature, allowing students in high school to complete two years of college concurrently with completion of a high school diploma (Chapa, Galvan-De Leon, Solis, & Mundy, 2014; SRI International, 2018). STEM academies emerged as a result, transforming schools into 21st century learning communities (Kennedy & Odell, 2014). The foundation of the Texas High School Project model is supported through four pillars: (1) effective teachers; (2) supportive and knowledgeable educational leaders; (3) learning systems (curricula, scheduling, and classroom design); and (4) streamlined data analysis of performance (Haney, Holland, Moore, & Osborne, 2013, p. 25). All Texas STEM (T-STEM) academies follow and implement the same *Design Blueprint,* serving as a guide to produce college and career-ready students. The University of Texas at Tyler University Academy extended the T-STEM Design Blueprint to encompass K-12 students, engaging students in PBL/PrBL/PhBL as the primary pedagogy for learning, resulting in a shift from a teacher-centered learning environment to a student-centered learning environment (Odell, Kennedy, & Stocks, 2019). 21st century STEM skills development is an important part of the school assessment model. Students not only receive grades for content achievement but are also evaluated on acquiring 21st century STEM skills.

Evaluation of dual enrollment in the Texas High School Project revealed students who completed college courses during their high school years, attended and completed an Associate's degree or higher during their college experience. These findings, along with findings of the Central Texas Student Futures Project, were particularly significant for completion rates held for traditionally marginalized groups, racial minorities, and students from low-income families (Cumpton & King, 2013; Struhl & Vargas, 2012) and showed that students from diverse backgrounds attending STEM academies outperformed their counterparts at traditional high schools (Kennedy & Odell, 2014). There are many STEM schools emerging in the U.S. and across the globe that provide access to any student interested in STEM. These schools implement 21st century pedagogies involving students in real-world projects.

Recommendations

The review of the preceding reports and STEM program descriptions are not exhaustive but provide enough evidence that there are clearly areas of commonality in the 21CS and learning documents, as well as from successful STEM school models from around the world. We are 20 years into the 21st century and we are still struggling to define 21CS with precision due to the dynamic nature of the skill sets necessary for success. These definitions will undoubtedly evolve as we move into the 22nd century.

In many countries, including the U.S., educational policies and accountability systems rely solely on standardized tests structured around memorization of facts and procedures. As long as accountability systems reward scores on achievement tests, schools will focus on maximizing test scores rather than on developing 21CS. Of particular importance today is the need to evaluate the actual level of penetration of 21CS into classroom instruction.

The U.S. Bureau of Labor Statistics estimates "that employment in STEM-related fields will increase by one million between 2012 and 2022" (Educational Research Center of America, 2016, p. 3). Teaching students to be creative producers of knowledge and innovation supports the development of 21st century STEM skills. Specialty Schools, especially those that focus on skills rather than test scores, can foster future STEM leaders and immerse students in high-quality STEM education aimed at developing 21CS. These environments, flexible in design, can also provide a venue to test teaching materials and provide professional development to prepare teachers ready to immerse their students in environments rich in PBL/PrBL/PhBL. In addition, open enrollment STEM schools, such as those described earlier, can serve as a transition to STEM majors and careers for all students.

The 22nd Century is just around the corner and we need to be forward-thinking. The constant reoccurrence of the Four C's implies that the most critical skills for students of all ages are represented through a continuing thread throughout most of the 21CS documents. Developing 21CS begins in the early years and progresses through secondary school, requiring educators to align how students learn through concept acquisition and development while promoting students to use the skills needed to survive and thrive in the modern workplace. Learning about scientific practice, through Problem and Project-based Learning, Phenomenon-based Learning, and other transdisciplinary STEM approaches to teaching science education, promotes student engagement in scientific inquiry for and by themselves. 21CS is a vehicle for promoting socio-political activism, assisting students to become active citizens in addressing science and technology issues at local, national and global levels. 21st century STEM skills are more important than ever as industry advances and places greater importance on preparing students for the world of work.

Summary

- The U.S. National Education Association and the Partnership for 21st Century Skills Four C's (critical thinking, communication, collaboration, and creativity) appear equivalent to international descriptions of desirable 21CS, as well as with the NRC and World Economic Forum competency categories.
- An examination of curriculum documents from multiple countries show the inclusion of 21CS, and that the goals encompassing 21st century competencies have been included in many standards and curriculum documents.
- Project-Based Learning (PBL), Problem-Based learning (PrBL), and Phenomenon-based Learning (PhBL), along with transdisciplinary approaches to teaching science education such as STEM (science, technology, engineering and mathematics) and STEAM (science, technology, engineering, the arts, and mathematics), break down the artificial barriers of disciplines enabling students to understand the connected nature of knowledge and utilize critical skills leading to success in the 21st century economy.
- Successful schools provide technology tools for classrooms, and equip teachers and students with the skills to use them in a meaningful context, thus promoting the development of the technical skills supporting 21st century STEM skills.
- There is broad recognition that 21st century STEM skills are important, but there is little research to indicate the level of implementation in the classroom.
- Classroom implementation of the 21CS, and the pedagogies that support them, may not come to fruition until standardized assessments are reformed to include measurements related to 21CS and professional development for teachers is designed to include 21CS across the primary and secondary school experience.

Further Reading on Model 21CS STEM Schools

Denver School for Science and Technology, Colorado: https://www.greatschools.org/colorado/
  denver/2427-Denver-School-Of-Science-And-Technology-Stapleton-High-School/.
High Tech High Charter School, San Diego, California: https://www.hightechhigh.org/.
Manor New Technology High School, Texas: https://mnths.manorisd.net/Domain/22.
University of Texas at Tyler University Academy: http://www.uttia.org/.

References

American Association for the Advancement of Science. (2015). *AAAS science assessment.* American
  Association for the Advancement of Science Project 2061 Science Assessment. Retrieved from
  http://assessment.aaas.org/pages/home.
American Montessori Society. (2018). Retrieved from https://amshq.org/Montessori-Education/
  History-of-Montessori-Education/Biography-of-Maria-Montessori.
Asia Pacific Economic Cooperation. (2008a). *Education to achieve 21st century competencies
  and skills for all: Respecting the past to move toward the future.* Retrieved from https://
  www.apec.org/Press/Speeches/2008/0115_cn_ednetserminarambcapunay   and   http://www.
  sociedadytecnologia.org/pages/view/94024/education-to-achieve-21st-century-competencies-
  and-skills-for-all-respecting-the-past-to-move-toward-the-future.
Asia Pacific Economic Cooperation. (2008b). *2nd APEC Education Reform Symposium: 21st Century Competencies.* Xi'an, China, Asia-Pacific Economic Cooperation Human Resources Development Working Group. Retrieved from http://hrd.apec.org/index.php/21st_Century_ Competencies.
Asia Pacific Economic Cooperation. (2008c). *Education to achieve 21st century competencies and skills: Recommendations.* 4th APEC Education Ministerial Meeting, Lima,
  Peru.   Retrieved   from   http://www.sei2003.com/APEC/forum_report_files/08_4aemm_018_
  Education_to_Achieve_21st_Century_Competencies_and_Skills.pdf.
Australian Education Council. (1989). *The Hobart declaration on common and agreed national
  goals for schooling in Australia.* Ministerial Council on Education, Employment, Training and
  Youth Affairs. Retrieved from http://www.curriculum.edu.au/verve/_resources/natgoals_file.pdf.
Beers, S. (2016). 21st century skills: Preparing students for their future. *21st Century Skills,* 1–6.
  ASCD. Retrieved from https://cosee.umaine.edu/files/coseeos/21st_century_skills.pdf.
Broadening Advanced Technological Education Connections. (2013, March 20). *March Newsletter*
  (D. Boisvert, Ed.) Boston, MA, USA: University of Massachusetts, Boston.
Bybee, R. W., & Fuchs, B. (2006). Preparing the 21st century workforce: A new reform in science
  and technology education. *Journal of Research in Science Teaching, 43*(4), 349–352. Retrieved
  from http://onlinelibrary.wiley.com/doi/10.1002/tea.20147/epdf.
C21 Canada. (2015). *Shifting minds 3.0: Redefining the learning landscape in Canada.* Canadians
  for the 21st Century Learning and Innovation. Retrieved from http://www.c21canada.org/wp-
  content/uploads/2015/05/C21-ShiftingMinds-3 pdf.
Carroll, V. (2015, February 6). Do DSST schools have an unfair advantage? *The Denver
  Post.* Retrieved from https://www.denverpost.com/2015/02/06/carroll-do-dsst-schools-have-an-
  unfair-advantage/.
Cernansky, R. (2013, April 14). *The very model of a modern STEM school.* Retrieved
  from the Smithsonian website https://www.smithsonianmag.com/innovation/the-very-model-of-
  a-modern-major-stem-school-23163130/.
Cernavskis, A. (2015, August 11). How teens move from innovative K-12 to college: With a
  new spotlight on San Diego's High Tech High, questions arise about its model. *U.S. News*

& *World Report*. Retrieved from https://www.usnews.com/news/articles/2015/08/11/how-teens-move-from-innovative-k-12-to-college.

Chapa, M., Galvan-De Leon, V., Solis, J., & Mundy, M. A. (2014). College readiness. *Research in Higher Education Journal, 25*(1), 1–5. Retrieved from https://files.eric.ed.gov/fulltext/EJ1055338.pdf.

Chu, S. K. W., Reynolds, R. B., Tavares, N. J., Notari, M., & Lee, C. W. Y. (2017). Twenty-first century skills and global education roadmaps. In Chu, et al. (Eds.), *21st century skills development through inquiry-based learning* (pp. 17–32). Singapore: Springer.

Crockett, L. W., & Churches, A. (2018). *Growing global digital citizens: Better practices that build better learners*. Bloomington, IN: Solution Tree Press.

Crockett, L. W., Jukes, I., & Churches, A. (2011). *Literacy is not enough: 21st century fluencies for the digital age*. 21st Century Fluency Project and Corwin Press: Thousand Oaks, CA.

Crown Copyright. (2003). *21st century skills realising our potential: Individuals, employers, nation*. Retrieved from https://www.gov.uk/government/uploads/system/uploads/attachment_data/file/336816/21st_Century_Skills_Realising_Our_Potential.pdf.

Cumpton, G., & King, C. T. (2013). *Trends in low-income enrollment and outcomes in Central Texas for school districts and campuses, 2008 through 2012*. Ray Marshall Center for the Study of Human Resources, Lyndon B. Johnson School of Public Affairs, The University of Texas at Austin. Retrieved from https://repositories.lib.utexas.edu/bitstream/handle/2152/25278/Trends%20in%20Low-Income%20Enrollment%20and%20Outcomes%20in%20Central%20Texas%20for%20School%20Districts%20and%20Campuses%2C%202008%20through%202012.pdf?sequence=3&isAllowed=y.

Disney. (2016). *About Imagineering*. Walt Disney Imagineering Imaginations: Dream, Design, Diversify. Retrieved from https://disneyimaginations.com/about-imaginations/about-imagineering/.

Educational Research Center of America. (2016). *STEM classroom to career: Opportunities to close the gap*. Educational Research Center of America. Retrieved from https://www.napequity.org/nape-content/uploads/STEM-Classroom-to-Career-report-FINAL.pdf.

Haney, J., Holland, T., Moore, N. N., & Osborne, C. (2013). *Stephen F. Austin High School*. Retrieved from http://www.edtx.org/uploads/research-and-reports/Stephen%20F%20Austin_Redesign.pdf.

High Tech High. (2016). *High Tech High: About Us*. Retrieved from https://www.hightechhigh.org/about-us/.

International Baccalaureate. (2010). *The IB Learner Profile*. Retrieved from https://www.ibo.org/globalassets/publications/recognition/learnerprofile-en.pdf.

Kennedy, T. J., & Odell, M. R. (2014). Engaging students in STEM education. *Science Education International, 25*(3), 246–258. Retrieved from http://www.icaseonline.net/sei/september2014/p1.pdf.

Lynch, S. J., Spillane, N., House, A., Peters-Burton, E., Behrend, T., Ross, K. M., & Han, E. M. (2017). A policy-relevant instrumental case study of an inclusive STEM-focused high school: Manor New Tech High. *International Journal of Education in Mathematics, Science and Technology, 5*(1), 1–20.

Mitchell, D., & Gerwin, D. (2007). *Survey of Waldorf Graduates, Phase II*. Research Institute for Waldorf Education, Wilton, New Hampshire. Retrieved from http://www.journeyschool.net/wp-content/uploads/Waldorf_Graduates_Gerwin_Mitchell.pdf.

Montgomery, S. L., & Kumar, A. (2015). *A history of science in world cultures: Voices of knowledge*. Routledge.

National Academy Press. (2012). *A framework for K-12 science education: Practices, crosscutting concepts, and core ideas*. Retrieved from https://www.nap.edu/catalog/13165/a-framework-for-k-12-science-education-practices-crosscutting-concepts#.

National of Academy of Sciences, National Academy of Engineering, & Institute of Medicine. (2007). *Rising above the gathering storm: Energizing and employing America for a brighter economic future.* Washington, DC: The National Academic Press. Retrieved from https://www.nap.edu/catalog/11463/rising-above-the-gathering-storm-energizing-and-employing-america-for.

National Academy of Sciences. (1997). *Preparing for the 21st century: The education imperative.* Washington, DC: National Academy Press. Retrieved from http://www.nap.edu/read/9537/chapter/2.

National Commission on Excellence in Education. (1983). *A Nation at risk: The imperative for educational reform.* Washington DC: United States Department of Education. Retrieved from https://www2.ed.gov/pubs/NatAtRisk/risk.html.

National Education Association. (2012). *Preparing 21st century students for a global society: An educator's guide to "the four Cs.".* Washington, DC: Author. Retrieved from http://www.nea.org/assets/docs/A-Guide-to-Four-Cs.pdf.

National Research Council. (2012). In J. W. Pellegrino & M. L. Hilton (Eds.), *Education for life and work: Developing transferable knowledge and skills in the 21st Century.* Committee on Defining Deeper Learning and 21st Century Skills, Board on Testing and Assessment and Board on Science Education, Division of Behavioral and Social Sciences and Education. Washington, DC: The National Academies Press. Retrieved from https://www.nap.edu/catalog/13398/education-for-life-and-work-developing-transferable-knowledge-and-skills.

National Research Council (U.S.) Committee on Highly Successful Schools or Programs for K-12 STEM Education. (2011). *Successful K-12 STEM education: Identifying effective approaches in science, technology, engineering, and mathematics.* Washington, DC: National Academies Press. Retrieved from https://www.nap.edu/catalog/13158/successful-k-12-stem-education-identifying-effective-approaches-in-science.

National Science Teaching Association. (2011). Quality science education and 21st-century skills. *NSTA Position Statement.* Arlington, VA, USA: National Science Teachers Association. Retrieved from https://www.nsta.org/about/positions/21stcentury.aspx.

New Zealand Ministry of Education. (2007). *The New Zealand Curriculum.* Wellington, New Zealand: Learning Media Limited. Retrieved from https://nzcurriculum.tki.org.nz/content/download/1108/11989/file/The-New-Zealand-Curriculum.pdf.

North America Reggio Emilia Alliance. (2018). Retrieved from https://www.reggioalliance.org/narea/.

Odell, M. R. L., Kennedy, T. J., & Stocks, E. (2019). The impact of pbl as a stem school reform model. *Interdisciplinary Journal of Problem-Based Learning, 13*(2). Retrieved from https://docs.lib.purdue.edu/ijpbl/vol13/iss2/4.

Organization for Economic Co-operation and Development. (2003). The definition and selection of key competencies: Executive summary. In S. D. Rychen & L. H. Salganik (Eds.), *Key competencies for a successful life and well-functioning society.* Göttingen: Hogrefe and Huber Publishers. Retrieved from http://www.oecd.org/pisa/35070367.pdf.

Organization for Economic Co-operation and Development. (2005). *The definitions and selection of key competencies.* Paris: OECD Publishing. Retrieved from http://www.oecd.org/pisa/35070367.pdf.

Organization for Economic Co-operation and Development. (2008). *21st century learning: Research, innovation and policy, directions from recent OECD analyses.* Paris: OECD Publishing. Retrieved from http://www.oecd.org/site/educeri21st/40554299.pdf.

Organization for Economic Co-operation and Development. (2014a). *Beyond PISA 2015: A longer-term strategy of PISA.* Paris: OECD Publishing. Retrieved from http://www.oecd.org/pisa/pisaproducts/Longer-term-strategy-of-PISA.pdf.

Organization for Economic Co-operation and Development. (2014b). *PISA 2012 results: Creative problem solving (volume V): Students' skills in tackling real-life problems.* Paris: PISA, OECD Publishing. Retrieved from https://doi.org/10.1787/9789264208070-en.

Özdemir, S. (2019). Soloborative learning: Solo thinking, collaborative tinkering. *International Electronic Journal of Elementary Education, 11*(3), 217–219.

Partnership for 21st Century Skills. (2003). *Learning for the 21st century: A report and mile guide for 21st century skills.* Retrieved from http://www.p21.org/storage/documents/P21_Report.pdf.

Partnership for 21st Century Skills. (2010). *21st century knowledge and skills in educator preparation.* Retrieved from http://www.p21.org/storage/documents/aacte_p21_whitepaper2010.pdf.

Partnership for 21st Century Skills. (2015). *Framework for 21st century learning.* Retrieved from http://www.p21.org/storage/documents/docs/P21_Framework_Definitions_New_Logo_2015.pdf.

Queensland Curriculum and Assessment Authority. (2017). *Queensland 21st century skills: Explanations of associated skills.* Retrieved from https://www.qcaa.qld.edu.au/downloads/senior/snr_syll_redev_21st_century_skills_associate_skills.pdf.

Reggio Emilia Australia Information Exchange. (2018). Retrieved from https://www.reggioaustralia.org.au/component/content/article/65.

Scott, C. L. (2015). *The futures of learning 1: Why must learning content and methods change in the 21st century?* United Nations Educational, Scientific, and Cultural Organization Education Research and Foresight, Paris [ERF Working Papers Series, No. 13]. Retrieved from http://unesdoc.unesco.org/images/0023/002348/234807E.pdf.

Silva, E. (2009). Measuring skills for the 21st century. *Phi Delta Kappan, 9*(90), 630–634. Sage Journals. Retrieved from http://journals.sagepub.com/doi/abs/10.1177/003172170909000905.

SRI International. (2018). *Evaluation of the Texas High School Project.* Retrieved from https://www.sri.com/work/projects/evaluation-texas-high-school-project.

Struhl, B., & Vargas, J. (2012). *Taking college courses in high school: A strategy guide for college readiness—The college outcomes of dual enrollment in Texas.* Jobs for the Future.

The Conference Board, Inc. (2006). *Are they really ready to work?: Employers' perspectives on the basic knowledge and applied skills of new entrants to the 21st Century U.S. Workforce.* The Partnership for 21st Century Skills.

UNESCO. (1996). *Learning: The treasure within. Report to UNESCO of the International Commission on Education for the Twenty-first Century* (pp. 1–46). Paris, France: UNESCO Publishing. Retrieved from http://unesdoc.unesco.org/images/0010/001095/109590eo.pdf.

Voogt, J., Erstad, O., Dede, C., & Mishrass, P. (2013). Challenges to learning and schooling in the digital networked world of the 21st century. *Journal of Computer Assisted learning, 29,* 403–413. https://doi.org/10.1111/jcal.12029.

Waldorf Education. (2018). Retrieved from https://waldorfeducation.org/waldorf_education.

World Economic Forum. (2015). *New vision for education: Unlocking the potential of technology.* Retrieved from http://www3.weforum.org/docs/WEFUSA_NewVisionforEducation_Report2015.pdf.

Teresa J. Kennedy, Ph.D. holds a joint appointment as Professor of International STEM and Bilingual/ELL Education in the College of Education and Psychology and in the College of Engineering at the University of Texas at Tyler, United States of America. Her research interests focus on STEM Education, international comparative studies, gender equity, and brain research in relation to second language acquisition and bilingualism.

Cheryl W. Sundberg, Ph.D. is a Research Scholar at The Ronin Institute (of the United States), with a substantial portion of her current scholarship centered on the use of Web 2.0 to facilitate science learning at all levels. Her research interests include interactive online teaching, teaching with emerging technology, and science teaching.

Correction to: The Bildung Theory—From von Humboldt to Klafki and Beyond

Jesper Sjöström and Ingo Eilks

Correction to:
Chapter 5 in: B. Akpan and T. Kennedy (eds.),
Science Education in Theory and Practice,
Springer Texts in Education,
https://doi.org/10.1007/978-3-030-43620-9_5

Chapter 5, "The *Bildung* Theory—From von Humboldt to Klafki and Beyond" was previously published non-open access. It has now been converted to open access under a CC BY 4.0 license and the Copyright Holder is "The Author(s)". The book has also been updated with this change.

The updated version of this chapter can be found at
https://doi.org/10.1007/978-3-030-43620-9_5

© The Author(s) 2021   C1
B. Akpan and T. Kennedy (eds.), *Science Education in Theory and Practice*,
Springer Texts in Education,
https://doi.org/10.1007/978-3-030-43620-9_33

Index

Note: Page numbers followed by *"f"* indicate figures; those followed by *"t"* indicate tables.

© Springer Nature Switzerland AG 2020
B. Akpan and T. Kennedy (eds.), *Science Education in Theory and Practice*,
Springer Texts in Education, https://doi.org/10.1007/978-3-030-43620-9

Printed in the United States
by Baker & Taylor Publisher Services

Printed in the United States
by Baker & Taylor Publisher Services